CULTURAL

ENVIRONMENTS • LANDSCAPES • IDENTITIES • INEQUALITIES

GEOGRAPHY

WILLIAM NORTON   **SECOND EDITION**

OXFORD

UNIVERSITY PRESS

OXFORD

UNIVERSITY PRESS

8 Sampson Mews, Don Mills, Ontario M3C 0H5
www.oupcanada.com

Oxford University Press is a department of the University of Oxford.
It furthers the University's objective of excellence in research, scholarship,
and education by publishing worldwide in

Oxford   New York

Auckland   Cape Town   Dar es Salaam   Hong Kong   Karachi
Kuala Lumpur   Madrid   Melbourne   Mexico City   Nairobi
New Delhi   Shanghai   Taipei   Toronto

With offices in

Argentina   Austria   Brazil   Chile   Czech Republic   France   Greece
Guatemala   Hungary   Italy   Japan   Poland   Portugal   Singapore
South Korea   Switzerland   Thailand   Turkey   Ukraine   Vietnam

Oxford is a trade mark of Oxford University Press
in the UK and in certain other countries

Published in Canada
by Oxford University Press

Library and Archives Canada Cataloguing in Publication

Norton, William, 1944–
Cultural geography : environments, landscapes, identities,
inequalities / William Norton. — 2nd ed.

Includes bibliographical references and index.

ISBN-13: 978–0–19–541922–1.–   ISBN-10: 0–19–541922–7

1. Human geography—Textbooks. I. Title.

GF41.N66 2005   304.2   C2005–902153–5

Cover image: Getty Images, PhotoDisc
Cover design: Brett J. Miller

5 6 7 – 12 11 10
This book is printed on permanent (acid-free) paper ∞.
Printed in Canada

Contents

Figures, Tables, Boxes, and Photo Essays

Tables

Boxes

Photo Essays

Preface

This is a substantially restructured and revised second edition of an ambitious textbook—ambitious because the intent is to offer comprehensive coverage of cultural geography as a subdiscipline of human geography. Unlike most other cultural geography texts, and there are several very good ones, this book is not restricted to cultural geography as it has developed since the 1970s in response to critiques of earlier work and to various conceptual developments. Rather, this book incorporates both the substantial volume of conceptual and empirical work initiated by Carl Sauer in the 1920s and known as the landscape school, and the more recently developed concepts and practice associated particularly with Marxist, humanist, feminist, and postmodernist thought. This inclusion of the relatively traditional and relatively new serves to highlight the continuity of the cultural geographic tradition, especially as this is evident in longstanding and ongoing interests: environments, landscapes, identities, and inequalities.

Why are there relatively few textbooks that address both traditional and recent interests in cultural geography? First, much traditional cultural geography has a strong empirical focus, and many practitioners have centered their energies in that arena rather than attempting to produce syntheses of empirical content. Second, the challenge of producing a text is a real one, not only because of the need to incorporate both traditional and newer themes, but also because of the related need to integrate North American cultural and European social traditions.

Necessarily, then, both planning and writing this textbook have been challenging. The aim here is to present cultural geography as a reasonably coherent subdiscipline concerned with making sense of people and the places they occupy. There is a focus on the geographic expression of culture in landscape and also a focus on the social and spatial constitution of culture. Links with other human geographic subdisciplines, with physical geography, and with other disciplines are regularly stressed.

This second edition is structured as follows. Following an introductory statement about cultural geography, there are two chapters that address conceptual issues. Chapter 2 covers what is sometimes called traditional cultural geography, with emphasis on ideas about humans and nature and the seminal writings of Carl Sauer, while Chapter 3 focuses on the many ways in which cultural geography has been rethought since the 1970s as Marxist, humanist, feminist, and postmodern theories have been applied to the traditional work.

Together, chapters 2 and 3 serve to inform the more empirical content of chapters 4, 5, 6, 7, 8, and 9, which deal, respectively, with environmental issues, landscape evolution, regional landscapes, global inequalities, other groups and their landscapes, and ordinary and sometimes extraordinary peoples and places. Although there are close links between the two conceptual and the six more empirical chapters, instructors using this textbook might choose to focus on the empirical chapters with only passing reference to the conceptual chapters as judged necessary. Chapter 10 offers some often tentative thoughts on the present and possible futures of the cultural geographic enterprise.

A number of colleagues and friends ensured that preparing this revision was a source of sustained pleasure. In particular, I thank Laura Macleod of Oxford University Press for her constant support of this second edition and Eric Sinkins for his constructively critical and meticulous editing. Among the many others at Oxford who deserve my thanks are Phyllis Wilson, Rachael Cayley, Lisa Rahn, and Euan White. Doug Fast expertly produced the new maps and diagrams and redrew those carried over from the first edition. Barry Kaye provided me with a wealth of information that, although sometimes distracting, has certainly enhanced this revision. Few authors can be so fortunate in their friendships. Pauline helped me in ways too numerous to mention.

1

Introducing Cultural Geography

Cultural geography is about understanding people and the places they occupy by analyzing cultural identities and cultural landscapes. It derives from a longstanding interest in culture—especially the geographic expression of culture in landscape—as a causal mechanism and from a more recent concern with cultural politics—especially the social and spatial constitution of cultures. It is concerned with the local and the global, and acknowledges that much of what is evident at the local scale is linked to global matters. Cultural geography incorporates both traditional and newer conceptual bases, and is closely related to other areas of geographic interest, especially social, economic, political, and physical geography; to other academic disciplines, especially anthropology, history, political science, psychology, and sociology; and to such interdisciplinary concerns as cultural, women's, and ethnic studies.

To better understand this explanation of cultural geography, consider four seemingly very different geographic phenomena that are discussed in this textbook: wheat fields, Buddhist shrines, gay neighborhoods, and segregated beaches. What these have in common is that they are all parts of cultural landscapes created and re-created over time by people as they relate, through their cultural identities, both to local and to more distant places. Initially, as people seek food and shelter, they change and add to local physical landscapes in an uncertain and accidental manner; through time, as cultures and technologies evolve, these processes of landscape change and addition become increasingly

directed. Culture, understood as the ability to take actions that result in intended consequences, facilitates the creation of desired landscapes, of what might be described as landscapes that ought to be. This objective of creating landscapes that ought to be is never wholly achieved, for at any given time there are examples both of intended landscapes and of unintended landscapes. To paraphrase a popular definition of geography, cultural geography examines what landscapes are located where, why they are where they are, and what the significance of place-to-place variations is.

It is not surprising, then, that cultural geography is a pragmatic concern. The role of place as it relates to the human condition was stressed by the distinguished cultural geographer Wagner (1990:41) when he asserted that 'a theoretically well-grounded, intellectually vigorous, and practically effective social and cultural geography might well assume, in time, a major role in guiding and guarding the evolution of humanity's environments'. Certainly, students of cultural geography are exposed to a diverse body of concepts and facts that, together, raise questions and offer insights into the condition of our world and its inhabitants. Indeed, within the larger discipline of human geography, cultural geography is recognized as an increasingly central concern, with Zelinsky (1995:750) claiming that cultural geography is a 'scholarly discourse that has shifted from the comfort of placid marginality toward the overheated vortex of ferment and creativity in today's human geography'. This shift began about 1970 and continues apace as cultural geography experiences a transformation from an essentially descriptive, empiricist enterprise to a more conceptually varied search for understanding and meaning of both peoples and places. In order to paint as comprehensive a picture as possible, this textbook reflects both the traditional and the newer emphases and, where appropriate, highlights the continuity of the larger cultural geographic enterprise.

This introductory chapter aims to provide you with the necessary broad context for tackling the chapters that follow. In this chapter you will find

- three examples of cultural geographic research, which are intended to provide at least a flavor of what is to come;
- a discussion of the larger scholarly context of cultural geography, both as an area of social science and as an interest within human geography. This discussion will help you appreciate that cultural geography does not function in a scholarly vacuum, but has origins and links to other academic interests;
- an introduction to key terms and ideas, together with an explanation of the varying interpretations of these;
- an outline of six prominent themes in the study of cultural geography along with the conceptual bases for these themes. It is acknowledged that multiple and ever-changing themes to a changing subject matter are welcome and necessary;
- a brief concluding section.

Doing Cultural Geography

The first of the three brief discussions designed to whet your appetite for what follows reflects the traditional approach to the study of cultural geography—what you will come to know as the **landscape school**. The primary objectives of this approach are to identify cultural groups, study the relationships these groups maintain with the natural world, delimit the geographic regions they occupy, and describe the landscapes they create. The second discussion reflects a more recently developed set of concerns that you will come to know as the **new cultural geography**, although, as this term first appeared during the 1980s, much of the new cultural geography is not so new any more. This approach focuses on matters of cultural identity, especially on how identities are formed with reference both to other groups of people and to places as these places are contested between groups. The third discussion addresses a topic not typically seen as part of cultural geography, namely reasons why places exhibit significant differences in quality of life. The example discussed is that of famine in Africa. This text argues that this type of problem straddles traditional and newer approaches and merits increased attention from cultural geographers. Indeed, all three discussions share a fundamental concern with how cultural groups create landscapes that, in turn, reinforce their cultural identities.

Identifying Groups and Describing Landscapes

A traditional concern of cultural geography is to identify groups, usually according to some criteria such as language or religion, and to describe the regional landscape created by these groups in the areas they occupy. Most work of this kind has treated both 'culture' and 'landscape' as uncontested terms, although as discussed later, both terms are complex and uncertain. One example of a distinctive regional cultural landscape is that part of the American intermontane west settled by members of the Church of Jesus Christ of Latter-day Saints, popularly known as Mormons. Who are the Mormons, and what is distinctive about the landscape they have created?

The Mormon religion originated in upstate New York in 1830 from the experiences, claims, and achievements of Joseph Smith, Jr. According to his own testament, Smith was visited by God and Jesus and told to found a new church. Converts accepted both the principle of prophetic revelation and Smith as a prophet, believed that Jesus had restored his Church through Smith, and accepted the *Book of Mormon*—an account of Hebrew immigrants to the western hemisphere around 600 BCE—as a companion to the Bible. Church growth was rapid, and a series of movements through Ohio, Missouri, and Illinois culminated in the 1847 migration to the Salt Lake Valley—the promised land. From this center

Mormons began a systematic and disciplined occupation of the surrounding valleys in Utah and neighboring states. Two characteristics of the church help explain why church members functioned as a group, creating a distinctive landscape in the areas settled. First, all members were expected to participate in the work of the church, including the work of settling the landscape around Salt Lake City that began in the years after 1847. Second, Mormons are a society of believers practicing co-operative effort and support for other members. The concept of community is evident in the favored pattern of Mormon village settlement.

The Mormon village and agricultural landscape contains distinctive features that are absent in surrounding areas. In towns, these features include especially wide streets, unpainted barns and granaries adjacent to residences, ward chapels, many I-style homes, and the use of brick for construction. In the surrounding agricultural area, roadside irrigation ditches, an open field landscape, crude unpainted fences, and wooden derricks used for loading hay characterize the landscape. This Mormon village and agricultural landscape is distinctive primarily because members of the Mormon cultural group share characteristics that affect their activities, and it is these activities, related to their cultural identity, that explain how the landscape evolved to become what it is today. Notwithstanding the worldwide reach of the contemporary church, for many Mormons the landscape created in the intermontane west continues to be a fundamental component of their cultural identity.

Dominant and Other Cultural Identities

A more recent concern of cultural geography is to describe and understand the tendency for some groups of people, who may or may not be

Located in a two-block area in downtown Salt Lake City, the Temple Square complex includes the Temple, Tabernacle, Assembly Hall, two visitor's centers, Church Office building, Joseph Smith Memorial Building, Church Administration Building, Lion House, Beehive House, and various monuments and memorials. Located close by is a Church Museum, Family History Library, historic log cabin, and Mormon handicraft store. This urban landscape speaks clearly of the Mormon presence and of Mormon power. (William Norton)

associated with a specific place and who may lack a clear identity, to find themselves in a struggle to maintain a common identity, often in opposition to some other dominant identity. Throughout much of Europe, Gypsy populations—also known as Roma—struggle to maintain their traditional lifestyle and to access places that facilitate that lifestyle.

Of uncertain origin—Gypsies were once thought to be from Egypt but are most likely descended from seventh-century migrants from India—and of uncertain number—estimates range from 4 to 15 million—Gypsies speak several dialects of a language, Romany, related to Hindi and have no common religion. Nevertheless, they are generally acknowledged as a group

at least partly because they prefer to live on the margins of dominant cultures. They lack interest in conventional wage labor, a characteristic that has many implications for lifestyle and for landscapes. They are very mobile, and this contributes to their being viewed and treated as outsiders when they stay in an area for a period of time. Unlike most ethnic groups, they are not associated with a particular territory.

Long viewed with suspicion by more settled populations, Gypsies have experienced much hostility and persecution. Many were forced into slavery in Romania during the nineteenth century, and between 250,00 and 500,000 were killed in the Holocaust. Gypsies continue to be stigmatized and ridiculed in ways that would be

This Gypsy encampment in Rome is typical of Gypsy encampments throughout Europe. Characterized by disorder and suggesting mobility rather than stability, the landscape offers a striking contrast to a suburban residential area. The main Gypsy camp in Rome is on the extreme eastern side of the city and is home to anywhere between 800 and 1200 Slavic and Romanian Gypsies. The homes are without electricity and drains, while running water is available only from two stand pipes. There are rats everywhere. Gypsies are widely perceived as threatening, especially to tourists. Travel literature frequently warns about Gypsies begging, stealing, or selling small items in public areas, such as main squares, subways, railway stations, and even outside hotel doors. (David Turnley/CORBIS Canada)

unacceptable if directed towards any other group. In November 2003, for example, the residents of a small village in southern England built a mock caravan with the license plate 'PIKEY' (a derogatory term derived from the old turnpike roads Gypsies traveled on), painted a Gypsy family in the windows, and then burned the caravan. Such incidents are not uncommon.

Hostility towards the group is compounded by a distaste for the landscapes they occupy. Gypsy landscapes are typically marginal. They may be waste areas inside cities or may be roadside areas. Regardless, they might appropriately be described as landscapes of exclusion, since they lie beyond the places occupied by the dominant cultural group. Further, these landscapes are different because they are characterized by disorder and are difficult to understand for those who do not belong to the group, especially when compared to the order of a suburban housing estate. Both Gypsies and their landscapes are in some sense outside of mainstream society and mainstream space, and this contributes to a negative image of both Gypsies and the places they occupy.

The future of the group, their lifestyle, and their landscapes is uncertain. Throughout much of Europe, governments are making it increasingly difficult for Gypsies to maintain their traditional lifestyle. In Britain, for example, a 1994 act released local authorities from the obligation to provide sites for mobile populations and introduced laws penalizing people who stopped in open areas without permission.

Unequal Worlds

Peoples and landscapes are clearly unequal, a fact that is fundamental to any understanding of cultural identities and cultural landscapes. Place-to-place variations in quality of life are evident at all spatial scales—local, regional, national, and global. At the global scale there is an all too easily identified distinction between less and more developed worlds, with the states of sub-Saharan Africa in particular demonstrating such characteristics of limited development as high fertility, high infant mortality, low life expectancy, low incomes, and an inadequate

economic and social infrastructure. To highlight some of these points, Table 1.1 provides 2004 data for Malawi and Canada.

The Canadian data are representative of countries of the more developed world; those for Malawi are unfortunately representative of many countries in sub-Saharan Africa. Most cultural geographers applaud place-to-place differences in peoples and landscapes as reflecting the activities of diverse cultures in diverse physical environments, but there is nothing to be applauded in these differences. Expressed simply, Malawi and Malawians are in a precarious state.

Compounding this dire situation is the fact that Malawi, along with much of sub-Saharan Africa including neighboring Zambia and Zimbabwe, is especially prone to famine and disease. In 2002 a state of national disaster was declared because of widespread famine, with reports that 70 per cent of the population was on the verge of starvation. Many of Malawi's poor are becoming poorer as they sell off their few possessions, with some parents selling their children to avoid the responsibility of feeding them. Locally, 2002 was described as *chinkukuzi*, meaning the 'year nobody will survive' in the local Chichewa language. One year later it was clear that the 2002

Table 1.1 Two Worlds

	Malawi	Canada
Total population (million)	11.9	31.9
Crude birth rate	51	11
Crude death rate	21	7
Infant mortality	121	5.2
Life expectancy (years)	44	79
Per capital income (US$)	570	28,930
% of population with HIV/AIDS	14.2	0.3

Source: Population Reference Bureau, *2004 World Population Data Sheet* (Washington, DC: Population Reference Bureau, 2004).

food crisis was the worst in living memory, with thousands dying.

Famines have various causes, including conflict, drought, and corrupt government. Recent factors contributing to the famine in Malawi include heavy flooding that destroyed bridges and a rail route, government mismanagement, and a reduction in aid from the West because of concerns about corruption and overspending. But famine in Malawi is more complicated because it is, effectively, an annual occurrence. Food shortages are routine during the 'hungry months' of December through March. This is the growing season, and the harvest from one year rarely lasts the twelve months needed before the following harvest. Famines in Malawi cannot be blamed on a cruel dictator, a damaging civil war, or disastrous weather. Famine in Malawi is simply endemic.

Tragically, chronic food shortages are not the only problem confronting Malawi. At an estimated 14 per cent, the incidence of HIV/AIDS is among the highest in the world, and yet Malawi has no effective prevention policies and minimal healthcare facilities, while discussions of sexual topics are largely taboo.

What this Book Is About

The three preceding vignettes identify some of the principal concerns of the cultural geography that is discussed in the following chapters. As these vignettes illustrate, cultural geography encompasses a diverse subject matter and diverse conceptual backgrounds. To accommodate these and other concerns, this textbook is organized as follows:

- Chapter 2 introduces the tradition of cultural geography, with emphasis on ways of understanding how humans relate to each other and to nature, and on the introduction in the 1920s by the doyen of cultural geographers, Carl Sauer (1889–1975), of the landscape, or Sauerian, school. This tradition, concerned especially with groups and related regional landscapes, continues to be important, especially in North America, where it has long been the dominant concern.

- Chapter 3 introduces some revisions to the landscape school tradition that derive from a variety of conceptual backgrounds, such as Marxism, feminism, and postmodernism, beginning in the 1970s. These revisions added to, perhaps even reinvented, cultural geography and still contribute to what is sometimes labeled the new cultural geography. In the era of new cultural geography, interest in the visible and material landscape as an object of study, evident in the tradition initiated by Sauer, becomes concern with the role of power relations and with symbolic and social identity. Further, there is recognition that landscape is not the sole object of study, which coincides with a growing interest in studying cultural identities as they are formed and reformed.

- Chapters 4, 5, and 6 are informed by the contents of chapters 2 and 3—especially 2—being concerned respectively with humans and nature, with landscape evolution, and with regional landscapes. In much of this work there is an underlying assumption that culture is some form of causal mechanism.

- Chapters 7, 8, and 9 are informed by the contents of chapters 2 and 3—especially 3—being concerned respectively with power and identity, with other voices and landscapes, and with imagining, writing, and reading landscapes.

- Chapter 10 provides an opportunity to evaluate the content of the text and to make some observations on the current and possible future status of cultural geography, emphasizing that the traditional and newer approaches to the discipline share some fundamental concerns about peoples and places.

Four Terminological Challenges

As is so often the case in social science, key terms tend to be understood differently by different practitioners and in different contexts. The following section highlights four such terms that are particularly important in the study of cultural geography.

CULTURE

One of the greatest terminological challenges facing cultural geographers is the very term

culture, which is the subject of considerable debate. One review stated that culture 'is one of the two or three most complicated words in the English language' (Williams 1976:87). Other compositional subdisciplines of human geography—such as political and economic geography—do not encounter this definitional problem to nearly the extent that cultural geography does. Quite simply, unlike the terms 'politic' and 'economy', the term 'culture' is contested, creating much uncertainty about an appropriate working definition of culture in geography. While it might be assumed that this uncertainty is a fundamental concern for cultural geographers, in fact the debate about the culture concept is, while important, not central to the study of cultural geography. To anticipate the conclusion reached in a discussion of the debate later in this chapter, multiple meanings of culture are legitimate, indeed valuable.

Cultural and Social

A second challenge comes from confusion about the closely related term '**society**' and about the differences between the words 'cultural' and 'social' as used in the study of cultural geography. Generally, this debate is rooted in the fact that there have long been uncertainties about these terms in other social sciences, especially anthropology and sociology. But there are also regional differences in preference, with 'cultural' being the term traditionally favored by North American geographers, and 'social' the term favored by British and other European geographers. As with the uncertainty about the word 'culture', this issue is addressed later in this chapter, where it is concluded that while these two terms refer to separate concepts, the debate about their precise differences is of no substantive intellectual merit for the contemporary cultural geographer, being essentially an accident of disciplinary history.

Four women search for water-lily bulbs to eat in the Elephant Marshes off the Shire river in southern Malawi. Picking and eating water-lily bulbs is considered a severe coping strategy, as this food is not very nutritious and not much liked in the region. These women continue the work despite the real risk of crocodile attacks, which have led to numerous deaths. The woman second from the left is four-months pregnant. When this photograph was taken in 2002 water-lily bulbs were one of the few sources of food remaining in the region. (Ben Curtis/CP)

In addition, it is important to acknowledge that cultural, social, political, and economic processes are interrelated. For this reason, the traditional insistence in human geography of teaching and researching these subdisciplines separately has some limitations. Indeed, the phenomenon known as the **cultural turn** can be seen as extending cultural geography into political, economic, and social spheres. The phrase 'cultural turn' refers to advances in philosophy and social theory—such as those of poststructuralism, postmodernism, and postcolonialism—that together encourage a greater appreciation of the importance of culture in studies of human life. Political geography, for example, is comfortable incorporating cultural as well as social and economic content as needed to explain and understand ways in which the political world and political life are changing. Similarly, much contemporary economic geography incorporates cultural as well as social and political processes. Finally, social geography is increasingly adopting a cultural perspective. This textbook recognizes both the importance of the cultural in other traditional subdisciplines of geography and the relevance of social, political, and economic processes in cultural geography.

Central to discussions of the contemporary political and economic worlds is **globalization**, a term that refers to the functional integration of internationally dispersed activities. It is also an issue that needs to be at the forefront of cultural analyses of people and place. Both identity and landscape are affected by globalization, not necessarily in the sense that it diminishes the importance of local identity or of place, but rather in that it produces a tension between the global and the local.

NATURE

The third terminological challenge centers on meanings of the word 'nature'. While culture is indeed a complex term, 'nature is perhaps the most complex word in the language' (Williams 1976:184). Given that one of the principal concerns of cultural geography is to analyze and understand relationships between nature and culture, this uncertainty about the meaning of nature poses real difficulties. At this stage of the textbook, the meaning of nature to focus on is that contrasted with culture; thus, nature is the material world excluding humans. This meaning is debated in Chapter 2, where alternatives are discussed.

LANDSCAPE

The fourth contested term is **landscape**. This term causes difficulties because of uncertainty about its initial meaning and, to a lesser degree, because of the various challenges to the mainstream Sauerian understanding of the term that has been adopted in cultural geography. The classic Sauerian interpretation of landscape refers to the cultural transformation of the natural world, emphasizing visible and material characteristics and close links between land and life—this is the landscape that is lived in. This interpretation will be expanded on later in this chapter and fully detailed in Chapter 2.

Uncertainties about the meaning of landscape stem mainly from the fact that landscape is the English rendering of a composite German word, *landschaft*. *Land* refers to the area used to support a group of people, and *schaft* refers to the molding of a social unity; *landschaft*, then, refers to group activities and experiences that occur in a particular place. Much early German and French cultural geography was concerned with locating and describing groups and their territories, and it is this interpretation of landscape that was introduced in North America by Sauer. But a second meaning of the term also became popular in the English language. This is the pictorial, visual meaning of landscape as scenery, referring to something that is beyond, rather than a part of, ourselves. This is the landscape that is looked at and that we are able to relate to in aesthetic terms. Both meanings of landscape serve to privilege vision.

Pause for Thought

As these opening remarks suggest, cultural geography today is an evolving subdiscipline with a changing, often contested focus. Is this situation to be applauded or deplored? On the one hand, cultural geography is an academic and practical concern that responds both to an increasingly sophisticated body of concepts and to the changing needs of society. Such ability and willingness to respond are indicative of a healthy and vibrant subdiscipline. On the other hand, because such responses are ongoing and may be difficult to place in a larger context, cultural geography is rather like a moving target, never quite as easy to pin down as we might like it to be. New concepts and a changing subject matter—are these circumstances suggestive of a dynamic or an uncertain subdiscipline?

Providing a Context

Cultural geography is one of many areas of academic and practical interest concerned with human activities and human identities, and this section outlines a larger intellectual and disciplinary context to help you understand the particular concerns of cultural geography. In order to appreciate this larger context it is important to understand the **social construction** of knowledge. All ideas, including theories, develop in specific social, cultural, and historical contexts. These contexts affect our ideas, such that all knowledge—including knowledge of what is considered to be real—is socially constructed. This important idea is closely associated with some areas of the new cultural geography and resurfaces on several occasions in this textbook.

Although the scholarly history of cultural geography is lengthy, with speculations about cultural and social matters—especially about human behavior—evident in Greek, Chinese, and Islamic civilizations, it was not until the mid-eighteenth century in western Europe that such issues received widespread attention. Throughout the extended period from classical Greece to the mid-eighteenth century, there were scholarly advances in both history and the physical sciences, but little discussion of social and cultural issues.

The Rise of the Social Sciences

By about 1750, social science—as opposed to the social sciences—was an established concern of intellectual elites, though it was not as yet the subject of academic study in universities.

Although there were various proposals concerning the goals and practice of social science, two basic assumptions prevailed. First, the aim of social science was the discovery of a few general principles that would form the basis for scientific inquiry. **Empiricism** was to prevail in this endeavor; this is an approach that asserts that all factual knowledge is based on experience, with the human mind being no more than a blank tablet (*tabula rasa*) before encountering the world. Second, a central aim of social science was to promote social progress through revealing truths about ourselves.

Together, these two assumptions reflected the prevailing philosophical ideas of the **Enlightenment** period. The subsequent rise of social science disciplines coincided with, and contributed to, changes that occurred with the onset of the **Industrial Revolution** and the social and economic system of **capitalism**. It is not surprising that this period of dramatic change spurred the establishment of disciplines, each of which functioned to serve society by identifying and proposing solutions to problems. Thus, it was during the nineteenth century that the various social science disciplines were differentiated and institutionalized in universities, where their various proponents began to articulate their particular methods and interests. This specialization was a trend that involved a rapidly increasing world of facts, the rise of universities, the appearance of specialized societies and journals, and the desire of groups of scholars to have their discipline placed on an equal footing with other emerging disciplines. All of these developments may be characterized as part of the rise of **modernism**. All involved some claims about the ability of the new social sciences to generate truths about humans in much the same way that the physical sciences were seen to generate truths about nature and in contrast to earlier approaches that focused on religion or **metaphysics**. Certainly, there were arguments against social science's acceptance of a physical scientific approach, arguments that were mostly based on the grounds that it was dehumanizing (that is, neglecting what it means to be human), but the dominant trend was clear.

By about 1900, the principal social science disciplines were firmly established, and since that time additions and changes have been limited. This is important because, as Johnston (1985:5) noted, academic disciplines, once created, 'have a defined existence and are invested in by individuals, who wish to protect their capital'. Furthermore, while it might seem surprising to suggest that the establishment of academic disciplines can be seen as 'a turning-point in human self-discovery', Smith (1997:575) argues that the very presence of social science disciplines is a 'visible sign of belief in and the special status of specialist knowledge and tech-

niques about human nature'. Indeed, in many ways, the late nineteenth century was the beginning of the modern study of social science.

In principle, each of the social sciences can be traced to the recognition of a group of interrelated questions concerning human behavior, a term that is intimately linked with our central concern—culture. Box 1.1 provides an overview of the history of attempts to explain human behavior in the Western world.

The Rise of Human Geography

What can be said of geography in the light of these general observations about the larger

Box 1.1 Explaining Human Behavior

A dominant tradition, beginning with Plato (c.428–c.348 BCE) and Aristotle (384–322 BCE), is that of metaphysical **dualism**. This theory, which asserts that substances are either material or mental, has typically been used to distinguish between a physical, material world, and a human, mental world.

Beginning in the sixteenth century with the rise of physical science, a **mechanistic** view of the physical world emerged, as scientists began to regard outside forces as responsible for physical motion. It proved to be but a short conceptual step to the assumption that human behavior might be similarly explained, an idea that was first advanced by René Descartes (1596–1650). He carried the dualistic tradition further by theorizing a distinction between voluntary and involuntary behavior. Descartes believed that voluntary behavior was governed by the mind, while involuntary behavior was mechanical. This Cartesian dualism, or **Cartesianism**—a distinction between mind and body—prompted the philosopher Gilbert Ryle (1900–76) to refer to the human mind as a 'ghost in the machine'. The materialist philosopher Thomas Hobbes (1588–1679), having shed any Cartesian inhibitions, departed from this dualistic position when he argued that all human behavior—in fact, the entire universe—was mechanical. Other empiricist philosophers, notably John Locke (1632–1704) and David Hume (1711–76), took the next logical steps by deriving physical laws assumed to determine human behavior and developing the argument that all human ideas could be explained by experience.

A principal dissenting argument was expressed by G.W.F. Hegel (1770–1831), who believed that each

historical period could be summarized in terms of some overarching theme, the *Geist*, or the spirit of the age. Hegelian thought emphasized meaningful behavior followed for voluntary reasons. Some other German philosophers, notably William Dilthey (1833–1911), pursued this **idealist** perspective, which is not dissimilar to a contemporary subjective concept of culture, as it suggests that behavior is related to something inside the person, such as emotions, feelings, or perceptions. Karl Marx (1818–83) developed this argument, favoring the more mechanistic approach and making a distinction between consciousness and social being.

Thus, broadly speaking, two conceptual poles have been used as the basis for understanding human behavior, with the dominant nineteenth-century view favoring the mechanistic argument based on physical science procedures over the idealist argument. Thinking about these two ideas along with text comments about the assumptions of Enlightenment philosophy and the rise of the social sciences can help us understand the intellectual climate during the nineteenth-century rise of social science disciplines. A principal feature of this intellectual climate was the assumption that all human behavior—the key subject matter of the emerging social science disciplines—could be explained in terms of a mechanistic physical science approach; that is, in terms of a cause-and-effect logic. This assumption of **naturalism** followed from the demonstrable success of physical science and received powerful support from Darwinian theory, which effectively validated the scientific study of

group of social sciences? The study of geography (as opposed to the more specific *human* geography) has a long and distinguished academic pedigree, with substantive contributions made by early Greek, Chinese, and Islamic civilizations, and a consistent history of growth from the fifteenth century onwards in Europe. Throughout, there were central concerns with mapping and written descriptions of lands and peoples. The seventeenth-century scholar Bernhardus Varenius (1622–50)—confronted both with the explosion in European geographic knowledge related to overseas activity and with the need to provide a framework for organizing

this knowledge—recognized that geography was a physical and also a human science involving studies of regions and systematic interests. In the late eighteenth century, the philosopher, Immanuel Kant (1724–1804), identified geography as essentially a concern with regions.

The growth of geography in the nineteenth century paralleled that of other social sciences. There was an ever-increasing factual base, evident most notably in attempts by two great German geographers, Alexander von Humboldt (1769–1859) and Carl Ritter (1779–1859), to produce comprehensive world geographies, a task that proved to be no longer within the reach of

humans. The principal nineteenth-century advances in social science thus confirmed what might be described as a 'privileging' of science.

Given this context, it is not surprising that the new social sciences, in order to explain their general subject matter of human behavior, introduced versions of physical science **determinism** at early stages in their disciplinary histories. In sociology, Auguste Comte (1798–1857) and Herbert Spencer (1820–1903) argued for a view of society as an integrated whole, comparable to a physical system that determined the behavior of all members, while Émile Durkheim (1858–1917) shared this superorganic view of society but saw society as altogether separate from the qualities of the individual members. In anthropology, a similar superorganic concept of culture was introduced by Alfred Kroeber (1876–1960) in 1917. Certainly, by the early twentieth century, both sociology and anthropology were seen as natural sciences taking the view that social or cultural forces constrained the behavior of individuals. Comparable viewpoints emerged in psychology in the form of various behaviorisms, and in geography in the form of environmental determinism, the idea that physical geography was the cause of human behavior. Each of these approaches is a particular version of the idea that the human world can be studied using cause-and-effect logic, a logic that, expressed simply, presupposes that a single principal variable explains the many and varied concerns of the social scientist. Thus, human behavior and the larger human world are: for the sociologist, explained by reference to society; for the anthropologist, explained by reference to culture; for the psychologist, explained by reference to a stimulus–response framework; and, for the human geographer, explained by reference to physical geography.

Although the approaches to the study of human behavior that derived from physical science were highly influential during the formative periods of the social sciences, they no longer represent dominant viewpoints. Rather, the favored approaches in social theory and research today are associated with other perspectives. The general but by no means complete rejection of physical science methodology in the social sciences occurred gradually during the twentieth century and is related to the rise of Marxist, humanist, cognitive, critical science, poststructural, and postmodern approaches. Each of these, to varying degrees, rejects the notion that the social sciences bear any methodological resemblance to the physical sciences. Indeed, it is not uncommon today to refer to these academic changes, along with the various cultural, economic, and political changes that began after the Second World War, as representing the decline of the modernist period—a period that had roots in the Enlightenment and the Industrial Revolution—and the onset of a postmodern period. In other words, this period can be seen as a transition from modernism to **postmodernism**.

Nevertheless, notwithstanding the current fall from favor of approaches inspired by physical science, an awareness of the great importance of these approaches during the formative phase of the various social sciences is important to our concern with the character of cultural geography as one interest within human geography.

a single scholar. Special interest societies, such as the Paris Geographical Society (1821) and the American Geographical Society (1851), were founded during this time, and there was a gradual recognition of geography as a university discipline with the creation of individual chairs, for example at the Sorbonne in Paris in 1809 and in Berlin in 1820. An even more important development was the establishment of several academic departments of geography, first in Prussian universities in 1874, then in universities in other European countries and in the United States beginning in 1903. Professional scholarly associations served to further validate the discipline.

This capsule account of the emergence of geography in the nineteenth century raises a key point for understanding the character of the new university discipline. Put simply, it is not appropriate to consider geography only in the context of the social sciences because geography was, of course, much more than human geography. Indeed, human geography is different from the other social sciences because of the explicit and longstanding associations with physical geography; works by Varenius and Kant clearly acknowledged the contributions of both physical science and human science to the study of geography.

This relatively early integration of physical and human geography distinguished the discipline from the other social sciences that were, by comparison, not as varied and had considerably less intellectual capital invested. But, more importantly, this integration helps account for the late nineteenth-century uncertainty about the identity of geography. As Martin and James (1993:164) noted, 'There was no professionally accepted paradigm to serve as a guide to the study of geography'. For this reason, the need to answer the question 'What is geography?' was paramount. Each of the other social sciences had to grapple with a comparable question, but they were not similarly disadvantaged by the quantity of intellectual baggage that accompanied geography and hence were able to respond to their question largely in the context of late nineteenth-century circumstances, specifically the rise of social science questions related to aspects of human behavior. By about 1900, four definitions of geography were evident: geography

as physical geography; the study of regions; the influence of physical environment on humans; and the study of the human landscape.

These definitions bear witness to the contradiction that geography faced, namely accommodating the legacy of the past with the pressing demands of late nineteenth-century science. Nowhere is this more evident than in the matter of the unity of physical and human geography, with most practitioners questioning whether it was possible and appropriate to include radically different concepts and methods in a single discipline. The consensus answer, institutionally speaking, was yes.

Thus, geography has a long history, such that during the late nineteenth-century period of formal discipline creation, geographers had much to consider when defining their academic field. Furthermore, this history was not one that could be easily accommodated into the emerging structure of disciplines that largely confirmed the acknowledged division between physical and social sciences. The majority of geographers asserted that geography continued to be a unifying discipline dealing with both physical and social facts. This collective 'decision' by geographers to maintain a focus on both physical and social science—or, perhaps more correctly, to focus exclusively on neither—is an ongoing source of discussion with debates about the need for and viability of an integrating discipline.

Human Geography Since 1900

THE REGIONAL APPROACH

The four definitions of geography noted above grew out of the need to secure an intellectual niche for the new university discipline. By the 1920s it was regional geography that assumed dominant status, a dominance that continued until the mid-1950s. Emphasis was on the delimitation and description of regions, and the underlying philosophy was a form of empiricism that saw the acquisition of geographical knowledge as a way to correct or verify facts about places and peoples. Typically, regional accounts of human-and-land relationships implicitly accepted the logic of **environmental**

determinism, which is the argument that the physical environment is a principal cause of human activity. Regional geography thus incorporated both physical and human subject matter, and for many practitioners geography was a bridging science that both straddled and linked physical and human sciences.

SPATIAL ANALYSIS

By the mid–1950s, some practitioners were expressing concern about human geography's lack of a scientific focus, a shortcoming blamed by some for the discipline's apparent failure to keep pace with advances in other social sciences. The response was to borrow heavily from economics in the introduction and application of what became known as spatial analysis. This approach involved both description and explanation, with particular focus on explaining why things are located where they are. Like some other social science interests, the spatial analysis approach derived from physical science, was objective in character, and had theoretical and quantitative content. The underlying philosophical support came from **positivism**, which was earlier accepted by most other social sciences as well as by the physical sciences. Human geography was late in accepting the logic of positivism, indeed so late that positivism was being rejected by the social science mainstream at the very time it was introduced into human geography. One consequence of this larger shift in conceptual emphasis was that spatial analysis was a major concern in human geography for only a brief period, from the late 1950s until about 1970, and it has never been of major importance in cultural geography.

THE LANDSCAPE SCHOOL

Although it is fair to characterize human geography from about 1900 until about 1970 as dominated, first, by a regional approach and, second, by spatial analysis, such a capsule account, from the perspective of the cultural geographer, omits the single most important tradition. In the early 1920s Sauer formulated the landscape school (also known as the Sauerian school), and this approach is central to the twentieth–century evolution of cultural geography. It has proven to

Carl O. Sauer (courtesy of the University of California, Berkeley)

be a longstanding approach that continues to be important today, although it is now enriched by additional concepts and ideas.

The landscape school developed on the premise that culture, operating on physical landscapes through time, was responsible for the transformation of those landscapes into cultural landscapes. It is therefore explicitly opposed to the premise of environmental determinism, although it can be seen as conceptually similar in that it incorporates a cause–and–effect logic. Foundations for this school are evident in the early nineteenth–century work of von Humboldt and Ritter (noted earlier), in the pioneering studies of human impacts on environment conducted by the American geographer George Perkins Marsh (1801–82), and in the work of such geographers as Paul Vidal de la Blache (usually known as Vidal) (1845–1918), Friedrich Ratzel (1844–1904), and Otto Schlüter (1872–1952). Much of the material discussed in this textbook, especially that relating to human and land relationships, to landscape evolution, and to regional landscapes, is informed primarily by this research tradition.

MARXISM AND HUMANISM

The demise of spatial analysis was partly a reflection of changes in other social sciences, especially the move away from objective analyses inspired by physical science toward a range of conceptual concerns that focused explicitly on human beings as subjects, not objects. In short, spatial analysis was criticized for being dehumanizing. In response, since about 1970, human geography has been a willing recipient of, and participant in, a wide range of concerns that, in various ways, place humans at the center of analysis. The two principal philosophical approaches that have embraced this emphasis are Marxism and humanism. Both are discussed more fully later in this textbook, and it is sufficient at this stage to emphasize that, notwithstanding the many differences between them (and, indeed, within each of them), they typically reject any suggestion that humans and human behavior can be studied in the objective terms of physical science.

Marxism and humanism proved to be refreshing additions to the conceptual arsenal available to all human geographers, including cultural geographers, and they continue to stimulate empirical work. Indeed, cultural geographers who were disenchanted with perceived inadequacies of the landscape school were among the first to apply these two methods to the study of human geography.

FEMINISM AND THE CULTURAL TURN

Most recently, since about 1980, at least four additional philosophical movements have added to the conceptual repertoire of the human geographer. That feminism, poststructuralism, postmodernism, and postcolonialism have had a particularly strong impact on cultural geography is a clear indication of the increasing importance of this subdiscipline. As with Marxism and humanism, each of these is explicitly opposed to any physical science analogies. But, more important, they combine to assert the relevance of an understanding of human cultural identity, to highlight the social construction of identity, to acknowledge that power relations are practiced through the expression of **discourse,** to emphasize the need to break through the conventional gender blindness of human geographic analyses, and most generally to insist on questioning the bases for the studies and interpretations carried out by cultural geographers.

What Is Culture?

Few scholars using the concept of culture have rigorously defined it. Here geographers have been exemplary rather than exceptional. Most users agree that the term resists simple definition. (Mathewson 1996:97)

The most important yet most confused concept in cultural geography is that of culture. Indeed culture is one of the most difficult concepts to interpret in all of social science. In other words, it is not only geographers who have difficulty with the term, but also anthropologists, sociologists, and others. Many of the difficulties encountered by cultural geographers stem from issues that arise in these other social sciences.

There are several good reasons for the continuing uncertainty about the meaning of this basic term. There was an early association of 'culture' with the biological term 'cultivation', the tending of natural growth, and a logical extension of this usage to refer also to human growth or development. By the mid-eighteenth century, culture was used to refer mainly to human organizations, gradually replacing the closely related term 'civilization' over the next hundred years. By 1871, culture, or civilization, was defined by the anthropologist Edward Tylor (1832–1917), as 'that complex whole which includes knowledge, belief, art, morals, custom, and any other capabilities and habits acquired by man as a member of society' (quoted in Kroeber and Kluckhohn 1952:81). By 1952, two leading anthropologists had combined to distinguish definitions proposed by 110 writers on the basis of 52 discrete concepts employed in those definitions (Kroeber and Kluckhohn 1952).

Culture and Society

Society and culture are closely related concepts but do relate to different phenomena, a point that has been generally accepted since 1958, when two leading figures, the anthropologist Kroeber and the sociologist Talcott Parsons (1902–79), together produced a statement that aimed to disentangle the terms and propose authoritative definitions. It is worth quoting their opening statements.

There seems to have been a good deal of confusion among anthropologists and sociologists about the concepts of culture and society (or, social system). A lack of consensus—between and within disciplines—has made for semantic confusion as to what data are subsumed under these terms; but, more important, the lack has impeded theoretical advance as to their interrelation.

There are still some anthropologists and sociologists who do not even consider the distinction necessary on the ground that all phenomena of human behavior are sociocultural, with both societal and cultural aspects at the same time. (Kroeber and Parsons 1958:582)

Kroeber and Parsons proceeded to explain the confusion they identified in terms of earlier uses of both terms. Thus, in the formative periods of disciplinary evolution during the late nineteenth and early twentieth centuries, leading anthropologists such as Franz Boas (1858–1942) defined culture as human behavior that was independent of genetic characteristics, while sociologists such as Spencer, Durkheim, and Weber (1864–1920) treated society in much the same way. 'For a considerable period', Kroeber and Parsons argue, 'this condensed concept of culture-and-society was maintained with differentiation between anthropology and sociology being carried out not conceptually but operationally' (1958:583). They proposed the following definitions, which have been generally accepted within the social sciences (1958:583).

- Culture refers to 'transmitted and created content and patterns of values, ideas, and other symbolic-meaningful systems as factors in the shaping of human behavior and the artifacts produced through behavior'.
- Society refers to the 'specifically relational system of interaction among individuals and collectivities'.

Thus, in anthropology it is usual to stress the non-genetic character of culture, referring only to those things that people invent, develop, or pass down through generations. In this sense, culture is extrasomatic—literally, beyond the body. Culture is conventionally seen as having three aspects: the values and abstract ideals that members of a human **group** hold; the norms and rules that they follow; and the material goods that they create. In sociology, society is understood as the system of interrelationships that connect individuals who share a common culture. This meaning of society is in accord with earlier usage that distinguished society (an association of free individuals) from the term 'state' (an association based on power relations). In response to the attractions of physical science modes of explanation, both disciplines during the formative stages of their development incorporated the idea of culture or society as a causal mechanism, although this idea is less important today. Anthropology and sociology offer many other interpretations that do not concern us at this time, though certain ones have carried over into cultural geography and are considered in later chapters as needed.

Cultural Geography, Human Geography, and Geography

It is fair to say that cultural geography, as a compositional subdivision of human geography, is concerned with some particular subject matter within the larger discipline. Note, however, that the flexibility of the term 'culture' has permitted two broader interpretations.

First, because culture can be equated with everything produced by humans, as distinct from everything that is a part of nature, there has been a tendency, especially in the United States, to treat cultural geography as synonymous with human geography, such that many courses and textbooks labeled as cultural are more appropriately seen as introductory human geography courses and texts.

Second, for many geographers, cultural geography, when broadly interpreted as the study of humans and land, is geography, both human and physical. This interpretation is supported by an alternative view of culture that includes nature. The editors of the *Companion Encyclopedia of Geography* articulate and endorse this view of geography explicitly: 'geography is, and at root always has been (despite excursions into spatial science and other exotic themes), about the interdependence of people and their environment, and about the evolving intercourse between humans and their earthly, and to a lesser extent celestial, habitat' (Douglas, Huggett, and Robinson 1996:ix). Identifying the study of human and land relationships as the core of geography is both a hindrance and an aid. It is a hindrance because it opens the cultural geographic door to such a diverse body of material that the subdiscipline almost disappears—it does in fact become all of geography. But it is an enormous aid because it obliges cultural geographers to acknowledge that cultural geography is not only about humans as individuals and as group members but also about the land they live on and with.

Both of these broader interpretations are at least partly a consequence of the uncertain status of the term 'culture'. Both are based on minor points concerning usage and need not concern us further, but they are instructive in illustrating the rather casual way key terms can be employed. Neither is a substantive issue because there has always been a majority interpretation of cultural geography as one part of human geography, and that interpretation is dominant today. Cultural geography, as defined earlier, focuses on peoples and places with particular reference to cultural identities and cultural landscapes. Human geography focuses on peoples and places more broadly with reference to culture, economics, and politics. Thus, human geography includes such subdisciplines as cultural, economic, and political geography. We now consider some of the more substantive issues surrounding the term 'culture' as it is used in cultural geography.

Culture in Cultural Geography

The meanings that cultural geographers attach to the word *culture* reverberate throughout this textbook, and it is inappropriate to attempt any detailed appraisal in this introductory chapter. Instead, the relevant meanings are discussed later, as needed, while the current discussion offers an overview, with details to follow.

CULTURE AND THE LANDSCAPE SCHOOL

Sauer introduced the dominant view of cultural geography in the 1920s in a series of writings that effectively set the agenda for the discipline in North America until the first serious reservations were voiced in the 1970s. Sauer regarded three factors as basic to an understanding of landscape: the physical environment, the character of people, and time. Sauer (1924:24) thus defined geography as 'the derivation of the cultural area from the natural area'. His classic statement of intent was perhaps the following: 'Culture is the agent, the natural area is the medium, the cultural landscape the result' (Sauer 1925:46). Although he saw culture as the 'shaping force', he viewed physical landscape as being 'of fundamental importance, for it supplies the materials out of which the cultural landscape is formed' (Sauer 1925:46).

For Sauer, then, the importance of culture to the study of geography was a given, such that the principal focus of his research was investigating the impact of culture on landscape, an impact evident in the creation of distinctive material landscapes that permitted the delimitation of cultural regions. But this concern with the material and visible landscape did not imply an exclusion of cultural identity, and practitioners frequently identified groups according to their way of life—what Vidal earlier labeled *genre de vie*—based on such characteristics as language, religion, and ethnic identity. Thus, culture was also viewed as a system of shared values and beliefs. This view, or perhaps more correctly these views, of culture remained largely uncontested in geography until the 1970s.

In this interpretation of culture there is an emphasis on culture as cause, on landscape outcomes, and on identifying characteristics. This view prompted a cultural geography—the landscape or Sauerian school—that functioned as a leading subdivision of human geography with three principal themes: landscapes as the outcomes of human and land relations; landscape change; and regional landscapes. (These three themes are addressed in chapters 4, 5, and 6 respectively.) Until relatively recently, most leading cultural geographers traced their intellectual antecedents to the ideas articulated by Sauer, while the first fully fledged textbook in cultural geography was unabashedly Sauerian in focus (Wagner and Mikesell 1962).

Questioning the Meaning of Culture

Challenges to the Sauerian view of culture, and hence to the landscape school of cultural geography, emanated from several directions, with the strongest being the claim that Sauer *reified* culture—that is, he treated culture as a reality and not as a concept or human construct. Expressed differently, critics raised concern at the use of culture alone as a cause of landscape without also focusing on the individual human decision-making behind it. This criticism was central to a seminal article by Duncan (1980), who argued that Sauer was reflecting the **superorganic** or cultural determinist view made popular in the anthropology of the 1920s and beyond by Kroeber. According to the superorganic perspective, any understanding of the cultural world had to be rooted in culture because it was there that decisions were made. Kroeber, along with the like-minded Robert Lowie (1883–1957), was a contemporary of Sauer at the University of California, Berkeley.

It is useful to see Duncan's argument as a major critique of the landscape school conception of culture and as contributing to the rise of the new cultural geography. Certainly, his argument struck a chord in a 1980s intellectual environment very different from that of the 1920s, such that the changes it sparked are but one component of the larger disenchantment with modernism in social science. The Sauerian tradition was criticized also for its rural and antiquarian emphasis and for its focus on material landscapes, as cultural geographers proposed a more complex and interpretive view of landscape as cultural construction.

Demands for a new concept of culture came not just from scholars outside the landscape school tradition with an agenda that was both cultural and social in inspiration. The landscape school was also criticized from within, as several major figures called for and proposed new ideas as early as the 1970s. For example, Mikesell (1978:13) asked practitioners to 'give more serious thought to how they wish to use the concept of culture', while English and Mayfield (1972:6) bemoaned the 'intense preoccupation with the visible material landscape' that 'led to an unfortunate neglect of the less obvious, invisible forces which in some cases form cornerstones in the explanation of spatial patterns of human behavior'. The important point here is that changes were being initiated by scholars immersed in the landscape school tradition, not only by those looking from outside of the tradition. Furthermore, it is rather misleading to imply that opposition to the Sauerian tradition and calls for new cultural traditions were evident throughout English-language geography. Writing with specific reference to New Zealand, Berg and Kearns (1997:1) noted that 'many cultural geographers here, but especially those with interests in "Maori issues", have always had a political edge to their work' such that 'the simple "old/new" binary constituted in representations of American cultural geography fails to capture the complexity of the situation' in New Zealand.

Culture and the New Cultural Geography

There is no doubt, however, concerning the impact on North American cultural geography made by scholars who objected to the Sauerian view principally because of the perceived tendency to reify culture. Uncertainty about how to correct this perceived flaw of the earlier view contributed to the emergence of several new understandings of culture, a reflection of the diverse theoretical landscape of the times. One particularly distinctive feature of this period was the interest shown in cultural geography by British social geographers; indeed, a 1987 conference entitled 'New Directions in Cultural

Geography' was organized by an organization called the Social Geography Study Group, which subsequently became the Social and Cultural Geography Study Group. Not surprisingly, then, several of the new conceptions of culture had a strong social flavor. The appearance of several varied concepts does not, however, disguise a fundamental similarity.

For the new cultural geography, culture is not seen as an explanatory variable, nor is culture approached as a cause. Rather, the new cultural geography emphasizes the plurality of cultures, defined as those values that members of human groups share in particular places at particular times. In brief, culture is seen as a medium or process rather than as an object; we might describe this as a change from an interest in culture to an interest in cultures. In an influential agenda-setting article, Cosgrove and Jackson (1987:99) defined culture as 'the medium through which people transform the mundane phenomenon of the material world into a world of significant symbols to which they give meaning and attach value'. This view of culture is not substantively different from earlier views. In accord with a general suspicion of what can be seen as oversimplistic causal arguments and a related preference for acknowledging cultural and spatial differences, it reflects a difference in emphasis rather than in substance.

The transformation of cultural geography that followed the expression of concerns about the Sauerian view of culture encouraged humanistic studies of behavior and landscape and paved the way for the interests in power and identity, in other voices and their landscapes, and in imagined landscapes and landscapes as text—the concerns of chapters 7, 8, and 9 respectively. This transformation also resulted in much questioning about the directions to be taken by cultural geography and the possible fragmentation of the sub-discipline. Regardless, there is little doubt that the new cultural geography was institutionalized by the end of the 1980s. This institutionalization is confirmed by the publication of cultural geography textbooks that chose to exclude much of the Sauerian tradition, focusing instead largely on certain aspects of the new cultural geography.

NEW CULTURAL GEOGRAPHY AND CULTURAL STUDIES

The understandings of culture in the new cultural geography are not simply critical responses to earlier understandings in the Sauerian tradition but are also reactions inspired by developments in the area of **cultural studies**. This is a term that could have been introduced earlier during our discussion of twentieth-century human geography, along with feminism, poststructuralism, postmodernism, and postcolonialism, but it is perhaps most appropriately seen as an umbrella term comprising all of these together with aspects of Marxism and humanism. It is an area of scholarly concern that is both rich and diverse, hardly surprising given the preceding statement and the complexity of the culture concept itself. As an interdisciplinary field, cultural studies embraces numerous concepts, mostly involving a rejection of modernism, and is represented in a number of schools of thought, four of which are noted briefly at this time.

First, a British tradition focused attention on working-class culture, on consequences of the Industrial Revolution, and on such new cultural forms as film and television. Much of the work conducted under the auspices of the Birmingham Centre for Contemporary Cultural Studies is central to this tradition. Second, an American tradition evolved from an anthropological concept of culture that centered on the meaningful ordinary lives of people from a culture other than that of the investigator. Anthropology from this perspective is thus an encounter with otherness. Third, a Marxist-inspired Frankfurt school of critical theory focused on a critique of modernity. A basic inspiration is the conviction that social science must not be the instrument of particular ideologies, notably those associated with dominant groups, including the political state; rather, social science must be emancipatory. Fourth, there are traditions, such as postcolonialism, feminism, and studies of racism, that are linked by their explicit concern with exposing the Eurocentrism that has been and often continues to be associated with the production of knowledge. For example, it can be argued that the Orient is really an invention of those who study it from outside, from the Occident, and that North America and Europe employ ideas from within their cultures—ideas such as freedom and individualism—to facilitate their conquest and domination of other regions.

The impact of cultural studies on cultural geography has been considerable, and the fact that its influence continues today is evidenced by studies of new cultural forms—of otherness, of ideologies of domination and oppression, and of Eurocentric attitudes. Perhaps the single most important cultural geographic idea derived from the cultural studies tradition is that of cultures as maps of meaning, as the 'codes with which meaning is constructed, conveyed, and understood' (Jackson 1989:2).

EVALUATING THE CULTURE CONCEPT

There are many views of culture within cultural geography, and Box 1.2 both summarizes and expands on the discussion so far by listing some representative definitions.

Clearly, understanding what is meant by the word 'culture' is not a simple task and undoubtedly provides both challenge and promise for cultural geography. There are at least three general reasons for this state of affairs. First, there has been much debate concerning the realness of culture, and this has contributed to confusion. The modernist approach has been to regard culture as a thing that is capable of explaining other things, an idea that is now rejected by most cultural geographers and other social scientists. Second, culture has never been easy to define separately from other related words, especially society, but also economy and politic. Third, the dominant current view is to acknowledge that culture has different meanings for different groups in different places and at different times.

Despite these comments, there are three basic statements with which most would agree:

- First, culture refers to our human way of life in the sense that it is what makes one group of people different from some other group.

- Second, culture is the principal means by which humans relate to nature and to each other; we are able to adapt to and change physical environments and are able to co-operate with and sometimes conflict with other groups.
- Third, culture is extrasomatic, distinguishing humans from other species; it is thus the reason humans are able to put other species on display in conservatories, aquariums, and zoos.

Pause for Thought

There continue to be differences of opinion concerning the claim that the culture concept used by landscape school geographers involved an almost unquestioning acceptance of the realness of culture whereas those concepts used by new cultural geographers are more flexible. Indeed it might be claimed that most concepts of culture, not only the Sauerian version but also those advocated within new cultural geography, involve reification. How do you react to these comments acknowledging uncertainty—indeed disagreement—about the single most important concept used by cultural geographers? Does this present a problem for you as you commence your studies, especially in light of the earlier uncertainty about the larger subdiscipline?

Most cultural geographers do not view this uncertainty with alarm, rather relishing in debates about past meanings and in the opportunities that various and varied definitions of culture present them as they strive to understand the ever-unfolding character of peoples and places. It is also useful to appreciate that similar situations prevail in other areas of social science. To cite just three examples, there are multiple meanings of society, behavior, and personality, as any study of sociology and psychology textbooks makes clear. Viewed from this perspective, some uncertainty seems both normal and desirable. Do you agree?

Human Identity—Who Are We?

So far in this text, the word 'culture' has been used to convey some idea of what cultural geographers study, namely the values, norms, behaviors, and material goods of humans. However, the degree of uncertainty about this term means that it has not been possible to convey a clear or comprehensive picture. One way to address this difficulty is to focus on socially constructed characteristics of individual and collective human identity—that is, characteristics that are made or acquired rather than inherited.

The following discussion addresses eight socially constructed characteristics that cultural geographers have identified: place, language, religion, ethnicity, nationality, community, class, and gender. A key concern of many social scientists today is the question of how it is possible to understand identity using group labels but at the same time avoid **essentialism,** that is, attributing any essential characteristics to a group. While it is typically necessary to group people, the resulting implication that groups

Box 1.2 Some Geographic Interpretations of Culture

This box details some representative geographic interpretations of the culture concept, organized chronologically. The intent is to provide an indication of the variety of definitions available and to indicate changes in definition through time, but also to suggest that, variety and change notwithstanding, there is an essential consistency.

Sauer (1925:30)

Culture is *'the impress of the works of man upon the area'.*

This is one of the classic statements in the seminal 1925 article by Sauer, emphasizing the idea of culture as a cause of landscape.

Sauer (1941a:8)

'Culture is the learned and conventionalized activity of a group that occupies an area.'

This less deterministic definition is perhaps closer to much of the empirical work accomplished by landscape school geographers than is the 1925 statement above.

Gritzner (1966:9)

'A culture is a human society bound together by a common complex of culture traits, each trait being anything which to the culturally-bound group has either material form and applicable function, or an expressed value.'

With this clear statement reflecting the consensus view and activities of the landscape school, Gritzner stresses that culture is transmitted spatially and temporally through the unique human ability to symbolize.

Spencer and Thomas (1973:6)

'Culture is the sum total of human learned behavior and ways of doing things. Culture is invented, carried on, and slowly modified by people living and

working in groups as each group occupies a particular region of the earth and develops its own special and distinctive system of culture.'

This textbook definition expands on the idea that culture is learned behavior.

Zelinsky (1973:70)

Culture is *'a code or template for ideas and acts'.*

This brief definition, taken from a larger and highly influential statement, is a clear assertion of culture as a causal mechanism.

Wagner (1974:11)

'Learned behavior is pretty much what we mean by culture.'

Wagner (1975:11)

'The fact is that culture has to be seen as carried in specific, located, purposeful, rule-following and rule-making groups of people communicating and interacting with one another.'

Together, the two statements by Wagner reiterate the importance of learning, stress that culture refers to human behavior, and emphasize the importance of communication and interaction between and among group members.

Jackson and Smith (1984:205)

'Culture, in the sense of a system of shared meanings, is dynamic and negotiable, not fixed or immutable. Moreover the emergent qualities of culture often have a spatial character, not merely because proximity can encourage communication and the sharing of individual lifeworlds, but also because, from an interactionist perspective, social groups may actively create a sense of place, investing the material environment with symbolic qualities such that the very fabric of landscape is permeated by, and caught up in, the active social world.'

have common traits or behavior patterns may be unwarranted given the often uncertain basis for defining group membership. The most popular alternative to essentialism is social constructionism, the idea that identities are formed and continually reformed through interaction with others and thus are fluid and negotiable.

This question surfaces frequently in the pages of this textbook.

PLACE

Cultural geographers of all persuasions, and increasingly other social scientists also, are sympathetic to the idea that place matters. Identity is

---

This definition, one of the first from the new cultural geography, incorporates elements of earlier definitions, notably Wagner (1975), but includes some additional ideas, notably a shift from a concern with behavior to a concern with meaning.

Jackson (1989:ix)

Culture is *'a domain, no less than the political and the economic, in which social relations of dominance and subordination are negotiated and resisted, where meanings are not just imposed, but contested'.*

This is one of the first statements to explicitly recognize that culture needs to be understood in the context of struggles between different groups of people. In both this and the preceding statement, there is explicit reference to the social world.

Shurmer-Smith and Hannam (1994:5–6)

Culture is *'that negotiated intersubjectivity which allows human beings as individuals to reach a tenuous understanding of one another, to experience each other jointly'.*

Rejecting a cultural geography concerned with mapping traits onto landscapes, this definition instead emphasizes ways of being in the environment.

McDowell (1994:148)

'Culture is a set of ideas, customs and beliefs that shape people's actions and their production of material artifacts, including the landscape and the built environment. Culture is socially defined and socially interpreted. Cultural ideas are expressed in the lives of social groups who articulate, express and challenge these sets of ideas and values, which are themselves temporally and spatially specific.'

After observing that culture is 'a notoriously slippery concept', McDowell (1994:148) suggests that 'a broad consensus' concerning the above definition 'might not be difficult to achieve'.

Jordan, Domosh, and Rowntree (1997:5)

Culture is *'learned collective human behavior, as opposed to instinctive, or inborn, behavior. These learned traits form a way of life held in common by a group of people.'*

This statement, taken from a highly influential introductory textbook, maintains the idea of culture as it evolved within the landscape school. This view remains popular with many contemporary cultural geographers, notwithstanding the undoubted achievements of the new cultural geography.

Mitchell (2003a:2)

'Culture is decidedly not everything. Indeed . . . *Culture needs to be understood as* no-thing, *which is to say that rather than an object, "culture" is an* on-going, struggled-over set of social relations *that give rise to social meaning, to differences within and across social groups and places, and to the exercise of power.'*

Mitchell is stressing that culture as way of life needs to be understood in political terms as it always serves particular groups and particular aims; thus, culture is both a component and a consequence of power relations.

Summary Comments

Although these selected definitions are but a few of many, they do combine to provide a meaningful overview of the approaches taken by geographers. No one single definition serves the purposes of this textbook, but it is appropriate to suggest that the final three reflect many of the current interests of contemporary cultural geography. However hard geographers and others try to achieve a definitive wording, such a goal is unlikely to be attained. As Jackson (1989:180) acknowledged at the end of a book-length study, 'the stuff of culture . . . is elusive.'

The agricultural Yakima valley in Washington is experiencing a significant increase in its Spanish-speaking population. In 1995, three customers were ejected from this tavern for speaking Spanish, and the three then filed a civil action lawsuit against the tavern's owner for discrimination. The sign hanging above the bar, which reads 'IN THE U.S.A. IT'S—ENGLISH OR ADIOS, AMIGO'—plainly states the position of the tavern's management. There is of course much at play here. The concern is with those who are seen as different to the rural norm of the region, and the incident can be interpreted in terms of language, ethnicity, discomfort with new arrivals, and opposition to change generally. The identities being objected to thus have several overlapping components—language, ethnicity, difference, and newness. (AP photo/Kirk Hirota)

not simply a matter of *who* we are, but also *where* we are. Humans in groups occupy areas and thus tend to create material and symbolic landscapes that might be susceptible to delimitation as regions. For most of us, place is an important element in our individual and collective identity.

LANGUAGE

North American cultural geographers, particularly those working in a landscape school tradition, typically distinguish groups of people according to what are viewed as primary cultural characteristics, notably language and religion. Such variables, which are similar to those used by anthropologists to classify peoples, are particularly appropriate to studies of tribal peoples and to areas of European overseas settlement. For many groups, language is the principal expression of culture and may embody a particular view of the world. In fact, during the nineteenth century, language was one of the principal ways for a population to assert a distinctive national identity. But attaching such cultural significance to language can be dangerous for a minority group whose language is dominated by that of a larger group in the same political unit; in such cases, the loss of language may mean the loss of the larger culture.

RELIGION

Religion, the second of the two traditional cultural universals, addresses matters of ultimate significance, such as questions about human meaning, about human presence in the world, and about the place of humans in the larger context of god, or gods, and the physical world. As a result, religion plays an important role in any consideration of human relations with nature and with the landscapes associated with particular groups of people. In particular, religion is also playing an increasingly important role in the formation of presumed ethnic identities.

ETHNICITY

Ethnicity refers to the perceived distinctiveness of one group relative to others, usually based on a shared history that may be partly mythical and that has political associations. Many groups strive to create ethnic solidarity and enhance ethnic pride by emphasizing origins, history, claims to a homeland, and distinctive material and other cultural traits. In landscape school cultural geography, ethnicity is often used instead of language and/or religion to identify a group.

Ethnicity is a problematic term. The assumption that ethnic groups exist typically implies that such groups have their own distinctive culture, which is another instance of reification. Ethnicity is also closely associated—and even confused—with the even more problematic term 'race'. As discussed in Chapter 7, the idea that there are human races is a myth. Unfortunately, this biological term is often treated as a synonym for ethnicity, or the impression is created that ethnic differences exist as a direct consequence of supposed racial differences.

NATIONALITY

Nationality is also a problematic term. Humans are divided into national groups (however artificial these may be), and also into political units known as nation states. But these two concepts are rarely congruent, and as a result, contrary to the aspirations of nineteenth-century nationalist thought, national groups are rarely, if ever, neatly packaged into a political unit, and political units rarely, if ever, comprise a single supposed national group. Regardless, the idea of national identity is central to many accounts of the contemporary world.

COMMUNITY

Unlike the word 'culture', community has never held a privileged position in the discourse of social science, although, like culture, it has been granted a great many different meanings. A term that has traditionally had such positive connotations as loyalty, shared concerns, and personal contact, community is today more frequently used as an alternative term for ethnic group.

CLASS

In the nineteenth century 'class' was used to refer to people with a common social or economic status. Karl Marx (1818–83) moved beyond classification by making a distinction between those who produce surplus (laborers) and those who appropriate it (capitalists) and concluding that each class has a collective consciousness and organization that places it in a perpetual class struggle. Marx used class consciousness in the sense that others used culture in the late nineteenth century. Today, class is a key variable in the area of political economy, which has contributed some ideas to cultural geography in the guise of the political ecology discussed in Chapter 4. Class is not, however, central to much other work in contemporary cultural geography.

GENDER

Gender refers to attributes that are culturally ascribed to women and men; it is a cultural construct subject to change. Thus, studies of gender emphasize the role played by male power and domination and assumed female subservience in the creation and maintenance of gender roles. It is usual to stress that there is no necessary link between gender and biological sex, as women may have masculine characteristics and men may have feminine characteristics. But not only is there recognition that gender is not an essential characteristic—that is, it relates to culture—there are also arguments that sexuality and sexual preference relate to culture in such a way that they also can be considered a social construction.

Pause for Thought

Do these eight aspects of human identity capture the spirit of that term as you might apply it to yourself or to others, or do you sense that identity is better captured by reference to certain other characteristics? Although these eight reflect much of the work in social science, are they reasonably comprehensive in the context of our contemporary world? It seems safe to say that some people choose to identify other characteristics as being important to themselves and to others. Some possibilities are membership of or identification with an environmental group or political party, a particular sexual identity, vision impairment, hearing impairment, or a personality type such as introversion or extroversion. Any one, or some combination, of these may be critical to identity.

Both the text list of eight characteristics and these suggested additions highlight the issue of essentialism as opposed to social constructionism. Most cultural geographers today emphasize the social construction of identity through life experiences and personal choices. Do you agree that identity is fluid and subject to change through interaction with others, or do you feel that there are important aspects of our identity that are effectively presented to us at birth and that are unchanging, indeed unchangeable? Perhaps you favor a view of identity that is capable of encompassing both essential as well as socially constructed characteristics?

Box 1.3 Some Organizational Frameworks

Although cultural geographers 'have not been prone to defend their interests or issue programmatic statements' (Mikesell 1978:1), there have been several efforts to clarify both subject matter and approaches. This box notes, in an abbreviated format, some of the principal efforts that have been made; further details of particular issues are included in this textbook as needed. Students seeking additional information are advised to refer to the sources noted below, to the progress reports on cultural geography and/or social geography published annually in the journal *Progress in Human Geography*, and to the assessments of cultural geography included in Dohrs and Sommers (1967), Mikesell (1967), Miller (1971), Salter (1972), Spencer (1978), Norton (1987), and Ellen (1988).

Wagner and Mikesell (1962)

In their pioneering and highly influential book of readings, Wagner and Mikesell (1962:1) proposed the following five themes as constituting 'the core of cultural geography':
- culture
- cultural area
- cultural landscape
- culture history
- cultural ecology

Gritzner (1966)

According to Gritzner (1966), cultural geography
- begins with the anthropological concept of culture;

- considers culture traits and groups in terms of development;
- studies culture/nature;
- interprets landscapes; and
- divides the world into culture regions and subregions.

Wagner (1972, 1974, 1975)

Wagner renounced the five themes outlined with Mikesell in 1962 and offered a revised focus based on institutions and communication.

Spencer and Thomas (1973)

Rather than specifying themes, Spencer and Thomas (1972) recognized the following four conceptual entities:
- population
- physical environment
- social organization
- technology

In this more process-oriented schema they also proposed six interoperative relationships, namely those involving
- population and environment,
- population and social organization,
- population and technology,
- social organization and technology,
- physical environment and social organization, and
- physical environment and technology

Themes in Cultural Geography

This textbook uses six themes to organize the ideas and work of cultural geographers. They are not formally institutionalized within cultural geography but are general themes that can be identified when tracing the development of cultural geography from about the 1920s onwards. Box 1.3 summarizes a selection of earlier attempts designed to impose order on the subdiscipline. Is this circumstance—a lack of evident and universally accepted research themes—peculiar to cultural geography? Of course not. Probably all of the principal interests in social science are appropriately viewed in this way.

There are at least two general reasons why the absence of a definitive list of themes is not uncommon in social science. First, research themes are not typically identified and then carefully adhered to. Rather, researchers conduct analyses that, although often prompted by earlier work, may be quite different in focus to that earlier work; they may even conduct research that is prompted by new ideas imported from other disciplines. Second, the nature of much social science work is such that commonalities often become apparent only as a body of work evolves, which is a principal reason for the difficulty of classifying new social science work; put simply, new work may require a new classification.

Mikesell (1978)

Mikesell (1978) identified seven persistent preferences:
- historical orientation;
- humans as agents of environmental change;
- focus on material culture;
- rural bias in North America, non-Western, or pre-industrial, bias elsewhere;
- links to anthropology;
- substantive research; and
- fieldwork.

He also identified three recent developments:
- environmental perception;
- cultural ecology; and
- focus on United States.

Norton (1989)

Norton (1989) proposed four themes:
- evolutionary;
- ecological;
- behavioral; and
- symbolic.

Murphy and Johnson (2000a)

In this tribute to Marvin Mikesell (subtitled 'enduring and evolving geographic themes'), the editors group papers into three categories:
- constructing cultural spaces;
- remaking the environment; and
- claiming places.

Buttimer (2003)

Buttimer (2003) outlined five principles:
- one earth inhabited by multiple cultural worlds;
- evolving geographical knowledge of nature and culture;
- inventions and conventions in texts and images;
- insiders and outsiders, mindscapes and landscapes; and
- self and other, dialogue towards mutual understanding.

Strohmayer (2003)

Strohmayer (2003) argued that in addition to the key concepts of culture and geography, the following should be prominent:
- landscape materiality;
- contestation;
- identity; and
- globalization.

Anderson, Domosh, Pile, and Thrift (2003a)

With the understanding that cultural geography is a contested and unfolding intellectual terrain, the editors identified five themes:
- culture as distribution of things;
- culture as a way of life;
- culture as meaning;
- culture as doing; and
- culture as power.

Logically, then, the six themes identified and used in this textbook could not have been easily identified, say, twenty years ago, and are unlikely to be appropriate, or at least sufficient, twenty years from now. In some cases the association of ideas and empirical work with a single theme is a necessary simplification designed to suit the purposes of this textbook, which assigns the work of cultural geographers to one of six themes. Other themes could be identified, and in some instances work could be assigned to more than one theme. In short, there is nothing 'correct' about the classification; it is simply one way to interpret the literature of cultural geography as it has developed and changed since about 1920. This cautionary statement does not serve to invalidate the themes identified but to make it clear that the classification must not be treated as sacrosanct.

It is important to note that the six themes, while they are different, are complementary. They do not compete with one another in some search for mythical universal truths. Rather, each one has made and is likely to continue making contributions to the fundamental goals of cultural geography, namely making sense of the real worlds of both people and place.

Six Themes

The first three themes are introduced in Chapter 2, in the discussion of the tradition of cultural geography, especially through the account of the landscape school. All three continue today as important and changing themes. The second three themes are more closely identified with developments since the 1970s than with the landscape school and are introduced in Chapter 3, in the account of the rethinking of cultural geography. These three are informed by different meanings of the culture concept and, more generally, by a rejection of the larger intellectual environment of the landscape school. Table 1.2 identifies the six themes and provides notations concerning conceptual inspirations, principal figures, and key concepts.

ENVIRONMENTS, ETHICS, LANDSCAPES

The first theme is concerned with relationships between humans and nature. Although there are important conceptual origins in the landscape school, this theme is also informed by new theory, especially in the form of political ecology and ecofeminism. Further, this is the most obviously practical area of cultural geography, being both environmentally and socially relevant. There are close links with physical geography, and it is this theme that comes closest to fulfilling the ambitions of many geographers to work within a unified physical and human science tradition.

LANDSCAPE EVOLUTION

Like the first theme, landscape evolution has origins in the landscape school. A great deal of work has been accomplished in this area, including work usually labeled as historical geography. The basic idea informing this theme is the recognition that, in accord with Sauer, cultures transform physical landscapes through time to create human landscapes. Thus, it is common for practitioners working within this theme to identify close links between the characteristics of a culture and changing landscapes.

REGIONAL LANDSCAPES

Studies of regional landscapes, especially as these are reflections of cultural **occupance** over time, are central to the landscape school tradition, and there are close links between this and the preceding theme in that regions can be seen as one outcome of evolution; the content difference between the two themes is one of emphasis. Regions are identified at various spatial scales, with global cultures and local, ethnically defined areas seen as parts of a single dialectic, not as opposites. There are also regional studies influenced by other concepts; for example, at a global scale, world systems theory explains the evolution of the modern political and economic system.

POWER, IDENTITY, GLOBAL LANDSCAPES

A principal concern of new cultural geography is the political dimension of culture, and this theme focuses on global issues as they relate to the control exercised by some groups over other groups. One consequence of this power imbalance is the uneven quality of life in different places, a circumstance that prevails despite globalizing processes. Links with landscape school interests are evident, with discussions of landscapes and regional differences, but the most important distinction between this theme and those based on the landscape school is the embracing of a political conception of culture.

Other Voices, Other Landscapes

The political dimension of culture is also the focus of this theme, which emphasizes the distinction between self and others as this is reflected in identities and landscapes. Recognition of the multiplicity and fluidity of identities in the contemporary world facilitates discus-

Table 1.2 Cultural Geography—Six Principal Themes

Theme	Conceptual Inspirations	Key Concepts	Principal Figures
Environments, Ethics, Landscapes	landscape school, ecology, anthropology, Marxism, political ecology	humans and nature, nature, culture, ecology, adaptation, human impacts	Barrows, Butzer, Mikesell, Mathewson, Peet, Sauer, Wagner, Zimmerer
Landscape Evolution	landscape school, history, *Annales* school, world systems	landscape, nature, culture, time, diffusion, acculturation, frontier	Carter, Clark, Conzen, Darby, Hoskins, J.B. Jackson, Kniffen, Lewis, Meinig, Sauer
Regional Landscapes	landscape school, regional geography, new cultural geography, culture worlds	culture, region, homeland, landscape, globalization, world system	Hart, Hartshorne, Hudson, Jordan, Meinig, Sauer, Shortridge, Zelinsky
Power, Identity, Global Landscapes	new cultural geography, Marxism, feminism, Postmodernism	power, identity, ethnicity, language, religion, nation, globalization	Blaut, Harvey, P. Jackson, Mitchell Taylor, Wallerstein,
Other Voices, Other Landscapes	new cultural geography, folk and popular culture, sociology, Marxism, feminism, postmodernism	social construction, discourse, difference, identity politics, others, gender, sexuality, resistance,	Cresswell, Domosh, Duncan, P. Jackson, Mitchell, Rose, Sibley, Valentine
Living in Place	new cultural geography, folk and popular culture, psychology, humanism, postmodernism	discourse, home, place, sense of place, landscape as text, identity, consumption	Carney, Cosgrove, Daniels, Duncan, Jakle, Ley, Tuan

Although incorporating some new content, the first three themes are traditional in that

- their principal conceptual inspiration is the landscape school;
- they include a view of geography as an integrating physical and human discipline.

The final three themes are more recent in origin. They introduce some new and different content to the ideas and practice of cultural geography and reflect

- generally close contacts between cultural geography and other social sciences;
- a concern with social theory and cultural studies;
- views of culture that explicitly reject the idea that culture is a causal variable.

sion both of self and others and of landscapes that are subject to different interpretations by different groups.

LIVING IN PLACE

Although the landscape school tradition considers both material and symbolic landscapes, it clearly places an emphasis on the visible material world. This theme extends that emphasis by focusing on identity and on the symbolic meanings attached to landscape. With principal conceptual inspirations from humanism and postmodernism, this theme explores landscapes as they are perceived and as they recreate culture and identity. It also examines folk and popular groups that can be identified by consumption preferences and related landscapes.

Concluding Comments

Cultural geography is a contested arena in that there is an abundance of approaches and research interests, with a particular distinction to be made between traditional and newer concerns. The Sauerian view of culture and the related landscape school, first introduced in the 1920s and linked to the larger modernist project, continues to dominate much contemporary research—particularly in North America—with ongoing interests in human and land relationships, landscape evolution, and regions. The new cultural geography that emerged in the 1970s has reservations about the larger modernist project, is critical of attempts to reify culture, emphasizes group and landscape inequalities, and considers symbolic as well as material landscapes.

From the 1920s to the 1970s a Sauerian view of cultural geography dominated in the United States but was absent in other English-speaking countries and in Europe, possibly because the superorganic interpretation of culture never appealed to either anthropologists or geographers outside the US. A comparable area of investigation in Britain was known as social geography.

The key concept of society was similar to that of culture in that it allowed for discussions of group ways of life and related landscapes.

There is no doubt that landscape school interests continue to be major areas of research endeavor. Indeed, even after the new cultural geography had been around for more than a decade, an American survey showed that 'mainstream cultural geography seems satisfied with the superorganic' (Rowntree, Foote, and Domosh 1989:212). These interests are more conceptually diverse than was the case early in the twentieth century, but the intellectual antecedents are clear. The new cultural geography interests are less established and have not generated as much research activity to date, but they are clearly the major growth areas.

But to emphasize the contested nature of cultural geography at the expense of the shared interests is misleading. As one part of the larger social scientific endeavor, cultural geography changes as the needs of society change and as additional ways of conceiving the subdiscipline appear. Differences between older and newer versions of cultural geography are fewer than has sometimes been asserted, and it may be that many of the purported differences result from the understandable tendency of those introducing new ideas to do so through a rejection of earlier ideas.

For this textbook the new cultural geography is seen as a welcome addition to, not a replacement for, the more traditional interests. Accordingly, this textbook

- acknowledges the presence of different ideas and approaches, but explicitly rejects any suggestion that there are grounds for excluding some ideas and approaches;
- reflects the tradition of cultural geography as initiated by Carl Sauer in the 1920s; and
- capitalizes on the reassertion of the cultural in geography that reinvigorated the subdiscipline in the 1970s and continues to stimulate important research today.

Further Reading

The following are useful sources for further reading on specific issues.

In a challenging philosophical text, Sack (2003) develops the idea of the human creation of landscapes that ought to be.

Gritzner (2002) outlines the study of geography as 'What is Where, Why There, and Why Care?'

Throughout this text there is a concern with both concepts and analyses—concepts shape cultural geographic knowledge while analyses—performed by cultural geographers—test these concepts. The need to consider both concepts and analyses is well established in social science generally and is demonstrated effectively for human geography by Hubbard, Kitchin, Bartley, and Fuller (2002).

Francaviglia (1978) provided the seminal description of the visible Mormon landscape.

Sibley (1995) discusses Gypsy identity and landscape, and Holloway (2003) explores the racialization of Gypsy identity and the contested construction of Gypsy places.

Morin (2003) provides a concise history and geography of landscape studies, while Olwig (2002) provides a counterpoint to scenic understandings of landscape.

Philo (1988) stresses the need to integrate cultural, social, economic, and political processes in human geographic research. Barnes (1996) comments on the cultural understood as materialized in the economic.

Knox and Agnew (1994), Berry, Conkling, and Ray (1997), Lee and Wills (1997), and Taylor and Flint (2000), are examples of textbooks that include cultural processes in their accounts of political or economic geography. Wheeler (2002) outlines the increasing cultural content of urban geography. Barnes (2003) discusses culture and economy in geography, and Gertler (2003) outlines a cultural economic geography of production.

Berger and Luckman (1966) discuss the social construction of knowledge.

Sopher (1972) outlines a spatial analytic cultural geography.

Feder and Park (1997:10) discuss the meaning of culture in anthropology, and Giddens (1991:35) discusses the meaning of society in sociology. Kuper (1999) provided a comprehensive account of the culture concept in anthropology.

Cosgrove (1978) and Jackson (1980) are important early statements of the new cultural geography; McDowell (1994) and Kong (1997) offer overviews; Price and Lewis (1993a) and Mitchell (1995) provide critical accounts.

Jackson (1989), Shurmer-Smith and Hannam (1994), Crang (1998), Anderson and Gale (1999), Cook, Crouch, Naylor, and Ryan (2000), Mitchell (2000), Shurmer-Smith (2002), Anderson, Domosh, Pile, and Thrift (2003a), Blunt et al. (2003), and Duncan, Johnson, and Schein (2004) are cultural geography books that choose to exclude much if not all of the Sauerian tradition.

Murphy and Johnson (2000a) edited a tribute to Marvin Mikesell that appropriately includes a wide range of cultural geographic studies.

Blaut (1993a) discusses cultural geography as a contested area with a particular distinction made between traditional and newer concerns.

Smith (2003) summarizes a survey of perceptions of cultural geography held by cultural geographers.

For accounts of changing approaches to cultural geography, see chapters 3, 4, and 5 in Duncan, Johnson, and Schein (2004).

R.J. Johnston (1997) is the fifth edition of an influential book, *Geography and Geographers*, which, for the first time, includes a chapter titled, 'The Cultural Turn'.

Hoggart (1957), Williams (1958), and Thompson (1968) are important works anticipating the British cultural studies tradition.

Geertz (1973) details the anthropological concern with the study of otherness.

Said (1978, 1993) identifies some of the biases that would be evident if a specifically European view of the world is adopted.

Philo (1989), Parr and Philo (1995), and O'Dwyer (1997) comment on particular identities and behaviors.

2

The Tradition of Cultural Geography

This chapter outlines the tradition of the cultural geographic enterprise, especially as it developed in the late nineteenth and early twentieth centuries as part of the larger institutionalization of geography as an academic discipline. The chapter focuses particularly on the geographic perspectives of environmental determinism and the landscape school, and on the ideologies through which humans understand their earth-transforming activities. In this chapter you will find

- a discussion of the reasons for viewing humans and nature as separate—what is often called the *dualistic preference*—and for distinguishing between presumed different types of human;
- an account of environmental determinism—that is, nature as it impacts on humans—as the principal example of dualistic thinking in the geographic tradition;
- a discussion of approaches that emphasize the human use of nature, rather than the impact of nature on humans;
- an outline of the landscape school approach to cultural geography, which was introduced in the 1920s and remained largely unchallenged until the 1970s. Possible links to the influential anthropological concept of culture as superorganic are stressed;
- a summary of alternative ecological traditions that favor attempts to integrate humans and nature—what is often called the *holistic preference*. Included here are the controversial claims of evolutionary naturalism;
- some concluding statements that reaffirm the value of understanding contemporary cultural

geography in both larger historical and disciplinary contexts.

An overriding concern in this chapter, as the outline of the chapter's contents suggests, is the broad issue of human relationships with nature and related landscapes, a concern that prompts three questions. First, when we refer to nature, are we limiting the concept to what geographers typically label the physical environment—geology, landforms, climate, weather, flora, and fauna—or do we understand humans as belonging to nature? Expressed simply, are humans seen as *apart from* or *a part of* nature? For many cultural geographers working in the classic landscape, or Sauerian, tradition, the separation of humans and nature was taken for granted. Indeed, the phrase 'humans and nature' immediately identifies a separation of the two, a dualism that is a traditional component of Western thought. Today, there are several perspectives on this important question.

Second, is human behavior affected, perhaps even determined, by physical environment, or does culture play a determining role? The emphasis on separating humans and nature resulted in what is perhaps an unfortunate tendency, that of privileging one at the expense of the other, and there are numerous examples in this chapter of this tendency towards **reductionism**. Both versions of determinism—environmental and cultural—played important roles in the history of cultural geography. Building largely on previous work, many early twentieth-century cultural geographers assumed that human behavior and cultural landscapes resulted primarily from characteristics of the natural world. Major challenges to this view were put forward by French geographers, especially Vidal, and by Sauer. Sauer's work has in turn received criticism for favoring a cultural determinist approach.

Third, what distinguishes humans from other living species, particularly from other primates, and why have some cultures often regarded other cultures—and also some groups within their own culture—as being closer to other living species than they are themselves? Most early twentieth-century cultural geographers simply ignored this question, as it was not seen as necessary to an understanding of cultures and landscapes. More recently, however, there has been much interest in uncovering the political component of culture and in identifying any unequal identities and landscapes found as a result.

Overall, responses to these three questions changed markedly as the larger cultural geographic enterprise changed, first with the development in the 1920s of the landscape school and later with the emergence, beginning in the 1970s, of the new cultural geography. Most notably, the various challenges to the landscape school initiated in the 1970s recognized that the earlier cultural geography had ignored some important matters. Consensus views are not yet evident, although one important development is that it is becoming increasingly popular to talk about the social construction of nature (as this term was introduced in Chapter 1), with the broader concern being the way in which we divide our world through our language as one means of naming, defining, and imposing order on that world.

Pause for Thought

Regardless of cultural background, it is likely that the phrase 'humans and nature' is part of your taken-for-granted world. For most people socialized in a Judeo-Christian tradition—which is the tradition that has had the greatest impact on the cultural geography considered in this book—it is usual to conceive of the two parts of the phrase as both separate and unequal entities, with nature subordinate to humans. But such a view of the world does not prevail in some other traditions, including Buddhism, Hinduism, and many indigenous cultures, in which humans are more characteristically seen as part of nature, neither separate nor superior. Is one view correct, and the other incorrect? Or is one view in some way better than the other? Or are the differences between cultural traditions best seen as simply different ways of thinking, different ways of structuring the world in which we live? These are challenging questions, and you will be asked to rethink this matter at the conclusion of this chapter when you have some additional bases for formulating a response.

Separating Humans and Nature

Humans have probably always wondered about their place in the world and about their relationship with nature. People have raised these questions in various cultural contexts and in many different ways that tend to reflect prevailing religious attitudes and scientific knowledge. One characteristic feature of Western thought is the presumed separation of humans and nature and the contrasting of one with the other to emphasize difference. The authoritative account of the changing views of humans and nature in the Western world from the classical period through 1800 is by Glacken, who began:

In the history of Western thought, men have persistently asked three questions concerning the habitable earth and their relationships to it. Is the earth, which is obviously a fit environment for man and other organic life, a purposefully made creation? Have its climates, its relief, the configuration of its continents influenced the moral and social nature of individuals, and have they had an influence in molding the character and nature of human culture? In his long tenure of the earth, in what manner has man changed it from its hypothetical pristine condition? (Glacken 1967:vii)

In a brief consideration of nineteenth-century thought, Glacken (1985) acknowledged the rise of a fourth question: how have humans interpreted nature in esthetic, subjective, and emotional ways?

Each of these four questions implies a separation of human and natural worlds. Although other traditions asked different questions that often stressed integrated rather than separated human and natural worlds, it is Western thought, as noted earlier, that has been most influential in the evolution of cultural geography. So, why is it that Western ideas about humans and nature have typically seen the two as separate entities?

Historical Perspective

GREEK AND CHRISTIAN THOUGHT

The first of Glacken's three questions concerns the idea of a designed earth, a view that reflects the doctrine of **teleology**. In accord with this idea, **Stoicism**, a philosophy that flourished in Athens after about 300 BCE and in Rome some 400 years later, held that because the natural world was clearly suited to human life, then nature must have been created for humans. All things had a purpose, with plants created for animals and animals created for humans. Many of these views, and similar ideas from Judaism, were absorbed into early Christian thought, and they have often been discussed in terms of the Biblical story of the Garden of Eden and the Fall, in which God grants humans dominion over nature: *Be fruitful and multiply, and fill the earth and subdue it; and have dominion over the fish of the sea and over the birds of the air and over every living thing that moves upon the earth* (Genesis 1:28). But as soon as

This seventeenth-century Flemish painting of Adam and Eve in the Garden of Eden is typical of many such representations. Although Adam and Eve are clearly in awe as they receive instructions from God, the animal world is depicted as subservient to the human figures while the luxuriant vegetation suggests fruitfulness and productivity. Such paintings both reflected and reinforced prevailing ideas about humans dominating and using the natural world. (Scala/Art Resource, NY)

Adam and Eve ate from the forbidden tree, Adam removed himself from the natural world of the garden, emancipating the human soul from nature. This separation was followed by God again granting humans dominion over all animals after the flood: *Every moving thing that liveth shall be meat for you* (Gen. 9:3). The account of the Fall in the Christian world and the resulting dualism of humans and nature can be contrasted with other **creation myths**, such as those of many indigenous cultures, that often involve seeing humans as a part of the natural world.

THE SCIENTIFIC REVOLUTION

Stoic, Jewish, and Christian traditions of a designed earth and the right of humans to rule over all other living things were widely accepted in Europe for many years. They were influential in the birth of modern science in the sixteenth century, in the philosophical movement of the Enlightenment in the seventeenth and eighteenth centuries, and in the rise of positivism in the nineteenth century (see Box 2.1). Each of these developments further contributed in some way to the idea that humans and nature were separate. Descartes, Hobbes, Locke, and Hume carried on the scientific method advanced by the empiricist philosopher Francis Bacon (1561–1626), which broadly speaking, had both a mechanistic worldview and a belief in human control of nature as central ideas. For Bacon, a key purpose of the scientific endeavor was to emphasize human dominion over nature, a dominion that some saw as partly lost as a consequence of the

Box 2.1 **Science and the Enlightenment**

The seventeenth century began with Bacon asserting the need to study nature scientifically and ended with Isaac Newton (1642–1727) vindicating this faith in the power of science with an explanation of the workings of the solar system, the *Mathematical Principles of Natural Philosophy*, which was published in 1687. Newton demonstrated the power of human reasoning as a vehicle to understand the world, a scientific revolution that facilitated the rise of the Enlightenment by ensuring that science had a philosophical partner.

The term 'Enlightenment' refers to an intellectual movement that began with the English philosopher John Locke (1632–1704) in the seventeenth century, although it was the French Revolution that effectively established the ideals of liberalism, equality, and popular sovereignty that characterized the 'age of reason'. Scholars today often contrast the Enlightenment with the earlier medieval period of presumed darkness, irrationality, and lack of reason. Among the principal doctrines of Enlightenment thought are the following ideas:

- that reason enables humans to think and act correctly;
- that humans are, by nature, both rational and good;
- that all humans are equal and all ways of life should be tolerated; and
- that local prejudices are devalued because they result from particularities rather than from reason.

Enlightenment philosophers initiated the modern age by looking forward, rather than back to Greek or Biblical thought. Together with the scientific revolution, the Enlightenment allowed scholars to believe that the world, both physical and human, was accessible and comprehensible. Humans could understand all things, and the growing understanding of physical laws and chemical laws fostered the belief that there must also be human laws. Enlightenment thinkers anticipated that this new knowledge would lead to increasing control over the environment, improvement in the human condition, and prediction of the future—all of these components of the emerging modernism.

But the late nineteenth century witnessed considerable strain on many Enlightenment ideals. It was, for example, all too clear that, contrary to expectations, much of the success of Europe in the modern age was built on the exploitation of others rather than on any principle of equality. And the confidence of Enlightenment thinkers in the ability of science and human reasoning did not go unchallenged. Practically, there was powerful opposition from both established religion and from many authoritarian European monarchs. Philosophically, there was opposition to applications of the naturalism of physical science to social science. As noted in Box 2.2, the principal philosophical opposition was from the idealist perspective.

Fall. Indeed, natural history, comprising both botany and zoology, was studied not only for reasons of scientific curiosity, but also for explicitly practical and utilitarian reasons.

ENLIGHTENMENT AND
NINETEENTH-CENTURY THOUGHT

The empiricist philosophy that guided the practice of this early science asserted that all factual knowledge is based on experience, thus denying the view that the human mind is provided with a body of ideas before it encounters the world. This empiricism has proven to be the dominant **epistemology** and method of science, including the early social sciences that established the credibility of the discipline in the nineteenth century.

As an outgrowth of the development of science, Enlightenment philosophy carried on the established empiricist tradition. Nature was seen as something that could be explained in terms of a system of universal laws, such as the law of gravity, rather than as a number of unrelated parts understood only by reference to God. Continuing Christian tradition, humans were seen as possessing immortal souls and as dependent on the natural world, a concept known as **materialism**. Thus, human behavior was explicable by reference to laws similar to those being discovered in physical science; this is the concept of naturalism as discussed in Box 1.1. Together, these closely related views (some philosophers argue that they are virtually synonymous) further contributed to the rise of positivist social science in the nineteenth century.

The positivism that came to the fore in nineteenth-century social science is associated with the writings of the French philosopher Auguste Comte (1798–1857), although the basic ideas were anticipated by the empiricism of Bacon, and this positivism has close links with both materialism and naturalism. Recall that the key assumption of empiricism is that knowledge is dependent on experience. Expressed simply, Comtean positivism claimed that all knowledge, human and natural, is derived from systematic study of the world and the formulation of laws about the world.

Not all scholars accepted that human observers had access to an objective view of nature, unaffected by subjective considerations. Friedrich Nietzsche (1844–1900), for example, insisted that it was not possible to discover nature, only to imagine and interpret nature. Nevertheless, the principal social science disciplines that developed in the nineteenth century—economics, sociology, anthropology, psychology, and, to a lesser degree, geography—became typically empiricist and positivist in emphasis.

The linking of human and natural worlds within one method of science did not lead to any real integration of the two subject matters. Although they were to be studied by the same method, they remained separate. Building on these comments, Box 2.2 summarizes two important philosophical debates that are important to cultural geographers but that remain unresolved.

What It Means to be Human

Although traditionally most Western thought took it for granted that humans were fundamentally different from other forms of life, ideas about the separation of human and natural worlds spurred interest in determining precisely what it was about humans that made them different.

Aristotle observed that only humans possess a rational or intellectual soul, and Christian theology insisted that to be human was to have an immortal soul. But such requirements did not make it easy to distinguish humans from animals and, from Aristotle onwards, much thought has been given to explaining the distinction. Regardless of detail, it has been usual to see the two as different in some fundamental way and to regard humans as superior. In addition to stressing anatomical distinctions, three specific characteristics distinguishing humans were proposed: speech, the ability to reason, and the possession of a conscience and religious instinct. Descartes added to the debate by proposing that, like animals, humans resembled machines in that many of their actions were instinctual, but that only humans also possessed an intellect. Ideas such as these served to confirm the separation of humans and animals and, more generally, of culture and nature. Animals were awarded low moral status, justifying such activities as hunting, domestication, the eating of meat, extermination of undesirable species, and the displaying of animals in enclosures for scientific, educational, and entertainment purposes.

DIFFERENT HUMANS

Ideas related to these sometimes prompted European thinkers, from the Greeks onwards, to relegate certain classes of people—different cultural groups, the insane, the poor, children, women—to the status of lesser animals. Thus, the various attempts to explicitly define *human* had implications not only for the way in which the natural world was treated but also for the way humans treated other humans. Further, it was usual in both Western and other traditions to regard nature as female. There were two contrasting components to this image. On the one hand, especially prior to the birth of modern science, nature was seen as a nurturing mother providing humans with their various needs. On the other hand, there was the wild aspect of nature, suggesting the need for control.

Women are devalued in most societies, a fact that has been interpreted in terms of this association of women with nature, where nature is something less than human, women are closer than men to nature, therefore women are inferior humans. Why are women seen as closer than men to nature? It may be because it is women who procreate and because of the related domestic social roles that women play. Men, on the other hand,

This illustration comes from Henry Sharpe Horsley's *The Affectionate Parent's Gift; and the Good Child's Reward* (1828), a collection of poems and essays designed to lead the tender 'Mind of Youth in the early Practice of Virtue and Piety'. The illustration accompanies a poem called 'A Visit to the Lunatic Asylum', which begins: 'Come child with me, a father said, / I often have a visit paid / To yon receptacle of woe, / For Lunatics.—Come, child, and know, / And prize the blessing you possess, / . . . / Come, shed a tear o'er those devoid / Of what you have through life enjoy'd: / See, in this mansion of distress, / The throngs of those who don't possess / Their reason.' These lines reflect the common view of those suffering from mental illness as somehow less than human.

have traditionally occupied a position closer to culture. An additional distinction that can be made is that traditionally men are concerned with the welfare of the social whole while women are concerned with more particularistic family matters. This is the classic distinction made between the male, public domain and the female, private domain.

It is difficult to exaggerate the significance of this 'transcoding of feminine qualities to Nature, and of naturalness to women' for the cultural geographic enterprise, both traditional and new (Rose 1993:69). Indeed, it might be argued that all of human geography is structured around the separation of humans and nature, especially as this is reinforced by the distinction between urban and rural ways of life that became increasingly apparent in the nineteenth century.

These various ideas concerning human authority over nature, the perceived variable human quality of particular people, and the linking of women with nature are themes that recur frequently throughout this textbook.

ADVANCES IN NATURAL AND SOCIAL SCIENCES

During the nineteenth century, developments in the natural and social sciences improved the methods used to distinguish humans from nature. Advances in geology and biology enabled natural scientists to clarify the **evolution** of the human species, while social scientists began to focus explicitly on the concept of culture as an exclusively human attribute. Although it is correct to note that these advances that provided new biological and cultural understandings did not allow scientists to make absolute distinctions between humans and nature, they did help to clarify some of the ideas previously used in this endeavor. More impor-

Box 2.2 Two Philosophical Debates

Students of cultural geography may well be forgiven for asking why certain philosophical concepts are being presented in this textbook. The answer is straightforward: the questions cultural geographers ask, the approaches they employ, the data they choose to discuss, the results they consider relevant—all are affected by philosophical preferences.

For our purposes, an understanding of twentieth-century cultural geography is enhanced by awareness of two fundamental but unresolved philosophical debates. Issues related to these two debates arise frequently in this chapter and generally throughout this book. Accordingly, this box provides brief summaries of these debates.

First, there is an epistemological debate about naturalism. Is it or is it not appropriate for social scientists to adopt a naturalist perspective? The success of the scientific revolution and of Enlightenment claims encouraged the idea that the methods of science could be applied to humans as well as to physical objects. As a result, many nineteenth-century social scientists began the search for laws of human behavior. This application of the mechanistic methods of physical science to social science is what is meant by the term 'naturalism'. The history of social science, including cultural geography, is full of examples of competing claims concerning the validity of a naturalist perspective. These debates are not resolved today, although naturalism is certainly out of favor relative to alternative approaches. This text has already touched on this issue in Chapter 1 with,

for example, references to deterministic approaches, which are naturalist, and references to versions of **humanism** that are opposed to naturalism.

Naturalism incorporates an empiricist focus, stresses the search for causes, and is typically positivist. Opponents of naturalism argue that humans cannot be treated as though they were akin to physical objects, favoring instead approaches that focus on **hermeneutics**, sometimes called *verstehen*. Hermeneutics refers to the study and interpretation of meaning in everyday life. There are many versions of hermeneutics, but all share a belief in the centrality of the mental quality of humans, asserting that, unlike physical objects, humans have needs and desires that affect their actions, and that these must be considered. From this perspective, human behavior is motivated by something inside the individual, such as feelings or perception. For many social scientists, this is *the* philosophical debate. Traditionally, naturalism has been strong in English-speaking areas, in the disciplines of economics, psychology, and sociology, and in the writings of Durkheim and Parsons. Hermeneutics has been more evident in Germany, in humanistically inclined social sciences such as history, and in the writings of Weber.

A second unresolved philosophical debate is an ontological debate about realism. **Realism** is the philosophical view asserting that the existence of material objects is independent of sense experiences. Thus, realism confirms what is essentially the common-sense view of the world, but it is consid-

tant, a consideration of these developments in the physical and social sciences reveals that both are relevant in any attempt to define what it means to be human.

Both geology and biology developed dramatically from the late eighteenth century through the nineteenth century. In the case of geology, these developments caused scientists to question seriously the Biblical account of the earth, which is based on a belief in **catastrophism** and which attributes the earth a maximum age of roughly 6,000 years. In 1774 the French scientist and pioneer of paleontology Count Buffon (1707–88) advanced the concept of **uniformitarianism**, the idea that changes to

the earth had occurred over time as a result of such observable physical processes as erosion and deposition, not because of a number of isolated and unobserved catastrophes. This perspective of the earth as having been shaped over time by natural, uniform processes was further developed by two Scottish geologists, James Hutton (1726–97) and Charles Lyell (1797–1875), whose influential texts, produced in 1788 and 1830 respectively, made it evident that the earth was much older than the 6,000 years previously believed.

These advances in geology were not the only developments to alter the established view of natural science at this time. Through

ered to be a philosophical position because an important case against it has been made. The case against realism is based on the system of thought known as idealism, a term that, in this context, refers to a group of philosophical doctrines claiming that what humans know of the world is fundamentally the product of the human mind. The idealist perspective does not oppose the view that material things exist, but it does oppose the claim that this material world is completely independent of the human mind.

This debate about realism may seem abstract and unimportant, and it is certainly not so easy to translate directly into the practice of cultural geography. Nevertheless, it does have a significant epistemological component concerning what role, if any, mental constructions play in our knowledge of the world. To be realist implies a rejection of the view that social science must involve a search for regularities and the testing of hypotheses, and a rejection of the view that social science must involve the interpretation of meaning. If this sounds a little like having your cake and eating it, note that in a sympathetic discussion, Yeung (1997) acknowledges a difficulty that a realist researcher faces, namely that realism is a philosophy that lacks a method.

In conclusion? These two philosophical debates are not easy to summarize in a few words because they reflect complex issues. However, the debates—and they are ongoing debates that have not been resolved by philosophers—are fundamental considerations to be taken into account by cultural geog-

raphers and others who attempt to develop ideas about culture and nature. Indeed, it is because of these unresolved debates that the authors of an introduction to a discussion of social theories stressed that the current lack of consensus might be endemic to social science: 'At the very least, whether there can be a unified framework of social theory or even agreement over its basic preoccupations is itself a contested issue' (Giddens and Turner 1987:1).

Examples of naturalist approaches mentioned in this chapter include environmental determinism, the superorganic, and sociobiology. But most approaches currently favored in cultural geography are unsympathetic to naturalism, choosing instead to adopt some version of the hermeneutic perspective, such as one of the several humanist approaches or postmodernism. Indeed, the challenge from hermeneutics has not left even the physical sciences unscathed, with some postmodernists arguing that both physical and social sciences involve an interpretive ordering of reality; such claims are evident in some feminist work. The position of cultural geographers in the debate about realism is less clear, partly because the philosophical issues are more difficult. Thus, some versions of realism can be naturalist in emphasis, while other versions can incorporate hermeneutics; indeed, critical realists insist that hermeneutics is the starting point, but not the sole concern, of research. But note that naturalism and realism, though they are both opposed to the idealist perspective, are not equivalent approaches.

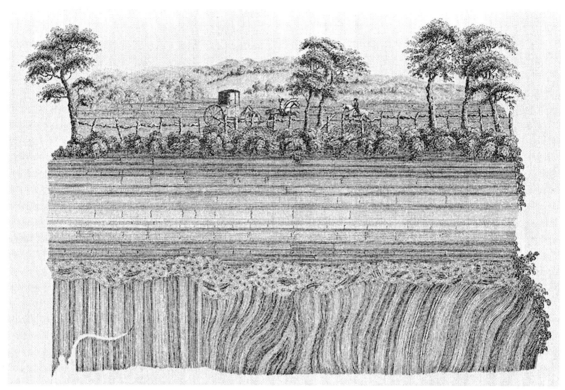

James Hutton presented his theory of the earth before the Royal Society of Edinburgh in March 1785. The first full version appeared in 1788, and his massive two-volume *Theory of the Earth with Proofs and Illustrations* was published in 1795. Hutton recognized that an unconformity—a discontinuity in the rocks, reflecting a gap between two episodes of rock formation—was the most dramatic evidence that changes to the earth had occurred over time because of such observable physical processes as deposition and uplift, not because of isolated and unobserved catastrophes. This illustration shows John Clerk of Eldin's famous conceptual rendering of the Jedburgh unconformity. The people are included as a means of contrasting the present with the past. (Reproduced by permission of Sir John Clerk of Penicuik)

developments in the study of plant and animal biology it was becoming clear that not only had the earth evolved slowly into its present form, but animal and plant species had also evolved. The first proposal for biological change was put forward by the French naturalist Jean-Baptiste de Lamarck (1744–1829). Lamarck correctly noted that animals and plants were adapted to their specific environments, but his explanation, known as **Lamarckianism**, as to how such evolution occurred—namely, that organisms develop into more complex forms— was incorrect. It was English scientists Charles Darwin (1809–82) and Alfred Russel Wallace (1823–1913) who, independently, proposed **natural selection** as the solution to the question of plant and animal evolution put forward

by Lamarck. Natural selection refers to the process by which nature effectively selects members of a species that are most suited to the environment; hence the most adaptive traits are reproduced and increase while the least adaptive traits decrease. Darwin published his monumental *On the Origin of Species by Means of Natural Selection* in 1859.

Darwinism as a theory is non-teleological in the sense that it explains how an event contributed to some result but does not explain the occurrence of that event. Thus, species variations occur randomly, but those variations chosen by the environment were not designed with some particular result in mind. For many, this absence of teleological explanation means that the basic logic cannot be directly applied to cultural change, an important consideration in assessing attempts to apply Darwinism to the social sciences. The Lamarckian evolutionary process, involving the idea that the organs and behavior of species can change and that new organs and behavior can be transmitted from individual species members to their offspring, may be more appropriately applied in a cultural context.

Building on these advances in natural science, natural and social scientists began to ask and answer questions about the evolution of one species in particular: humans. The natural science answer—still incomplete—was based on biological developments in such areas as brain size and the ability to walk upright. One important approach addressing this issue is sociobiology. The principal social science answer was based on the always uncertain and contested concept of culture and on the assumption that cultural evolution had occurred. An alternative social science answer argues that we have an understanding of nature, but not of ourselves. These three bodies of ideas are now discussed.

Sociobiology

Human beings have a great deal in common both biologically and culturally, and it is critical that cultural geographers not lose sight of this essential human unity. But humans also have much in common with some animal species, such that we can say that humans are a part of nature. At the same time humans possess culture, and this represents a difference from nature. In short, humans are both natural and cultural beings. It is this recognition that, for many, justifies the argument that most studies of humans are too **anthropocentric** and that much can be learned by treating the human cultural experience as but one part of the larger biological continuum.

Sociobiology is a particular expression of the relationship between biological and cultural or social evolution that argues for a reduction of human behavior to biological concepts. In *The Descent of Man*, Darwin studied social organization generally by applying evolutionary theory combined with the biological knowledge he

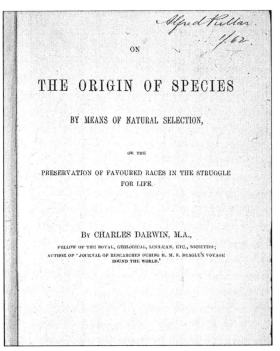

Charles Darwin's classic, *The Origin of the Species*, published in 1859, maintained that the process of natural selection tends to favor the survival of those who are best adapted to their environment. (Thomas Fisher Rare Book Library)

derived from observing both animal and human behavior. This early linking of biology and human subjects was criticized from religious quarters, and also by sociologists such as Durkheim, who argued for the autonomy of sociology and hence for separation of the human and the natural. Versions of this debate have resurfaced in recent years with a number of sociobiological theories being proposed.

Several sociobiological theories make explicit use of Darwinian concepts to describe a process of natural selection of cultural characteristics that produces, through time, an increasingly complex, increasingly well-adapted, and hence improved culture. Changes in human culture may be seen as similar to organic mutations, with parallels between biological and human cultural evolution along with the added idea that cultural evolution is a form of learning process leading to ever-improving human circumstances. This is a particular interpretation of behaviorism as developed in psychology. But the most elaborate interpretation of biological evolution in human terms incorporates the idea that cultural change is often intentional and purposive, as in the Lamarckian evolutionary process, rather than accidental, as in Darwinian theory. The temptation of relating biological and cultural evolution has also proven irresistible for some biologists, who accept the idea that the process of cultural evolution is Lamarckian. One particular view is based on a set of theories that assert that human behavior is a product of the co-evolution of human biology and culture.

Not surprisingly, sociobiology has been subject to criticism. Most notably, it represents a major challenge to much contemporary social science because its emphasis on a naturalist perspective puts it out of tune with the prevailing social science view of humans as active subjects. Further, feminists have expressed concern with its male construction of science: 'The patriarchal voice of sociobiology is less the effusive sexism that ripples over the whole plane of the text than it is the logic of domination embedded in fashioning the tool of the word' (Haraway 1991:74). Given the generally critical appraisal of sociobiology from the social sciences, as well as related internal disagreements,

it seems unlikely that the approach will make a major impact on the contemporary understanding of culture and culture change regardless of the validity of the arguments. Certainly, the principal cultural geographic initiative in this direction—namely, the habitat and prospect-refuge theories discussed in Chapter 9—have not been widely accepted. At the same time, such ideas should not be ignored, as they reflect a clear concern with the idea that humans are a conventional animal species.

Cultural Evolution

The evolutionary ideas advanced in eighteenth- and nineteenth-century geology and biology were received and applied with great enthusiasm by many nineteenth-century social scientists. Specifically, the thesis that cultures evolved and thus display only limited ties to nature—in other words, that culture changed while nature remained unchanging—was a dominant feature of the early disciplinary experiences of both anthropology and sociology. Although by no means a new idea—indeed it was evident in classical Greece and became especially important during the period of European overseas expansion—the concept of cultural evolution became the subject of numerous theories in the period after about 1860. But despite the number and diversity of evolutionary theories, all of them have three features in common: first, they identify a number of distinct cultural stages; second, they recognize that the probability of remaining in a stage is greater than the probability of returning to an earlier stage; third, they identify a probability of moving through the sequence of stages.

Three great social thinkers of the nineteenth century, Comte, Marx, and Spencer, adopted evolutionist perspectives, while the anthropologists Edward Burnett Tylor (1832–1917) and Lewis Henry Morgan (1818–81) portrayed culture and cultural change in terms of **unilinear evolution**—a series of stages from savagery to civilization. Together, Tylor and Morgan established the comparative method of inquiry, collecting ethnographic data from different cultures in order to demonstrate the increasing complexity of culture through time. Their study reflected an acceptance of the Enlightenment doctrine of progress, and it was implicitly assumed that European civilization represented the apogee of evolution, while primitive peoples were both biologically and culturally inferior. As one component of evolutionist thought, the anthropological typologies employed new data produced by European overseas expansion to support their theories of unilinear evolution, which postulated that all cultures would, by means of independent invention, pass through a series of increasingly superior stages.

In *Primitive Culture*, first published in 1871, Tylor (1924) proposed three stages of cultural evolution and their characteristics as follows:

- savagery—hunting and gathering, limited technology
- barbarism—agriculture, settled villages and towns
- civilization—writing

More generally, Tylor noted:

On the whole it appears that wherever there are found elaborate arts, abstruse knowledge, complex institutions, these are results of gradual development from an earlier, simpler, and ruder state of life. No stage of civilization comes into existence spontaneously, but grows or is developed out of the stage before it. This is the great principle which every scholar must lay firm hold of if he intends to understand either the world he lives in or the history of the past. (Tylor 1916:20)

Although the three stages are distinguished essentially in terms of technology, the greatest emphasis was placed on language, religion, and myth. Further, Tylor employed the concept of cultural survivals, components of a culture that, for reasons of tradition, persist from one stage to a later stage.

In similar fashion, Morgan (1974), in his 1877 book *Ancient Society*, proposed seven stages of cultural evolution and their characteristics as follows:

- lower savagery—beginnings of human life, gathering of wild fruit
- middle savagery—eating of fish, origins of speech, and use of fire

- upper savagery—use of bow and arrow
- lower barbarism—use of pottery
- middle barbarism—agriculture, irrigation
- upper barbarism—use of iron tools
- civilization—writing

Morgan distinguished the stages in technological terms, although he wrote primarily about the evolution of social and political forms and about a transition from communal to private ownership.

A lack of interest in evolutionism in the early twentieth century ended with a revival prompted by Childe in the 1930s and continued by White in the 1940s and Steward in the 1950s. Childe used evolutionist thought in accounts of the origins of agriculture and subsequent development of an urban way of life. White employed a superorganic concept of culture, specifically a form of technological determinism, to argue that technology determined the availability of energy and the efficiency of its uses, and further, that the structure and successful functioning of culture was the direct result of the amount of energy produced and the manner in which that energy was used. Technology, through energy, was therefore the cause of cultural characteristics and change. Steward, a student of both Kroeber and Sauer, was critical of earlier unilinear evolutionist ideas, especially their often grand claims and gross generalizations. He favored instead a more modest **multilinear evolution**. The aim of multilinear evolution was to discover the laws that determined cultural development by identifying parallels in the development of different cultures.

Boas initiated arguments against evolutionism in anthropology with strong support from Kroeber, Lowie, and Wissler. He based his criticism on the belief that the crucial cause of cultural change was **diffusion** and related culture contact, not independent evolution. He also raised objections to the racism implicit in evolutionism's emphasis on progress, to the lack of objectivity in evolutionary theory, and to the rigidity of the sequences proposed by Tylor and Morgan. Boasian anthropologists emphasized diffusion and contact as these related to the creation of cultural areas. He described human history 'as a sort of "tree of culture", with fantastically complex branching, intertwining and budding off—each branch representing a uniquely different cultural complex, to be understood in terms of its own unique history rather than compared in cultural complexes in other world regions in some grand scheme of "stages of evolution'" (Pelto 1966:24).

Although the evolution of culture is seen as a central component of the landscape school methodology—it is one of the five themes identified by Wagner and Mikesell (1962) (see Box 1.3)—it is nevertheless an interest that has been ignored by most cultural geographers (but see Newson 1976). It appears that cultural geography's interest in landscape outcomes of cultural occupance largely precluded any detailed concern with cultural change, and, further, that such discussion was seen as more properly belonging to the discipline of anthropology. Indeed, the principal evidence of evolutionist thought in human geography is in association with environmental determinism.

Understanding Ourselves

The traditional way of defining *human* was to distinguish between humans and other animals in order to identify some presumed distinctive characteristics, such as having been made in the image of a Christian God, possessing a soul, or being spiritual as well as corporeal. But at the same time, humans have always been intrigued by the possibilities of intermediate species, of creating human life from non-living things, or of creating machines that are akin to humans. The basis for distinguishing humans and other animals lessened with mechanistic logic, which provided explanations of the whole material world—humans and nature alike—and with the advances made by Darwin that showed that human biological evolution was subject to laws similar to those governing the rest of the natural world. To offer a conclusion on this matter is difficult; it may be that a persuasive definition of what it means to be human is impossible. One perspective on this matter is discussed in Box 2.3.

Environmental Determinism

Nineteenth-century thinking about humans and nature incorporated some contradictory elements, but the dominant view was to see the two as separate. Viewing humans and nature as sep-

Box 2.3 Achieving Effective Knowledge of Ourselves

According to Gellner (1997), effective knowledge of nature exists, but effective knowledge of humans and their institutions does not. During the long period between the initial domestication of plants and animals and the scientific and industrial revolutions, European culture was based on the production and storage of food, on a rather unchanging technology, and on minimal advances in science. Accordingly, a Malthusian situation prevailed, with resources remaining essentially constant and any increases in population or decreases in food supply creating stress; this is, of course, the first stage of the demographic transition model. Under these circumstances, lack of sufficient food for the least powerful was an ever-present possibility, causing the social hierarchy to function essentially as a queue to the storehouse. The principal concern for individuals and groups was, therefore, not with the issue of total food production, but rather with their ability to exert power over others in order to enhance their social status, or, more simply, to move up in the queue. Accordingly, the prevailing cultural values were those that enhanced personal and group authority and strength, and not those related to technological progress and innovation.

Under these conditions, humans lacked any real knowledge of either nature or themselves. There was necessarily some understanding of parts of the natural world, but the integration of ideas was not encouraged in circumstances where the most powerful assumed ownership of new knowledge and innovation in order to further enhance their status. So how did this situation change? The answer lies in the rise of northwestern Europe beginning in the seventeenth century, which involved a transition from coercive or martial society to productive or commercial society. In Marxist terms, this is a change from a feudal to a bourgeois order. There had been similar transitions at earlier times that proved unsuc-

arate did not, however, prevent scholars from holding the view that both were susceptible to study by the same scientific method expressed in the philosophy of naturalism and the associated idea of positivism. For the discipline of geography, the separation of nature and humans was most evident in the explicit distinction typically made between physical geography and human geography, but it was apparent also in the acceptance and application of the idea of environmental determinism, an approach to the study of geography that was closely related to naturalist epistemology.

The doctrine of environmental determinism asserts that the physical environment—nature—determines or at least influences the human world, both culture and cultural landscape (see Figure 2.1a). Thus, all of our human actions are to be understood primarily if not exclusively in the context of the physical world of climate, landforms, soils, and vegetation. This idea was often treated as though it were a law of science rather than an assumption. Recall that concern with physical geographic impacts on humans was one of the three questions identified by Glacken (1967) as characterizing Western thought about humans and nature from Greek times through to about 1800.

This concern is one particular interpretation of the larger philosophical position of determinism—the idea that all events are effects of earlier events. In physical science, determinism was the dominant view for at least 200 years after Newton, although it has now been replaced by the indeterminism of contemporary physics. As noted, the social science disciplines found determinism very appealing at the time of their foundation, as such arguments appeared to give credibility to their specific disciplinary subject matter and to render their enterprise scientific in the nineteenth century sense of science as empiricist and positivist.

Pause for Thought

Determinism implies that events have causes and that, in principle, it is possible to uncover those causes and perhaps even to formulate laws that relate causes and effects. Applied to human behavior, this idea has profound implications concerning free will and moral responsibility for our actions. If the argument that all of our behaviors are predetermined by prior events is accepted, then does this

---

cessful in changing the larger cultural circumstances. This particular transition was successful because it occurred in conjunction with a second development, namely the rise of science.

Accordingly, since the seventeenth century, success in European cultures has not been based only on power, but also on a sound knowledge of nature, on how to increase production, on progress. The contemporary world thus has a sound grasp of nature and an ability to change nature, but no equivalent grasp of ourselves or how to change ourselves in desirable ways—a process some call *social engineering*. There are five possible reasons for what we might regard as our continuing failure to achieve an understanding of ourselves comparable to our understanding of nature. These are:

- the inherent complexity of the human world;
- the idea that human behavior is affected by human ideas and interpretations (refer to Box 2.2);
- the idea that everything humans do is linked to everything else that human do;
- the idea that cultural evolution is Lamarckian;
- the role played by chance in human affairs.

In discussing the prospects for advances in understanding ourselves comparable to advances that we have made in understanding nature, Gellner (1997:16) concludes that

we simply do not know whether the social sciences will have their 17th century, whether the breakthrough and the subsequent fall-out will occur. What we do know is that *if* it did occur, it would, once again, completely change the rules of the game, as radically—perhaps more so—than was the case when technologically based natural science made possible Perpetual Growth and the vision of World-as-Progress.

mean that humans cannot be held responsible for their actions? If this is the case, do you agree that this is unacceptable? Are you surprised that social sciences were attracted to determinist logic, especially as applied at a group rather than an individual scale? Moreover, are you concerned to hear that the social sciences used determinist logic without offering any meaningful critique of the larger philosophical context? It might be helpful, when considering environmental determinism as developed and applied within geography, to have preliminary responses to these critical questions in mind.

Historical Perspective

Environmental determinist thought has been supported throughout its long history by countless scholars, many of them writing from very different philosophical positions. Indeed, some combined thoughts about environmental influence with thoughts about human impact on the earth, the two being divergent but not contradictory ideas. The following sections trace the history of environmental determinist thought.

GREEK AND CHRISTIAN THOUGHT

Early references to environmental determinism in classical Greece are found in the writings of Hippocrates (fifth century BCE), as Glacken (1967:80) explains: 'From early times there have been two types of environmental theory, one based on physiology . . . and one on geographical position; both are in the Hippocratic corpus.'

The physiological theories derive from the idea that human health varied according to specific locations, and could be affected by such factors as altitude and proximity to water. Such theories tended to be raised in discussions of, for example, the best locations for houses. More significant for our purposes are the geographical position theories that derive from comparisons of large areas and concern the character of groups of people rather than the health of individuals. In this category are discussions of environments and related human characteristics, focusing especially on the contrast between the well-off populations of the eastern Mediterranean and the more penurious populations of Europe north of the Mediterranean area. In similar fashion, Aristotle described

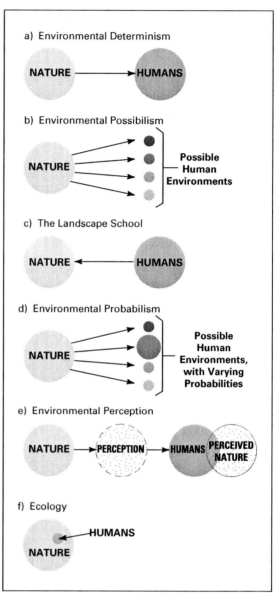

Figure 2.1 Geographic Approaches to Human-and-Nature Relationships. This figure summarizes five of the approaches to the study of humans and nature detailed in this chapter. These schematic diagrams necessarily simplify the very real complexity evident in the various debates. It is not appropriate to interpret these approaches as different from one another in any absolute sense; rather, each lies somewhere on a continuum ranging from the dualism most evident in extreme environmental determinist arguments through to the holism evident in some ecological arguments.

Source: Adapted from D.N. Jeans, 'Changing Formulations of the Man–Environment Relationship in Anglo-American Geography', *Journal of Geography* 73 (1974):37.

some eastern Asiatic populations living in warm areas as thoughtful and skillful but lacking in spirit and courage, in contrast to the less intelligent but braver and more warlike northern Europeans. According to Aristotle, the inhabitants of Greece, living in an intermediate environment, combined the better—interestingly, not the worst—qualities of both these populations.

Many other Greek scholars, especially Posidonius (c.135 BCE–51 BCE), a Stoicist who also discussed the influence of the position of the sun and stars on both physical geography and humans, continued this tradition of thought. In this sense, early environmental determinist thinking does not necessarily imply a separation of humans and nature, because all things are related. Using more contemporary terminology, much of this early thought was ecological in character in that all aspects of the physical and human worlds were seen as linked; humans grow in an environment and are affected by that environment.

Strabo (c.64 BCE–20 CE), a member of the later Stoic school, adopted a broader cultural perspective in discussing the geography of the known world. Although environmentalist overtones are evident in much of his writing, especially concerning the physical geographic reasons for the rise of Rome, there is also recognition of the human ability to overcome environmental challenges. The writings of Strabo highlight the important general point that many scholars who contributed environmental determinist arguments also recognized other human-and-nature relationships.

After the demise of the Roman empire in the fifth century, many of the Greek and Roman contributions were unavailable to European scholars, and the most studied and influential book for about the next thousand years was the Bible. The Biblical design argument was most influential and contributed to the number of accounts of human use of the earth; original environmental determinist ideas were less common, although there were contributions from such scholars as Albertus Magnus (c.1200–80) and St Thomas Aquinas (c.1225–74). Accounts of humans and nature played an important role in Islamic scholarship, with ibn Khaldun (1332–1406), one

of the greatest philosophers of the period, emphasizing the impact of environment on social organization and economic life.

THE SCIENTIFIC REVOLUTION

In Europe, the onset of the scientific revolution and the beginning of significant overseas exploration combined to produce additional examples of the presumed impact of physical environment, especially climate, on humans. The new era of sea travel meant that the basic methodology of cause and effect, reinforced by advances in science, could be applied to an increasing number of locations in the expanding world known to Europeans. One twentieth-century environmental determinist was fond of quoting Bacon, who said: 'True knowledge involves the study of Causes' (Taylor 1928:3). Certainly, this way of thinking flowered in many European works during this period. The French philosopher Jean Bodin (c.1529–96) built on earlier work to argue that cultural differences were a result of astrological, latitudinal, and local factors. One conclusion he reached was that people in temperate climates were both more talented than those in the colder north and more energetic than those in the warmer south. However, even such an enthusiastic proponent of environmental determinism as Bodin acknowledged as well the role of other factors, such as the consequences of cultural contacts related to war and migration.

ENLIGHTENMENT AND NINETEENTH-CENTURY THOUGHT

Montesquieu (1689–1755) restated many of the earlier positions on environmental determinism using new information that was becoming available because of European overseas movement. He used climate to explain cultural differences—he posited that people in cold areas were strong and courageous while those in warm areas were more suspicious and cunning—and to explain cultural persistence, as in the presumed unchanging cultures of Asia. By this time, environmental determinism was much more than simply a way of explaining culture: it also justified cultural relativism. Thus, Montesquieu, though he was opposed to slavery, understood that it could develop where a population was so indolent that slavery was the only means available by which that population could be used as labor. Again, as with Bodin, Montesquieu did not adhere to environmental determinism at the expense of recognizing other aspects of the human world; he included in his writings discussions of human impacts on environment.

The idea of natural controls is also evident in the writings of Thomas Malthus (1760–1834), who argued that it was not possible for a genuinely happy society to evolve because of the universal circumstance that population numbers increased more rapidly than did the supply of food, which was limited by nature. It was in his famous 1798 essay that Malthus first referred to the 'struggle for existence'.

One of the most explicit statements of environmental determinism during the Enlightenment came from the French philosopher Victor Cousin (1792–1867):

Yes, gentlemen, give me the map of a country, its configuration, its climate, its waters, its winds, and all its physical geography; give me its natural productions, its flora, its zoology, and I pledge myself to tell you, *a priori*, what the man of that country will be, and what part that country will play in history, not by accident but of necessity; not in one epoch, but in all epochs; and, moreover, the idea which it is destined to represent! (quoted in Febvre 1925:10)

The second half of the nineteenth century provided many examples of the apparent relevance of the new ideas about biological evolution for human beings. Darwinian theory served to confirm the largely non-scientific arguments of earlier writers, with the result that many versions of environmental determinism—such as that proposed by the English historian H.T. Buckle (1821–62), which explained the onset of the scientific age in Europe in terms of climate, soil, and configuration—began to assume increased prominence. One proposal for a positivist social science incorporating environmental determinism was especially important, namely that associated with the evolutionary perspective most fully developed by Spencer. As noted earlier, the second half of the nineteenth

century was the heyday of evolutionary theories, with major contributions also from the anthropologists Morgan and Tylor. But it was Spencer, in his articulation of what is usually labeled **social Darwinism**, who argued for an analogy between animal organisms and human groups with regard to the struggle to survive in a physical environment. The logic of social Darwinism is straightforward, hypothesizing that cultural groups evolved in accordance with their ability to adjust to physical environments. Accordingly, Spencer, for example, saw history as a steady move from warmer to colder areas.

During the nineteenth century, then, environmental determinism continued to be popular, and in fact became a component of other conceptual schemes, such as cultural evolution and social Darwinism. But the century culminated in the adoption of the approach in a quite traditional, straightforward form by the newly institutionalized discipline of geography. This adoption was notwithstanding the fact that several highly influential mid-nineteenth-century geographers had expressed alternative views. Environmental determinism appealed to geographers for a number of reasons, including the following:

- it had a long scholarly tradition;
- it offered a way to account for ever increasing evidence of cultural differences in different physical environments;
- it belonged to the body of mechanistic science, which continued to be popular;
- it was supported by Darwinian theory and related ideas that applied to cultures.

Geography as the Study of Physical Causes

THE IMPACT OF RATZEL

The German geographer Friedrich Ratzel (1844–1904) accepted many of Spencer's ideas concerning the similarities between human societies and animal organisms, strongly emphasizing that humans were subservient to nature. Indeed, Ratzel is often seen as the originator of modern environmental determinism, proposing the approach as a specifically geographic concern in 1882 in the first volume of *Anthropogeographie*. As restated by Ratzel, environmental determinism proved an immensely popular approach to the study of humans and nature even though various alternatives emphasizing other relationships, including a unity of humans and nature, were available at the time. It is clear that Ratzel regarded this concept only as a generalization, not as a scientific law, and in the second volume of *Anthropogeographie* he substantially modified the view. But the early Ratzel appears to have been the more influential, probably because it was the early Ratzel whose views mirrored the needs of geography at the time, namely a subservience of human to physical geography and an apparent scientific emphasis.

In addition to the various reasons for the attraction of environmental determinism already noted, it bears mentioning that one compelling reason for other geographers' positive reception of Ratzel was the fact that his approach gave credibility to the geographic enterprise, stressing, as it did, the role of physical geography in human affairs. At a time when the various social sciences were competing to generate explanations, the practitioners of a discipline typically seized upon any approach that highlighted their discipline at the expense of others. A related reason for the use of environmental determinist logic was its implicit incorporation in the dominant regional approach to the study of geography.

The writings of Ratzel played a critical role in what was to become cultural geography, such that it is difficult to exaggerate the importance of environmental determinism during the early years of the newly institutionalized geographic discipline. Thus, the leading British geographer at this time, Halford J. Mackinder (1861–1947), accepted the view that human geography was derivative of physical geography. Similarly, in the United States, the environmental determinist view was articulated by the founder of the Association of American Geographers, William Morris Davis (1850–1934), and, more generally, became an integral component of the new university discipline. But the principal examples of environmental determinist thought are contained in the works of two American geographers, Ellen Semple (1863–1932) and Ellsworth Huntington (1876–1947).

SEMPLE AND HUNTINGTON

Semple visited Germany in the 1890s to study with Ratzel and subsequently introduced his understanding of physical influences on humans to American geographers. In *Influences of Geographic Environment*, Semple famously began:

Man is a product of the earth's surface. This means not merely that he is a child of the earth, dust of her dust; but that the earth has mothered him, fed him, set him tasks, directed his thoughts, confronted him with difficulties that have strengthened his body and sharpened his wits, given him his problems of navigation and irrigation, and at the same time whispered hints for their solution. She has entered into his bone and tissue, into his mind and soul. (Semple 1911:1)

These opening sentences set the tone for a book that contains determinist explanations throughout. Many of these explanations support the longstanding view of climate as a principal cause of human character, human activities, and human landscapes. Semple's work also reflects the prevailing thought about racial differences between groups of people. The following brief quotation is not atypical of the thrust of the 1911 book: 'The northern peoples of Europe are energetic, provident, serious, thoughtful rather than emotional, cautious rather than impulsive. The southerners of the sub-tropical Mediterranean basin are easy-going, improvident except under pressing necessity, gay, emotional, imaginative, all qualities which among the negroes of the equatorial belt degenerate into grave racial faults' (Semple 1911:620).

In similar fashion, Huntington accepted the basic logic of environmental determinism but used the idea to focus on the supposed links between climate and the growth and decline of civilization. The basic theme was that areas of stimulating climate encouraged the growth of civilization, whereas the monotonous tropical climate was especially inimical to such growth:

. . . a prolonged study of past and present climatic variations suggests that the location of some of the most stimulating conditions varies from century to century, and that when the great countries of antiquity rose to eminence they enjoyed a climatic stimulus comparable with that existing today where the leading nations now dwell. In other words, wherever civilization has risen to a high level, the climate appears to have possessed the qualities that today are most stimulating. (Huntington 1915:4)

But Huntington did not simply produce general arguments; he also provided details of climatic change, ways of measuring civilization, maps of civilized areas, and implications of climatic controls for humans. As an example of his measures of civilization, Huntington (1945:6) noted, 'intermediate peoples, such as the Chinese, occupy intermediate places, with more cars than the pygmies, less than the Bulgarians.' His concern with the present human condition is well stated as follows: 'The climate of many countries seems to be one of the great reasons why idleness, dishonesty, immorality, stupidity, and weakness of will prevail. If we can conquer climate, the whole world will become stronger and nobler' (Huntington 1915:294).

APPRAISAL

Certainly, the single most disturbing aspect of environmental determinism as an approach for the newly institutionalized discipline of geography is that it quickly proved to be embarrassingly simplistic. Nevertheless, although environmental determinism may well be an old and essentially discredited idea, it continues to reverberate in many discussions about humans and nature.

Generally speaking, those who applied the approach assumed the argument to be correct and sought examples rather than approaching geographic facts in a more objective and open-minded manner. Thus, the attraction of environmental determinism was well expressed by Griffith Taylor (1880–1963): 'as young people we were thrilled with the idea that there was a pattern anywhere, so we were enthusiasts for determinism' (Taylor, in Spate 1952:425). Indeed, Taylor made an important contribution in arguing for a modified determinism: 'Man is able to accelerate, slow or stop the progress of a country's development. But he should not, if he is wise, depart from the directions as indicated by the natural environment. He is like the traffic-

controller in a large city, who alters the rate but not the direction of progress; and perhaps the phrase "Stop and Go Determinism" expresses succinctly the writer's geographic philosophy' (Taylor 1951:479).

Although advocates such as Semple, Huntington, and Taylor acknowledged the need for cautious applications and also recognized the role played by humans, it is fair to say that in many geographic studies of the late nineteenth and early twentieth centuries, culture, technology, and economics were subservient to physical environment. Further, a substantial body of literature, including much earlier geographic literature arguing for the unity of humans and nature, was ignored.

One useful way to think about environmental influence is in the context of human technology. Expressed very simply, it is possible to distinguish between two types of nature, benign and difficult, and between two types of technology, high and low. The outcomes of different pairings of humans and nature are identified in Table 2.1. The table suggests that the variable 'determining' the outcome of a particular relationship between humans and nature is the human variable of technology, not nature. This is because when the human variable changes from one level of technology to another, the outcome, the nature of the relationship, changes. When nature changes, from one type to another, the outcome is unaffected.

Pause for Thought

The preceding account of environmental determinism hopefully makes it clear that the approach was extremely popular among geographers from the late-nineteenth century until the mid-twentieth century. This is the case despite the fact that it seems appropriate to view the approach as fundamentally flawed, except perhaps when employed in very broad terms to help uncover general relationships, such as that between global physical geography and population density and distribution. Consider the following five observations and ask yourself if you agree with them. First, interpreted as geography, environmental determinism neglects much earlier work, defines the discipline in terms of one specific causal relation, neglecting alternative forms of interpretation, and typically excludes subject matter that does not accord with the approach. Second, it is overly simplistic, as perhaps are all versions of determinism when applied to human behavior; the complexities of human behavior are ignored in the desire to identify a cause-and-effect relationship. Third, although the physical environment is an important consideration in much decision-making—for example, there are obvious links between behavior and diurnal and seasonal cycles and between behavior and such dramatic physical events as floods and earthquakes—it remains the case that all such behavior still needs to be understood in terms of human environmental contexts, such as those of culture, society, politics, and economics. Fifth, as is readily apparent, all human landscapes change through time notwithstanding the fact that physical environments are relatively unchanging. After considering these five questions, are you inclined to conclude that, at best, the approach is only superficially attractive given that it does not stand up to rigorous scrutiny?

Human Use of Nature

Recall that many scholars from classical Greece onwards combined thoughts about environmental influence with thoughts about human impact on the earth, the two ideas being divergent but not contradictory. Indeed, the two ideas share a conceptual focus in that both imply the presumed separation of humans and nature. Nevertheless, it is not usual to find scholars before the eighteenth century addressing the question of human impacts in the same detailed philosophical terms as they discussed physical influences. For this reason, and because the larger philosophical context is covered in the earlier discussion of environmental determinism, this discussion of writings about human use of the earth is relatively brief.

Historical Perspective

GREEK AND CHRISTIAN THOUGHT

The idea that humans were able to change the physical environments in which they lived is a logical consequence of human life itself. Scholars in classical Greece typically justified this on philosophical grounds by recognizing that humans, being capable of reasoning, were different from nature and as such were orderers, or organizers, of the natural world. This rationale was an important component of the designed earth argument: God created the earth for humans, who were then granted authority and dominion over it to change or indeed complete it as they saw fit. As Europeans became aware of environments outside of Europe, they viewed these new lands as simply awaiting their finishing touch.

These classical Greek and Judeo–Christian ideas about the inherent responsibility of humans to organize, change, and complete nature provided ample justification for the effects of such human activities as irrigation, drainage, mining, and animal and plant domestication. As there was no real evidence of undesirable changes resulting from these activities, they were seen in a positive light. Indeed, many

Table 2.1 Human-and-Nature Relationships

Relationship	Outcome	Example
High technology/ benign nature	Humans not limited by nature—various activities	Temperate areas today
High technology/ difficult nature	Humans not limited by nature—various activities	Hot, dry deserts today
Low technology/ benign nature	Intimate human–nature relationship—foraging	Temperate areas 12,000 years ago
Low technology/ difficult nature	Intimate human–nature relationship—foraging	Hot, dry deserts 12,000 years ago

of the scholars who commented on the role of physical influences—such as Montesquieu and Sebastian Münster (1489–1552), whose 1544 book *Cosmographia Universalis* was the authoritative volume on world geography for about the next 100 years—commented also on the positive contributions humans were making in the sphere of environmental change. This broad interpretation of human impacts prevailed until the nineteenth century.

THE SCIENTIFIC REVOLUTION

Ideas about human impacts were central to the thoughts of the first scientific philosophers, Bacon and Descartes, and to the tradition that they initiated in that there was new confidence in the ability of scientific knowledge to control physical environments. This confidence combined with prevailing religious understanding of human dominion over nature contributed in the seventeenth century to the general sense that humans were both improving themselves and improving nature through their impacts on nature.

ENLIGHTENMENT AND NINETEENTH-CENTURY THOUGHT

It was in the eighteenth century, beginning with the work of the French naturalist Buffon, that the favorable view of human impact on nature assumed real importance. Buffon acknowledged that humans needed to change environments in order to enhance their civilization and stressed the improvements that had been achieved through orderings of nature. He wrote extensively about animal and plant domestication, the removal of forests, changes to soils, and the creation of cultural landscapes in general. Central to his arguments were distinctions between humans and animals, which focused on human intelligence, creativity, and adaptability.

The first suggestions by European scholars that humans might be abusing rather than using, or indeed improving, physical environments surfaced between the eleventh and fourteenth centuries, a period of agricultural expansion that involved the transformation of forest and marsh, the establishment of settlements, and increased mining activity. But it was the American geographer Marsh who made the first real advances

in this area with the 1864 publication of *Man and Nature*. He portrayed humans as essentially a destructive power and noted their hostile influences on environments: 'man is everywhere a disturbing agent. Wherever he plants his foot, the harmonies of nature are turned to discords. The proportions and accommodations which ensured the stability of existing arrangements are overthrown' (Marsh 1965:34). In addition, he presented considerable evidence to demonstrate the frequency, magnitude, and deleterious effects of human activity, including detailed examples concerning plants, animals, forests, water bodies, and coastal areas. Overall, Marsh provided a compendium of factual information that focused on the theme of human activities and their detrimental consequences for nature.

Cultural Geographic Interpretations

POSSIBILISM

Ratzel published the first volume of *Anthropogeographie* in 1882, and the determinist arguments contained therein were accepted and further developed by Semple and others. The second volume, published in 1891, advocated a different approach to human geography. Rather than focusing on the influences of physical environment on humans, Ratzel emphasized the role of humans in their cultural groupings. As we have seen, this approach, like that of environmental determinism, had a long scholarly precedent, and it was also favored by the contemporary German geographer Alfred Kirchoff (1838–1907).

It was this approach, advocated in the second volume of *Anthropogeographie*, that was further developed by the eminent French geographer Vidal. In this tradition, the physical environment was regarded not as a determinant of human activities but as a factor that set limits on the range of possible human options in an environment, and it is this emphasis on environmental possibilities that prompted the label 'possibilism' (see Figure 2.1b). More generally, possibilism is one component of *la tradition Vidalienne*, a tradition that also includes concerns with local regions (*pays*), with immediate physical surroundings (*milieu*), and with the culture or way of life of a group of people (*genre de vie*). For Vidal and those who followed him, such as Lucien Gallois (1857–1941), Emmanuelle de Martonne (1873–1955), and especially Jean Brunhes (1869–1930), *genre de vie* was the key to understanding which of the various possibilities offered by the environment was selected for use by a human group. Vidal was also a key influence in the development of the *Annales* school of history in France, which included such early members as Lucien Febvre (1878–1956) and Marc Bloch (1886–1994), all of whom consistently studied the relevance of environment in human history.

Vidal and his followers—often referred to as Vidalians—recognized close relationships between humans and environment, explicitly rejecting the idea of environmental determinism. The approach continued the dualism evident in much of the previous thought, although the emphasis on the separation of humans and nature is less explicit than is the case with environmental determinism. Further, possibilism arose in direct opposition to the concept that social phenomena are explicable only in terms of other social phenomena, an approach associated with the French sociologist Durkheim. For Vidal, neither determinism—environmental or social—was acceptable. For a fascinating account of the practical relevance of these questions, see Box 2.4.

VARIATIONS ON A THEME
Although, the basic logic of possibilism was generally accepted by many cultural geographers, it remained the subject of debate during the first half of the twentieth century, and a small number of geographers proposed modifications. In addition to the stop-and-go version of determinism proposed by Taylor, Spate argued that neither environmental determinism nor possibilism was adequate; he coined the term 'probabilism' as an appropriate middle road (Spate 1952:422). If one accepts the basic logic of possibilism, then this variation makes sense. Probabilism simply recognizes that the various possibilities have varying probabilities of occurrence (see Figure 2.1c), a logic that was indeed implicit in possibilism.

At the same time, there was also a continuing concern, rooted in the tradition initiated by Marsh, with human impacts. Sauer was one of the principal instigators of a seminal volume of readings, the motivation for which was well stated by Thomas:

> Every human group has had to evaluate the potential of the area it inhabits and to organize its life about its environment in terms of available techniques and the values accepted as desirable. The identification, use and care of resources is in the end a problem of human values and behavior. Cultural differences in techniques and values, and hence in utilization of the physical biological environment and its conversion into a human habitat, have distinguished one human group from another. The effects of man on the earth are geographically varied and are historically cumulative. (Thomas 1956:xxxvi)

Another variation on the possibilist approach involved a more explicit acknowledgment that the way groups of people perceive, and therefore behave in, environments varies according to their particular group characteristics (see Figure 2.1d). Differences between real and perceived environments are emphasized in this approach, which became popular during the 1960s (see chapters 5 and 9).

Controlling Nature and Controlling Others
'Human use of nature': the very phrase conveys the impression that nature is something to be manipulated according to human needs and wants. Returning to the Christian ideas of the separation of humans and nature and of humans being granted dominion over all other creatures, and recognizing the long tradition of subsequent religious and scholarly thought, it becomes evident that one feature of the human experience, especially in the Western world, has been the quest for ever greater control over the natural world. Further, this ambition to control has been extended to include human control of other humans, notably those who are perceived, for whatever reason, as perhaps less than human. These are, then, old ideas that have long been acknowledged, but they are also ideas that are subject to various interpretations.

As noted, in the Jewish, Greek, and early Christian worlds, humans were generally enthusiastic to establish hegemony over nature. Hierarchies were established with Aristotle, for example, speculating about human nature and developing a sliding scale with human males at the top and human females second, followed by ranks of other living beings. The principal Christian view was to see humans as masters of animals, with wild animals there to be hunted and domesticated animals there to perform labor. Thus the experiences of hunting and of animal and plant domestication served to confirm human domination of the natural world.

By the end of the eighteenth century, the growth of science had given rise to a rational rather than a religious justification for human domination of the natural world and for identifying different groups of humans. **Slavery**, forced labor, **feudalism**, the isolation of the mentally ill and of lepers were all instances of humans identifying differences within the human species and exercising control over those seen to be inferior. If to be human was to have some qualities defined by those in positions of power, then it was inevitable that some humans would be regarded as in some way less than other humans. Indeed, just as there had been those who had claimed that plants could be domesticated or removed and that animals needed to be domesticated, hunted, or displayed in zoos, so there were those who claimed that humans needed to possess culture—European culture, that is—or else be exterminated, enslaved, or forcibly moved.

An especially important phase in the development of these ideas occurred during the period of European overseas expansion, when several European powers effectively carved up and divided a large portion of the globe among themselves. During this period the idea that humans, specifically European males, could also dominate others became paramount. In some respects, as noted in Box 2.1, this European achievement was the culmination of the Enlightenment, involving a triumph of progress and reason. The domination of others was considered legitimate because Europeans considered them-

Box 2.4 **A Historic Encounter**

Geography's concern with the relative merits of different ways of conceiving relationships between humans and nature is not, of course, only an academic issue. A fascinating early example of the practical significance of the differing viewpoints of environmental determinism and possibilism involves the debate about the prospects of settling the arid Australian interior after the First World War. Taylor, a proponent of stop-and-go determinism, was teaching geography in Australia at the time and was an important figure in the debate. Openly critical of settlement proposals, Taylor referred to 'Nature's Plan', identifying 'optima', meaning climatic conditions ideally suited to settlement, and 'limits', meaning the minimum climatic conditions suited to settlement, and published a textbook that was banned by the Western Australian education authorities because of its pessimistic views on settling the interior. On the other side of the debate were those who favored the expansion of settlement, including a number of politicians and the Canadian explorer and ethnologist Vil-

hjalmur Stefansson (1879–1962), who was well known in Canada for his accounts of time spent in the Arctic, in which he expressed his belief in the potential of that area for human habitation and development. Stefansson visited Australia in 1924.

The very different perspectives of Taylor, the determinist, and Stefansson, the possibilist, prompted very different assessments of the Australian interior, assessments that were widely reported in the press. Stefansson optimistically referred to the prospects of the area and, without any real justification, compared it favorably to the American West. His expansionist views were precisely what the Australian politicians and public wanted to hear. Taylor subsequently described Stefansson as a 'dangerous anthropogeographer'; he became an isolated figure and left Australia in 1928 (Powell 1980:181).

Who was right? Certainly, Taylor underestimated the ability of humans to adapt to difficult environments, but, just as certainly, Stefansson greatly exaggerated the potential of the area.

selves superior; indeed, the conquerors were replacing the perceived impoverishment of others with the values of justice and reason. The 'White Man's Burden' was to civilize the savage. This logic was reinforced by the evolutionary biology of Darwinian theory, and especially by social Darwinism.

It was not unusual for groups that were relatively powerful, such as the English establishment, to view groups that were relatively powerless—such as the Irish, Africans, infants, women, the poor, the mentally ill—as inferior, and even to describe them in animal terms. Once seen in animal terms, people were liable to be treated in analogous ways—by being domesticated, for example, or even hunted. (Racism and sexism are treated in greater detail later on, especially in chapters 7 and 8.) Another way to express this general point is to note that it became possible to measure human progress, advances in civilization, in terms of the domination of nature and of others.

The Landscape School

Initially, the possibilist approach was closely identified with, and indeed essentially restricted to, a group of French geographers. But two closely related concerns were initiated outside of France during the early part of the twentieth century, both asserting the primacy of culture over environment. In 1906 the German geographer Schlüter built on the ideas of Kirchoff and proposed the study of a cultural landscape, *Kulturlandschaft*, which, he suggested, was a landscape created by a cultural group from a previous physical landscape, *Naturlandschaft*. It was this idea of landscape transformation that attracted Sauer and that provided impetus for the highly influential landscape school. Expressed simply, Sauer proposed that geography be concerned with the development of the cultural landscape from the natural landscape. The emphasis was thus on the evolution of the visible material landscape.

Blainey (1983) described the Australia of the early twentieth century as *A Land Half Won*. That description remains apposite today, with most of the arid or semi-arid interior unsuited to cultivation and even challenging for arable activity. This illustration dates from 1910 and shows plowing of marginal land. In any given year the success of such an endeavor was closely related to rainfall. (Mary Evans Picture Library)

It is helpful to interpret the approaches of Schlüter and Sauer as a further move away from an environmental determinist argument in that, with their explicit incorporation of culture, they allowed for an ongoing process of environmental change and adaptation rather than an acceptance of certain environmental options. Thus, in a textbook inspired by the landscape school, Carter (1968:562) noted that it is 'human will that is decisive, not the physical environment. The human will is channeled in its action by a fab-ric of social customs, attitudes, and laws that is tough, resistant to change, and persistent through time'.

In 1925, Sauer wrote the article, 'The Morphology of Landscape', that is usually viewed as the seminal statement of the landscape school. The substantial impact of this one piece on most work in cultural geography prior to about 1970 and on much work since that date is undeniable, and yet there are a number of intriguing questions surrounding the article:

The Atlantic slave trade began in the fifteenth century, when the Portuguese, followed by the Dutch, French, English, and others, explored the African coast. The market for slaves expanded rapidly, first in South and Central America and then in the United States, starting in 1619 when a Dutch trading ship first brought African slaves to Virginia. In 1807, the British Parliament abolished slavery, but it was not until 1833 that the trade was abolished throughout the Empire, and it was 30 more years before President Lincoln issued the Emancipation Proclamation freeing slaves in the US. Inspired by a friend's poems, George Morland painted *The Slave Trade* in 1791. The painting is set on the West African coast and shows an African couple being separated by traders. (Bildarchiv Preussischer Kulturbesitz/Art Resource, NY)

- Was it an attempt to delimit a subfield of human geography—which is what the article effectively did—or an attempt to set a course for all of human geography?
- Was it influenced more by earlier European work, such as that of Ratzel in volume 2 of his *Anthropogeographie*, of Schlüter, and of Vidal, or more by the anthropological arguments about the meaning of culture as elaborated especially by Kroeber?
- Did Sauer himself, in his considerable body of empirical work, adhere closely to the methodological principles outlined in the 'The Morphology of Landscape'?

The discussion of these questions as part of this larger account of the landscape school enriches understanding of the identity of the cultural geography that built on the ideas of Sauer.

Origins

At the time when Sauer was expressing his ideas about landscapes, the larger discipline of geography had an uncertain identity. Following the first establishment of geography departments in Prussian universities in 1874, there were various claims and counterclaims about the identity, methodology, and subject matter of the newly institutionalized discipline. The debate took place against a complex intellectual backdrop. Geography, both physical and human, was a long-established area of both intellectual and more general concern, with traditional interests in exploration, mapping, and written descriptions of places and peoples. Scholars writing about geography before 1874 contributed to many important debates, such as those about humans and nature, about populations and resources, about human differences, and about changes in ways of life. Further, the late nineteenth century was a period during which two seemingly contradictory trends were evident: there was an increasing separation of physical and human sciences as the social sciences became institutionalized, and yet there was much sympathy in both areas of endeavor for cause-and-effect analyses. Geography occupied a distinctive position in the organization of knowledge at this time because it was both a

physical and a social science, and most geographers who achieved leadership roles found that it was incumbent on them to define the discipline of geography.

Principal statements about the identity of geography included the declaration by Ferdinand von Richthofen (1833–1905) in favor of a geography centered on the regional concept, and the two ideas promulgated by Ratzel, namely geography as environmental determinism, in volume 1 of *Anthropogeographie* (1882), and geography as the study of human landscapes, in volume 2 of *Anthropogeographie* (1891). Vidal followed the later Ratzel in advocating the study of human landscapes. For most American and British geographers at this time the central concern was the study of earth and life relationships, but they saw two additional tasks as particularly important: determining the influence of physical environment on human geography and human history, and delimiting and analyzing regions. Certainly, both the physical cause and the regional approaches to geographic study were widely accepted in the English-speaking world. The formative physical cause work was accomplished especially by Semple and Huntington, while the formative regional statements were contributed by Mackinder and others in Britain and by Davis in the United States. It was into this somewhat confused and quite uncertain scholarly landscape that Sauer ventured in the 1920s.

Sauer taught at the University of Michigan from 1915 to 1923, after which he moved to the University of California at Berkeley. He remained at Berkeley until 1957, when he began an active 23-year period of retirement. It was during his early years at Berkeley that Sauer made his contribution to the debate about the identity of geography. In the period from 1924 to 1931, Sauer succeeded both in articulating a subject matter, namely landscapes as created by humans in their capacity as members of cultural groups, and in outlining an approach to that subject matter, an approach that emphasized relations between cultures and the physical world, landscape evolution, and region delimitation. The discipline's interest in physical and cultural landscapes was anticipated in earlier writings;

the evolutionary concern was similar to earlier ideas of Vidal and Schlüter; the regional concern was in accord with the dominant view of the discipline. Sauer's methodological achievements were to treat geography as one of the social sciences, to synthesize and expand upon earlier geographic works, to express this synthesis in clear and forceful language, and to influence the activities of the many younger geographers who were attracted to work with him.

The considerable literature about Sauer and the landscape school includes a variety of interpretations about precisely what Sauer and those who followed him believed. Varied interpretations are perhaps inevitable when a scholar writes often and at length, contributes to knowledge over an extended period, and proves to be highly influential. But it is important to note that two evaluations of Sauer's philosophical stance arrived at quite different conclusions. First, Entrikin (1984:387) stated: 'Culture history became his model of social science, and its underlying methodological and philosophical basis was drawn from natural science rather than from historiography'. This view is supported by Leighly, Sauer's closest associate at Berkeley, who identified geomorphology and plant ecology as principal influences and noted that, on one occasion, Sauer described himself as an earth scientist. Second, a rather different view, argued by Speth (1987), identifies a historicist perspective as Sauer's principal inspiration, an opinion that has been accepted in broad terms by humanist geographers who have traced their roots back to the landscape school. This disagreement is, of course, but one particular version of the larger philosophical debate addressed in Box 2.2. Notwithstanding these two different interpretations, six basic ideas promulgated by Sauer can be identified.

Six Key Ideas

1. *First, Sauer was adamant in his rejection of environmental determinism at a time when both Semple and Huntington were major figures in American geography and when much regional geography was sympathetic to the approach.* For Sauer, accepting environmental determinism involved

needless generalization. This generalization resulted in a loss of objectivity, since advocating a particular viewpoint typically produced results that were inferior in value to results that were achieved without some initial assumptions having been made. Thus, facts were the first priority, to be followed— not preceded—by interpretation, with interpretation proceeding without either stressing or ignoring physical geography. Sauer's rejection of environmental determinism was based on flaws he perceived in the argument and not on any downplaying of the importance of physical environment. He consistently referred to possibilism, as discussed by Vidal and Febvre, in sympathetic terms: 'Within the wide limits of the physical equipment of area lie many possible choices for man, as Vidal never grew weary of pointing out' (Sauer 1925:46).

2. *Second, rather than either of the two preferred approaches to geography, namely environmental determinism and regional geography, Sauer favored the study of the facts of landscape as the subject matter of the discipline:*

    > It then becomes the task of geography to grasp the content, individuality, and relation of areas, in which man comes in for his due attention as part of the area, but only in so far as he is areally significant by his presence and works. This is a unitary and attainable objective. The landscape is constituted by a definite body of observational facts that may be studied as to their association and origin. . . . A phrase that has been much used in German literature, unknown to me as to origin, characterizes the purpose perfectly: 'the development of the cultural out of the natural landscape'. (Sauer 1927:186–7)

    Although a regional emphasis was implicit in this landscape methodology, the term 'landscape' was thought to be a clearer alternative to either of the then favored terms, 'area' and 'region'.

3. *Third, Sauer was also critical of the regional approach to geography that centered on the question of 'where':*

'The geographer is hardly likely to prosper by giving his main attention to all kinds of distributional studies' (Sauer 1927:185). Sauer expressed concern that focusing on the distribution of any and all phenomena would result in geographers studying some subject matter in which they lacked competence. He did, however, acknowledge the importance of regions as areas that exhibited a characteristic expression as a result of cultural occupance.

4. *Fourth, Sauer saw a focus on landscape as constituting both physical geography (the physical landscape) and the impacts of humans on physical geography (the cultural landscape).* Sauer consistently recognized the unity of physical and cultural components of landscape.

5. *Fifth, Sauer stressed that cultural impacts on landscapes prompted landscape change, and also that occupying cultures underwent change:* 'An additional method is therefore of necessity introduced, the specifically historical method, by which available historical data are used, often directly in the field, in the reconstruction of former conditions of settlement, land utilization and communication' (Sauer 1931:623). According to Sauer, neither cultural landscape nor culture was fixed, and he made reference to a succession of landscapes with a succession of cultures. This evolutionary emphasis was similar to the concept of sequent occupance, which related changing landscapes to changing cultural occupances (Whittlesey 1929). Sauer outlined the evolution of both natural and cultural landscapes, the latter being summarized as follows: humans, through their culture, transform the natural landscape (see Figure 2.2). Cultural landscape, not culture, is the object of analysis. Recall from Chapter 1 that, although Sauer (1925:46) referred to culture as 'the agent' and the 'shaping force', he referred to the physical landscape as 'fundamental'.

6. *Sixth, as part of the opposition to environmental determinism, Sauer regularly stressed that culture functioned as a cause of landscape.* The anthropological impetus for this claim, known as the superorganic concept of culture, is discussed in Box 2.5, and the extent to which Sauer and his followers embraced this culture concept is debated in Box 2.6.

These six methodological statements combined to encourage three related themes that have been central to most subsequent landscape school research (Sauer 1941a):

- Cultural landscapes result from occupance of a physical landscape; hence there is an ecological component to cultural geography.
- Cultural landscapes change through time, as indeed do the cultures occupying the landscape; hence there is an evolutionary component to cultural geography.
- Cultural activities result in the creation of a relatively distinct cultural landscape; hence there is a regional or cultural area component to cultural geography.

Impact

The impact of Sauer's early writings was both immediate and long-lasting. In 1934, a major overview of geography as a social science quoted the 1925 article at length and included discussion of the cultural landscape in regional analysis: 'wherever man enters the scene he immediately alters the natural landscape, not in a haphazard way but according to the culture system which he brings with him, his house groupings, tools, and ways of satisfying needs' (Bowman 1934:149–50).

GEOGRAPHY AS THE STUDY OF LANDSCAPE

The landscape school was one of five attempts made to define geography between about 1900 and 1960; the other four were environmental determinism associated with Ratzel, possibilism associated with Vidal, regional geography associated with Hartshorne, and spatial analysis associated with Schaefer. Cultural geography has often served as a synonym for human geography, and in fact Sauer was perhaps attempting to determine an agenda for this larger discipline rather than generating a subdiscipline. And although landscape school cultural geography never really dominated the larger

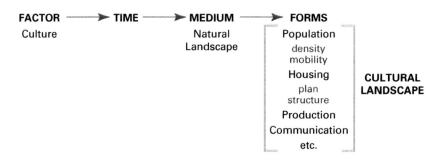

Figure 2.2 The Morphology of Landscape. This is the classic diagram included in the seminal 1925 article by Sauer, 'The Morphology of Landscape'. It is, in fact, the second of two very similar diagrams, the first outlining the established research area of the morphology of the natural landscape, which Sauer (1925:41) described as 'the proper introduction to the full chorologic enquiry which is our goal'. Thus: 'The natural landscape is being subjected to transformation at the hands of man, the last and for us the most important morphologic factor. By his cultures he makes use of the natural forms, in many cases alters them, in some destroys them' (Sauer 1925:45).

Although 'cultural morphology might be called human ecology' (Sauer 1925:45), the term morphology was favored over the alternative term ecology at least partly because, following the assertion by Barrows (1923) to the effect that geography was human ecology, ecology was identified with the idea of environmental determinism. Referring to the diagram, Sauer (1925:46) wrote:

> The cultural landscape is the geographic area. . . . Its forms are all the works of man that characterize the landscape. Under this definition we are not concerned in geography with the energy, customs, or beliefs of man but with man's record upon the landscape. . . . The cultural landscape is fashioned out of a natural landscape by a cultural group. Culture is the agent, the natural area is the medium, the cultural landscape the result. Under the influence of a given culture, itself changing through time, the landscape undergoes development, passing through phases, and probably reaching ultimately the end of its cycle of development. With the introduction of a different, that is, alien culture, a rejuvenation of the cultural landscape sets in, or a new landscape is superimposed on remnants of an older one.

This diagram, and the related commentary, provide a succinct summary of the six key ideas of the landscape school, namely that

- physical landscape is not a determining factor;
- cultural landscape is the object of study;
- studying cultural landscapes involves more than a concern with distributions;
- physical and cultural geography cannot be separated;
- both cultures and cultural landscapes change through time; and
- culture can be understood as a a cause of landscape.

Source: C.O. Sauer, 'The Morphology Of Landscape', *University of California Publications in Geography* 2 (1925):37. (Reprinted courtesy of the Department of Geography, University of California, Berkeley)

field of human geography, it has proven to be an especially enduring set of ideas and practices, initiating a compositional subdiscipline of human geography that has made major contributions from the 1920s through to the present.

The prevailing consensus is that the legacy of Sauer is a concern with cultural landscapes, although this is an association that is critically discussed on several grounds. For instance, along with the concept of landscape, Sauer and his followers focused on links with anthropology, on regional studies in Latin America, and on ecological analyses. Further, many geographers identified with Sauer—Denevan and Peet, for example—centered their studies on topics other than landscape. At the same time,

it is worth noting that some other cultural geographers not directly associated with Sauer—such as Hart, Jakle, Lewis, and Meinig—stressed the landscape concept. Among the geographers directly associated with Sauer who did focus on landscape are Kniffen, Parsons, Salter, and Zelinsky.

REVISIONS OF THE APPROACH

Prior to the first indications of a major revision of the agenda of cultural geography in the 1970s, the landscape school concepts introduced by Sauer were subject to four revisions, both by Sauer and by associated scholars.

The first revision was a decreasing emphasis on the evolution of cultural landscapes from physical landscapes, which was owing at least in part to the considerable amount of specialized knowledge required to analyze physical landscapes. The second revision to landscape school concepts was a diminishing emphasis on maintaining a regional focus, as implied in the concept of sequent occupance, and an increasing emphasis on visible landscape. The third revision was the close integration of historical and cultural analyses, prompted by concern with landscape evolution. The importance of time for Sauer and the landscape school may have been influenced by the work of anthropologists, who saw time as continuous, a view to which Sauer had fully subscribed by 1941. The fourth revision concerned the decreasing emphasis on culture as cause—an implicit rejection of the superorganic and its unproven assumptions.

Such transformations were relatively minor and did not radically change the character of cultural geography as a concern with the evolution and description of visible material landscapes—landscapes as repositories of **material culture** (see also **non-material culture**). Meaningful criticisms of the historical landscape tradition first appeared in the 1960s and centered on the perceived emphasis on visible, material landscapes, and the related neglect of values and beliefs. Such criticisms appeared notwithstanding Sauer's explicit proposal to the effect that: 'Human geography, then, unlike psychology and history, is a science that has nothing to do with individuals but only with human institutions, or cultures' (Sauer 1941a:7).

Pause for Thought

Given your understanding of geography by the 1920s, do you believe that Sauer's ideas were innovative and merited adoption by others? Interestingly, the reaction from other American geographers at the time was mixed. Some of Sauer's contemporaries followed his lead and practiced the cultural geographic study of landscape, but many others, including most of the older generation of geographers, were largely unmoved, preferring instead to concentrate on regional geographic studies. Further, Sauer's ideas had little impact outside of North America, with most European cultural geographers preferring what is best described as a possibilist emphasis in their studies.

Regardless, it is difficult to exaggerate the importance of Sauer's contribution to the development of cultural geography as this was pursued in North America, and much of the content of chapters 4, 5, and 6 derives from the methodological principles enunciated by Sauer.

Toward Holistic Emphases

Rationale

Consider the following four statements:

- 'It is wrong to say that the western tradition has emphasized the contrast between man and nature without adding that it has also emphasized the union of the two. The contrasting viewpoints arise both because man is unique and because he shares life and mortality with the rest of living creation' (Glacken 1967:550).
- 'The old dichotomy between man and Nature, the view that environment is an antagonist that must be conquered, or to which one must passively submit can only lead to disaster or stagnation' (Tatham 1951:162).
- 'May I remind you that much faulty thinking on the part of historians, philosophers and others is due to the fact that they always will set man against nature as if they were two distinct categories. Man versus Nature instead of man and his environment being regarded as a single complex' (Taylor, in Spate 1952:425).
- 'The danger is in setting up a false duality and so involving ourselves in insoluble or unnecessary questions of the chicken–and–egg order' (Spate 1952:419).

These four quotations introduce an approach to the study of humans and nature that aims to think in terms of unity rather than separation, in terms of **holism** rather than duality. There is a long scholarly tradition of attempting to view humans as a part of nature, although, unlike the theme of separation, this tradition has never been dominant. This is not surprising given the prevalence of separation ideas in the major philosophical traditions as well as in Christianity. Many of the scholars associated with a holistic approach, although they argued for a consideration of humans and nature together, found it necessary to use these terms separately, thereby implicitly acknowledging a division between the two in accord with prevailing thought. Indeed, the historical perspective presented below is similar in overall emphasis to that contained in the earlier section on human use of nature.

Historical perspective

GREEK AND CHRISTIAN THOUGHT

The themes of interdependence among all things and of the unity of all living things have been a part of human thought since at least the time of classical Greece. Both Herodotus (484–*c*.425 BCE) and Plato observed that all life functioned as a unity and served to maintain a stable universe. (This is the tradition of a mother earth, or Gaia, discussed more fully in Chapter 4.) A holistic theme is evident also in the work of Posidonius, referred to earlier in the discussion of environmental determinism. Philo (*c*.20 BCE–*c*.50 CE),

known especially for achieving a broad synthesis of Greek and Jewish thought, viewed humans as a part of nature in the sense that all were created by God. In his view, the ranks of different species in nature was determined by the order of their creation, with the superior humans created last. Indeed, there are two somewhat different Christian creation stories in Genesis, with that referred to earlier in this chapter having humans created first and all other creatures subsequently.

It can be suggested, then, that from the onset of the Christian tradition, humanity is both apart from and a part of the community of all living things. St Augustine (354–430), the greatest of the Latin church fathers, similarly saw nature as a continuum and ranked living things above those without life, and living things with intelligence—humans—above living things without intelligence—animals. Again, the Jewish philosopher Maimonides (1135–1204) observed that humans

Box 2.5 The Superorganic Concept in Anthropology

Until the mid-twentieth century, the most influential way of thinking about humans as members of cultural groups was the superorganic interpretation of culture. Recall that this essentially mechanistic and determinist way of thinking was developed within anthropology as one component of cultural evolution logic, and that there were similar developments in sociology, with both Comte and Spencer conceiving of society as an integrated entity, comparable to a physical system and entirely determining the behavior of the people within it. The superorganic concept flows from this idea, seeing societies as wholes created either by growth from within or by absorption of groups from without. Similarly, Durkheim argued for the separation of sociology from both biology and psychology, using the term *social organism* to refer to the social as a unique phenomenon, a process separate from the qualities of individual members of a society and one that cannot be explained by reference to psychology and biology. Both the social organism of Durkheim and the superorganic of Spencer see the social or cultural as a force constraining individual behavior.

There are close conceptual parallels between what was happening in geography, namely the

approach of environmental determinism, and what was happening in sociology and anthropology, namely the approaches of social and cultural determinism. All these approaches were versions of larger determinist logic, all were mechanistic, all were reductionist, and all were privileging one part of the world to the essential exclusion of other parts. The superorganic concept is especially important to cultural geographers because of the argument that Sauer's own acceptance of the approach was such that it became a central part of the landscape school.

As used by Kroeber, the term 'superorganic' referred to the non-organic human product of societies, cultural institutions, modes of production, and levels of technology. Although the term is borrowed from Spencer, it was similar to Spencer's only in the sense that it referred to non-biological aspects of societies. Kroeber's superorganic concept made a distinction between social processes and biological or organic processes, emphasizing that in contrast to biological evolution or changes in organic structure, human societies did not use the principle of heredity to transmit new adaptations to other members of the species. Further, 'the distinction between animal and

needed to understand their place in nature rather than believing that nature was created for humans. Related to these ideas about humans being one part of a hierarchy, other scholars, such as Henry More (1614–87), noted that nature was unthinkable without humans.

Thus, in addition to the many arguments concerning a separation of humans and nature are numerous arguments, again often based on religion, that emphasized a harmonious relationship between or integration of humans and nature. Such ideas continued to be stated until the late eighteenth century, by which time the first of a group of scholars, often with explicitly geographic concerns, provided a new set of arguments about the unity of humans and nature.

GEOGRAPHICAL THOUGHT

From the Renaissance to 1800, a handful of scholars took a holistic approach to the study of geography. Johann Reinhold Forster (1729–98), who traveled around the world in the 1770s, studied the world as a coherent whole, while Anton Friedrich Büsching (1724–93) wrote a geography of Europe that incorporated the idea that all nature, affected by humans or otherwise, was one. Two of the leading scientists of the late eighteenth century, the geologist Hutton and the biologist Lamarck, both conceived of humans as a part of nature.

But it was the nineteenth-century geographers von Humboldt and Ritter who made the greatest contributions to the holistic theme. Both were scholars of tremendous importance and influence in the first half of the nineteenth century, achieving prominent positions in science and society. 'Never before or since have geographers enjoyed positions of such prestige, not only among scholars but also among educated people all around the world' (Martin and James 1993:112–13).

---

man which counts is not that of the physical and mental, which is one of relative degree, but that of the organic and social, which is one of kind. . . . in civilization man has something that no animal has' (Kroeber 1917:169). Although humans may be biological organisms, they are social animals, and it is through language and social and cultural institutions that they learn the values of the civilization in which they live:

> All civilization in a sense exists only in the mind. Gunpowder, textile arts, machinery, laws, telephones are not themselves transmitted from man to man or from generation to generation, at least not permanently. It is the perception, the knowledge and understanding of them, their ideas in the Platonic sense, which are passed along. Everything social can have existence only through mentality. Of course civilization is not mental action itself; it is carried by men, without being in them (Kroeber 1917:189).

Advocates of the superorganic opposed explanations that focused on the natural world and employed environmental determinism, and also opposed explanations that focused on individuals and employed psychological logic. According to the argument against psychological explanations, the length of time an individual was involved in culture precluded the possibility of a paramount role for individuals as initiators of human activities.

It is possible to exaggerate the specific meaning of this concept. Although many subsequent commentators stressed the explicit cultural determinism evident in the superorganic, others take a rather different view. In a major overview of anthropological theory, Harris (1968:342) concluded that Kroeber was not a cultural determinist. Indeed, Kroeber (1917:205) himself noted that to infer 'that all the degree and quality of accomplishment by the individual is the result of his moulding by the society that encompasses him, is assumption, extreme at that, and at variance with observation. . . . [N]o culture is wholly intelligible without reference to the non-cultural or so called environmental factors with which it is in relation and which condition it.' This is what Mikesell (1969:231) called a 'cautious philosophy roughly comparable to the geographic concept of possibilism'.

Their contributions to geography and to the advance of knowledge in general are considerable. Our concern is with their shared commitment to a view of the earth as an organic whole, and to the idea that all things on the earth's surface are related—a concept known as *zusammenhang*, meaning literally 'hanging together'.

It is important to observe that such a commitment did not necessarily exclude the idea that there could be some environmental impacts on humans: writing in 1808, von Humboldt acknowledged that climate and configuration could affect agriculture, trade, and communications. Yet for both von Humboldt and Ritter, the interrelationship of humans and nature was clearly apparent and could not be ignored. To quote von Humboldt (in Tatham 1951:44): 'The earth and its inhabitants stand in the closest reciprocal relations, and one cannot be truly presented in all its relationships without the other.

Hence history and geography must always remain inseparable. Land affects the inhabitants and the inhabitants the land.' Similarly, Ritter (in Dickinson 1969:37) stated: 'My aim has not been merely to collect and arrange a larger mass of materials than any predecessor, but to trace all the general laws which underlie all the diversity of nature, to show their connection with every fact taken singly, and to indicate on a purely historical field the perfect unity and harmony which exist in the apparent diversity and caprice which prevail on the globe, and which seem most marked in the mutual relations of nature and man.' Ritter, unlike von Humboldt, expressed a teleological philosophy, believing that the earth was designed to be the home of humans.

Several other nineteenth-century geographers held views similar to those of von Humboldt and Ritter in that they clearly favored a holistic emphasis, especially rejecting any sug-

Box 2.6 A Superorganic Concept of Culture in the Landscape School?

In light of some later commentaries on the work of the landscape school, it is important to consider whether or not Sauer was proposing an approach that incorporated a cultural determinist or superorganic perspective. There are three obvious reasons why this might not be the case.

First, Sauer was highly critical of environmental determinism because he saw it as 'based on a belief that a single natural law can explain the social order' (Sauer 1927:173); it is reasonable to assume that Sauer would not replace an environmentally centered law with a culturally centered law that shared a similar failing. Second, Sauer consistently acknowledged the importance of physical landscape in all of his methodological writings as well as in many of his empirical studies. Third, the European sources that Sauer cited favorably were largely concerned with questions about the interactions between humans and nature from a possibilist or similar perspective. Indeed, there was a debate between Vidal and Durkheim on precisely this issue, with Durkheim objecting to the inclusion of physical landscapes in attempts to explain human behavior.

On this basis, it seems reasonable to conclude that Sauer was not any version of determinist, environmental or cultural. But there is an alternative argument that has support from at least two areas. First, Sauer did refer to culture using terms such as 'cause'. In this context, Figure 2.2, the classic landscape school diagram, has culture acting on physical landscapes to produce cultural landscapes. Second, there is no doubt that Sauer was intellectually close to both Kroeber and Lowie, who were in the anthropology department in the same university as Sauer. Leighly (1976:339–40) emphasized these close ties: 'Basic to Sauer's course in world regional geography was the concept of "cultures", now used in the ethnological sense, which he learned from the anthropologists at Berkeley.' Sauer (1974:192) also acknowledged this association and influence: 'Anthropologists were our tutors in understanding cultural diversity and change. Robert Lowie in particular introduced us to the work of such geographers as Edward Hahn and Ratzel as founders of an anthropogeography that I had not known.' The intellectual association between Sauer and both Kroeber and Lowie is significant in that both

gestion of environmental influences. Thus, the French geographer Élisée Réclus (1830–1905) emphasized the processes by which cultures used environments to satisfy their needs, and the Russian geographer Peter Kropotkin (1842–1921) saw nature as a dynamic whole that included humans and that was always changing.

Geographers were not the only scholars reacting to the traditional conceptual separation of humans and nature. In addition to the idea of a human–and–nature separation, nature has been seen as external in that it is beyond the realm of humans, and as universal in that there is both external nature and human nature. A universal view implies that humans are as natural as are the external components of nature. Although Marx can be read in various ways, for some he introduced a different way of thinking that rejected the external/universal dichotomy and that emphasized the continuity between humans and nature, with humans viewed as part of nature. Some geographers have recognized this contribution of Marxist thought referring to inner actions within nature rather than to interactions between humans and nature.

ECOLOGICAL EMPHASES

Perhaps the most developed contribution to the idea that humans and nature are best treated in a holistic fashion, focusing on the unity of the two rather than on their separation, is contained within the broad theme of ecological thought (see Figure 2.1e). Indeed, much of the work previously noted, especially that by such geographers as Marsh, von Humboldt, and Ritter, is often interpreted today as being within the ecological tradition (a tradition discussed more fully in Chapter 4).

The word 'ecology' is derived from two Greek terms: *oikos*, meaning 'place to live' or

anthropologists were sympathetic towards a superorganic concept of culture.

Was Sauer an explicit superorganicist, proposing explanations of cultural landscape exclusively in terms of culture as cause? A balanced answer seems to be no, given the considerable weight that Sauer accords to physical landscapes. The aim of cultural geography, as he defined it, was to 'explain the facts of the culture area, by whatever causes have contributed thereto' (Sauer 1931:623). The close links between the landscape school and *la tradition Vidalienne* suggest a shared concern with broadminded interpretations of the human-and-land relationship that were incompatible with either the superorganic of Kroeber or indeed the similar social organism concept proposed by Durkheim. For both Kroeber and Durkheim, the major concern, respectively, was with cultural and social causes of human behavior, but in cultural geography both culture and society were neglected, relatively speaking, in favor of visible material landscapes as objects of study. Arguments about possibly excessive influences from Kroeber also ignore the fact that Sauer was intellectually close to Bolton, a Latin American historian at Berkeley. More generally, a scholarly association does not necessarily result in indiscriminate acceptance of ideas.

On the other hand, there is little doubt that several later landscape school geographers either explicitly or implicitly accepted the superorganic concept. Duncan (1980) details compelling examples. Thus Zelinsky (1973:70) claimed: 'The power wielded over the minds of its participants by a cultural system is difficult to exaggerate.' In similar fashion, Carter (1968:562) stated: 'It is human will that is decisive, not the physical environment. The human will is channeled in its action by a fabric of social customs, attitudes, and laws that is tough, resistant to change, and persistent through time.' More generally, many cultural geographers referred to Kroeber and other anthropologists with superorganicist inclinations in their discussions of the concept of culture. Although there is no unequivocal final word on debates of this type, it is perhaps worth noting that, in a retrospective comment, Sauer (1974:191) reflected: 'The morphology of landscape was an early attempt to say what the common enterprise was in the European tradition.'

'house', and *logos*, meaning 'study of'. Literally then, ecology is the study of organisms in their homes. The key impetus for the development of ecology was the third chapter of *The Origin of Species*, which centered on the various adaptations and interrelationships of organisms, implicitly including humans, and environment. Haeckel first used the term in 1869, although it was Darwin, in *The Descent of Man* (published in 1871), who explicitly included humans. The broad theme of ecology proved to be of compelling interest in both physical and social science. Thus, along with the specifically physical approaches, such as plant ecology and animal ecology, a variety of human and cultural ecologies appeared in the various social sciences. Details of and differences between these approaches are included in Chapter 4.

In principle, these various ecological approaches rejected the dominant late nineteenth-century concern with the separation of humans and nature. In particular, these approaches repudiated any idea of human dependence on nature or human independence from nature. Instead, they encouraged studies of humans and nature without privileging one at the expense of the other, thereby avoiding a reductionist perspective. Ecology is based on the premise that things cannot be studied except in context, such that relations with other things are critical. In practice, however, many specific applications of ecology maintained a separation, at least in part because much of the terminology of the approach originated in physical science. As a result, several ecological approaches—including the approach incorporated in the landscape school as well as the cultural ecology initiated by the anthropologist Steward, which was probably the most influential ecological approach—incorporated a view of humans as members of cultural groups dominating nature. It seems reasonable to suggest that humans are both constrained and enabled by nature, but also affected by their culture.

Evolutionary Naturalism

There is little doubt that much contemporary cultural geography is comfortable with the various critiques of earlier ways of thinking about humans and nature, but there is another interpretation of the scholarly tradition in social science. For Hutcheon (1996), the problem lies not in physical science's seduction of social science, but rather in the failure of social science to fully articulate a position based on the twin scientific pillars of evolution and naturalism. According to her interpretation, the social sciences have largely failed to take advantage of the scientific approach, with the various attempts to do this being steps in the right direction that have never been adequately pursued. Recall that naturalism implies the universality of cause and effect, while evolution is seen to apply to all things. Evolutionary naturalism, as Hutcheon describes it, is a way of thinking based on 'the premise that human beings are continuous with all of nature, and that all of nature is continuously evolving' (Hutcheon 1996:vii). This is an important argument that merits consideration in our discussion of humans and nature.

According to the evolutionary naturalist argument, two currents of philosophical thought have been established in the Western world since about the sixth century BCE. The dominant tradition is based on a metaphysical dualism, a separation of the physical and the spiritual, or the natural and the supernatural. This implies that there are two categories of experience—namely secular and sacred—and two ways of knowing—namely experiential and mystical. Application of the scientific method, of causal logic, has been essentially restricted to the physical realm. The minority tradition, described as evolutionary naturalism, sees both humans and the universe as natural, and therefore encourages the study of human behavior and human institutions using procedures similar to those applied in physical science. Roots of this minority tradition are evident in Buddhist thought as well as in anti-teleological Greek Epicurean philosophy.

It was Descartes who provided a developed logical base for the dominant dualist perspective stressing the separation of science from religion, and it was Hobbes, Locke, and Hume who, although sharing the concern with a scientific approach, countered dualism with the argument that humans could also be studied scientifically. Hutcheon (1996) traced the history of social scientists' efforts to employ the methods of physical science, specifically in the form of evolutionary naturalism, and argued that such efforts have always been countered by approaches explicitly accepting that humans need to be studied differently from the physical world. Principal examples of social scientists who have included elements of the evolutionary naturalist perspective in their work include Spencer, Durkheim, George Herbert Mead (1863–1931), and B.F. Skinner (1904–90). Hutcheon (1996) also identified a series of links between contemporary physical and social science, evident especially in the sociobiological emphasis outlined earlier in this chapter.

In a related interest, the cultural geographer Wagner (1996:1) introduced the admittedly speculative claim that 'human beings are innately programmed to persistently and skillfully cultivate attention, acceptance, respect, esteem, and trust from their fellows', an idea that forms the basis for a materialist and evolutionary selectionist perspective on human behavior. This idea, more fully discussed in Chapter 9, is in close accord with the basic argument of evolutionary naturalism since both theories involve an explicit rejection of the separation of humans and nature.

Pause for Thought

Based on your reading so far and on your understanding of other social sciences, are you inclined to agree with Hutcheon that the social sciences have failed because they have not fully pursued a physical science approach? Certainly, the argument in favor of applying evolutionary naturalism in contemporary social science is a minority view, but it does merit consideration. On the one hand, it is out of tune with most current thought because it rejects the idea that the study of humans requires different concepts to those used in physical science. On the other hand, it is in tune with most current thought because it is an argument against separating humans and nature in that it advocates a common approach to physical and human phenomena.

Concluding Comments

This chapter first introduced a number of philosophical concepts and associated scholars in the attempt to clarify our understanding of the meanings of the terms 'humans' and 'nature', and the character of relationships between the two. Necessarily, the philosophical concepts have been discussed in simplistic and selective terms. As noted in Box 2.2, the principal reason for incorporating these concepts in the discussion is that they provide relevant larger contexts for an understanding of many of the issues that concern cultural geographers. Unfortunately, their introduction can also cause confusion, and it is with this thought in mind that the Glossary provides basic explanations and brief discussions of key philosophical and other terms. A complete understanding of these issues is made difficult by the existence of various interpretations of these terms and of several unresolved issues involving the status of reality and the merits of competing claims.

This chapter also details several variations of the presumed relationship between humans and nature, which are relevant to an understanding of the human world, versions of which continue to be evident in contemporary cultural geography. Views that argue for cause–and–effect relationships remain attractive for many cultural geographers. The idea of physical geography as cause, while no longer a widely accepted argument, remains important in some specific contexts. Of course, cultural geographers cannot avoid consideration of the physical environment, as physical processes necessarily have an impact on the surface of the earth and on human activities. The view that culture is a principal determinant of human activities also continues to be an important component of some cultural geography. However, as has been stressed in both this and the previous chapter, much contemporary cultural geography downplays cause–and–effect approaches in general, whether physical or cultural, favoring instead approaches that advocate some holistic perspective. The most notable such approaches are those that are ecological in emphasis.

From a European perspective, the emergence of the modern world, beginning in the sixteenth century, involved and encouraged the application of the new methods of physical science to the social sciences—including the derivation of universalist impersonal processes, objective forces, and dominant ideas—as means of explaining the human world. Culture was one such process, moving from simple to increasingly complex forms in an essentially determinist manner.

But these ideas were not without contemporary critics. Idealist thought was an important if perhaps less influential tradition, rejecting the claim that it was possible to reflect the reality of the physical and human worlds; they were not there to be discovered and explained, but rather to be interpreted. During the late twentieth century a series of developments, notably postmodernism, raised further questions about the claims that there is a real and knowable world and that humans can objectively study their own world. Further, both physical and social science have been accused of being nothing more than politically motivated practices designed to provide a privileged few (usually White males) with power while simultaneously denying power to others.

One transition in the cultural geographic approach to the study of humans is, by now, very clear. There has been a substantial redirection of scholarly work concerned with human behavior and cultural change from the search for causes, whether the causes are physical or cultural, to the search for meaning (this is the first philosophical debate in Box 2.2). Although both endeavors have always been present, the search for causes achieved paramount status during the period from the seventeenth century until about the 1970s; since then, the search for meaning has become increasingly central. It is significant that the search for causes was especially influential during the formative period of the social sciences in the late nineteenth century, and this emphasis was therefore able to exercise a significant impact on the development of cultural geography at a key stage.

A related transition concerns the general rejection of arguments that focus on nature as cause in favor of arguments that focus on culture and society, not as in the superorganic concept, but rather as providing the context for our understanding of the world. Stressing the presumed dependence of humans on nature involves a privileging of the natural world, while stressing the presumed dependence of humans on culture involves a privileging of the cultural world. Recently, there has been a movement toward a rather different privileging of the human, as in the accounts of nature as socially constructed (this is discussed in Chapter 4). It is useful to recognize that all of these emphases can be interpreted as examples of reductionism, in the sense that they seek understanding in terms that are restricted to what is viewed, *a priori*, as most important.

There are clearly many unanswered questions in the above discussion, such as what nature is, and what it means to be human. Our discussion of nature considered separate from humans sought to explain how this separation came about and what it implies. Perhaps the easiest way to think about this is to recognize that it simply results from defining nature as everything that is not human. If we are uncomfortable with this approach and prefer to conceive of humans as a part of nature, then an alternative definition is needed, one that views nature as everything that there is in the world of experience. According to such an alternative definition, to be natural is to be a part of the world.

Pause for Thought

Have you revised your understanding of humans and nature in light of the material introduced in this chapter? Certainly, it is difficult not to be impressed by both the quantity and quality of the cultural geographic and other work on this topic. Indeed, you will encounter additional material relevant to humans and nature throughout this book. At this stage, you may wish to conceive of nature in two rather different ways: first, there is nature without humans present, wilderness; and second, there is nature with humans present, the cultural landscape. In both cases nature has meaning to us as it is a part of our perceived world. Does it seem sensible to suggest that humans can be understood as both apart from and a part of nature? We are apart from nature in the sense that we have culture and are able to do one thing and not another according to our judgment. But we are also a part of nature in that it is not possible to conceive of humans without there being a natural world that sustains us.

Further Reading

The following are useful sources for further reading on specific issues.

The history of human-and-nature relationships is discussed in Teich, Porter, and Gustafsson (1997) and in Coates (1998).

E. Graham (1997) outlines philosophical debates in human geography.

Bhaskar (1989) and Sayer (1992a) discuss merits and demerits of the realist position.

Kuznar (1997) discusses naturalism in anthropology.

Ortner (1972), Ortner and Whitehead (1981), Fitzsimmons (1989), and Johnson (1996) review possible links between women and nature.

Flew (1978) explains the difficulty of defining what it means to be human.

Because geography followed social Darwinism rather than Marxism, Peet (1985) claimed that the discipline was prevented from achieving a meaningful approach to humans and nature.

Olwig (1996a:86) argues that the concept of nature was 'tainted by its use in the context of a nature/culture dichotomy in which nature was seen to have some form of determinant relation to culture—a relation that had to be, and largely was, erased from academic geographic discourse'. See also Olwig (2003).

For more on 'stop and go determinism' see Taylor (1937:459).

Rostlund (1956:23) claimed that environmental determinism was 'not disproved, only disapproved.'

Berdoulay (1978) and Andrews (1984) discuss the debate between Vidal and Durkheim, with Berdoulay noting that neither environmental nor social determinism was acceptable to Vidal.

Olwig (1980), Bowen (1981), and Breitbart (1981) discuss nineteenth-century geographers sympathetic to an integrated humans-and-nature perspective.

Anderson (1995, 1997) outlines the human interest in displaying animals in zoos.

Whatmore and Thorne (1998) discuss geographies of wildlife.

Tuan (1997:6) examines the acknowledgment by Humboldt that physical geography could affect human geography.

Mathewson and Kenzer (2003) provide a valuable volume of readings about the work and legacy of Sauer.

Mikesell (1968, 1969), Hooson (1981), Entrikin (1984), Solot (1986), Macpherson (1987), Martin (1987a), Bowen (1996), and Mathewson (1996) comment on the methodological writings of Sauer.

Leighly (1976) and Macpherson (1987) focus on the specific Berkeley context for the growth of geography and related disciplines.

Sauer (1924, 1925, 1927, 1931) outlines the basic arguments of the landscape school.

Platt (1962:39) and Parsons (1979:13) consider links between Sauer and the anthropologists Kroeber and Lowie.

Duncan (1993, 1998), Price and Lewis (1993a, 1993b), and Penn and Lukermann (2003) discuss the superorganic in landscape school cultural geography.

Aschmann (1987:137) identifies links between Sauer's methodological and empirical writings.

Rowntree (1996:130–1) identifies the diverse character of landscape school cultural geography.

Brookfield (1964) and English and Mayfield (1972) note the possible exclusion of human values and beliefs in landscape school cultural geography.

Young (1974:8) notes: 'Human ecology may be defined (1) from a bio-ecological standpoint as the study of man as the ecological dominant in plant and animal communities and systems; (2) from a bio-ecological standpoint as simply another animal affecting and being affected by his environment; and (3) as a human being, somehow different from animal life in general, interacting with physical and modified environments in a distinctive and creative way.'

Rossi and O'Higgins (1980), Wright (1983), Sanderson (1990), and Trigger (1998) outline evolutionist arguments.

Childe (1936), White (1949), and Steward (1955) are examples of evolutionist writing

Langton (1979) outlines parallels between biological and human cultural evolution.

Cavalli-Sforza and Feldman (1981) interpret biological evolution in human terms.

Durham (1976:89) claimed that 'both biological and cultural attributes of human beings result to a large degree from the selective retention of traits that enhance the inclusive fitness of individuals in their environments.'

Leaf (1979) and Heyer (1982) discuss the superorganic.

Rossi and O'Higgins (1980) consider various ways of thinking about humans as members of cultural groups.

3

Rethinking Cultural Geography

The story of human geography from 1900 to the present is one of changing conceptual emphases and related changing empirical interests. Building on a substantial tradition, the regional approach to human geography, with emphasis on delimiting and describing parts of the earth's surface, prevailed from about 1900 until the late 1950s. Growing in response to recognition that regional geography was unscientific, spatial analysis, with the principal empirical concern of explaining why geographic facts are located where they are, became the leading approach between the late 1950s and 1970. However, beginning about 1970 many human geographers saw spatial analysis as a dehumanization of the discipline and, finding inspiration instead in humanist and Marxist concepts, initiated new studies concerned with inequalities and subjectivity. As each of these developments occurred, the discipline became more and more conceptually and empirically diverse, and these tendencies were exacerbated around 1980 with an explosion of conceptual and empirical interests related to feminism and the cultural turn. Accordingly, human geography today is best described as an eclectic discipline without any one dominant approach.

As these changes in human geography unfolded, the landscape school proved to be enduring and fruitful, persisting until 1980 with only minor modifications. Its stability was owing at least in part to the fact it was the regional approach that was challenged by spatial analysis, and it was spatial analysis that was challenged by Marxism and humanism. The only meaningful link between spatial analysis and the landscape school was the development of a spatial behavioral focus that interested some cultural geographers, while the key link between the landscape school and the Marxist and humanist approaches was that some cultural geographers played leading roles in the introduction of the new approaches. But both feminism and the cultural turn that began around 1980 differed as they incorporated explicit challenges to the landscape school, especially concerning the superorganic concept of culture and the seemingly conservative character of landscape school work. These criticisms contributed to the rise of what is often labeled the new cultural geography. Nevertheless, despite challenges, Sauer's influence continues to stimulate much of the current work in cultural geography.

The principal aim of this chapter is to tell the story of cultural geography as it has been affected by and taken part in the several transformations that the larger discipline of human geography experienced. To help place the transformations in context, recall that landscape school cultural geography derived from a clear statement about the evolution of landscape, tended to see culture as a causal mechanism, emphasized historical and usually rural landscapes, included both material and symbolic aspects of culture, involved field and archival research, and notwithstanding the explicit 1925 and several subsequent statements, was essentially empirical and not theoretical. This agenda has been interpreted by some critics as inherently conservative, and the distinction with those approaches introduced from about 1980 onwards is striking, with new interests in places, landscape interpretation, symbolism, human identity, difference, and inequality. The perspectives and approaches introduced through the newer bodies of thought have clearly had considerable impact on the concepts and practice of all areas of cultural geography, and indeed on all areas of social science. All are complex, interrelated, and even contested bodies of thought that are continually evolving. Accordingly, the discussions in this chapter do not attempt to reflect the full richness and diversity of these ideas. Rather, what is attempted here is a series of capsule accounts with emphasis on material that is relevant to the cultural geographic studies discussed in the following chapters, especially chapters 7, 8, and 9. In this chapter you will find

- a discussion of the rise of spatial analysis and its limited links with cultural geography;
- a description of the rise of several humanist approaches;
- an outline of five principal psychological approaches to the study of behavior;
- an outline of two different geographic approaches to studies of behavior in the context of what is described as the model of humans;
- a discussion of geographic interest in Marxist social theory;
- an outline of feminist approaches, with emphasis on the oppression or subordination of women and other oppressions. The challenges that feminist thought poses for the foundations of the geographic enterprise are identified;
- a discussion of the cultural turn, with reference to a variety of innovative conceptual contributions including poststructuralism, postmodernism, and postcolonialism;
- a discussion of some implications of the cultural turn, including its critique of the previously accepted mode of representation, a concern with human difference, and a concern with the social construction of knowledge;

- a comment on the introduction of culture into traditional social geography;
- concluding comments, which aim to place these various interests into a larger context and indicate how they inform the contents of later chapters.

Spatial Analysis

Regional geography did not place any priority on studying culture and can be baldly characterized as empiricist, descriptive, atemporal, and implicitly if not explicitly environmental determinist. The growing discontent with this approach was most effectively articulated in a landmark article that argued for a more scientific approach to geographic subject matter—for a **nomothetic** rather than an **idiographic** emphasis (Schaefer 1953). This discontent resulted in a major revision of the geographic enterprise with the onset of what can be described as the quantitative revolution and ushered in the approach known as spatial analysis. Table 3.1 lists basic differences between regional and spatial approaches.

Why was it that the subdiscipline of cultural geography was essentially unaffected by spatial analysis? The general answer is that cultural geography was separate from the dominant tradition of geography that was challenged by spatial analysis, namely regional geography. The cultural geographic interest in evolution and in the study of process through time was an especially important difference between cultural and regional geography. Challenges to the regional concept were aimed explicitly at regional geography's concern with regions as the core of geography, not at cultural geography's concern with regions as outcomes of evolutionary processes. The demise of the region as the core geographic concept was of major significance to such interests as economic geography, which lacked a substantive conceptual basis other than that provided by regional geography; it was of considerably less significance to cultural geography because of its firm foundations within the landscape school. Further, because spatial analysis maintained the atemporal emphasis

Table 3.1 Regional Geography and Spatial Analysis

Regional Geography 1900 to Late 1950s	Spatial Analysis Late 1950s to 1970
is idiographic	is nomothetic
asks *what?* and *where?*	asks *what?*, *where?*, and *why?*
aims to describe	aims both to describe and to explain
is atemporal	is atemporal
focuses on facts	includes both factual and conceptual content
is essentially a classification	uses theories, hypotheses, and quantitative methods
is implicitly environmental deterministic	rejects environmental determinism
is linked to empiricist philosophy	is linked to empiricist philosophy
declined about 1955 because it was seen as not scientific	declined about 1970 because it was seen as dehumanizing

The above is a basic summary of some differences between the traditional regional and the spatial analytic approaches as practiced in twentieth-century geography. The dates are approximate.

of the regional school, cultural geographers saw little to attract them.

Geographers who advocated spatial analysis directed their critical attention especially to the areas of economic and physical geography, which seemed to offer the greatest potential for practicing theory construction and quantitative methods in a manner guided by a positivist philosophy, specifically **logical positivism**. The deductive procedures involved in theory creation were certainly unattractive to cultural geographers, who were often intent on analyzing the particular and not the general. For a research tradition concerned primarily with questions of history and culture, a positivist perspective was necessarily of limited appeal, arguing as it did for objective research and for research strategies adopted essentially from the physical sciences. (This despite the fact that the superorganic content of the landscape school could be interpreted as reflecting the 'behaviorist claim that habit should be construed not as thought but as activity' (Duncan 1980:194–5).)

In one respect the failure of spatial analysis to affect cultural geography is surprising, because the discipline of history was undergoing some methodological debate and developing an increasingly positivist concern at about the same time. Further, both the French *Annales* school of historical method and the mostly North American new economic history embraced some theoretical and quantitative content. Although these developments in history, which were pursued by some cultural geographers, are comparable to spatial analysis, they did not lead cultural geography any closer to adopting a spatial analytic approach. Interestingly, toward the end of the 1960s, when spatial analysis was being challenged by other approaches, the deficiency of excluding historical work was recognized and a process-to-form methodology proposed. Process-to-form ideas involve identifying causal processes and linking these directly to spatial outcomes. Although this methodology is explicitly historical, it was not welcomed by most cultural historical geographers.

Overall, the failure of spatial analysis to affect cultural geography, while unsurprising, is regrettable. Although spatial analysis was quickly challenged by other approaches, it benefited most other subdisciplines of geography by introducing new ideas and initiating new methods of analysis.

Humanisms

The substantive philosophical debate introduced in Box 2.2—concerning whether or not the study of humans can be modeled on approaches developed in physical science—applies to the naturalist perspective of spatial analysis, discussed in the preceding section. From a humanist point of view, the idea of a physical science approach to studying humans is untenable: to treat humans objectively would be to treat them essentially as objects and not as thinking, feeling individuals. For many cultural (and other) geographers, the fundamental problem with spatial analysis was its inability to accommodate individual human characteristics, and this is one reason that various humanist approaches became popular beginning around 1970.

Three key aspects of humanist philosophy are relevant. The first is the view that humans are ontologically and epistemologically irreducible. This means that claims about knowledge cannot be derived from physical science. This humanist tenet involves emphasizing the uniqueness of such human phenomena as creativity and understanding: 'human beings present a unique reality, irreducible to the order of animal behavior' (Wertz 1998:53). The second relevant aspect of humanist philosophy is its focus on human experiences and symbolic expression rather than abstract principles, and its acknowledgment of many different truths. The third aspect is humanism's respect for individual freedom and dignity.

In responding to the ideas of humanist philosophers, especially Dilthey and Edmund Husserl (1859–1938), many scientists argued that the subject matters of physical and human science necessitated different concepts and methods. The most obvious justification for this separation was the fact that humans are free to act as they choose. But a further justification was based on the fact that human phenomena are internal, not external, to the experience of the scientist. This is a critical observation, and Wertz (1998:49) expressed it this way: 'Because physical phenomena and their laws of interconnection are not directly given, they must be explained by constructed models, theoretically derived hypotheses, and experiments revealing functional relationships among isolated factors. Human phenomena, because they are immediately given to knowledge and internally related to each other, need not be hypothesized but are to be described and understood in their meaningful connections.'

Humanists acknowledge the need for hypothetical systems in physical science, but reject their application in the human sciences, arguing that the reason human sciences often fail in their tasks is that they adopt physical science methods that are totally unsuited to their subject matter. Expressed more simply, humanists demand a model of humans that is quite different from the one that results from the use of physical science methods. Intriguingly, however, as discussed in Box 3.1, some scholars interpret the larger humanist tradition quite differently.

Concepts and Procedures

In geography, it was primarily cultural and historical geographers who introduced humanist ideas, which offered a powerful critique of spatial analysis. Box 3.2 outlines the three principal humanist interests advocated by cultural geographers—phenomenology, existentialism, and idealism. The practice of humanist geography is discussed especially in Chapter 9, in the context of accounts of imagining, writing, and reading places.

Most work in humanist geography is associated with phenomenology and stresses human subjectivity and free will, the relevance of participant rather than observer research, and the need for a hermeneutic or interpretive focus. For Bunkse (1996:361), the key humanist element in cultural landscape creation is imagination—'the unique human ability to step outside the immediate context of life, to, as it were, absent ourselves from the here and now and to enter into other realities or to create entirely new realities'. Perhaps surprisingly, the attraction of humanist ideas in the 1970s for certain geographers, including many cultural geographers who were disillusioned with the positivist spatial analytic tradition, did not involve close links with psychology. Rather, as noted in Box 3.2, humanist geographers turned directly to some of the philosophical sources. Regardless, the interpretation of humanism made by psychologists and by cultural geographers was fundamentally similar.

Despite some questions concerning the preferred individual social scale of analysis and also a specific focus, humanist approaches have generated some important concepts. Most generally, it is customary to talk about the **taken–for–granted world**, sometimes called the lifeworld, of everyday living and thinking. This is the intersubjective world of lived experiences and shared meanings, including previously hidden aspects of our personal geographic environment. Intersubjectivity refers to the shared basis of experi-

ence in everyday life, to the understanding that individuals have of each other, and to their similar perceptions of the world. Our human experience of this intersubjective or taken–for–granted world comprises the facts that the phenomenologist explores. Studying the taken–for–granted world involves recognizing the centrality of mundane, everyday experiences, and one task of the humanist cultural geographer is to grasp the dynamism of this world.

Other humanist concepts are used to discuss parts of the surface of the earth as these are understood, perceived, lived in, and experienced. Thus, the need to focus on **place** and not space became the key feature of early humanist geography and continues to be a paramount concern; indeed, much humanist geography is devoted to illustrating and clarifying this term. The meaning of a place is inseparable from the consciousness of those who inhabit it, such that place is not so much a location as a setting, not a thing but a relationship. Closely tied to this focus on place are three related ideas. The first,

Box 3.1 An Alternative Humanist Perspective

In a sensitive series of writings, Hutcheon outlined a 26-century history of humanist thought that has one great idea at heart—namely, the philosophical premise of naturalism (recall the discussion of evolutionary naturalism in Chapter 2). This idea that humans are grounded in nature is seen as compatible with the recognition of humans as a distinctive species—distinctive because we have developed critical consciousness and culture. Thus:

> Humanists believe that we humans are the only species thus far to have evolved the capacity for constructing reliable knowledge of our surroundings, and about ourselves. Therefore, we no longer need to resort to myths of revelation from on high, or to fictions about mysterious intuitive messages from unknowable forces beyond what is accessible to human experience. (Hutcheon 1995:31)

This alternative humanism also has an overriding focus on morality and social activism, but it is moti-

vated by philosophy rather than politics. There are two key points to consider: first, this is an interpretation committed to the scientific method, including the operation of cause and effect in human behavior; second, this is an interpretation arguing that the longstanding foundations of this humanism are occasionally attacked from within by those who become attracted to some version of subjectivism, such as existentialism or postmodernism.

Because this alternative humanist interpretation has not been adopted by either humanist psychologists or cultural geographers—who, as we know, have been attracted instead to versions of humanism that are subjectivist, opposed to the scientific method, and unsympathetic to naturalism—it is not considered further here. This should not be seen as a criticism of the interpretation but rather as something that is necessary in a textbook designed to reflect the practice of cultural geography. If this interpretation were accepted, then the debate about naturalism would be resolved.

topophilia, refers to love of place, to the affective ties between humans and environment. As people live in and become comfortable with a particular place, they develop a **sense of place** that serves as a basic emotional underpinning for everyday life. The idea of **placelessness**, referring to locations that lack meaning, has not proven very useful, possibly because any location that can be identified necessarily has meaning to someone.

Box 3.2 Phenomenology, Existentialism, and Idealism

Much of the support for the humanist movement in cultural geography as it emerged in the early 1970s originated in opposition to the model of humans embraced by positivist spatial analysis. Humanism contends that nature can be explained, but that humans, social life, and individual behavior need to be understood. Necessarily, then, humanists consider the scientific method inappropriate for research into human behavior, because humans have intentions; this is the fundamental principle of subjectivity. Further, humanist thought maintains that social phenomena are not entirely external to the researcher, and this has led to the development of a hermeneutic tradition, that aims to reveal expressions of the inner life of humans by *verstehen*. Ley and Samuels (1978:9) describe this 'principal aim of modern humanism in geography' as 'the reconciliation of social science and man, to accommodate understanding and wisdom, objectivity and subjectivity, and materialism and idealism'.

The three humanist approaches referred to in this box are well established in the larger philosophical literature. Although it has been customary to use this tripartite classification in geography, there is much overlap and integration of the three approaches given the common interest in subjectivity and individuals.

Phenomenology, as introduced especially by Tuan and Relph, has prompted numerous variants. For example, although it was initially articulated as a study of individuals, phenomenology has been adapted to the social scale by Alfred Schutz. A key concern of the perspective is the contention that there is not an objective world independent of human experience, and hence there is a focus on the individual lived world of experience and the understanding of meaning and value. The implications for cultural geography are varied, but phenomenology certainly presents cultural geographers the major task of reconstructing individual worlds.

Existentialism was proposed especially by Samuels, although 'the boundary between existentialists and phenomenologists cannot be drawn precisely. . . . existentialists concern themselves with the question of the nature of "being" and understanding human existence' (Entrikin 1976:621). With existentialism as with phenomenology there is concern with individuals and their relationship both with the world of things and the world of others. There is little evidence to suggest that this philosophy is emerging as central to cultural geography.

Idealism as advocated by Guelke is one component of the larger idealist alternative to naturalism discussed in Box 2.2. The Guelke version derives specifically from historical idealism as proposed by the historian Collingwood. Idealism insists that phenomena are only significant when they are a part of human consciousness; to comprehend the world, it becomes necessary to rethink the thoughts behind actions in order to discover what decision makers believed rather than why they believed it. Although this approach was cogently advocated, and although there have been several applications, most notably by Guelke (especially 1982), idealism conceived in this manner has not proven to be especially attractive to other cultural geographers.

Each of these three approaches stresses human subjectivity and free will. Each shares the view that a spatial analytic geography is dehumanizing because it treats humans as objects, not subjects. But most contemporary cultural geographers do not accept the claim that the appropriate scale of analysis is that of the individual, as they recognize that the behavior of the individual is subject to many cultural constraints. Jackson and Smith (1984:21), for example, noted that a problem of humanist philosophies was that of 'building an effective bridge between individual cognition, perception, and behavior, on the one hand, and an appreciation of man's place in society, on the other'. The tendency of humanism to focus on individuals, to explain behavior in terms of individual mental processes, has led to its being criticized as psychological reductionism.

Although often interpreted in complex conceptual terms, place is essentially a very simple idea. Each of us is able to identify locations that have meaning to us, and it is these locations that we call places. Places may be large or small, may be warm and welcoming or cold and hostile, and may have significance for individuals or groups. This flexibility means that many parts of the earth's surface qualify as places, making place an especially difficult concept to illustrate. The location shown here, French River on the north shore of Prince Edward Island, is selected because the inlet setting and the hills behind suggest comfort, enclosure, and security; it seems evident that this location may represent a place for those who live or have visited here. (Bill Lowry/Ivy Images)

Unlike positivists, who generally adhere to a scientific method, humanists do not employ a particular method in their studies, as there is general agreement that humanism cannot be defined as a set of procedures or techniques. Humanist geographers are generally hostile to statistical techniques and opposed to formal reasoning; they tend to prefer creative writing. Some use conversational language to capture the nuances of everyday experience, while others adopt a more self-conscious literary style to uncover profound experiences in everyday life. Probably the most popular procedure humanist geographers use is participant observation, in which the researcher openly empathizes with the researched. This method, which had become important in both anthropology and sociology by the 1970s, was quickly adopted by humanist geographers, who favored its explicit recognition that people and their lives do matter. An informal procedure humanists use is phenomenological reduction, which involves suspending usual ways of knowing, such as language or academic expertise, in order to achieve a fundamental intuition of the world. Although proposed, this procedure has not been satisfactorily explained or used in cultural geography. Overall, humanists use qualitative methods that involve researcher observation of and involvement in everyday life, which they regard as central to an understanding of humans and human landscapes.

One problem with much humanist geography is that it typically prioritizes individuals at the expense of groups. Such an emphasis is not in accord with much current work that sees a concern with individuals as a reflection of Western, middle–class values. Indeed, notwithstanding many arguments to the contrary, one leading humanist geographer has argued that no real humanist tradition in geography has developed, and that even some ten years after being introduced, the approach consisted of 'little more than a few expressions of possibilities' (Relph 1981:134). If this claim is correct, humanist geography can be unfavorably compared with corresponding interests in sociology and psychology, in which scholars have succeeded in developing ideas on the basis of substantive research activities. Regardless, humanism has contributed some important concepts and is closely allied with some aspects of postmodernism.

Behavioral Geographies

Following landscape school principles, the approach traditionally preferred in cultural geography was to see individual behavior as subsumed within a larger cultural context, so that cultures were understood to be the best means of explaining landscape change—this is one component of the superorganic concept of culture. But, by about 1970 two revisions to this landscape school interest were evident. The first involved the rise of a positivist spatial approach to the study of behavior, and the second involved a more humanist and individually based approach.

In formulating these two behavioral geographies, geographers turned to psychology, focusing in particular on those psychological concepts that appeared to relate directly to their traditional concerns with space and environment. Thus, geographers largely ignored the many difficult issues that psychologists confronted. Of course, the way these concepts were interpreted depended on which particular geographic approach was used, and it is for this reason that two different behavioral geographies, each incorporating a different model of humans, emerged. In both cases it is helpful to consider the underlying psychological theories prior to outlining the behavioral geographies.

Understanding Human Behavior

Knowledge and learning are divided into disciplines that change through time, such that the contemporary social sciences are social constructions with boundaries that are neither necessary nor unchanging. Nowhere is this more evident than in attempts to understand human behavior. Two points are relevant here. First, although long an area of scholarly interest, it is only in the twentieth century that the study of human behavior has become identified particularly with the disciplines of psychology and sociology; prior to this time, such work was conducted primarily by philosophers and historians. Second, the current division of interest between psychology and sociology may be unfortunate given the shared interest in human behavior. The division reflects certain trends in the way disciplines were formed in the nineteenth century, especially the desire of Comte, and later of Durkheim, to establish sociology as an independent science of society. Hence, while any full understanding of human behavior requires consideration of both individuals and collectivities, most theoretical formulations are derived from disciplinary origins and are limited to one or the other. Box 3.3 provides a brief overview of this question of social scale.

During the nineteenth century, psychologists introduced two approaches to the study of the human mind, in the original Greek tradition of psychology. Some argued that the mind could best be understood through **structuralism**, analyzing its parts or structures, with the principal research method being that of introspection. Others argued that the mind was to be understood in terms of ongoing thought processes that are responsible for human learning; this approach, known as **functionalism**, was influenced by Darwinian ideas of evolution and adaptation.

From these bases, twentieth-century psychology introduced many varied ideas about how to understand and explain behavior. It is not possible to do full justice to the diversity of psychological interest in human behavior, but the principal schools of thought are readily identifiable, and those to which geographers have turned for inspiration are relatively few. The approaches of psychoanalysis, behaviorism, Gestalt psychology, humanism, and cognition are outlined in Box 3.4, which shows that one of the major sources of disagreement among the different approaches was how to define the subject matter of psychology. Was psychology the study of the human mind, of human behavior, or of both of these? As the contents of Box 3.4 suggest, behavioral concepts developed in psychology are substantial and varied, and there is an abiding tension as psychologists have often desired the objectivity of the natural sciences but have been concerned that they do not and cannot possess such objectivity because some of what they study, such as mental states, seems necessarily subjective and incapable of measurement.

Three other complex unresolved issues that psychological theories address are as follows. First, is it appropriate to regard human behavior as possessing what are often labeled purposive or teleological qualities? That is to say, are some phenomena best explained in terms of ends (what they have become or what they achieve) rather

Box 3.3 The Question of Social Scale

Individuals or groups? This is a recurring issue in social science.

Because individuals live in societies, and because societies comprise individuals, there is no clear distinction between the two, and it can be claimed that they must be considered together. Indeed, it was only in the nineteenth century that social scientists established disciplines distinguished on this rather flimsy basis. Certainly, psychology as the study of individuals, and both sociology and anthropology as the study of groups, suffer from poorly defined key terms; recall the discussions of culture and society in Chapter 1. Even today, the distinction between individual and group is less than clear, with university courses on social psychology often included in both psychology and sociology departments. Psychological social psychology typically studies social relations from the individual outwards by reference to such individual characteristics as attitudes, and the approach is often derived from naturalism. Sociological social psychology typically employs group identities such as class and community, and is more receptive to methods related to the idealist perspective.

What is the appropriate social scale for analyzing the kind of human behavior that is of interest to cultural geographers? There is no simple answer to this question, since different approaches favor different scales. Indeed, the question itself is misleading, as a group can range from two individuals to the global population.

Of course, Sauerian cultural geography focused explicitly on culture: 'Human geography, then, unlike psychology and history, is a science that has noth-ing to do with individuals but only with human institutions, or cultures' (Sauer 1941a:7). Further, most contemporary cultural geography accepts the need for a group scale. For example, Marxist approaches discussed later in this chapter contend that humans are more than individuals, and that there is a need to consider the intersubjectivity of social life. From this perspective, the question becomes one of who does something—an individual or a class, an institution or a culture? The Marxist answer is that individuals need to be defined within the wider cultural context; a focus on individuals is a concern for their beliefs or ideas, which need to be defined culturally. On the other hand, as discussed in Box 3.2, humanist geographers, because of their concern for subjectivity and free will, employ philosophies that typically encourage an individual scale of analysis.

Not surprisingly, then, cultural geographers today display a willingness to study people at the full range of social scales. Most of those who continue to work in an essentially Sauerian tradition focus on cultural groups, as do those whose inspiration is from Marxist or feminist thought. The many cultural geographers whose work is informed by postmodern and related thought show less commitment to a particular social scale, preferring instead to employ the scale that seems most appropriate for the topic at hand, although it is fair to say there is a particular concern with cultures understood as fluid, changing identities. Only humanist geographers are, on the whole, committed to the individual scale. On balance, it is the group scale that is most popular in the practice of cultural geography today.

than in terms of causes? Certainly, some theories view purpose and striving for goals as integral components of individual behavior, while other theories see these factors as being of marginal relevance, arguing that they accompany behavior but are not causes of behavior. Second, what is the relative importance of conscious determinants of behavior and unconscious determinants? There is general agreement in psychology that there are unconscious determinants, but there is much debate about their importance. Third, how important are the anticipated consequences of behavior, especially rewards, as causes of that behavior? Some argue for cause and effect while others argue against any such direct relationship.

These are three principal issues that psychological theories address, though there are certainly others. Among the many other issues that psychologists debate are the importance of learning to behavior, the question of individual uniqueness, the relatedness of acts of behavior, and the degree of relationship between behavioral acts and environmental contexts.

Cultural geographers have not made much use of these complex and divergent theoretical formulations. Rather, they have been attracted particularly to those approaches, or parts of those approaches, that appear to be in accord with prevailing geographic perspectives or that appear to be relevant to geography. Of course, this may not be such a bad thing if Hall and Lindzey (1978:69) were correct in stating, 'The fact of the matter is that all theories of behavior are pretty poor theories and all of them leave much to be desired in the way of scientific proof.'

Pause for Thought

The different psychological approaches to behavior may not be quite as different as they appear. Consider, as one example, attempts to integrate behaviorism and cognition. The distinction between these two is clear in principle, as behaviorism studies observable behaviors and cognitive psychology studies internal mental processes. However, the distinction is not quite so clear in practice, as one school of thought, sometimes called cognitive behaviorism, claims that cognitions are best thought of as behaviors that occur inside the body. Following this logic, the fact that a cog-nition—for instance, a thought—cannot be seen does not make it any different from an observable behavior, such as waving a hand. For the cognitive behaviorist, both types of behavior are subject to the same underlying principles. This example of the overlap and integration between two approaches that are often thought of as quite different from each other is but one instance of how uncertain the contemporary conceptual landscape of the understanding of behavior can be. There are also some interests shared between cognitivists and humanists, as both object to the behaviorist exclusion of mental life, focus on perceived meaning as the basis of behavior, and argue that humans are quite unlike animals.

Two Models of Humans

SPACE AND BEHAVIOR

In the late 1960s a behavioral geography developed from spatial analysis. At this time, many social scientists considered human geography to be one of the social sciences, an understanding that was in accord with the claim that all social sciences study social life or 'patterns of conduct that are common to groups of people', such that it is 'not their subject matter but their approach to it that differentiates the various social sciences' (Blau and Moore 1970:1). Indeed, geography was routinely included in every 'list of the social sciences' with each discipline having 'its own subject matter, its own part of the study of man' (Senn 1971:60). Also, during the 1960s heyday of spatial analysis, many human geographers viewed their discipline as a social science concerned with some aspect of behavior, with behavior often used as an umbrella term to refer to human activities in general. Thus, Ginsburg (1970:293) noted that human geography was concerned with 'questions of human behavior to the same degree, though not necessarily in the same way, that the other social sciences are'. Similarly, a leading cultural geographer asserted: 'Regardless of the tradition they follow or the methodology employed in the definition of their role, most human geographers have no hesitation in identifying themselves as social scientists' (Mikesell 1969:227).

This conjunction of circumstances sparked a geographic interest in the study of behavior that subscribed to a model of humans with close links to spatial analysis and some tenuous links with behaviorism and cognition as discussed in Box 3.4. This behavioral geography was unconcerned with human creativity and understanding, and the model of humans its exponents employed saw humans as responding to particular stimuli but paid only minimal attention to questions of human freedom and dignity. In addition to this rather narrow view of humans as rational decision makers, which was borrowed from economics and used typically in economic geographic analyses, behavioral geography held a more flexible view of humans as optimizers. This idea was in accord with some work in economics that viewed humans as boundedly rational animals—that is, as following simple step–by–step rules given constraints of limited knowledge and time. A classic study of farming practices in Sweden

further established that humans adopted strategies designed to find a satisfactory outcome, and this led to the introduction of the term **satisficing behavior** (Wolpert 1964). Many of the studies conducted in this tradition studied human behavior and inferred cognitive characteristics such as **perception**. Major geographic analyses informed by these ideas included work on natural hazards.

PLACE AND BEHAVIOR

Most cultural geography concerned with human behavior favored a different model of humans, one that incorporated humanist concerns such as subjectivity. It is for this reason that geographers were attracted to such psychological concepts as cognitive or mental maps and perceived environments that involve identifying a psychological environment, or what might be called the subjective frame of reference. This focus on sub-

Box 3.4 Psychological Approaches to the Study of Human Behavior

Psychoanalysis as introduced and developed by Sigmund Freud (1856–1939) is a deterministic approach claiming that human behavior is beyond the control of the individual such that the role of free will is minimal. Proponents of this approach believe that because it is the presence of a dynamic unconscious that explains why people behave as they do, the task of a psychoanalyst is to uncover the forces that operate in the unconscious. But this unconscious cannot be accessed directly, and there is accordingly an emphasis on uncovering this hidden world through other means, such as the analysis of dreams. Despite the controversial nature of these ideas within psychology, the complexity of the concepts, and the implicit determinism, versions of psychoanalysis are attractive to some cultural geographers (Kingsbury 2004).

Behaviorism as introduced by John B. Watson (1878–1958) in 1913 claimed that the subject matter of psychology was behavior, not the human mind, and that behavior was to be studied by means of objective procedures as practiced in the physical sciences. Radical behaviorism was developed by Skinner, who introduced the concepts of antecedent

conditions, operants, and consequences, with the key idea being the law of effect—that is, that consequences influence behavior. Few cultural geographers are attracted to these ideas (Norton 2003a).

Gestalt psychology developed in the early twentieth century from the idea that human behavior is determined by a psychophysical field analogous to a gravitational or electromagnetic field. Thus, human perception of a particular object is explained by reference to the larger field within which the specific object is contained. The human mind and perception are one whole and not a number of parts. This way of analyzing conscious experience is different from those proposed by both psychoanalysis and behaviorism. The emphasis is on how the world is perceived rather than on how individuals behave because it is perception that determines behavior. Thus, behavior can be understood by uncovering the perceived world of individuals rather than by studying the objective world. The principal method of analysis used by Gestalt psychologists is a phenomenological approach involving the study of individual perception and subjectivity. Some geographers have been attracted to these ideas because of the empha-

jective environments involves the recognition that the physical world can affect individuals only in so far as it is perceived or experienced. Thus, in much geographic work, objective reality is not seen as a cause of behavior; rather, it is objective reality, as perceived or assigned meaning by the individual, that is judged important. Some work in this tradition is discussed in chapters 5 and 9.

While several geographers were devoting themselves to the substantive body of work centering on perceived environments, one geographer adopted the transcendental phenomenology of Husserl. According to this view, humans have misunderstood the character of the world by believing that a scientific, mathematical, and objective view of the world is also the world itself. Rejecting such a view, Husserl claimed that it was not appropriate to interpret events in our lifeworld scientifically; what was

more important was to attempt to understand the general structures of meaning of the lifeworld. To do this, Husserl explained, it is necessary to suspend our preconceptions by putting aside any suggestion that the real world is naturally ordered. With these ideas as a basis, Hufferd (1980:19) argued that 'In order to explain why an object (human or not) acts or reacts in whatever fashion, it is necessary to understand it in relation to the perspective resulting in its action'. According to this argument, any attempt to understand human behavior needs to proceed as follows. First, the researcher must record the accounts offered by numerous participants as precisely as possible until major themes become evident and no additional new themes are forthcoming. Second, the researcher must list the frequently mentioned topics, classify them, and compare the different interpretations. Third, the researcher must summarize the top-

sis on perception and subjective environments, the concern with individuals as members of groups, and the explicit acknowledgment of the role played by external, non-psychological environments.

Humanism rejects the scientific tradition of conducting research that is replicable and verifiable by other researchers, seeking instead an understanding of the world as it is appreciated by the individual or group being studied. It asserts that behavior results from subjective understandings—perceptions—of the objective world, and all versions of humanist psychology maintain a central concern with human experience and personal growth. Hall and Lindzey (1978:229) describe the position of humanism in relation to psychoanalysis and behaviorism thus: 'Humanistic psychology opposes what it regards as the bleak pessimism and despair evident in the psychoanalytic view of humans on the one hand, and the robot conception of humans portrayed in behaviorism on the other hand.'

Cognition is an approach arguing that mental processes, especially thinking but also perception, memory, attention, pattern recognition, and problem solving, are principal causes of behavior. Cognitive psychology claims that the presence of mental

factors means that any attempt to explain all human behavior in terms of physical laws is necessarily inadequate, and that while it may be the case that behavior is shaped by consequences, it is still the case that cognition must be studied in its own right. With its emphasis on information processing, it is related to the cognitive science approach, which aims to explain cognitive processes in terms of the means by which information is handled by the mind, usually proposing that the mind works in a series of logical steps rather like a computer. Some cognitive psychologists have borrowed concepts from such areas as computer science, communications science, and neuroscience. Additionally, cognitive psychology has a related geographic interest in the computational process modeling of spatial cognition and behavior.

All too clearly, psychologists do not agree on the best approach to study of human behavior. Each of the approaches noted has some merit, and it is not possible to definitively state that one is in some way inherently superior to the others. A balanced response to the above overview may be to suggest that different approaches are suited to different specific problems.

ics and identify key features so that the perception held by the group can be reconstructed as authentically as possible. Arguably, this reconstruction forms the basis for an understanding of human activities.

ONE OR TWO GEOGRAPHIES OF BEHAVIOR?

Behavioral geography as it developed from the spatial analytic tradition was not closely associated with work in cultural geography, unlike the humanist-flavored behavioral geography promoted by cultural geographers. But this simple distinction in principle, specifically involving different inspirations and different models of humans, was much less clear in practice as geographers struggled to introduce behavioral analyses into the discipline. A classic early study intentionally and effectively integrated positivist and humanist geography to explain and understand American inner-city life with specific reference to an inner-city area of Philadelphia with a high concentration of African-American residents. For some contemporary cultural geographers, such an integration of approaches is necessarily flawed, although Ley later insisted that objective measurement and subjective interpretation were compatible:

> For while one may conduct a multilocale ethnography of more than one place, an upper limit is quickly reached. In moving from two sites to 20 some standardization of accounts—usually measurement—becomes necessary. I see nothing but intellectual gain from incorporating the harvest of an intensive interpretation of one or a few sites with an extensive analysis of a large number. (Ley 1998:79)

Overall, however, geographers undertaking studies of behavior favored either a positivist or a humanist perspective, and thus were making a decision about naturalism, specifically about what might be called the model of humans. They either considered it appropriate to view humans as objects and to work towards the development of laws governing behavior, or else they preferred to view humans as subjects, active participants in their worlds. In retrospect, it is clear that most behavioral work favored the humanist perspective. Thus, research focused on the idea that humans are subjects, emphasizing presumed links between perception, behavior, and the creation of landscape, and often building on the assumption that perception caused behavior, which in turn caused landscape. The idea that humans imaged environments as something other than reality emerged as a fundamental basis for research.

The Inference Problem

It is not satisfactory to approach an understanding of the cultural landscape solely from the rather naive proposition that perception causes behavior causes landscape. All too clearly, factors other than perception influence behavior, and factors other than behavior influence landscape. Nevertheless, much useful and innovative work has centered on the role played by perception. But should cultural geographers organize their research around the logic that perception causes behavior, or should they infer perceived environments on the basis of actual behavior? Both of these research frameworks involve difficulties.

On the one hand, using **image** to explain behavior is difficult in that it can be argued that perceived environments cannot even be objectively measured. Further, a close correspondence between perception and behavior is often difficult to demonstrate. On the other hand, inferring images from behavior also poses logical problems, as it is one example of the inference problem—that is, assuming the identity of causes on the basis of an analysis of effects.

On balance, the second concern is the more problematic. Inference is an uncertain form of discovery, for there are no guarantees that the inferred cause is indeed a true cause. To demonstrate the dangers of inference in this context, consider the following simplified circumstance. Farmer A perceives that either crop x or crop y will produce a satisfactory income, while farmer B perceives that either crop y or crop z will produce a satisfactory income; both choose to cultivate crop y, and therefore their two landscape decisions are identical, despite being based on different perceptions of reality.

Pause for Thought

Although several geographers writing around 1970 anticipated that a behavioral approach might prove a panacea for perceived ills of the discipline, this has not been the case. But why not? One reason might be that two quite different versions of behavioral geography developed, based on quite different conceptions of what it means to be human. Viewed retrospectively, it seems clear that in both cases the underlying philosophical impetus, positivism in one instance and humanism in the other, proved unacceptable to the majority of geographers. Further, Walmsley and Lewis (1984:4) identified the confused character of behavioral geographies by referring to such features as a focus on individual decision-making units, the relevance of both acted-out and mental behavior, an emphasis on the world as it is rather than as it should be under some theoretical constraints, an interest in economic location topics, and use of models. This list is suggestive of the mixed inspirations and varied practice of behavioral geographies. It will be interesting to see if cultural geographers pursue further the behavioral or humanist approaches to the study of behavior, or indeed if they choose to pursue some alternative approach.

Marxisms

As with humanism, versions of Marxism were introduced to human geography around 1970, prior to the cultural turn of the 1980s. Indeed, in some respects Marxism was one of the approaches that the cultural turn reacted against because of the emphasis that Marxism placed on grand theory—the claim that there is necessarily some right answer to a problem—and also because of the essentialist character of the Marxist approach. Rather like humanism, Marxism has contributed some important concepts to cultural geography and is linked to some aspects of postmodernism. Most notably, Marxism informs ideas about human identity, difference, power, inequality, and commodification.

Marxism, as developed by Marx and Friedrich Engels (1820–95), is both a body of social theory and a political doctrine. During the twentieth century the political doctrine was implemented in modified form in a number of countries, notably the former Soviet Union and China. The concern in the present context is with the social theory.

Marxist Social Theory

It is often suggested that in developing a social theory Marx followed Hegel and employed a **dialectic** as an alternative to formal logic, although it seems more appropriate to suggest that the dialectic used by Marx was ontological, referring to the fact that societies are organic wholes that evolve only to eventually collapse as a consequence of their internal contradictions. The three components of Marxist social theory are as follows.

The first principal component of Marxist social theory involves the identification of principal types of human society in a historical context. Examples of societies are slavery, feudalism, capitalism, and **socialism** (all four terms are included in the Glossary). Marx discussed each society as a **mode of production**, comprising both **forces of production**—that is, the raw materials, implements, and workers that produce goods—and the **relations of production**—that is, the economic structures of society. It is the forces of production that produce goods, while the relations of production determine the way in which the production process is organized. The most important relation concerns matters of ownership and control. Two insights emerge from this component of Marxist social theory: there is a historical transition from one mode of production to another, and social classes play a key role in social formation and social change. Further, society can be differentiated into **infrastructure** (or **base**)—another term for relations of production—and **superstructure**, the legal and political system and forms of consciousness. These terms are helpful because infrastructure can be interpreted as a determinant of superstructure, meaning that human thinking results from material conditions. This is effectively a form of economic determinism.

The second principal component of Marxist social theory is the transition from one type of society to another explained in terms of two related processes, namely technological change

This larger-than-life statue of Marx (seated) and Engels is in a small park with paths, trees, and shrubs. One of Berlin's most famous statues, it is located in Marx-Engels-Forum in East Berlin, capital of the former East Germany. The statue was once a symbol of the ideological roots of the communist state, but today it is better known as a place for tourists to have their photographs taken. (Dave Bartruff/CORBIS Canada)

and **class** struggle. Marx emphasized the emergence of new classes and the fact that an existing dominant class was able to limit the opportunity for further social change. Because a dominant class typically favors maintaining things as they are, Marx recognized the need for class struggle as the means for replacing one mode of production with another mode that comprised a different set of classes or that lacked classes altogether.

The third principal part of this theory is its analysis of nineteenth-century capitalism, which anticipated a transition to socialism. Marx was especially critical of capitalism because the dominant class (owners) was concerned primarily with profits—or what he called the ceaseless accumulation of capital. Further, he saw owners as exploiting the dominated class (workers). Although such exploitation of one class by another was by no means new, Marx recognized

that what was previously open in modes of production such as slavery and feudalism was disguised in the capitalist mode of production. Marx envisaged capitalism as the last of the class-based societies. Much of the political doctrine of Marxism is concerned with the various mechanisms by which classless socialist societies would eventually replace capitalism.

The Importance of Division of Labor

Marxist social theory is concerned with the material basis of society and aims to understand society and social change by referring to historical changes in social relations. This approach is termed **historical materialism**. Marx conceived of history in materialistic terms, and, as we have seen, a central feature of Marxist social theory concerns the division of labor and the formation of classes.

For Marx, the first human historical act is the production of means to satisfy material needs, and fulfillment of these needs leads, as population numbers increase, to other needs. The first form of social relation, the family, is required to fulfill the initial material needs; when the family proves unable to satisfy these new needs, a new social formation develops. In this sense, it is appropriate to identify each mode of production with a particular social formation that emerges to satisfy the material needs of society. This is what Marx intended when he said that society has a materialist basis. Thus, in tribal societies there is an elementary division of labor based on age and gender, but with the development of agriculture, civilization, and industry the division becomes increasingly complex. In the industrial world of the nineteenth century, the capitalist mode of production involved a separation and conflict of industrial and agricultural interests, a separation and conflict of capitalist and worker interests, and a separation and conflict of individual and community interests. The implications of these divisions of labor, from the elementary forms found in the family to the much more complex forms found in capitalism, are enormous.

Thus, Marxist social theory maintains that there are unequal distributions of labor, of the products of labor, and of private property. Further, the distinction between individual and community interests leads to the creation of states that serve the community interest. When human relations are reduced to commercial relations—a form of commodification—the result is an estrangement or alienation of social activity. Marx was concerned both with the material conditions of workers and with the way in which their lives assumed an alien form.

Versions of Marxism

Marx was a prolific writer whose works included numerous uncertainties and contradictions as his ideas developed and transformed through time. Accordingly, there have been numerous interpretations of both Marx's social theory and his political doctrine. Certainly, there is no one correct reading of Marx. Two principal versions of Western Marxism that appeared during the twentieth century are outlined here.

The Frankfurt school of critical theory emphasized the humanist aspects of Marxist thought and added elements of social psychology. From this perspective, Marx was a voluntarist, seeing human history as reflecting human intentional activity. This school criticized bourgeois society for being dominated by a form of technological rationality and social science for its prevailing scientific orientation.

A second version of Marxism rejected the humanist aspects and focused on its structuralist character. A popular approach during the 1960s, structuralism emphasized that understanding a social system required an appreciation of the structural relations between the component parts. From this perspective, human problems were essentially structural in character, as they were rooted in some specific economic system, and humans were not considered as autonomous active subjects as they were in the Frankfurt school.

In addition to these two versions, there have also been attempts to read Marx in terms of rational choice theory in economics—this is known as analytic Marxism—while Marxist theory has also been influential in some feminisms and in cultural studies generally. Regardless of the specific version, Marxism is always concerned with analyzing inequality, oppression, and subordination, and with identifying some means to overcome these. The basic solution for Marx was, of course, the creation of a socialist society involving the abolition of class, of private ownership of the means of production, and of commodity production.

Marxisms in Geography

As the account of Marxist social theory has shown, there is no single Marxist tradition to which geographers could turn, and because of this contested situation, any Marxist approach adopted in geography has been challenged not only by those geographers who are generally unsympathetic to the Marxist approach but also by other Marxist geographers who favor an alternative version. The most serious debate has been between those geographers inclined to the humanist-oriented Frankfurt school of thought and those inclined to structural Marxism. The disagreement between these two is one part of the larger 'structure and agency' debate—that is, the dispute about the ability of human beings to function under the constraints imposed on their behavior by social structures. We have already encountered this debate in the guise of what was called the social scale of analysis (see Box 3.3).

Not surprisingly, humanist geographers in particular have objected to structural Marxism on the grounds that, like positivism, it adopts a passive view of humans. Yet some human geographers have favored structural Marxism because of its emphasis on the presumed crucial influence of the infrastructure—meaning the economic structure of capitalist society—on the superstructure of the human geographic world, which includes the landscape. Other geographers have acknowledged the importance of both the more humanist Frankfurt school and structural Marxism, noting with approval that Marxist thought is open to a variety of interpretations, ranging from an active view of people as the makers of their own history to a more passive conception of human development as the determined product of relatively autonomous structures.

Within cultural geography specifically, there are two important claims. First, it has been asserted that 'Marxism and cultural geography share important basic presuppositions concerning the significance of culture, but in different ways and for different reasons both have failed to sustain those presuppositions in their practice and have not developed a dialogue with each other' (Cosgrove 1983:1). Cosgrove's attempt to correct this failing recognized a common focus on the historical aspect of human and land relations and stressed conceptual similarities such as a shared Marxist and Vidalian view of the relationship. Second, it has been argued that a Marxist cultural geography corrects perceived deficiencies in the discipline, such as cultural geography's failure to recognize that culture has a political component and its failure to recog-

nize that cultures are divided into classes. Further, according to Blaut (1980), a Marxist view is one that avoids two of cultural geography's humanist tendencies: the tendency to set individuals and culture in opposition, and the tendency to neglect external constraints from society at large. The most substantive recent work along these lines sees cultural geography as necessarily involving a conception of culture that focuses on questions of power and authority (Mitchell 2000).

Pause for Thought

Marxism informs many cultural geographic studies because it explicitly addresses the fact that the world is characterized by significant human and landscape inequalities, inequalities that traditional landscape school studies have rarely considered. The need to study such inequalities is clear. Agreed? But from a student perspective, does it seem unfortunate that Marxism is riddled with internal uncertainties, meaning that there is often disagreement concerning how best to employ Marxist concepts and principles in particular analyses?

Feminisms

Although there are many versions of **feminism**, all share the fundamental belief that **sexism** prevails, is wrong, and needs to be eliminated. **Patriarchy** is usually seen as the fundamental wrong, although feminists are concerned not only with the oppression or subordination of women but also with other oppressions. This discussion provides an appropriate basis for the discussions in Chapter 8 of the social construction of **gender** and of landscapes of sexuality.

Three Versions

The basic tradition of feminist thought, dating back to the late eighteenth century and known as the first wave or liberal feminism, aimed to obtain equal rights and opportunities for women. Note that from some later perspectives, such a seemingly meritorious approach can be seen as fundamentally flawed because, in aspiring to equal status with men, it supports the view of the male model as the norm.

The feminisms that appeared during the 1960s and continued into the 1980s are often described as the second wave of feminist thought. These tended to assume that it was possible to identify a specific cause of the oppression of women. Suggested causes included men's ability to control women's fertility and the capitalist requirement of a pliable workforce. Versions of cultural feminism, with a focus on technologies and ideologies developed by men and the idea of essential relationships, belong in this second wave category. These more radical feminisms often have the general intent of moving beyond the redistribution of rights and resources pursued by liberal feminists in order to initiate changes to the structure of society. Fur-

ther, many of these radical feminisms contend that the oppression of women is so deeply rooted in both psychic and cultural processes that much more than superficial change is necessary.

During the 1980s, the third wave of feminist thought appeared. This involved a critique of the earlier assumption that it was appropriate to attempt to universalize the experiences of women (usually White and middle-class) in the more developed world. Feminists thus developed a fuller concern for the multiple oppressions that structure our gendered identities. These multiple oppressions are based on social class, skin color, income, religion, age, culture, and geographic location. Third-wave feminism also involves an expanding interest in sexual difference as well as in gender, an interest stimulated by psychoanalytic theory. There

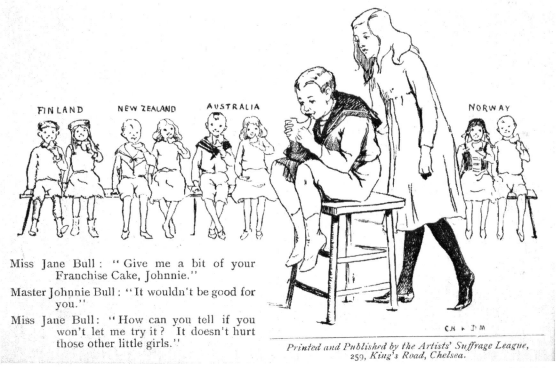

FINLAND   NEW ZEALAND   AUSTRALIA   NORWAY

Miss Jane Bull : " Give me a bit of your Franchise Cake, Johnnie."

Master Johnnie Bull : " It wouldn't be good for you."

Miss Jane Bull : " How can you tell if you won't let me try it ? It doesn't hurt those other little girls."

C.H + J.M

Printed and Published by the Artists' Suffrage League,
259, King's Road, Chelsea.

This British cartoon from 1908, with the simple appeal from Miss Jane Bull for equality with Master Johnnie Bull (John Bull personifies England much like Uncle Sam personifies America) neatly captures the essential and compelling logic of the first wave of feminist thought. In 1893, New Zealand became the first self-governing nation to grant the vote to all adult women and to allow women to stand for election. The other countries represented in the cartoon—Finland, Australia, and Norway—granted the women the right to vote and/or to stand for election in 1906, 1902, and 1907 respectively. (The Women's Library/Mary Evans Picture Library)

is an increasing realization that to insist on equality is to deny the importance of sexual difference. Finally, it has involved an engagement with aspects of poststructuralism and postmodernism that includes a concern with the social construction of gender. Most current work in feminist geography is identified with this third wave.

Note that for most versions of contemporary feminist theory, sex is a natural category based on biological differences between women and men (but see the discussion of sex and gender in Chapter 8), while gender, although based upon the natural category of sex, is a social construction that is developed over time and that refers to what it means to be female or male.

Locating Feminism in Geography

Unlike many of the other conceptual terms that have entered geographical discourse in the past twenty years (Marxism, structuralism, postmodernism), feminism provides neither a methodology nor a theory for human geography, although methodological and theoretical differences have been derived from it. Rather, feminism as it was originally introduced to geography had a far more radical goal—it provided a political basis for a critique of the practices of geography, practices that had allowed for the invisibility of women as both practitioners in the field and objects of enquiry. (Domosh 1996:411)

In light of this assessment, it is unsurprising that, despite a considerable body of both conceptual and empirical literature by geographers and others, feminism does not merit any significant explicit mention in many standard reviews of geographic approaches. The contrast with the ways that Marxism and postmodernism were received by the discipline is striking. It does appear that geography is less than certain about how to accommodate a body of thought that not only includes new perspectives and approaches and implies new subject matter, but that also questions the discipline of geography as a legitimate enterprise. Cloke, Philo, and Sadler (1991:xi) (three male geographers) explained their decision not to include a chapter on feminism in their account of approaches to human geography as follows: because 'the issues involved are more important than this "ghettoising" might imply', because of the belief that feminism is something 'more than just "another" approach', and because 'such discussion is most ably pursued by women'.

In addition to this uncertainty, there are debates about feminisms and feminist geographies within the feminist geography literature itself. Some have identified areas of shared feminist and geographic interest and argued for integration, while others express concern about the fundamental masculinism of geography that might effectively prohibit meaningful integration. The term 'masculinist' refers to work that claims to be exhaustive but that ignores the existence of women and that accordingly is concerned only with the position of men. Another important debate arose in the writing of a feminist geography textbook aimed at introductory-level students (Women and Geography Study Group of the Royal Geographical Society with the Institute of British Geographers 1997). This book was written by a collective that made the decision ('not an easy one') to exclude men from the writing team (Rose et al. 1997:3).

Feminist Challenges to Geography

The concerns of feminist geography have changed through time in general accord with changes in the larger feminist context. In the early 1970s and in accord with the logic of liberal feminism, there was concern on the part of some feminists about women in geography, especially since the discipline of geography, like many other disciplines, was dominated by men. There was also an attempt to incorporate gender as a variable into geographic studies, along with more usual variables such as culture and ethnicity.

By the 1980s, studies of oppression had begun to consider oppression both within the discipline of geography and within the larger world, and many of these studies adopted a Marxist orientation. The early concern about the lack of women in the discipline had expanded to include concerns about the essential masculinism of geography, especially because of geography's traditional link with exploration and its continuing involvement with fieldwork. Most recently, a more radical tradition with links to postmodernism has emerged. This tradition attempts to construct theory in response to the idea that geography is a masculinist discipline that claims to speak for everyone but that only speaks for White, middle-class, heterosexual males. Feminist geographers and others have begun to further explore the social construction of human gendered identities, both femininity as well as masculinity.

Probably the single most important concept in feminist geography is that of gender, along with the derivative ideas of **gender roles**, **gender relations**, and patriarchy. Gender roles are sets of behaviors deemed to be socially appropriate or inappropriate solely on the basis of sex. Few, if any, of these behaviors rely on inborn sexual differences. Most important, gender roles vary culturally and are, accordingly, best thought of as learned differences in behavior. The concept of gender relations recognizes that gender roles are not necessarily accepted; rather, gender roles are sometimes contested, and they change through time. Both of these characteristics of gender roles refer to the idea that gender involves power relations between women and men, specifically the relationship of patriarchy, whereby men dominate women. A similar argument about sexual identity is included in Chapter 8.

It is clear that feminism poses several challenges—perhaps better thought of as invitations—for geographers to consider. Since about the 1960s, forms of feminism have vigorously challenged virtually all of the belief systems and institutions of dominant patriarchal cultures, including geography. Feminism has questioned contexts for knowledge as well as basic concepts such as reason and logic. For geography this has involved questioning the legitimacy of approaches such as positivism, humanism, and Marxism on the basis of what is perceived as their sexist bias. Indeed, feminists were among the first to recognize the importance of the social construction of knowledge, meaning that those who create knowledge are responsible for determining the research problems, the data to be used, and the methods to be employed—the ear-

lier discussion of Marxism clearly demonstrated the validity of this recognition.

Feminists, along with postmodernists, use a number of concepts that, together, highlight the need to assess precisely what is happening during the research process. Most fundamentally, there is a need to acknowledge the importance of **situated knowledge**—that is, to recognize the discourse within which knowledge is produced. Similarly, there is a need for **reflexivity**, meaning self-reflection at all stages of research, in order to ensure that the particular characteristics of the researcher are not influencing the research activity in an inappropriate way. A third related concept is that of **gaze**, referring to, for example, the need to avoid a particular way of looking at things, such as a masculinist or heterosexist perspective.

Pause for Thought

It is perhaps inevitable that the discussion of feminism in geography is much more than a mere academic discussion. As with Marxism, feminism has a powerful political dimension, and it appears that cultural and other geographers are unsure how to accommodate the combination of academic and political challenges. Since most feminist geography is conducted by women, is it possible that questions about who is able to study the experiences of women have contributed to what might be described as a ghettoizing of feminist interests? And, if so, is this an unfortunate circumstance given that feminist philosophies have clearly added significantly to the body of concepts available to cultural geography?

The Cultural Turn

A Context

It is usual to suggest that both the social sciences and the humanities have undergone a cultural turn in discussions of humans and human activities. This cultural turn refers to placing increased importance on culture while downplaying other considerations such as economy and politics. An important aspect of this cultural turn is the emphasis on language, or linguistic turn, a term that is sometimes used as an alternative to cultural turn. Specifically, there is increasing realization that language is important, since it enables the communication of meanings, but that it is also constraining because of the limitations resulting from the meanings we attach to words; we do not know the world for what it is, but rather only as it is mediated through language and other symbolic systems. Expressed simply, the linguistic turn challenges the ability of language to function as a neutral conveyor of information.

This discussion considers the cultural turn from four interwoven perspectives, namely those of cultural studies, poststructuralism, postmodernism, and postcolonialism. This is a less than perfect way to categorize material that is by definition fluid and dynamic, but imposing some order—however artificial it might appear to be—is important to the student of cultural geography at this time because many of the ideas being raised are new to the subdiscipline, are of uncertain status, and are contested. However, it is appropriate to stress that the very spirit of the cultural turn is unsympathetic to classification and the imposition of order. This is an idea that resurfaces on several occasions in this discussion, but it is also one that proves difficult to adhere to at times.

Another point of information is in order. Although it may not be difficult to understand that there could be a cultural turn in some other disciplines, it may sound confusing to claim that the turn is affecting cultural geography—how can there be a cultural turn in cultural geography? The short answer is that the meanings of culture involved in the cultural turn are different from the meaning incorporated within Sauerian cultural geography—refer back to the various definitions of culture included in Box 1.2, and recall the deterministic implications of the classic Sauerian definition in contrast to the understanding of culture as a negotiated intersubjectivity implied in some of the more recent definitions.

Cultural Studies

Cultural studies is perhaps most appropriately seen as an umbrella term incorporating a wide range of philosophical and social theoretical ideas, including feminism, poststructuralism, postmodernism, postcolonialism, and aspects of Marxism and humanism. Central to cultural studies is the claim that established disciplines are unable to cope with the complex circumstances of the cultural world such that the advantages of an interdisciplinary approach are stressed.

Box 3.5 provides an overview of some of the more important authors and ideas that are emerging as influential in cultural geography. The information in this box is designed to promote a general understanding of the cultural turn.

Cultural studies literature typically rejects modernism in addition to the Enlightenment concerns with an empiricist epistemology and with a search for universal truths. Together, the varied bodies of thought that are labeled 'cultural studies' constitute a key feature of the cultural turn in the social sciences and humanities generally. Principal components of this work include several British studies of working-class culture (for example, Thompson 1968), the concept of culture introduced by Geertz (1973), and the Frankfurt school of Marxist critical theory. Also important in the current context are several advances in social theory associated with such writers as Gramsci, Derrida, Foucault, Baudrillard, and Deleuze, as these are discussed in Box 3.5, along with several traditions—such as feminism, postcolonialism, subaltern studies, and studies of racism—that are linked by their concern with exposing the Eurocentrism associated with the production of knowledge in the Western world since the Enlightenment.

Understandably, different bodies of thought that are gathered under the cultural studies umbrella emphasize different matters: feminism, for example, is critical of much of the gender blindness of traditional Marxism, while Marxism is a major source of inspiration for many postmodern social theorists. But together these works stress the importance of culture as a dynamic and primary force that is not necessarily always predictable from political, economic, and social forces. Further, all see culture as a site of negotiation and conflict in societies that are dominated by power and that are splintered especially in terms of ethnic identity, gender, and class. Much of this work can be described not only as part of the cultural turn, but also as part of the linguistic turn because of the emphasis on the role played by language.

Pause for Thought

Cultural geography clearly welcomes the importation of ideas from elsewhere, including from the cultural studies tradition, but there may be a real difficulty in such movement because cultural studies is seen by many of its practitioners to be a postdisciplinary and critical practice. Thus, from a cultural studies perspective, does it make sense that ideas be exported to disciplines and subdisciplines and then be employed within precisely those disciplinary and subdisciplinary frameworks that cultural studies challenges? It might be suggested that cultural geographers seem to want the best of both worlds—both the critical theories developed in cultural studies and the traditional subdiscipline of cultural geography in which to apply these critical theories. But for some practitioners of cultural studies the two are irreconcilable, not least because disciplines and subdisciplines are seen as something other than neutral when it comes to discussions of such topics as racism and sexism. Cultural studies does not take the existence of disciplines for granted, whereas most new cultural geography implicitly accepts the existence of a body of knowledge labeled cultural geography. Simply put, there is a very real challenge involved in attempting to employ theories derived from an avowedly postdisciplinary tradition in a disciplinary context. Indeed, for many in cultural studies, it is a challenge that should not be taken up.

Poststructuralism

Structuralism is one of several bodies of thought that view culture as a system of ideas based on advances in linguistic theory that see a sound and the object it represents as entirely arbitrary. In other words, structuralism's chief concern is with relationships rather than with those things that create and maintain relationships. For example, a structuralist interpretation of gift-giving in primitive societies focuses on the act of giving rather than on what is given, or by whom or to whom it is given. Structuralism aims to remove the difficulties of cross-cultural comparisons by showing that elements of all cultures are the product of a common single mental process. This idea has proven attractive to some anthropologists, but despite the interest it appears that many of the arguments are highly impressionistic. Two of the principal insights of structuralism—both antihumanist—are important, in modified form, in **poststructuralism**.

The first insight, following the linguistic philosopher Ferdinand de Saussure (1857–1913), is that language is a medium for understanding social organization. Thus, meaning is not conveyed through language but rather produced within language. Simply put, the facts do not speak for themselves. The second insight, following the Marxist philosopher Louis Althusser (1918–90), concerns the interpretation of subjectivity and claims that notions of individuality are seen as ideological constructs. Thus, humans cannot act autonomously; rather, their actions are constrained by various structures.

Poststructuralism is a broad term. Its central ideas reflect an expansion of, rather than a break with, structuralism. Many of the most important concepts, including those modified from structuralism, are often included as one component of the cultural studies tradition and are, as noted, also identified with postmodernism.

Postmodernisms

Box 3.6 provides a basis for outlining the key arguments and content of **postmodernism** by offering distinctions between four terms, namely modernity, modernism, postmodernity, and postmodernism. According to many (but by no means all) observers, the contemporary world is increasingly displaying the characteristics of postmodernism. Although this discussion offers a series of generalizations about postmodernism, it is not intended to convey the impression that there is some clearly agreed upon set of ideas. Postmodernism is a highly contested arena with numerous competing ideas that together constitute a real challenge to established theories and understandings of social science.

Postmodernism effectively rejects all other attempts at social theorizing because of the emphasis they place on establishing foundations for knowledge. Social theories developed within modernism (Marxism, for example) are seen as privileging science and the related foundational bodies of knowledge that discuss matters in terms of truth and falsity. Postmodernism, on the other hand, is antifoundational, opposing attempts at grand, all-embracing theories and favoring instead more local and particularistic studies that emphasize difference.

Postmodernism also questions—even denies—the ability of science and theory to establish truth, and rejects the claim that any such truths, were it possible to attain them, could be liberating. These are challenging ideas based on the argument that all knowledge, including truth claims, is socially created. Postmodernism proposes that the myths and ideologies of modernism serve only to promote ethnocentrism, specifically Eurocentrism. This postmodernist way of looking at human identity encourages studies of disadvantaged groups such as women, the working class, ethnic minorities, and gays—groups that were previ-

Box 3.5 Cultural Studies—Scholars and Ideas

Although many of the scholars and ideas associated with cultural studies are also closely identified with feminism, postmodernism, poststructuralism, and postcolonialism they are discussed together in this box for reasons of convenience. Specifically, because much of the empirical material in subsequent chapters (especially 7, 8, and 9) relies on the ideas presented in this box, it aims to serve as a reference informing many of those studies.

Antonio Gramsci (1891–1937) was a Marxist theorist and political activist whose most influential conceptual contribution was to elaborate the idea of **hegemony**. According to Gramsci, any class that achieved a position of economic importance and desired to attain power in a larger society first needed to achieve a degree of cultural and intellectual hegemony. This hegemony allowed the class to express its worldview, to structure social and other institutions so that they accorded with the class's aims, and most generally to create a context sympathetic to the class.

Gramsci also introduced the term **subaltern** to refer to those socially subordinate groups that lacked both the unity and the organization of more dominant groups that are able to exercise **authority** and control. The term has been popularized by Indian Marxists and generally refers to those groups considered to be in some way socially inferior to other groups.

The work of Michel Foucault (1926–84) both crosses and challenges disciplinary boundaries and is especially difficult to classify. One of his important contributions concerns the closely related ideas of discourse and **episteme**. Foucault saw subjectivity as constructed within and through discourses, with discourse defined as a system, comparable to a language, that enables the world to be made intelligible. (The related idea of episteme refers to the world-views or structures of thought that a society holds at a particular time.) All terms, then, need to be discussed within the specific context of a given discourse, as the meaning of any term varies from one discourse to another.

Discourses are important because they serve to legitimize a particular view of the world that then becomes part of the taken-for-granted world. Discourses define our world for us, as Foucault (1980:52) explains: 'the exercise of power perpetually creates knowledge and, conversely, knowledge constantly induces effects of power.' Thus, power is practiced through discourse. It can be argued, for example, that the discipline of geography evolved within the late nineteenth-century discourses of imperialism, colonialism, and racism. Foucault stressed the necessarily partial and situated character of any particular discourse. In short, language cannot 'serve as a perfectly transparent medium of representation' (Duncan and Ley 1993:5). Expressed simply, language can be seen as constructing rather than conveying meaning. What Foucault achieved in this respect was a reversal of the idea that knowledge is power, with the claim that only those in power have the right to say what is knowledge. To quote a few simple examples, consider what is involved in renaming rape as sexual assault, jungle as rainforest, and swamp as wetland—in each case the identical subject is being placed in a different context for evaluation.

ously marginalized or excluded because of their lack of power. In addition to encouraging studies of the disadvantaged and oppressed, postmodernism asserts that the voices of such groups are as important as the voices of so-called authorities. Authors and authority, as identified in modernism, are not privileged in postmodernist theory. For a postmodernist, individuals are repositories of cultural systems and are therefore unable to speak for themselves; rather, they speak for the cultural system. It should be clear from these comments that any

adoption of postmodernist claims has some significant implications for cultural geography.

Postmodernism emphasizes the idea of social construction. Postmodernists share the feminist concern with gender, class, and ethnic identity, recognizing that these are social constructions rather than established realities and, accordingly, that they can be changed. Postmodernists further acknowledge, with feminists, that academic disciplines such as geography are social constructions that exist only because they were created (and can therefore be dismantled)

Jean Baudrillard (b. 1929) is a particularly controversial postmodern writer. As with many other twentieth-century theorists, Baudrillard challenged Marxism both for being deterministic and for reducing culture to a secondary status at the expense, particularly, of the political and the economic. Baudrillard sees contemporary society as a **consumer culture** based on consumption and not on production. He also introduced the idea of the simulacrum, arguing that the contemporary world is dominated by signs, images, and representations, obliterating the real and rendering any search for truth and objectivity fruitless.

The French philosopher Jacques Derrida (1930–2004) has demonstrated a continuing concern with the taken-for-granted character of much philosophical thought and, according to some postmodernists, has effectively shown that philosophy has no privileged access to meaning and truth. Derrida is best known for the introduction of **deconstruction** and the related idea of **text**. Essentially a school of philosophy and literary criticism, deconstruction has had a significant impact on many of the traditional social science and humanities disciplines. However, it is by no means universally accepted, especially by philosophers, at least in part because it is such a radical set of ideas. Most generally, it has been used to challenge some of the various structures of the dominant European cultural tradition.

Gilles Deleuze (1925–95), a philosopher, and his occasional collaborator Felix Guattari (1936–92), a political theorist and psychoanalyst, are best known for their attempts to integrate the ideas of Marx and Freud. One of their joint efforts has been described as a 'vast, chaotic rag-bag of a book' (in Honderich

1995:183), and although they have been welcomed into cultural geography, especially by Shurmer-Smith and Hannam (1994), the long-term significance of their work remains to be determined.

Edward Said (1935–2003), a literary and cultural theorist, contributed an influential body of ideas—**Orientalism**—contending that the Orient is an invention of those who study it from outside, and that North America and Europe employ ideas from within their cultures, ideas such as freedom and individualism, to facilitate their conquest and domination of other regions. Thus, the Orient is a necessary European image of the **other**, a construct of a dominant European discourse, as that term is discussed above. For Said, culture is best seen as a hegemonic environment with specific prevailing modes of thought. The contemporary importance of these ideas is evident in the claim that the United States sees the Arabic world in dehumanized terms, a tendency that is exacerbated by terrorist activities and by the ongoing conflict between Palestinians and Israelis.

Much of the work that is considered part of the cultural studies tradition, including feminism and studies of racism, qualifies as critical theory—a term that refers to the conviction that both the social sciences and the humanities need to be emancipatory and not ideological instruments. Critical theories are fundamentally different, epistemologically, from theories in physical science in that they are reflective and not objectifying. These critical theories have origins in the Frankfurt school of critical theory, especially in the work of the German philosopher and sociologist Jürgen Habermas (b. 1929).

by society. Note that this general idea, without the accompanying philosophical argument, was central to the account of the discipline of geography included in Chapter 1. Also, like feminists, postmodernists stress the importance of positionality and the situated character of knowledge. The claim that language is at the heart of all knowledge, in the sense that social constructions are reflected in and often main-

tained through language, means that it is necessary to uncover the limitations that language places on our thinking and understanding. For this reason, a key feature of the postmodern agenda is the philosophy and method of deconstruction (see Box 3.5). Deconstruction seeks to uncover what is not said in a text and what cannot be said because of the constraints imposed by language. Further, deconstruction

Box 3.6 Modernity and Postmodernity—Modernism and Postmodernism

Postmodernity refers to a new phase of history and a new cultural system, while postmodernism is a narrower term referring to a set of ideas, a scholarly movement that opposes the intellectual logic of modernism. Much academic literature describes both the historical phase and the scholarly movement using the one term 'postmodernism'; however, it is certainly useful to distinguish the two in principle, although this proves to be rather more difficult in practice. By extension, modernity and modernism can be similarly distinguished from each other (Duncan 1996).

Modernity, the modern period, was initiated by the eighteenth- and nineteenth-century processes of industrialization, urbanization, secularization, and nation-state creation. Durkheim identified the increasing division of labor as the key to the onset of the modern period, while Tönnies identified the transition from **Gemeinschaft** to **Gesellschaft**. In broad terms, the contemporary world continues to display the basic characteristics of modernity. However, many observers recognize an emerging phase of postmodernity that is gradually replacing modernity.

The principal characteristics of postmodernity are those of a postindustrial economy accompanied by globalization processes, a loss of national identities, and a corresponding rise of more local ethnic identities. Harvey (1989) analyzed postmodernity as a condition of society. For many geographers, one appeal of the concept of postmodernity is the interest shown in spatial difference as compared to the emphasis on progress through time. Jameson (1991) described postmodernity, using Marxist terms, as a mode of production, a period of **late (taken-for-granted) capitalism** that is characterized by multinational capitalism and by the commodification of culture.

Modernism, as a scholarly movement, is associated with the Enlightenment ideals of progress and rationality, with an empiricist epistemology, and with advances in scientific knowledge (see Box 2.1). It was in this intellectual context that the various academic disciplines were established in universities, a process of institutionalization described in Chapter 1.

Postmodernism is a scholarly movement based on a loss of faith in the values and ideas that support the modernist project. Postmodernism emphasizes a blurring of disciplinary boundaries and the rise of new interdisciplinary movements and, in some instances, even argues for a postdisciplinary academic context.

Conflating these ideas, as many scholars do, it can be argued that such a significant set of changes is occurring today that the modern Western world is in a state of crisis, reflected in such circumstances as the general weakening of Western political traditions, the lessening of authority of some social institutions, and the resurgence of religious fundamentalism.

Note that, unfortunately, the terms 'modernity' and 'modernism' are inherently flawed because what is modern at one period of time is not necessarily modern at some later period. It is for this reason that the rather clumsy terms 'postmodernity' and 'postmodernism' can be described as following, subsequent to, and even in opposition to modernity and modernism respectively. Note also that the terms 'postmodernity' and 'postmodernism' are problematic because they appear to preclude the possibility of a coexistence of modernity with postmodernity and of modernism with postmodernism. The use of the prefix 'post', meaning after or behind, is usually interpreted to mean that one circumstance is replaced by another.

regards all aspects of a text as necessarily inter-textual in the sense that meaning is produced from one text to another rather than being produced between the world and any given text. More generally, this **intertextuality** refers to the idea that all things are related. Any reading of a text, meaning also any view of reality, is as good as any other reading. As applied in the context of social science generally, a key purpose of deconstruction, which postmodernists prefer to theory construction, is to reveal the interests served by theories.

Postcolonialism

Closely related to postmodernism is **postcolonialism**, which is concerned with theories of identity, the role played by power, and the production of culture, especially in the less devel-

Opened in 1911, the remarkable Transporter Bridge (*above*)—the only working bridge of its kind in Britain—spans the River Tees in Middlesbrough, northeastern England. The Bridge is a monument to nineteenth-century engineering and to the iron-and-steel origins of the town, and it remains today as a relict of an industrial past—both a visible and a symbolic landscape. Passengers and vehicles are lifted across the river on a large platform suspended from an overhead gantry. This photograph looks east across an extensive complex of chemical works between Middlesbrough and the North Sea. The area upstream was the location of iron and steel works that closed by the 1980s, leaving a desolate wasteland and a heavily polluted river. During the 1990s, however, the industrial structures were demolished and the river cleaned. Today, the area is a post-industrial recreational landscape, used for canoeing, whitewater rafting, and similar pursuits (William Norton). As one component of the modern-to-postmodern transition, many old industrial sites are being transformed from landscapes of production to landscapes of consumption. The experience of Granville Island in Vancouver (*below*) is typical. It was a shipbuilding center during the Second World War, derelict land by the 1960s, and today is a commercial and recreational centre with a large market place, galleries and shops, a hotel, and theaters (Courtesy Hotson Bakker Boniface Haden Architects).

oped formerly colonized world. Postcolonial geographic writing explicitly reflects an appreciation of the consequences of colonial activity as it has caused or contributed to circumstances of human and landscape inequality. The impacts of colonialism are profound, and a postcolonial perspective endeavors to uncover issues surrounding the loss of identities and places that characterizes the lives of those affected.

It is clear that the larger academic discipline of geography, along with the other social sciences, was not born innocently, but rather was a product of—and was complicit in—larger nationalist, colonial, imperial, racist, and masculinist ventures. The close association between Western ideas about exploration and discovery on the one hand and geography on the other hand cannot be denied. In much the same way that feminism challenges geography to rethink some fundamental assumptions about what is knowledge, so postcolonialism provides a compelling critique of the foundations of the geographic enterprise. Most notably, postcolonialism challenges the Eurocentric perspective of much geography; this challenge was evident in the work of Said, who was especially critical of the efforts made by colonizers to fix the identities of the colonized (see Box 3.5). Much postcolonial work stresses the Western tendency to see the world in dualistic terms, such as self and other, civilized and savage, and seeks to understand how these ideas came into being and how they can be dismantled.

The Mode of Representation

The task of cultural geographers is to represent the world to others, but this raises the question of what is an appropriate form for this **representation**. Prior to the emergence of the cultural studies tradition, the answer to this question was taken for granted. Today, as much of the content of this chapter has implied, it is debated, such that there is what might be called a crisis of representation.

The answer that was previously taken for granted is that any form of representation can be used that claims both to be universally valid and to result in an accurate understanding of the world. In this form of representation, 'the task of writing is the mechanical one of bolting words together in the right order so that the final construction represents the thought or object modelled' (Barnes and Duncan 1992:2). Both the Sauerian approach and the occasional attempts to employ a positivist approach are seen to belong in this category. Most important, the Sauerian approach can be seen as an attempt, usually based on observation, to accurately describe and classify the subject matter of cultural geography. Much of the material included in chapters 5 and 6 reflects this mode of representation.

But there are other modes of representation that reject the suggestion that it is possible to produce universally valid and accurate representations of the world on the grounds that any representation is actually an interpretation. This idea is central in humanist studies and explicit in several of the more influential modes of representation introduced in the context of the new cultural geography. Both feminism and postmodernism, for example, claim that accurate representations are not possible. More broadly, the hermeneutic method explicitly acknowledges the role of interpretation (see Box 2.2).

As discussed by Barnes and Duncan (1992:2–4), three critical implications arise from accepting the argument that it is not possible to produce accurate representations of the world. The first two of these implications were touched on in the accounts of feminism and postmodernism earlier in this chapter.

First, if there is no preinterpreted reality for writing to reflect, such that it is not possible for writing to mirror the world, then what does writing reflect? The general answer that has been proposed is that writing reflects earlier texts; this is the idea of intertextuality. From a postmodern perspective, this also implies that all truths are inside, not outside, texts.

Second, all writing reflects a particular and necessarily local vantage point and is inevitably marked by its origins. Bondi stated the feminist argument clearly:

According to feminist philosophers, dominant conceptions of knowledge are 'gendered'. This claim is elaborated in various ways but, of central importance, is the notion that western intellectual traditions operate through interrelated dualisms, such as reason and emotion, rationality and irrationality, objectivity and subjectivity, general and particular, abstract and concrete, mind and body, culture and nature, form and matter. In each case, the terms are defined as mutually exclusive opposites but are not equally valued: the first occupies a superior position and is positively valued; the second is subordinate and negative. And intertwined within this system of hierarchical dichotomies is a distinction between masculine and feminine. It is through this intertwining that dominant knowledge systems are 'gendered': the superior terms in the dualisms are associated with masculinity, the subordinate terms with femininity. (Bondi 1997:245–6)

More generally, it is possible to claim that because all texts are a reflection of a particular and not a universal viewpoint, it becomes critical to understand the position of the author of the text and also the intent of the text. The ideas of situated knowledge and positionality assert that authors are unable to speak for themselves; rather, they speak for the cultural system of which they are a part. Thus, all knowledge is partial. Some postmodernists move beyond these claims to assert that no one text is more privileged or in any sense a better interpretation than any other; this is clearly a claim that is especially open to debate. Ley (1998:80) noted that to abandon authority 'leads to the abandonment of responsibility, to a tentativeness that in its reluctance to achieve any closure embraces, sometimes celebrates, ambiguity and indeterminacy'. Further, it may be that the tendency for any representation to be **ethnocentric** need not necessarily imply that political projects ought to be excluded from academic work.

The third implication is that there is a need to employ a variety of writing strategies, tropes, and especially rhetorical devices to help convey meaning. In attempts to mirror the world, the principal writing strategies were those of objectivism, often involving a narrative. Metaphors are increasingly employed to describe the

unknown using the language of the known. The basic argument here is one component of the philosophical debates discussed in Box 2.2, and is a fundamental one that is reverberating through cultural geography.

Conducting Research

In addition to the challenges raised by Marxism and feminism, a cultural turn in cultural geography has introduced new research norms to the traditional study of cultural geography. Together, these three philosophical movements have taken the subdiscipline in new directions. Two principal developments are noted here.

First, Marxism, feminism, and the cultural studies tradition have encouraged studies of human identity, human **difference**, and the politics related to these matters. Many contemporary cultural geographic studies are characterized by a persistent questioning of concepts, classifications, and categories, a situation best described as one of **constructionism**. Identity, for example, is no longer taken to be something that is pre-given but, rather, is acknowledged to be socially constructed and therefore open to change. This has led cultural geographers to develop an increasing sensitivity to culture, especially concerning relations between dominant groups and other groups, and also an increasing awareness of the politics of difference. It is for this reason that the role played by power relations and the way power is practiced through discourse are central features of many of the approaches discussed in this chapter. There is also much evidence that contemporary cultural geography is including, even welcoming, the voices of those previously excluded.

A second development sparked by these three movements is that any discussion of ideas and facts is now interpreted with a recognition of the specific discourse being employed; this is the idea of **contextualism**. Cultural geography now demonstrates an increased awareness of and sensitivity to how knowledge is constructed, as well as by whom and for what purpose. As a result, much contemporary cultural geography, especially that informed by feminism and post-modernism, explicitly acknowledges the positionality and situatedness of the author.

But these developments, especially those associated with the cultural turn, are not without their critics. Indeed, Duncan (1994a) characterizes the debate between Sauerians and new cultural geographers during the 1980s and early 1990s as a 'civil war' within cultural geography. Further, concerning the effect that the cultural turn has had on social geography, Gregson (1993:527) observes that 'on the one hand social geographers move to deconstruct the socially constructed categories which form the basis for social differentiation (in so doing, frequently criticizing these categories as essentialist), on the other they find themselves having to recognize that these categories are the very ones through which people make sense of themselves and of social life.'

There have also been more general expressions of concern about the seemingly ceaseless introduction of new ideas into geography, a process that might be explained in terms of the commodification of knowledge, with competition resulting in flexible specialization. Berry (1995:95) refers to a 'combination of self-defeating trendiness and the calculating acceptance of a mainstream agenda by academics more concerned about creating successful niches than about principle', and suggested that postmodernism in particular is fundamentally flawed because a geography without a rational and science-based core cannot accommodate specialties.

The current philosophical and theoretical background that underpins cultural geographic research is multifaceted, complex, and dynamic, perhaps even uncertain and unstable. It has certainly been responsible for a wealth of empirical analyses.

Pause for Thought

As you read through the account of postmodernism, are you inclined to think that there is an intriguing ambiguity in the postmodern emphasis on social construction on the one hand and the continuing concern with academic disciplines and disciplinary status on the other

hand? Particularly puzzling is that, as noted earlier, cultural studies sometimes aspires to disciplinary status. Thus, Benko and Strohmayer (1997:xiii) noted that 'more than any other disciplines, with the possible exception of Cultural Studies, it was Human Geography that came away from this encounter with a renewed sense of mission, vindication, even of pride.' The importance of disciplines is evident in this assertion, with the identity of those disciplines reinforced by the use of capitals. Somewhat ironically, others have referred to 'the delineation both of a tradition and a future trajectory for cultural studies as the Academy's most upwardly mobile discipline' (Carter, Donald, and Squires 1995:vii).

Studying Society

It is clear that cultural geographers are now actively involved in working with, and in some instances even advancing, social theory. This is interesting because, traditionally, social theory was almost exclusively the property of the discipline of sociology and, accordingly, lacked spatial content. This circumstance changed beginning in the 1980s, when Giddens (1984:368) was able to claim that human geography was a social science, based on the fact that 'there are no logical or methodological differences between human geography and sociology.' Further, Giddens (1987:9) asserted that the 'social sciences developed as a family of disciplines such that the different "areas" of human behavior that are covered by the various social sciences form an intellectual division of labor which can be justified in only a very general way.' Geographers, too, stressed the social theoretic recognition of the importance of space, with Massey (1985:12) asserting that 'space is a social construct—yes. But social relations are also constructed over space, and that makes a difference', and Agnew and Duncan (1989:1) identifying a 'recent revival of interest in a social theory that takes place and space seriously'. The inclusion of place or space in social theory, previously almost unheard of, is now quite usual.

Social Theory and Space

The first substantial work in this area was the **structuration theory** of Giddens. The key feature of this theory is its focus on the links between human agents and the social structures—both local social systems and larger social structures—within which they function. Structuration theory also explicitly recognizes links between social relations and spatial structures, with space viewed as a medium through which social relations are produced and reproduced rather than as a mere backdrop against which social relations unfold. The term **locale** is used to refer to the settings in which everyday social interactions occur.

Although several geographers have endorsed and used structuration theory, others have questioned its value, and there is little evidence that it will play a major role in contemporary cultural geography. Sympathetic accounts by human geographers tend to be critical of geography for viewing place as inert when it ought to be viewed as a historically contingent process, as a transformation of space and nature that cannot be detached from the transformation of society. But cultural geographers might well resist any suggestion that they have viewed place as inert. The many regional studies by Meinig discussed in Chapter 6 may not be conceptually sophisticated by structuration standards, but they are explicit analyses of related spatial and social change.

Another body of work to which geographers have been attracted is concerned with what might be called the production of space, a phrase that refers to the fact that the spaces within which social life occurs are socially produced. In this approach to social theory, a rational and scientific interpretation of space is replaced with an interpretation that sees space as experienced or imagined.

But the most important advances in contemporary social theory for the cultural geographer are those that strive to build on ideas that are explicitly opposed to the Enlightenment concern with an empiricist epistemology and with the search for universal truths. Accordingly, cultural geographers have shown much interest in the works of some critical social philosophers. These theoretical contributions were considered earlier in the context of the discussion of cultural studies and related interests as a source of con-

cepts for cultural geographic analyses of both identity and difference.

Social Geographies

But what of the subdiscipline of social geography? Until the 1980s, social geography could be regarded as the British counterpart to North American Sauerian cultural geography. The different experiences are explained in terms of different intellectual origins. Thus, although cultural geography was stimulated by leading European geographers such as Humboldt, Ratzel, and Schlüter, it was really initiated by Sauer and never fully incorporated into geography curricula outside of North America. The British social geographic tradition has origins in French sociology, in British human geography, and in the Chicago school of human ecology.

The term 'social geography' was used by French sociologists of the Le Play school in the 1880s and by Réclus in the 1890s, and a detailed proposal was made by Hoke (1907). However, none of these developments had a significant impact on geography, and it was not until the 1930s that social geography emerged as a subdiscipline following a programmatic statement by Roxby (1930), the publication of social geography books by members of the French school, and the completion of some of the sociological human ecological work that emphasized social values (referred to in Chapter 4). The result was a British social geography essentially concerned with the regional distribution of population, occupation, and religion. There was a focus on identifying different regions according to various social phenomena, and Pahl (1965:81) described the subdiscipline as being concerned with the 'processes and patterns involved in an understanding of socially defined populations in their spatial setting.'

By the 1960s, this British social geography was an uncertain field because of the varied intellectual stimuli. During the 1960s and 1970s three transformations occurred: a behavioral perspective was introduced; quantitative methods were used; and a focus on social problems, often stimulated by Marxist thought, developed. But social geography continued as an uncertain subdiscipline with much relevant material regarded as belonging to other compositional subdivisions such as population geography and urban geography.

At the same time, in North America cultural geography had a much clearer identity, with a single dominant school of thought, the Sauerian school, focusing on ecology, evolution, and regions. Certainly, British social geographers, like sociologists of the time, tended to see culture as the concern of the anthropologist, not the geographer. A competing North American movement promoted a social geography built on the ideas of symbolic interactionism (see Box 3.7), with social geography defined as the study of human spatial behavior and related human landscapes from the point of view of society as the sum of a population's symbolic interactions. Proponents of the school favored the label 'social geography' over 'cultural geography' because of its implied emphasis on communication processes through which symbolic interactions take place and its related claim that individuals are constrained in their behavior by the character of the groups to which they belong. The cultural geographer Wagner (1972, 1974) likewise saw distinct landscapes and societies evolving in response to communication processes often generated by social institutions. Neither of these interests has prompted substantial research activity.

Since the 1980s the status of social geography has been especially uncertain. Reflecting this uncertainty, Dennis and Clout (1980:1) began their textbook with the following concern: 'Of all the varieties of geography, "social geography" is probably the most elusive and poorly defined.' Similarly, Cater and Jones (1989:viii) began their textbook with a discussion of 'social geography's identity crisis'. A few years earlier, Eyles and Smith (1978:55) concluded their review of the status of social geography on a cautionary note: 'Social geography's greatest achievement may yet turn out to be the acceleration of its own destruction.' On the other hand, Hamnett (1996:3) is more certain about the identity of social geography: 'The approach I take is that social geography is primarily concerned with the study of the geography of social structures, social activi-

ties and social groups across a wide range of human societies.'

Converging Cultural and Social Geographies

The differences between social and cultural geography were emphasized during the 1960s, when cultural geography remained outside of, and social geography was involved in, spatial analysis. But a convergence of the two interests was initiated in the 1970s when both subdisciplines found inspiration in humanism and Marxism. According to some observers, this convergence has proceeded apace, especially since the mid-1980s (see Figure 3.1).

Nevertheless, it is misleading to suggest that the two have in some way become one. Indeed, the message delivered in simplified form in Figure 3.1 is readily apparent given the material presented in chapters 1 and 2: specifically, the extent to which cultural and social geography have experienced a genuine integration is open to much debate. This fact is evidenced by both the form and content of the various Progress Reports written for the journal *Progress in Human Geography*. Certainly, the content of these reports indicates that the viability of a meaningful integration of cultural and social geography is questionable. For example, Gregson (1995:139) noted 'the ascendance of cultural, as opposed to social, theory in social geography (and in human geography more generally)', and wondered whether a discussion of progress in social geography was really an 'obituary'. Hamnett (1996:4), on the other hand, claimed that the 'revival of the "new" cultural geography in Anglo-America has been very influential in the revival of social geography'. Finally, Peach (1999:286) maintained that work in social geography continued 'to have more in common with the spatial side of sociology than with the cultural side of geography'.

Pause for Thought

Does this debate about what social geography is and is not seem of marginal relevance? Perhaps in some respects the details are somewhat marginal, but at the same time, the larger debate highlights a fundamental issue because it is about what is at the core of the subdiscipline, not

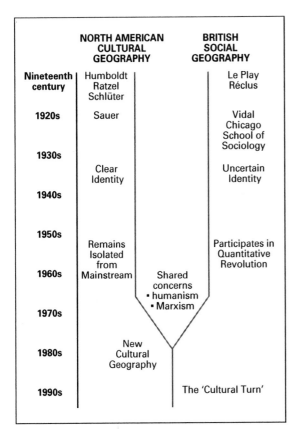

Figure 3.1 Cultural Geography and Social Geography—Origins, Concerns, and Convergence. This schematic diagram highlights the separate origins, different concerns, and convergence (beginning around 1970) of cultural and social geographic traditions.

about what is at the margins. The issue probably arises because of a complicated intersection of circumstances, including the quite separate European social and North American cultural geographic traditions and the labeling of the post-1980 developments as a cultural, not a social turn. Recall from Chapter 1 that the distinction between these two terms has never been especially clear, and it seems likely that confusion and uncertainty will continue for some time.

Concluding Comments

Consider the following three circumstances. First, as stated in Chapter 1, the new cultural geogra-

phy is a welcome addition to, rather than a replacement for, the Sauerian approach. Indeed, although the cultural turn in particular is most correctly seen as a set of new developments, the Sauerian antecedents of some aspects of these new developments are evident. Further, there is no clearly definable new cultural geography because the term does not refer to an approach but rather to several varied conceptual advances that may have some common content but that also demonstrate substantive differences. This is an important point to stress at this stage of the book—much of chapters 7, 8, and 9 especially reflects the new cultural geography, but this does not mean that the content of these chapters reflects some monolithic approach. None of this should be surprising, of course—the Sauerian approach itself was not and is not a single monolithic approach.

Second, because of the relative newness of much of the conceptual material, there is less certainty about what is and what is not important; the consensus judgments made today are unlikely to be identical to those made tomorrow. Any academic discipline travels many conceptual pathways, all interesting at the time perhaps, but not all equally profitable in terms of the empirical analyses they prompt or the contributions they make to knowledge. Indeed, several geographers have expressed concern about the seemingly indiscriminate, and perhaps even self-indulgent, borrowing of ideas from other disciplines, and also about the short lifespan of many of the concepts borrowed. More generally, other geographers worry that 'human geography is to be subsumed within a broadly defined social science' with a resultant loss of disciplinary identity (Unwin 1992:210). It is certainly the case that some recent conceptual advances embraced by some human geographers have further added to what was already a rather uncertain discipline, seeming to further remove the 'geo' from human geography.

Third, many of the recent conceptual advances in cultural geography are more difficult to understand and apply than is the Sauerian approach. This is because they are often associated with complex and disputed philosophical and/or social scientific origins, with the result that there are typically multiple versions of each of the newer conceptual advances. It is also because many of these advances are linked to postmodernism, a term that refers to a varied corpus of ideas but that characteristically insists upon the need to take into account non-traditional sources and the need to question established wisdom. Finally, it is because the cultural turn revolves around writing styles and vocabulary that makes it seem exclusionary to some and pretentious to others.

These three issues combine to make cultural geography one of the most exciting, most challenging, and, yes, sometimes most difficult areas of contemporary scholarly analysis. Added to

Box 3.7 Symbolic Interactionism

Symbolic interactionism refers to a group of related social psychological theories, inspired by the ideas of the social philosopher George Herbert Mead and subsequently developed especially by Herbert Blumer, that have close links to the humanist philosophy of **pragmatism**. There are three fundamental premises:

- Individuals learn the meaning of things primarily through interactions with others, meaning that we learn how to define the world through our experiences of social interaction.
- Although we derive meanings of things from social interaction, we do not simply accept those meanings; they serve only as a basis for our understanding.
- Our behavior results from our interpretations of the meanings presented to us.

In addition to these three fundamental premises, symbolic interactionists agree on an additional seven assumptions:

- Humans are unique in their ability to use language.
- Other unique human traits arise because of language.
- Human development is dependent on social interaction.
- Humans are purposive and rational animals.
- Humans seek rewards and avoid costs.
- Humans are conscious actors.
- Humans are active.

All major versions of symbolic interactionism accept the three fundamental premises and the seven additional assumptions, although there is often some disagreement concerning additional conceptual content.

A human ecology built on these ideas is different from other versions advocated by sociologists.

Specifically, symbolic interactionism stresses that humans are different from other animals in some very important ways, whereas many of the other versions of ecology, explicitly derived as they are from physical science, emphasize similarities between humans and other animals rather than differences. The most important difference identified by symbolic interactionists concerns the efficiency of human language and the implications for what it means to be human, given that effective communication is the key to complex behavior. Other important differences include the interaction with others that characterizes human life, the purposiveness of human behavior, and the high degree of awareness that humans possess. Ericksen (1980) opted to employ a symbolic interactionist interpretation of human ecology because of its emphasis on interaction, which is central to any ecological approach.

Although the human ecology proposed by Ericksen (1980) has not been adopted by cultural geographers, the symbolic interactionist perspective has been discussed and applied. Most of the relevant work is associated with the cultural geography that criticized the perceived superorganic concept of culture employed by Sauerians. As Duncan (1978:269) noted: 'The self is largely a product of the opinions and actions of others as these are expressed in interaction with the developing self.' The idea that spatial and social behavior are entwined was accepted in a prominent social geography textbook, which had as its central concern 'human spatial behavior and the derived geographical patterns from the point of view of society: the summation of a population's symbolic interactions' (Jakle, Brunn, and Roseman 1976:7). Note also that the definition of culture offered by Jackson and Smith (1984:205) and included in Box 1.2 is based on this approach.

and sometimes building on Sauerian landscape foundations, these conceptual advances and related empirical work are making major contributions to both academic and practical knowledge.

But there is a more critical interpretation of the rethinking of cultural geography. Specifically, an understanding of the various and varied approaches to cultural geography reviewed in this chapter is enhanced by appreciation of one particular criticism of the landscape school: writing from a South African and postcolonial perspective, Dodson (2000:140) noted that it is an 'unpalatable truth that the tools of traditional cultural geography were also the very tools by which the colonial and apartheid states were created and maintained'. It is difficult to dispute this observation, although it is important to acknowledge that all academic work needs to be appraised in the context of the dominant discourse of the time and place. Indeed, the criticism applies to the larger discipline of geography, and to other disciplines, especially anthropology, as these flowered in the late nineteenth-century climate of imperialism, bellicose nationalism, and racism.

Following the review of spatial analysis and the behavioral geography that it prompted, this chapter introduced a series of conceptual advances that are linked by a shared suspicion—in some cases an outright rejection—of science and the scientific method. More particularly, and with explicit reference to cultural geography, there is a general concern about the Sauerian tradition, which is now considered by many to view culture and cultures in an inappropriate manner. A crucial addition to our conceptual arsenal is Foucault's idea that power is exercised through the production of knowledge and truth in the context of a particular discourse, meaning that knowledge and truth are uncertain, even changeable. Certainly, a popular view of culture emphasizes that culture is, as noted in Box 1.2, 'a domain, no less than the political and the economic, in which social relations of dominance and subordination are negotiated and resisted, where meanings are not just imposed, but contested' (Jackson 1989:ix). Of course, such an interpretation is not universally accepted, as evidenced by the body of ongoing cultural geographic literature, a literature that continues to be informed by a Sauerian or modified Sauerian view.

Further Reading

The following are useful sources for further reading on specific issues.

Hubbard, Kitchin, Bartley, and Fuller (2002) provide a history of geographic thought that includes clear accounts of the transformation reviewed in this chapter. The edited text by Shurmer-Smith (2002) includes accounts of transformations from 1970 onwards.

Dodson (2000) provides an interesting commentary on new cultural geography as it developed at least partly as a reaction to the conservative agenda of the landscape school.

Sopher (1972) suggests possible links between cultural geography and positivism.

Amedeo and Golledge (1975), Chisholm (1975), and Norton (1984) describe process-to-form methodology as one way to incorporate time in spatial analysis.

Daniels (1985:144) identifies Vidal and Sauer as inspirations for humanist cultural geography.

Rose (1981) discusses some ideas of Dilthey as they relate to humanist geography.

Pioneering contributions in humanist geography are those by Relph (1970) and Tuan (1971), both of whom identified phenomenology as a legitimate approach.

There have been partially successful attempts to base humanist geography on an idealist philosophy (see, for example, Guelke 1974), on an existential philosophy (see Samuels 1978), and on a critical encounter with historical materialism (see Cosgrove 1978, 1983).

Tuan (for example, 1972, 1975) and Pickles (1985) outline phenomenology and geography; Samuels (1978, 1979, 1981) outlines existentialism and geography; Guelke (1971, 1975) outlines idealism and geography. For critical commentaries on idealism, see Watts and Watts (1978) and Harrison and Livingstone (1979).

Tuan (1974), Relph (1976), Ley (1977), and Seamon (1979) discuss some humanist concepts introduced by geographers.

Gold (1980:7–15) and Martin (1991:6–11) provide brief overviews of psychological approaches to the study of behavior.

Spencer and Blades (1986) outline some links between psychology and geography.

Philo and Parr (2003) provide a useful introduction to what they descibe as 'psychoanalytic geographies.'

Watson (1913) is the pioneering statement on behaviorism and psychology, while Skinner (1969) details the most popular version of behaviorism, namely radical behaviorism.

Homans (1987) and Lamal (1991) provide guidelines for group–based behavior analyses.

Neisser (1967, 1976) outlines cognitive psychology.

Smith, Pellegrino, and Golledge (1982) discuss the geographic modeling of spatial cognition and behavior; Golledge (1987) outlines environmental cognition, and Golledge (2003) reflects on recent cognitive behavioral research.

Jensen and Burgess (1997) provide evidence for misrepresentations of behaviorism within psychology, while Wertz (1998) provides evidence for misrepresentations of humanism.

Norton (1997a) identifies various geographic confusions about behaviorism. Some cultural geographers have misinterpreted both the practice of behavioral geography and the intentions of behavior analysis. For example, Relph (1984:209) stated: 'Since I have never been able to establish just what "behavioral geography" is and how it distinguishes itself from other sorts of geography, I have assumed it to be a version of B.F. Skinner's behaviorism somehow transferred from psychology to geography.' Similarly, Pile (1996:36) described behavioral geography as behaviorist and also incorrectly described both the Watsonian and Skinnerian versions of behaviorism as premised on the logic of stimulus and response. The incorrect association of Skinner's radical behaviorism with a stimulus–response framework is repeated by Griggs (2000:19).

Golledge (1969) and Kitchin, Blades, and Golledge (1997) discuss the varied relations between geography and psychology.

Werlen (1993:15) outlined an original argument for a human geography that was oriented to society and action on the grounds that 'no activity can be explained by reference to psychological factors only, but must be explained in relation to social context.'

Christensen (1982) debates the positivist–humanist split.

Burton (1963) and Brookfield (1969) debate the merits of behavioral geography as a corrective for perceived flaws.

Cullen (1976) and Bunting and Guelke (1979) describe what they see as unfortunate links between spatial analysis and behavioral geography.

Schmitt (1987) discusses Marxism generally, and Sayer (1992b) focuses on the importance of the division of labor.

Habermas (1972) outlines the Frankfurt school of critical Marxism, and Althusser (1969) outlines structural Marxism.

There are accessible discussions of Marxism in geography in Cloke, Philo, and Sadler (1991), Unwin (1992), Peet (1996a), R.J. Johnston (1997), Robinson (1998), Blunt and Wills (2000), and Mitchell (2004). Quani (1982) provides a book-length treatment. As discussed by Peet (1977a, 1977b), Marxist social theory was introduced into human geography as a part of an emerging radical geography during the 1960s—the radical journal *Antipode* began publishing in 1969—with the first major Marxist-inspired work being the book *Social Justice and the City* (Harvey 1973).

Sayer (1979, 1982) considers radical approaches and social scale.

Peet and Thrift (1989) explain the structure and agency debate.

Duncan and Legg (2004) provide one of the few cultural geographic commentaries on social class.

Gregson et al. (1997:67–8) discuss a feminist geographic analysis of the term 'patriarchy.'

Overviews that provide relatively accessible introductions to feminist geography include Hanson (1992), Rose (1993—generally acknowledged as a classic text), Domosh (1996), Monk (1996), McDowell (1997), Women and Geography Study Group of the Royal Geographical Society with the Institute of British Geographers (1997), Robinson (1998), Domosh and Seager (2001), Moss (2002), Pratt (2004), and Sharp (2004).

McEwan (2003) outlines debates between Western and other feminisms.

Hanson (1992) discusses areas of shared geographic and feminist interest; Bondi and Domosh (1992) and Longhurst (1994) explain the masculinism inherent in most geography. Recognizing that there are multiple histories of an academic discipline, Monk (2004) examines the careers of women geographers in the United States prior to about 1970.

Dowling and McGuirk (1998) review some of the principal themes reflected in research on gendered geographies.

England (1994) and Rose (1997, 2001) provide an understanding of feminism and the situated character of knowledge.

Duncan (1980), Ley (1981), Berry (1995:95), and Dunn (1997) outline some concerns about cultural geography borrowing ideas from other disciplines.

Chambers (1994:121–2) discusses cultural studies and disciplinary identities.

There is little doubt that the various stylistic tactics employed by some of the cultural geographers who are deeply involved in postmodernism are not always well received. Walker (1997:172–3) expressed it this way: 'I become easily irritated with the posturing of the postmodernist and the mannered style of discourse that glibly condemns the linear, logical, and evidentiary essay in favor of fragments of literary allusion and freely tossed Lacanian word salads, which leave a faint and convoluted trail of simulacrumbs for the poor reader to follow'.

Gregson (1993, 1995) and Badcock (1996) discuss some concerns about writing styles associated with the cultural turn.

Barnett (1998a, 1998b) examines the cultural turn in human geography.

Rosenau (1992) looks at differences within postmodernism; Berg (1993) assesses the claims that even the differences between modernism and postmodernism are based on false logic.

Accessible accounts of postmodernism by geographers include those of Dear (1988), Webb (1990–1), Cloke, Philo, and Sadler (1991), Unwin (1992), Duncan (1996), Edwards (1996), R.J. Johnston (1997), Robinson (1998), Dear and Flusty (2000), and Minca (2001).

Challenging commentaries on poststructuralist geographies are provided by Doel (1999, 2004).

Nash (2004) reviews postcolonial geographies in an attempt to answer the question: 'Can the spatial discipline of geography move from its positioning of colonial complicity towards producing postcolonial spatial narratives?' Other discussions of postcolonial geography include Sidaway (2000), Nash (2002), and Ryan (2004).

Barnes and Curry (1983), Duncan and Ley (1993), and Duncan and Sharp (1993) examine modes of representation in cultural geography.

Soja (1989:Chapter 1) looks at space and social theory; Gregory (1994) used the work of Lefebvre (1991) to facilitate discussion of the historical production of modern capitalist spatiality; other instances of the incorporation of space into social theory include the ideas of habitus (Bourdieu 1977) and of normalizing enclosures (Foucault 1970).

Gregory (1981), Pred (1984, 1985), and Gregson (1987) discuss geography and structuration theory.

Discussions of cultural and social geographies are provided by Jackson (2000a) and Peach (2002).

Watson (1951) and Dunbar (1977) chronicle the history of social geography.

Jakle, Brunn, and Roseman (1976) and Dennis and Clout (1980) are examples of social geography textbooks that employ a symbolic interactionist focus.

Hamnett (2003) worries that contemporary human geography as it is involved in the cultural turn is becoming increasingly detached from social issues and concerns.

Clifford and Valentine (2003) include a series of chapters designed to help students conduct geographic research. Much of the content is directly relevant to work in cultural geography.

Pain and Bailey (2004) summarize recent trends in British social and cultural geography, noting that there are increasing efforts to wed theoretical and action research.

4

Environments, Ethics, Landscapes

Cultural geography has a considerable and sustained interest in human-and-land relationships such that the subdiscipline, and for some even the larger discipline of geography, can be broadly interpreted as the study of humans and land. This chapter develops this interest both conceptually and practically.

Of the various approaches to the study of humans and land discussed in Chapter 2, it is several versions of the ecological approach that have attracted increasing interest in recent years. Indeed, in addition to the continuing importance of traditional cultural ecological concerns—evident in some nineteenth-century geography and in the landscape school—there are other areas that have been the subject of more recent focus. For example, there is an increasing concern with human impacts on environment, a concern that has given rise to an abundance of new literature in physical and social sciences. There is also a concern based on recent advances in biology that explicitly acknowledges the variability rather than the stability of nature. Further, and not surprisingly, developments in social theory have prompted more sophisticated versions of ecological thought, such as ecofeminism and political ecology. This chapter considers both the traditional and these newer interests. The sequence of material is as follows:

- a discussion of approaches to the study of ecology, stressing the subject's nineteenth-century origins in physical science, the interests in it shown by other social sciences, and the use made of its systems concepts. It is clear that ecology is a multifaceted arena that, although sometimes referred to as a discipline, is more appropriately thought of as an approach open to multiple interpretations and applications;
- a discussion of principal geographic interpretations of human ecology, namely those of Barrows and Sauer;
- discussions of the idea of nature as socially produced, of the 'new ecology' that acknowledges the ever-changing character of the natural world, of environmentalism, of ecofeminism, and of political ecology. Ecofeminism was first formulated in the 1970s and is concerned with the multiple oppressions that structure gendered identities. Political ecology came to the fore in the 1980s in the context of attempts to understand problems of soil degradation in the less developed world, and the many applications attest to both the popularity and success of the approach, especially in less developed world settings;
- a discussion of environmental ethics that includes accounts of various attitudes and values that relate to our behavior in environment;
- an overview of cultural and environmental change through time, centered on the concept of natural resources as cultural appraisals and on population and technological change;
- an acknowledgment of the changing character of ecological approaches and a consideration of the prospects for ecological approaches as used by cultural geographers.

Ecology: A Unifying Science?

It is helpful to begin by dispelling two popular misconceptions. First, it is a mistake to equate ecology with concern for environmental **conservation** and preservation; ecology was not introduced in response to concerns about human use of land. Indeed, in nineteenth-century thought there were numerous interpretations of the human–and–nature relationship, including two diametrically opposed views of nature: from one perspective, nature was there to be exploited for the benefit of humans, while from another perspective, nature needed to be conserved in order to minimize the damage being inflicted by humans. Ecology came to the fore in this contradictory intellectual environment; indeed, some scholars regarded ecology as a means of facilitating the continuing exploitation of nature by pointing out ways in which the resultant damage could be minimized.

Second, notwithstanding some attempts in this direction, it is a mistake to think of ecology as an academic discipline. The origins of ecology as a scientific approach can be traced to the third chapter of Darwin's *The Origin of Species*, which considered the various adaptations of organic beings to environment, although the more general context was biologists' attempt to understand how the distribution of species is related to environment. The first use of the term 'ecology' is generally attributed to Haeckel in the 1860s, and by the end of the nineteenth century ecological approaches were well established in both plant and animal science. There has never been a single discipline of ecology; rather, ecological concepts, notably the concept of a plant or animal community, were applied to a number of existing disciplines: as Bowler comments (1992:365), 'many scientists could do "ecology" while retaining their primary disciplinary loyalty elsewhere.' This multidisciplinary aspect of ecology is fundamental.

Ecology has its specific origins in physical science, though there are versions of human ecology in several of the social sciences. As noted in Chapter 2, the term 'ecology' is derived from two Greek words, *oikos*, meaning 'place to live' or 'house', and *logos*, meaning 'study of'. Accordingly, the central concern of ecology as a science is with the ways in which living things interact with each other and with their environments or homes. Ecology is, then, best thought of as an approach to the study of relationships between organisms and their environments and not as a particular body of subject matter, and it is for this reason that there are ecological approaches in many of the physical and social sciences.

Following the introduction of the term 'ecology', the first significant advances in understanding the term were made by Clements, an American botanist, who developed the idea that a plant community was a distinctive superorganism that had a life of its own. This idea, which derived its intellectual inspiration from the work of Spencer, had parallels with the superorganic concept of culture. Thus, Clements thought of a plant community as something more than the sum of the individual species and as responding to laws that operated only at the community level. The Spencerian idea of social progress was reflected in the concept of climax vegetation, which Clements viewed as the logical outcome of a process of evolution. Clements's *Research Methods in Ecology* (1905) was the first textbook to describe this new methodology of ecology as applied to plants.

By the early 1900s, ecology was an established approach in biology, botany, and marine science. The British Ecological Society was founded in 1913, and the Ecological Society of America in 1915. Six principal related concepts were used in early ecological analyses in physical science:

- *Community* refers to a group of related organisms in a particular area.
- *Competition* refers to the struggle between member species of a community.
- *Invasion* refers to the ability of some species to take over areas occupied by another species.
- *Succession* refers to the sequence of change as a community moves toward the finished or climax situation.

- *Dominance* refers to the fact that one or more species in a community is normally able to seek out and occupy the most favorable environment.
- *Segregation* refers to the spatial patterning of species.

Ecologies in Social Science

Although ecology began as a study of nature without humans—or, certainly, without privileging humans—versions of human ecology soon appeared. The first direct reference to humans in an ecological framework was by Darwin in 1871 (*The Descent of Man*), but it was Spencer who was the principal intellectual inspiration for the early advances in human ecology, and it was plant ecology that provided the basic concepts. Recall from Chapter 2 that Spencer, in his articulation of what is usually labeled social Darwinism, argued for an analogy between animal organisms and human groups based on the common struggle to survive in a physical environment. The logic of this social Darwinism—a form of naturalism—is straightforward, with cultural groups evolving in accord with their ability to adjust to particular physical environments. Indeed, it was Spencer who introduced the oft-quoted phrase, 'survival of the fittest'. These ideas are related to views of biological evolutionism as an inexorable form of social progress and, of course, to the conception of society as a superorganic entity. The following sections discuss ecological approaches in sociology, anthropology, history, and psychology, especially as these aid in understanding human ecological approaches in geography.

SOCIOLOGY

Sociology was the first of the social science disciplines to develop a coherent human ecological approach. This development took place at the University of Chicago, where several members of the geography department (including both Goode and Barrows, discussed below) were also expressing interest in human ecology. The key sociological figure was Park, one of the leaders of early American sociology. In conjunction with others—notably Burgess, McKenzie, and Wirth—Park successfully articulated and employed a human ecological approach to the analysis of urban areas.

For Park and his colleagues, usually described as the Chicago school of sociology, human ecology centered on the distribution of and interactions between humans and human institutions. Their approach involved mapping these distributions and explaining them by reference to the various concepts developed within plant ecology.

Park treated the concept of an animal or plant community as virtually synonymous with society. He viewed humans as competing for space, a competition reflected in differential urban land values and therefore varied land uses. Segregation—the presence of urban areas usually distinguished by land use or ethnic identity—was one outcome of competition. Park used the terms *invasion* and *succession* to refer to land use changes through time, and *dominance* to refer to the ability of particular land uses to locate in the most desirable area. Within the city, this most desirable area was the central zone, and the dominant land uses were those belonging to the financial and retailing sectors. The concepts as they were applied, especially in the Chicago context, allowed sociologists to suggest a concentric zone model of intraurban land uses.

Park's accomplishment was a human ecology explicitly modeled on plant and animal ecology—a human science derived from physical science with humans seen as organic creatures affected by the laws of the organic world. But these conceptual advances were stimulated not only by physical science ecologies, but also by personal experiences. 'I expect that I have actually covered more ground tramping about in cities in different parts of the world, than any other living man,' Park explained (1952:5). 'Out of all this I gained, among other things, a conception of the city, the community, and the region, not as a geographical phenomenon merely, but as a kind of social organism.'

This is precisely the type of approach that geographers might have been expected to have adopted, but their interests typically lay elsewhere—in environmental determinism, in the regional approach, and, of course, in the emerging landscape school. Indeed, sociologists placed great emphasis on the intellectual separation of sociology and geography, stressing especially that geography was an idiographic science whereas sociology was nomothetic. For Park and his colleagues, the role of geography was limited to supplying facts that the human ecologist could explain. Park also regarded geography as a discipline concerned with humans and their physical environments, whereas sociology was concerned with humans and their social environments.

The early human ecological approach in sociology was successful. The conceptual and empirical initiatives, mostly in midwestern American cities, sparked considerable interest. This was an interest that geographers did not share until the rise of a spatial analytic approach in the 1960s, by which time the approach was largely out of favor in sociology.

A principal criticism of human ecology from within sociology, readily understandable in the context of the first of the two philosophical debates introduced in Box 2.2, concerned the tendency to see and describe human groups in physical science terms. Hence, a major revision occurred with the introduction of the idea that various symbolic associations and sentimental attachments could be identified with city areas, and that these variables were able to counter the ecological processes (Ericksen 1980). Rather than treating the ecological process as dominant, proponents of human ecology emphasized humans themselves as leaders, not followers, of the ecological process. The conceptual inspiration for these ideas is the approach to sociology known as symbolic interactionism (see Box 3.7).

ANTHROPOLOGY

Although the sociological interest in human ecology had little impact on the work of cultural geographers, the anthropological interest is another matter, a fact that reflects the close ties between these two disciplines since at least the 1920s. Anthropological human ecology relies on the same physical science foundations as does the sociological counterpart, but anthropologists have developed their contributions more fully.

Steward, who had un undergraduate degree in zoology and did graduate work with Kroeber,

provided the seminal ecological statements in anthropology in the 1930s in a series of empirical analyses, in this way introducing ecology into anthropology by application rather than programmatically. The first explicit outline of an ecological approach came much later, and Steward (1955:30) is generally credited with introducing the term used to describe it: 'In order to distinguish the present purpose and method from those implied in the concepts of biological, human, and social ecology, the term **cultural ecology** is used.'

For Steward, the various human ecologies had failed as a result of their proponents' insistence on attempting to define each one as a new subdiscipline. Steward believed that the ecological approach was better seen as precisely that: an approach, a means to an end and not an end in itself. The aim of the approach was to explain particular cultural features, not to derive general laws. Steward included culture as a superorganic factor but drew a distinction between the behaviors of humans and the behaviors of other organisms. This was not, however, the extreme superorganic concept, as he (1955:36) also referred to the 'fruitless assumption that culture comes from culture'.

Steward rejected the concepts of community, competition, invasion, succession, dominance, and segregation (as introduced by physical scientists and as adopted by Park and others) on the grounds that they were inappropriate for humans. Rather, he saw culture as the key to understanding human relationships with nature. He acknowledged, for example, that competition might be present, but suggested that it could be interpreted only in cultural terms and needed to be approached through an understanding of culture and not as some independent process. Further, he saw human culture as extending beyond the range of local communities.

Steward's rejection of physical science concepts and emphasis on culture explain the choice of name for the approach—cultural ecology. The role of physical environment was limited in Steward's approach, which contained the implicit assumptions that 'cultural and natural areas are generally coterminous because the culture represents an adjustment to the particular environment', and that 'different patterns may exist in any natural area and that unlike cultures may exist in similar environments' (Steward 1955:35). Steward's approach was one of environmental adaptation that took into account the character of the culture. These ideas are comparable to the possibilism introduced by Vidal.

Three fundamental procedures are evident in Steward's cultural ecology, namely analysis of technology-and-environment relations, analysis of behavior patterns affecting environment, and analysis of relations between these behavior patterns and other aspects of culture. Numerous other anthropologists continued Steward's pioneering work, often in amended form, but there have also been substantive criticisms and developments of the approach. Five are noted:

- First, the superorganic implications have been criticized, particularly in light of some of the larger changes in social theory.
- Second, there have been arguments for a unified science of ecology instead of a specific cultural ecology, thus reintroducing the idea that human and nonhuman behavior could be studied in similar ways.
- Third, proponents of cultural materialism proposed, as an alternative to cultural evolution, a theory of sociocultural evolution based on the doctrine of natural selection and focusing on the demographic, economic, technological, and environmental causes of cultural evolution (see Box 4.1).
- Fourth, Barth (1956, 1969), refining the cultural ecology approach for analyses of ethnic group distributions, introduced the ecological **niche**.
- Fifth, some other ecological anthropoligists argued for 'numerous unifying concepts and principles' (Hardesty 1977:vii), including those of adaptation, human energetics, and the ecological niche.

This fifth development is pursued in greater detail in the account of adaptation later in this chapter.

Both sociological and anthropological ecologies have intellectual origins in plant and animal ecology, and both added humans to the ecological equation; however, the specific concerns of the two approaches, both conceptually and empirically, are clearly different. Further, reflecting their parent disciplines, the sociological human ecology was urban and Western in orientation, while the anthropological cultural ecology was rural and preindustrial. It is perhaps significant to an understanding of the differences between the two to note that Steward was also influenced by a legacy of earlier anthropological work by Kroeber and others, whereas Park created a human ecology with minimal inspiration from within larger sociology.

History

In history the introduction of ecological approaches was related to the belated appreciation that human behavior and environment are closely related—as recently as 1984, Worster (1984:2) found it necessary to propose 'the development of an ecological perspective in history', while according to Crosby (1995:1180), historians of earlier generations 'could not see what they were not ready to see'. What they did not see was the significance of the ecological analyses conducted by geographers such as von Humboldt and Marsh, nor did they see the relevance of population growth and technological change as factors basic to historical analyses. For most historians, the significance of environment was merely as backdrop to—and not as a meaningful player in—human affairs. Even at the global scale, both Toynbee and Spengler discussed civilizations with only minimal reference to environment.

In the 1930s and 1940s, two contributions concerned with settlement of the Great Plains focused explicitly on physical environment and culture (Webb 1931; Malin 1947). Both treated Great Plains settlement in the context of human-and-land relations through time, with natural resources such as soil and water discussed in the context of technology and culture. The fact that these important American studies largely failed to generate substantive analyses for other times and places was a reflection of historical scholarship at the time.

Ecological history is now inspired by the geographic work of Réclus, Ratzel, Semple, and Huntington, by the anthropological work of Kroeber, Wissler, Steward, Rappaport, and Harris among others, and by the historical work of Webb, Malin, and the German historian Wittfogel. Following Wittfogel, and the argument that the 'fundamental relation underlying all social arrangements . . . is the one between humans and nature', Worster (1984:4) provided a powerful argument for including more nature in the study of history. Ecology is now a thriving approach in history with two rather different thrusts, namely concerns with natural history and with the history of ideas about nature.

PSYCHOLOGY

In psychology, an ecological approach was initiated by Lewin (1944:22–3) as one part of a concern with the role of non-psychological inputs into human behavior: 'Any type of group life occurs in a setting of certain limitations to what is and what is not possible, what might or might not happen. The non-psychological factors of climate, of communication, of the law of the country or the organization are a frequent part

Box 4.1 Cultural Materialism

Cultural materialism is an ecological research strategy that is 'based on the simple premise that human social life is a response to the practical problems of earthly existence' and that 'opposes strategies that deny the legitimacy or the feasibility of scientific accounts of human behavior' (Harris 1979:ix). At the outset, cultural materialism distinguishes between

- thoughts and behavior, acknowledging that both can be studied from the perspective of either the observers or the observed, and
- emic and etic operations, with the former allowing those observed to determine the legitimacy of any analysis and the latter allowing the observer to so determine.

While humanists favor an emic perspective, cultural materialists emphasize an etic perspective: 'The starting point of all sociocultural analysis for cultural materialists is simply the existence of an etic human population in etic time and space' (Harris 1979:47). Analyses of etic human populations recognize an infrastructure that includes modes of production (such as technologies of food production) and modes of reproduction (such as technologies of population change), a structure that organizes production and reproduction at both domestic and political social scales, and a superstructure that includes activities such as rituals and science.

With these points as a basic framework, cultural materialism asserts that 'etic behavioral modes of production and reproduction probabilistically determine the etic behavioral domestic and political economy, which in turn probabilistically determine the behavioral and mental emic superstructures' (Harris 1979:55–6). This is the key principle of infrastructural determinism, claiming the priority of etic and behavioral conditions and processes over their emic and mental equivalents, and claiming the priority of infrastructural conditions and processes over their structural and superstructural equivalents. This is not a form of monocausal determinism, as infrastructure comprises a host of demographic, technological, economic, and environmental variables.

Behavior that creates human landscapes can be treated as an instance of human adaptation to natural and social environments, and much of that behavior is directed at ensuring successful adaptation, or survival. Human survival is dependent on certain kinds of behavior. Cultural practices, group behaviors that aid the survival of the group, are themselves likely to survive. Infrastructure, as defined by cultural materialism, is the principal interface between culture and nature. This means that an understanding of human behavior can be achieved by reference to its natural consequences. Put differently, cultural practices result from material causes.

Because cultural materialism argues that behaviors, as responses to environmental variables, precede mental rationalizations about the reasons for responses, it is subject to criticism from many contemporary social scientists—including many cultural geographers—who prefer to think of **cognition** as a more appropriate approach to the understanding of human behavior.

of these "outside limitations". Lewin (1951) thus referred to behavior as a function of individuals and their psychological environment, which together make up the life space, and of the larger non-psychological environment. Together, these three components constitute what is called a field (see Figure 4.1).

This approach is not clearly related to comparable developments in other social sciences. While the basic principle—namely, that humans are parts of environments and not objects in and separate from environments—is similar, the psychological contribution is very clearly a part of the parent discipline. Perhaps the principal sharing of ideas concerns the psychological interest in the way that humans adapt to their surroundings, an interest that includes considera-

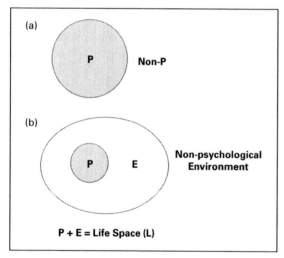

Figure 4.1: Lewin Field Theory. In 4.1a, the individual person, P, is formally separated from the rest of the world, non-P. The reason Lewin represents this distinction in graphic form and not simply in writing is that a spatial representation is susceptible to mathematical manipulation. The shape and size of the figure enclosing P are irrelevant; all that matters is that there is a bounded area completely surrounded by a larger area.

In 4.1b, a second step is taken in the representation of psychological reality. In this case, a second bounded figure is drawn around the bounded figure that encloses P. The area between the boundary of P and the boundary of the second figure is E, the psychological environment. The total area of both P and E is labeled the life space. Finally, the area outside the boundary of the second figure is the nonpsychological environment that includes physical, cultural, and social facts.

tion of culture as a 'group's adaptation to the recurrent problems it faces in interaction with its environmental setting' (Berry 1984:87).

Human Ecology and Geography

'GEOGRAPHY AS HUMAN ECOLOGY'

Although geography might be regarded as the most convivial disciplinary home for human ecology, given the traditional geographic concerns with both physical and human worlds, geographic ecological approaches have a checkered history, possibly because they are not associated with a specific seminal statement and do not have a coherent body of empirical research. Most of the suggestions that geography adopt a human ecological perspective have come from geographers sympathetic to an integrated physical and human geography.

The first use of the term 'human ecology' in the context of geography was in 1907 by Goode, a geographer at the University of Chicago. Following Clements and other plant scientists, Goode envisaged human ecology as an organizing concept for the discipline of geography. Such an idea was popular at the time, with several ecologically inspired papers presented at meetings of geographers, while Huntington became the second president of the Ecological Society of America. The culmination of this interest in a human ecology was the seminal methodological statement by Barrows (1923), 'Geography as Human Ecology', initially delivered as a presidential address to the Association of American geographers in 1922. It is useful to think of this statement as another in the series of attempts to claim a particular identity for the discipline of geography.

The specific aim of Barrows's statement was to define geography as the study of mutual relations between humans and the natural environment and thus as human ecology. More generally, Barrows intended to carve a niche for geography separate from other academic disciplines: 'Geography finds in human ecology, then, a field cultivated but little by any or all of the other natural and social sciences. Thus limited in scope it has a unity otherwise lacking, and a point of view unique among the sciences which deal with humanity' (Barrows 1923:7). He emphasized human adjustment to, as opposed to influence of, physical environment, and distinguished geography from geology and other physical sciences on the grounds that it was not concerned with the origins and distribution of physical features—landforms, climates, vegetation types, soil types, and so on—but only with human adjustment to those features. Barrows also made an explicit distinction between geography and history, suggesting that the historian began 'with what our remote ancestors saw' while the geographer began 'with what we ourselves see' (Barrows 1923:6). In sum, Barrows argued that geography as a discipline needed to concentrate on human ecology interpreted as human adjustment to physical environment.

Overall, geographic contemporaries of Barrows had good reason to be somewhat perplexed at the thrust of this statement. Given the established concerns of geography at the time, some objected to the explicit exclusion of the traditional geographic content of physical geography, while others were concerned at the lack

of reference to the traditional regional focus, and still others were dismayed at the rejection of environmental determinism. Overall the statement was not well received, and it is not surprising that Barrows's statement created little positive impact on the discipline of geography; in retrospect, it can be seen as an idiosyncratic attempt to forge an identity for the discipline. In terms of its immediate impact, Barrows's statement inspired an ecological account of the geography of Chicago (Goode 1926), and two university-level textbooks, *Geography: An Introduction to Human Ecology* (White and Renner 1936) and *Human Geography: An Ecological Study of Society* by the same authors (White and Renner 1948). Both textbooks saw geography as the study of human society in relation to the earth (see Box 4.2). Overall, however, most subsequent statements about geography as human ecology were not sympathetic to the Barrows interpretation.

HUMAN ECOLOGY AND THE LANDSCAPE SCHOOL

It is appropriate to think of the Sauerian tradition as containing three cultural geographic emphases—human-and-land relationships or

---

Box 4.2 Adjusting to Environment—A Human Ecological Theory

White and Renner (1948:635–7) outlined a human ecological theory for geography that built on the ideas of Barrows. The theory can be summarized as follows:

- A human society is made up of living organisms with a particular social structure and patterns of activities.
- On encountering a new environment, a group adapts that environment to its needs and habits so far as is feasible, but will also adjust its previous institutions and behaviors as necessary.
- A human group in an environment is thus always defined both by its past practices and by its responses to the environment.
- The adjustments to environment comprise three types of relationship, namely *use relationships*, *control relationships*, and *ideological relationships*,

all of which are continually changing in response to technological change.
- The outcome of these relationships is the creation of a cultural landscape.

White and Renner (1948:636) summarized their position thus: 'It is the geographer's contention, therefore, that human society can be understood only when its culture, make-up, and behavior are viewed against the background of its location, the space it occupies, and the resources which it utilizes.'

This formulation has not been significantly developed or applied by other geographers. Nevertheless, it is interesting to note that it is similar in broad outline to much of the cultural geography discussed in both chapters 5 and 6 concerning the evolution and regionalization of cultural landscapes.

ecology, landscape evolution, and regional landscapes. The ecological interest is evident in the concern with human–and–land relationships (see Figure 2.2) and was regularly stressed. For example, Sauer (1924:33) noted: 'Land is passing out of economic use in this country more rapidly than new land is being occupied. Timber devastation, soil erosion, overgrazing, soil depletion, failure of irrigation, dry farming, and drainage, here and there causing a pitiless revaluation of areas. Land is increasingly becoming a restricted economic good.' Again, Sauer (1927:192) claimed: 'To what extent is man as a terrestrial agent, that is by his areal expressions of culture, living harmoniously in nature (symbiotically), and to what extent is he setting narrowing limits for future generations by living beyond the means of the sites that he occupies? Man . . . appears, periodically, to effect his own ruin. . . . He has certainly been an engine of unparalleled destruction.' From the outset Sauer expressed concerns about human impacts, and several of his students completed major ecological analyses, including some studies of vegetation and the exemplary cultural geographic studies by Wagner (1958a) and Mikesell (1961).

The ecological content of the landscape school is not as readily acknowledged as are other aspects of the tradition, at least partly because Sauer rarely used the terms 'ecology' or 'human ecology'. This is perhaps because Sauer took exception to the statement by Barrows on the grounds that it involved an inappropriate limitation on the discipline of geography, rendering it simply the study of humans in relation to natural environment. Like the majority of geographers at the time, Sauer was less than enthused with the human ecology proposed by Barrows.

Pause for Thought

Why was the landscape school statement by Sauer so influential while the human ecology statement by Barrows had minimal impact? Here are three suggestions. First, Sauer not only outlined his position, he also restated and clarified that position several times, conducted related analyses, and attracted students to work with him. Bar-

rows on the other hand progressed little beyond the initial statement. Second, Sauer built on a rich European tradition of geographic thought and explicitly incorporated physical geography, unlike Barrows, who was seen as proposing something that lacked relevant intellectual foundations in the discipline. Third, Sauer's position was both clear and forceful, while Barrows's ideas were interpreted in two contradictory ways, with some seeing his position as a diluted form of environmental determinism and others seeing it as excluding physical geography. Passing informed judgment on this detail of disciplinary history is not easy, but it is an intriguing circumstance in light of the reception that human ecological approaches experienced in sociology and anthropology.

OTHER CONTRIBUTIONS

Several other geographers issued appeals for an explicit concern with human ecology; four are noted here. Thornthwaite (1940:343), a student of Sauer, claimed that 'human ecology must transcend all of the present academic disciplines, and . . . the development of a science of human ecology must involve cooperation of geography, sociology, demography, anthropology, social psychology, economics, and many of the natural sciences as well.' Such a view of a unifying human ecology, a co–operative enterprise, bears little resemblance to either the earlier statement by Barrows or the Sauerian cultural tradition. Morgan and Moss (1965) argued for an ecological approach that focused on the community concept, drawing parallels between biological and human communities. This account acknowledged that any analogy between animal organisms and human groups, in the Spencerian or social Darwinian tradition, was subject to criticism, but argued the case for viewing human communities in two ways: '1) Ecologically, as aggregates of different species focusing on man, including all the animals and plants which depend on man and all those on which man in turn depends. 2) Socially, as aggregates of one species, consisting of individuals with a variety of interests and functions which, joined together, make possible one social and economic unit' (Morgan and Moss 1965:348). In both cases, the key feature is the uniting of individual community members. In a different vein, Chorley (1973) aimed to show how human ecology provided a unifying link between physical and human geography, while Norwine and Anderson (1980:vii) built on the idea of the environment as a factor 'impacting man's destiny'. None of these four statements has had meaningful impact on the discipline of geography.

Summary

Ecology is an approach applied in several physical science disciplines, and human ecology is an approach applied in several social science disciplines. Young (1983:2) described human ecology as 'a strange nonfield, a hybrid attempt to understand the ecology of one species, the being *Homo sapiens*'. At first thought, it seems reasonable to suppose that if ecological analyses including humans are to have one logical disciplinary home, then that home should be in geography. But this has not proven to be the case. Put simply, geography, despite extending several invitations, has not proven to be an especially congenial abode for human ecology. Notably, the views of such mid-nineteenth-century geographers as von Humboldt and Ritter centered on human-and-land relations, and these relations were also a concern in both Vidalian and Sauerian traditions, but ecology *per se* was not stressed. It is interesting to ponder how different the history of geography might have been if one of the leading late nineteenth-century geographers had explicitly highlighted an ecological focus rather than the study of regions or environmental determinism.

From Barrows onwards, human ecology in geography has been characterized more by proposals of method than by demonstrations of method. Two observations highlight this circumstance. First, reviewing human ecology in geography between 1954 and 1978, Porter (1978:15) characterized the advances as 'a decade of progress in a quarter century'. Second, Zimmerer (1996a:161) referred to ecological concepts as being both 'persistently foundational and yet doggedly problematic in their application' in human geography. Overall, twentieth-century geographers identified both a physical sci-

ence–human science division in their discipline and a division between ecological approaches and other dominant approaches, first regional and later spatial analytic; the failure of geography to center on an ecological approach is at least partly a consequence of these internal identity uncertainties.

Most cultural geographic work with ecological content has been inspired by the landscape school. For Mathewson (1999:268), ecology's place in many of the cultural landscape studies conducted by landscape school geographers is 'implicit and pervasive', and indeed there does not seem 'to be any end in sight for this tradition, despite the emergence of new varieties of cultural geography whose practitioners jejunely and routinely issue uninformed dismissals of what has preceded them'.

Pause for Thought

Given the general failure of human ecology to become established as a leading concern in any one of the social sciences, you may be interested to learn of a new development taking place in a few North American universities: 'Human ecology is a new academic discipline which seeks, through a study of the interaction of man with his environment, to improve the near environment and the quality of life. It focuses knowledge of technical advances and improving coping skills on strengthening the ability of people to cope with the environment and to improve human–human and human–environment relations' (Edwards, Brabble, Cole, and Westney 1991:3). Most of the content of this proposed discipline is borrowed from elsewhere, especially from sociology, anthropology, and home economics. While the prospects for this purported new discipline seem limited, the fact that such a claim can be made reflects the uncertain status—and potential—of human ecology generally.

Ecology and Systems Analysis

As evident from the preceding overview, ecology is a unifying approach that has not been widely adopted. For most interested scholars, especially those approaching problems from a specific disciplinary perspective, the challenges of adopting an ecological approach are simply too great, given that a unified science needs to integrate both physical and human phenomena and be genuinely interdisciplinary. It has not even proven possible to integrate the various human ecologies and, indeed, the evidence points to increasing diversity of approaches rather than to integration. But there is one principal exception to these statements: the application of systems analysis in an ecological context. Many ecological analyses in various disciplines turned to systems concepts to support their work, such that these concepts represent the closest that researchers have come to a genuine integration of ecological approaches.

Rather surprisingly, the term **ecosystem** was not introduced until 1935, when it was coined by the English botanist Tansley. Sometimes called an ecological system, an ecosystem is any self-sustaining collection—an interacting system—of living organisms and their environment. Although this is a simple definition, ecosystems are both complex and dynamic.

An ecosystem has four principal properties. First, it is monistic, meaning that it integrates living things and their environment, allowing for recognition of the interactions between elements. In a general sense, this property is in accord with the ambitious aims of geographers such as von Humboldt and Ritter, who advocated a holistic view of humans and nature. Second, an ecosystem has a structure that can, in principle, be investigated. Third, it functions, in that there are interactions occurring within the system. Fourth, an ecosystem is similar in principle to any other system, such as a hot water system or the cardiovascular system. In addition, certain generally held beliefs about ecosystems make them an attractive basis for study. First, it is generally recognized that ecosystems exist at a multitude of scales, from the world to a single pond. Second, the ecosystem is seen as emphasizing the complexity of interrelationships between all member organisms Third, it is commonly acknowledged that an ecosystem is in a dynamic balance, with the environment constantly adapting in response to changing conditions.

These properties and features of an ecosystem along with the difficulties of conducting ecological analyses using traditional methods prompted use of the procedures of systems analysis as a means of identifying key variables and regulating factors. Systems analysis is a scientific approach to complex phenomena that proceeds through five stages (see Figure 4.2):

- In the first stage, known as *systems measurement*, objectives are outlined and data collected.
- The second stage, *data analysis*, involves calculating relationships between variables in order to determine the key variables given the objectives of the study.
- *Systems modeling* is the third stage, during which models are built to serve as a theoretical interpretation of the system.
- The fourth stage, *systems simulation*, involves manipulating the theoretical models in order to assess consequences of specific changes.
- Finally, during the stage of *systems optimization*, the aim is to develop strategies for achieving desired objectives.

The use of modeling and simulation is often essential given the size and complexity of many of the systems analyzed. As an example, consider attempts to study the effects of an increase in humans' carbon dioxide emissions on our global ecosystem. Similarly, one of the reasons many attempts to predict future population growth have been so spectacularly unsuccessful is that the complexity of factors involved has rarely been adequately acknowledged. In many cases, futures are predicted on the premise that relevant current trends will continue without significant change, which is rarely a reasonable assumption. In brief, systems analysis, which does at least allow for the complexity of a system to be recognized, is often the only feasible way to gain some understanding of ecosystems and their changing character.

In principle, analysis of an ecosystem that includes humans focuses on human modification of environment, whether accidental or deliberate, and on the consequences of modification. Typically, human involvement in an ecosystem leads to species destruction and loss

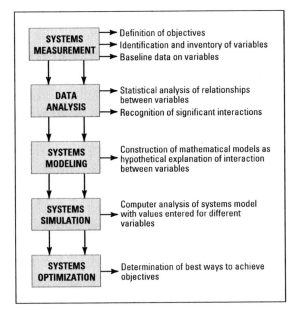

Figure 4.2 Stages in a Systems Analysis. This diagram out-lines five stages through which a typical systems analysis of an ecosystem might proceed. Details of each stage nec-essarily vary according to the specific aims of the analysis; for example, the fifth stage is appropriate only if the aim is to solve a particular problem. The procedure can be modi-fied if the objective is to identify the consequences of a par-ticular ecosystem change, such as the location of an industrial plant.

Source: C.H. Southwick, *Global Ecology in Human Perspec-tive* (New York: Oxford University Press, 1996):45. (Reprinted by permission of the author)

of equilibrium. Most analyses of this kind have been conducted from within the physical sci-ences, with humans seen as but one component of the system. A typical systems analysis with a social science emphasis employs the ecosystem concept in broad terms, rarely following the stages outlined in Figure 4.2. However, there are a number of favorable overviews and applica-tions of the approach by both geographers and anthropologists. Certainly, the attraction of a systems framework for the cultural geographer seems clear, with the explicit recognition of human–and–land relations and of humans as a part of nature. Nevertheless, following an outline of the approach, Foote and Greer-Wootten (1968:89) noted that they had 'presented "old wine in new bottles"', with the vintage being Sauer in the 1920s.

Pause for Thought

Although there is, at least on the surface, a compelling logic to the systems approach, it has not been widely employed by cultural geographers. Why not? Two reasons seem possible. First, analyses of complex systems using sophisticated analytical skills typically require input from teams of scientists, and team projects are not part of the tradition of cultural geography. Second, many cultural geographers are somewhat uncomfortable with seeing humans as merely one component of a system, preferring instead to see humans as a distinctive and perhaps sepa-rate component. This second observation is especially rel-evant today with the increased interest in humans as active decision makers.

Adaptation

A useful and yet much debated concept used in cultural ecological analyses, including those with a systems analysis focus, is **adaptation**. Defined loosely as changes in behavior that are designed to improve quality of life, or rather more for-mally as the strategies adopted to achieve eco-logical success, the meaning of the concept is clouded, especially by the recurring issue of environmental as opposed to cultural control and by the choice of social scale of analysis. Not surprisingly, adaptation is a concept introduced, interpreted, and applied in a number of disci-plines, including psychology, anthropology, and cultural geography.

In Psychology

In psychology, there is concern with the way in which individuals adapt, and an ecocultural psy-chology centers on adaptation to the recurrent challenges posed by environments, with the argument being that individual behavior must be considered in some complete ecological and cultural context. From this perspective, adapta-tion is interpreted in terms of the reduction of dissonance within a system. Ecological balance or harmony is promoted through three types of adaptation. The first and most usual form of adaptation is adjustment, which involves making changes in behavior to reduce conflict with environment. The second type, reaction, involves a form of retaliation against the environment to force the environment to adjust. The third type of adaptation, withdrawal, involves conflict reduction through human removal from the source of conflict. Recognition of these three versions of adaptation is comparable to earlier work in psychology that recognized movement toward, against, or away from a stimulus. This perspective sees all adaptations as individual solutions to particular environmental circumstances.

This psychological concern with individual behavior highlights questions about the social scale at which adaptation occurs, with anthropologists and cultural geographers usually favoring a group scale, although acknowledging the need for a variety of scales. Both individual and group behaviors can be seen as adaptations, although the details of the processes are different according to social scale. What may be a sound adaptive response for an individual may not be so for the group, and further, sound adaptive responses for both individual and group may not be so for the environment. Group adaptation may be interpreted as the state of management of physical resources.

In Anthropology

In anthropology, adaptation is closely linked to cultural evolution, with three levels distinguished: behavioral, physiological, and genetic. Anthropologists are primarily concerned with the behavioral level. According to Hardesty (1977:23), behavior 'is the most rapid response that an organism can make and, if based upon learning rather than genetic inheritance, is also the most flexible'. From this perspective, two types of adaptive behavior may be distinguished, namely idiosyncratic (that is, unique individual responses as studied by psychologists) and cultural (that is, shared responses as studied by anthropologists and cultural geographers). Group cultural adaptation occurs because of changes in technology, organization, and ideology, all of which are aspects of culture. How do these changes promote adaptation? The usual response is that they provide solutions to the problems posed by environment. But

changes in these cultural variables may also help by improving the effectiveness of solutions, providing adaptability, and providing awareness of environmental problems.

Adaptation is thus characterized by continual changes in human-and-environment relations. A simple adaptation model building on these ideas has three components (Hardesty 1986). First, there is focus on individuals and their fitness to adapt, with fitness linked to cultural baggage and learning. Second, there is a recognition that space is culturally meaningful, with ideology as the key determinant as to what is meaningful. Third, there is a recognition that adaptation itself frequently occurs because of some revolutionary, as opposed to evolutionary, change in ideology.

In Cultural Geography

In cultural geography, some cultural ecological work has modified the adaptation concept as it has been traditionally employed. Denevan (1983) and Butzer (1989b) argued that a version of cultural adaptation could be a key explanatory procedure for the cultural geographer, with human survival dependent on an available supply of adaptive responses, the appropriate scale of analysis being the ecological population, and the research aim being to account for the sources of variation and the processes of selection. There has also been an increasing recognition of the importance of ethnicity and of social and political power. Gade (1999) provided a series of analyses in the Andean context, including an intriguing review of rats and humans in Guayaquil, while Knapp (1991) favored a concern with adaptation as human behavior at the individual scale in a study of pre-European wetland agriculture in Ecuador, with emphasis on coping behaviors such as decision making and problem solving.

Box 4.3 offers a particular geographic interpretation based on the idea that all human behavior is social and therefore learned, such that adaptation is appropriately viewed as one part of the matrix of a society. According to this argument, 'explanations about variations in adaptive processes and about adaptive innovations . . . are to be looked for in variations of other elements of sociological systems, and not in changes of either ecological or biological systems' (Mogey 1971:81). Regardless of specific interpretation, the central question addressed using this concept concerns why humans behave as they do in the environments they

occupy. More generally, Butzer described the concerns of cultural ecology as follows:

It focuses upon how people live, doing what, how well, for how long, and with what environmental and social constraints. It emphasizes

Box 4.3 Adaptation as Social Adjustment

It may be surprising to discover that both cultural geographers and anthropologists continue to debate the relative importance of physical environment and humans to an understanding of the human-and-land relationship in the specific context of adaptation. While the extreme of environmental determinism is no longer seriously argued, the complexity of the relationship and variations from place to place encourage analyses that consider a variety of particular situations. Both Malin and Webb emphasized human adaptation to fixed and stable environments, rarely recognizing that the human mind was a crucial variable. Speth (1967), meanwhile, in an analysis of the intermontane Mormon landscape, argued that an adaptation focus lay midway between the two extremes of environmental and cultural determinism. Similarly, Porter (1965:419–20) concluded a study of subsistence activities in Kenya by noting: 'A model which seeks to describe culture as adaptive, through subsistence, to environmental potential cannot ignore man himself as a causative agent of environmental change.' Certainly, discussion of adaptation has renewed debate about determinism, with Mogey (1971), in a penetrating but infrequently referenced argument, proposing that all human behavior is social.

In order to maintain themselves, all societies have social mechanisms that enable them to adapt to environment, mechanisms that include both techniques and objectives and that are learned through time. There are different sets of social mechanisms in specific societies, and these are subject to ongoing change. With these claims in mind, Mogey (1971:80) explicitly argued against any form of environmental determinism, noting that even major natural disasters do not lead to social change with many societies adapted to specific disasters. From this perspective, all adaptation is social. Community was favored as the appropriate social unit for analysis because it is at this social scale that adaptation

occurs, with different communities employing different adaptive strategies. The specific differences among communities are based on two factors, technology and values. Following this argument, Mogey advanced six propositions:

- Community is the unit of social structure through which environments affect human behavior. It is this process that is labeled adaptation.
- Adaptive innovations require modification of the value system and of the techniques of the community if they are to be acceptable.
- In equalitarian communities, rates for the appearance of adaptive innovations are low.
- The mere presence of a status system that supports a set of leaders does not itself guarantee the ready acceptance of innovations.
- Adaptive innovations are frequent and acceptable in those communities with bureaucratic organizations, because they have values based on achievement and a status hierarchy that demands performance. However,
- Bureaucratic organizational systems have their own internal rigidities, which may impede their performance in the process of adaptation. Hence, more work on the interrelations of value systems and techniques in political, religious, and family systems is needed.

These are propositions, not conclusions, and do not imply a version of sociological determinism. As noted, they have not been widely tested or adopted by geographers in either cultural or social traditions. The final point made by Mogey implies that adaptation does not occur simply because it is needed. Certainly, many of the institutions of human society are constructed precisely because they facilitate cultural continuity—religion is a prime example—and such institutions are a 'prison de longue durée' (Braudel, quoted in Tarrow 1992:179).

that human behavior has a cognitive dimension and is dependent upon information flow, values, and goals. Finally, cultural ecologists recognize that actions are conceived and taken by individuals, but that such actions must be examined and approved by the community, in the light of tradition and the prevailing patterns of institutions and power, before decisions can be implemented. (Butzer 1989b:192–3)

Note that, in this tradition, culture is not superorganic; rather, there is a concern with cognition, and individuals are viewed as members of groups. Conceptually, there are closer links with symbolic interactionism (Box 3.7) than with cultural materialism (Box 4.1). Box 4.4 summarizes one example of an analysis of the evolution of landscape.

Rethinking Ecological Approaches

The fact that the origins of ecology were largely unrelated to what is now described as the environmental, or green, movement was stressed earlier. Today, however, the two are closely related—at least in the public imagination—such that ecology is often interpreted as a concern with environment. Certainly, two things are becoming increasingly clear: first, human activities pose ever-increasing threats to environments, and second, human activities in one part of the world can have impacts as far away as the other side of the globe. Our awareness of these facts is related to our growing interest in broad, even global, social issues, and to physical science's acknowledgment of population-and-environment relationships.

Box 4.4 Modeling the Evolution of the Cultural Landscape

The term 'cultural ecology' is commonly used to refer to a series of stages of human-and-nature interactions. In an analysis of northern European changes in vegetation, Emanuelsson (1988) stressed the roles played by both physical and human factors and outlined a model to study human impacts on nature that focused on the interaction of both factors. The three components of the model are discussed here.

The first component is based on the following argument. Human history is characterized by a series of inventions that have allowed humans to support ever increasing population numbers with a fixed resource base. Each of these inventions, by increasing the carrying capacity of an area, allows for a stepwise change in the human impact on landscape (see Figure 4.3).

Emanuelsson proposed four such steps over five levels of technology; these are outlined in terms of specific impacts on landscape as reflected by population density characteristics for southern Sweden, although the principles are argued to apply generally.

- The first level is that of a foraging economy, which can typically provide food for population densities between 0.2 and 0.8 per square mile (0.5–2.0/km²); the vegetation in the immediate vicinity of the group is affected.

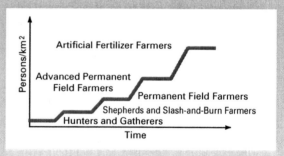

Figure 4.3 Hypothesized Relationship Between Population Size and Technology. The logic behind this diagram is that humans increase the carrying capacity and therefore the population of a given area through technological advances. These increases occur in a step-like fashion as technology changes. In this example, five levels of technology are identified—hunters and gatherers, shepherds and slash-and-burn farmers, permanent field farmers, advanced permanent field farmers, and artificial fertilizer farmers.

Source: Adapted from U. Emanuelsson, 'A Model for Describing the Development of the Cultural Landscape', in *The Cultural Landscape: Past, Present and Future*, edited by H.H. Birks, H.J.B. Birks, P.E. Kaland, and D. Moe (New York: Cambridge University Press, 1988):112.

- The second level is that of either shifting cultivation or pastoralism. This supports about 8 per square

Ecology is a term referring to the study of relationships and not, as noted, a term with any specific ideological implications. Tansley, the botanist who introduced the term 'ecosystem' in the 1930s, argued against the idea that ecology necessarily involved questioning human exploitation. He recognized instead that human activity was but one component of an ecosystem, and that it was not easy to identify those human actions that were destructive. Ideas linking ecology and concern for the environment are also in opposition to the claims of Clements concerning the supposed superiority of natural systems, claims that had their origins in Darwinian evolutionary ideas.

This section begins by acknowledging two important ideas about nature. First, nature may not be quite so natural after all, but rather may be more correctly interpreted as a social construction, meaning that when we talk about nature we are saying as much about ourselves as we are about the thing we are describing. Second, nature changes regardless of human activity, and this fundamental circumstance is important to any understanding of human impacts, although it does not undermine concerns about what impacts humans are having on environments.

Hybridity

The idea that nature is part of culture, meaning that our human experience of nature is, and always has been, mediated through cultural lenses, was a central feature of the classic account by Glacken (1967). For example, the image of nature implied by environmental determinism was mediated through religious

---

mile (20/km²). In the case of shifting cultivators, vegetation is substantially affected in the areas close to temporary settlements, while in the case of pastoralism, vegetation over a larger area is affected.

- The third level is that of permanent farming. This supports about 19 per square mile (50/km²), with substantial and permanent vegetation impacts in the areas occupied.
- The fourth level is that of a more efficient application of permanent farming, reflecting advances in the use of natural fertilizers, in cropping rotations, and in the selection of crops. These advances allow population densities to increase to 70 per square mile (200/km²) and again causes permanent vegetation change.
- The fifth level is that of permanent farming that is dependent on artificial fertilizers. This supports 386 per sqare mile (1,000/km²) and results in permanent vegetation change.

The second component of the model recognizes that developing a cultural landscape does not necessarily reflect a smooth transition through these five stages, for there are periods during which population densities decline because of overexploitation of soil or because of catastrophic events such as wars or disease. Such events are likely to result in a loss of ecological control, or what might be called a loss of the established equilibrium between humans and nature.

The third component of the model incorporates the fact that specific land uses and related human impacts are different in different parts of the world, owing mainly to variations in physical geography and, especially, to climate. For example, in northern Europe the climatic gradient, which, going from west to east, moves from a maritime to a continental climate, is very significant. In some other areas an extreme climate may make the transition from one stage to the next particularly difficult. Emanuelsson also noted that within any given climate region, specific land use activities are related to variations in other aspects of physical geography, notably soils and landforms.

Clearly, although this is described as an evolutionary model formulation, the primary concern is the significant ecological content. It is a formulation that deals with long spans of time and, further, that says little about the details of the human landscape, emphasizing rather the consequences of human and physical geographies for vegetation change. The model provides a helpful example of one aspect of the basic Sauer dictum—that humans operate on physical landscapes through time to affect change in those landscapes. In this case, the changes considered are not those with which cultural geographers are usually concerned, namely human additions to the landscape, but with specific changes in the physical landscape.

(specifically Christian) and philosophical (specifically empiricist and naturalist) beliefs and values. Nevertheless, it has typically been assumed that nature and culture are separate entities. Only recently has it been widely acknowledged that nature is manufactured by humans to the extent that nature and culture are not separate entities, a circumstance that prompts the concept of the hybridity of nature (Whatmore 1999).

The key idea behind this concept, evident especially in some humanistic arguments and in postmodern philosophy, is that nature is a social construction. According to earlier naturalist scientific views, it is possible to achieve an understanding of nature that is independent of the individual observer; however, more recent views stress the impossibility of any such objective interpretation, arguing that individuals are capable of constructing many different worlds, with the supposedly objective nature understood by science being but one of these. From a humanist perspective, it is necessary to interpret individual lifeworlds, the worlds of our direct experiences, not to observe objective nature.

Many observers stress the need for a culture of nature that is sympathetic to concerns expressed by environmentalists and others about negative human impacts on nature. In a discussion of the North American landscape, Wilson (1992:13) noted that 'the whole idea of nature as something separate from human experience is a lie. Humans and nature construct one another. Ignoring that fact obscures the one way out of the current environmental crisis—a living within and alongside of nature without dominating it.' From Wilson's perspective, what humans need is a new culture of nature based on a revised set of power relations. Box 4.5 discusses some related ideas concerning the invention and reinvention of nature.

If nature is best seen as a part of culture, then we might also argue that humans are a part of nature, an idea put forward in several recent arguments. First, there is the concept, building on earlier ecological ideas, of a global ecosystem. This is a holistic idea in that it refers to all things, organic and inorganic, and to the web of relationships between all parts. For many geographers and others, this concept is proving

to be an effective stimulus for what is at present the relatively new science of global ecology. Second, the controversial Gaia hypothesis, introduced by Lovelock in 1979, asserts that the environment and all life combine to form a single unified system, and have done so for about the past 3.8 billion years. Further, it claims that life itself has a profound impact on the environment, especially on the composition and temperature of the atmosphere, so that life maintains an environment that is favorable to life. This hypothesis opposes Darwinism in that it can be interpreted as having a teleological component, implying some action that is purposefully directed toward a specific goal.

In addition to these two holistic arguments, there are also views that strive to reconcile the need for integrating humans and nature and the need to acknowledge the distinctiveness of humans. According to this logic, any analysis of humans and of human landscapes must be **ecocentric**, focusing on the larger environmental context, including humans. But at the same time, analyses need to be anthropocentric, centering on humans because humans are different from other parts of the ecosystem, including other animals, in that they have culture. It is in this sense that humans might be seen as both a part of and apart from the rest of nature. A conceptual framework that aims to move beyond dualism to offer a new way of thinking about society–nature–technology–animal relations is that of actor–network theory. Derived from social analyses of sciences, these concepts attempt to respond to evidence that social theories are becoming increasingly anthropocentric at the expense of the various intimate relations between humans and animals.

New Ecology

A basic idea behind much traditional ecological thought, including the systems approach, is that nature strives to be in balance, and undergoes a process of evolution as it moves to a state of balance. The rise of the new ecology can be understood in the larger context of reactions against the intellectual traditions within which this initial ecological approach arose, especially the scientific tradition of mechanistic analysis and the Enlightenment tradition of continuing progress. Proponents of the new ecology maintain that the volatility of the natural world challenges the conventional ecological wisdom that depicts nature steadily progressing toward equilibrium. As a result, a central feature of new ecology is a non-equilibrium view of nature. Originating in biology, this new ecology is especially critical of systems analysis, which is seen as implying that nature can be understood in mechanical terms.

Zimmerer (1994:109), explaining the new ecology, noted that 'this new perspective calls attention to the instability, disequilibria, and chaotic fluctuations that characterize many environmental systems as it challenges the primordial assumption of systems ecology, namely that nature tends towards equilibrium and homeostasis.' Meanwhile Blumler (1996:32), referring with modest humor to the concepts of vegetational succession and to an eventual climax vegetation as expounded by Clements,

remarked: 'I have my doubts, given the highly dualistic nature of feminist writings, but it does seem possible that Clements' succession model represents a case of unconscious male psychosexual wish fulfilment: vegetation proceeds from small and flaccid to hard woody and upthrusting, becoming ever bigger and better until the ideal climax is reached.'

Although discussions of new ecology focus on issues with implications for ecological analyses, such as the origins of agriculture, there is, as Cronon points out, a larger issue involved in the idea that nature is inherently unstable, namely our human image of nature:

Recent scholarship has demonstrated that the natural world is far more dynamic, far more changeable, and far more entangled with human history than popular beliefs about the 'balance of nature' have typically acknowledged. Many popular ideas about the environment are premised on the conviction that nature is a stable, holistic, homeostatic community capable of preserving its

Box 4.5 Inventing and Reinventing Nature

There have been two contrasting images of nature as female, namely nature as a nurturing mother providing humans with their needs, and nature as wildness that needs to be tamed or controlled. The latter image became dominant during the scientific revolution: 'As Western culture became increasingly mechanised in the 1600s, the female earth and virgin earth spirit were subdued by the machine' (Merchant 1980:2). Such a mechanistic image encourages human use of the earth—the manipulation of nature—and can be contrasted with an alternative social construction of nature, an ecological image that sees nature as a whole and humans as a part of that whole. These two worldviews are founded on very different ontological and epistemological premises; the mechanistic view is both masculinist and exploitative of nature as female, while the ecological worldview acknowledges the need to break from these social constructions that involve human domination of nature and male domination of female.

Arguments such as these have contributed to a substantial and growing set of ideas on what might

be called the invention and reinvention of nature. These ideas center on the way nature is constructed in terms of prevailing ideas about race, sexuality, gender, nation, family, and class, and the dominations and inequalities that are contained within these. Such arguments require a re-evaluation of much earlier thought, especially concerning the centrality of female representations of nature. Feminists have initiated discussions of the character and power of the knowledge that passes for science, stressing that such knowledge—including the idea that there is a difference between man (not humans) and nature—was constructed by men at a time when the domination of nature by men was considered desirable. From this perspective, there is a need to reinvent a true human-and-nature relationship, a new scientific perspective that stresses human unity and human unity with nature. Haraway (1991:77) expressed the overall critical concern: 'Facts are theory-laden; theories are value-laden; values are history-laden.' These matters are one part of the philosophical debates referred to in Box 2.2.

natural balance more or less indefinitely if only humans can avoid 'disturbing' it. This is in fact a deeply problematic assumption. (Cronon 1995:24)

It is especially problematic because abandoning our belief in an essentially unchanging nature is psychologically disturbing. How are we to judge our actions without such a benchmark? We have a double difficulty here. First, nature is socially constructed. Thus, our descriptions of it reflect not just nature itself, but also ourselves, our values, and our assumptions, implying that we cannot in any objective sense 'know' the natural world. Second, nature changes regardless of our human activities.

What does all this mean for us when we ask ethical questions of ourselves about how we should treat nature? There are not, of course, any clear responses to such a question, but the double difficulty noted does provide a useful backdrop for the remainder of this chapter.

Pause for Thought

Interestingly, although the implications of new ecology are as yet barely touched upon, the basic argument may prove attractive to some cultural geographers because equal emphasis is placed on both nature and culture. Whereas much physical science emphasizes nature and much social science emphasizes culture, essentially each area of interest knows relatively little about the other. Cultural geographers, in classic geographic fashion, have always incorporated both concerns. Blumler (1996:26) wrote: 'I am more inclined to the Sauerian viewpoint which although strongly opposed to environmental determinism, nevertheless recognizes that nature and culture were equal players, interacting with each other in non-linear fashion. . . . In this sense, the Sauerian approach is consonant with the new ecological paradigm.'

Environmentalism

A classic argument proposes that current environmental problems are rooted in Jewish and Christian traditions because God was seen as giving the earth to humans for them to use as they chose, even if choosing meant destroying.

Some commentators reject this argument, claiming there is a substantive difference between the Biblical concept of dominion and the concept of environmental domination. The subsequent debate around this issue includes discussions of religions other than Judaism and Christianity to suggest that, even where religions traditionally see humans as a part of nature, there is evidence of ruthless exploitation of nature.

Major challenges to the Western tradition that saw humans as different from and superior to nature generally and to other animals in particular came from several and different directions. Malthus envisioned disastrous social consequences because of excessive and apparently uncontrollable population growth, relative to increases in the **resource** base, and John Stuart Mill (1806–73), echoing his concerns, warned that increases in population and in resource consumption could not simply continue. Henry David Thoreau (1817–62) viewed all creatures as a part of the community of nature, while John Muir (1838–1914), the founder of the Sierra Club, urged preservation of selected natural environments. These four thinkers provided pioneering accounts that stressed the fragility of our human use of earth.

Although it was Darwin who provided scientific support for the idea that humans were a part of nature, geographers have long been concerned with deleterious human impacts on their environment, and the contributions made by such major scholars as von Humboldt, Ritter, and Marsh are especially noteworthy. The Sauerian interest in such matters was a principal spur to a 1955 conference that laid the groundwork for the landmark volume *Man's Role in Changing the Face of the Earth* (Thomas, Sauer, Bates, and Mumford 1956). Aldo Leopold (1886–1948) urged society to adopt a new land ethic aimed at protecting the environment by rejecting the idea of land as mere property, and it is frequently claimed that this argument represents the true philosophical beginnings of the contemporary environmental movement. Certainly Leopold (1949) outlined a non–anthropocentric, holistic environmental ethic. But widespread popular interest in the environment dates from the 1960s, and especially from the 1962 publication of *Silent Spring* (Carson 1962)—a powerfully critical account of the effects of pesticides on human life. It was this work that ushered in the age of environmentalism, loosely defined as a concern with changing human attitudes and behavior regarding environment.

The interweaving of the physical and human—cultural, political, and economic—aspects of this topic is undeniable. 'It is now taken for granted that the global environmental crisis and a renewed concern with global demography (the return of the Malthusian specter) are inseparable from the terrifying map of global economic inequality' (Peet and Watts 1996:1). Certainly, current concern about environment cannot be detached from concern about people, as the United Nations–sponsored conferences on Environment and Development in Rio de Janeiro in 1992, on Population and Development in Cairo in 1994, and on Women in Beijing in 1996 made abundantly clear. Today, it is common to portray concern for the environment as being in opposition to progress, science, technology, and population growth, an attitude that might be explained by reference to the close association between the expansion of European empires and related environmental impacts.

Pause for Thought

It is instructive to consider whether the typical Western attitude towards human impacts on environment is changing today. There is a useful distinction to be made between an earlier anthropocentric view and a more current, but by no means dominant, ecocentric view. This transition can be seen as from a dominant emphasis on material well-being toward greater emphasis on overall quality of life. Pinkerton (1997) described the ecocentric view as one of enviromanticism, politically useful but practically questionable. Certainly, contemporary environmental movements are complex social and political forces, ranging from those favoring preservation of landscapes to those with more radical ideas, such as dismantling capitalist society and economy. It is not difficult to see how environmental movements can translate into formal political movements.

Ecofeminism

As discussed in Chapter 3, there is general agreement among the many versions of feminism that sexism prevails, is wrong, and needs to be eliminated. In the broadest sense, something is a feminist issue if an analysis of it helps one understand the oppression or subordination of women in any place and at any time.

In addition to this basic argument, academic feminism recognizes that the liberation of women cannot occur until all women are freed from the many oppressions that structure our gendered identities. Included in these multiple oppressions are those based on social class, skin color, income, religion, age, culture, and geographic location. But, despite these general premises, there is not one feminism. Nor is there one environmental philosophy and, accordingly, there is not one ecofeminism or one ecofeminist philosophy. Given the above, an explicit statement of interest is as follows:

> According to ecological feminists ('ecofeminists') important connections exist between the treatment of women, people of color, and the underclass on one hand and how one treats the nonhuman natural environment on the other. Ecological feminists claim that any feminism, environmentalism, or environmental ethic that fails to take these connections seriously is grossly inadequate. (Warren 1997:3)

These ideas first appeared in the mid-1970s through an integration of some feminist ideas and the green movement—from feminism comes the idea that humanity is gendered in such a way that women are subordinate and exploited, and from the green movement comes the idea that humans are damaging the natural world.

There are two fundamental versions of ecofeminism. Cultural ecofeminism focuses on the links traditionally made between women and nature, and the related circumstance that men oppress both. Cultural ecofeminism is thus a response to the feminist claim that Western culture has identified women and nature as belonging together and has devalued both.

Human nature is seen as grounded in human biology in that sex/gender relations have resulted in different power bases. As such, cultural ecofeminists see the world as dominated by technologies and ideologies developed and controlled by men, and they seek to elevate and liberate women and nature, usually by means of direct political action. Relations between women and the environment are seen as essential.

Social ecofeminism rejects the biological determinism implied in the previous position, focusing instead on nature as a social and political construct and not as a natural construct. Thus, social ecofeminism advocates liberating women by overcoming the constraints imposed by marriage, the nuclear family, patriarchal religion, and capitalism. There is a rejection of the essentialist idea that the linking of women and nature results from biology in favor of the idea that the relations between women and the environment are constructed. Because they are constructed, these relations are not universal; rather, they vary from place to place and through time.

Landscapes of waste and dereliction are all too commonplace inside cities, reflecting and reinforcing some of the less palatable aspects of contemporary urban society. Such damaged environments are sources of danger to children, breeding grounds of disease, and visually distasteful. (Getty Images/PhotoDisc)

IMPLICATIONS

The fundamental implication of ecofeminism is best described in the larger context of feminism in general. Since the 1960s, the various forms of feminism have effectively called into question virtually all of the established and cherished belief systems and institutions of dominant patriarchal cultures. The implications of ecofeminism for the larger ecological perspective are enormous. Many ecofeminists regard ecocentric views such as deep ecology as inadequate correctives, though there are others who see ecofeminism as a union of deep ecology and feminism. Ecofeminism asserts

- that the dominant development paradigm sees the earth as a resource to be exploited and developed for human benefit,
- that there are compelling connections between the treatment of nature and the treatment of women, and

• that there is a larger domination of the world by the values of a Western cultural tradition.

Accordingly, ecofeminists strive to move beyond the various dualisms, such as nature–culture and female–male, in order to create a new consciousness of people and the earth in harmony. One way to demonstrate some of these matters, and also to clarify the difference between cultural and social versions of ecofeminism, is to review the example of the Chipko movement, the most often discussed of the many struggles that symbolize the relationship between women and environment.

THE CHIPKO MOVEMENT

Chipko, meaning 'embrace' or 'hug' in Hindi, is a grassroots, non-violent movement initiated in 1974 by women in an area of northern India who expressed concern about the removal of broad-leafed indigenous trees. Since then it has spread remarkably and is credited with saving perhaps 4,634 sqare miles (12,002 km²) of trees. Why the concern? Because these trees are an essential resource for the rural poor, who depend on the trees for fuel, food, fodder, building materials, and household utensils, and also for limiting soil erosion. Why was it women who initiated the movement? One reason, certainly, is that women are more dependent than men on trees, being the ones who collect fuel and fodder and perform household tasks, and so it is women who would be the principal sufferers if trees were removed. Meanwhile, replacing the indigenous trees with pine or eucalyptus (as the government planned to do) would benefit men by providing them with employment. But a cultural ecofeminist would answer this question differently, focusing on the idea that ancient Indian cultures worshipped tree goddesses and that movements similar to the Chipko movement were in place over 300 years ago. On the other hand, a social ecofeminist might answer that it is because women are marginalized members of society and have consequently created a link with nature that they are the ones to initiate the movement.

This example can be interpreted in a way that supports the cultural ecofeminist essentialist claim that women are closer to nature and

Women of all ages, and often their children, joined in an organized resistance that became known as the Chipko (Hindi for 'tree hugger') movement. Led by Sunderlal Bahuguna, one of the best-known early leaders of this remarkable environmental movement, the protesters placed themselves between the trees and the contractor's axes to save the trees from destruction. (Photo reprinted courtesy of the Food and Agriculture Organization of the United Nations, from 'A Field Guide for Project Design and Implementation: Women in Community Forestry,' FAO, Rome, 1989)

are, therefore, necessarily more committed to caring for the environment. But it can be interpreted rather differently following the social ecofeminist constructionist argument that women in the less developed world view their relations with nature differently from women in the more developed world.

To summarize this discussion of ecofeminism is not easy because generalizations are difficult, but the following statements reflect a moderate ecofeminist position:

- Nature does not need to be ruled over by humans.
- Humans need to recognize their intimate relationship with nature.
- From a religious perspective, God is neither male nor anthropocentric.
- Mutual interdependency replaces the traditional hierarchies of domination that have God above humans above nature, or perhaps have God above men above women above children above animals above plants.
- Prevailing patterns of interdependency based on sex, gender, skin color, class, culture, income, religion, age, and geographic location need to be reconstructed to create more equitable situations.

Pause for Thought

Much of the content of this section so far is based on the idea—not the fact—that nature is not so natural after all, but is better interpreted as a social construction. For postmodern philosophers, this view is one part of a larger argument that insists on a lack of givens in most areas of inquiry. A contrary argument about nature asserts that 'Social construction is necessary but not sufficient for our being. Some values on earth are not species-specific to homo sapiens' (Rolston 1997:62). The key point for our purposes is that philosophers continue to debate the logic of constructionist claims. Cultural geographers need to appreciate that different opinions on this important matter are valid. This point is especially important since culture and nature are seen by many landscape school geographers as essential categories and by many others as constructions.

Political Ecology

Political ecology 'combines the concerns of ecology and a broadly defined political economy. Together this encompasses the constantly shifting dialectic between society and land-based resources, and also within classes and groups within society itself' (Blaikie and Brookfield 1987:17). Political ecological analyses recognize and incorporate various social and spatial scales and stress the need to understand a problem in a broad cultural, social, political, and economic context. Thus a political ecologist may conduct an analysis of a local area and of a small group but may also try to understand the problem at other scales; further, while the problem may be specifically economic, the political ecologist will incorporate other human and physical considerations in the analysis.

THE APPROACH IN CONTEXT

Suggestions that political and economic processes need to be considered in human or cultural ecological analyses first appeared in the 1970s. Arguments focused on a socialist human ecology as the best means of practice, since it is concerned with environmental histories, human adjustments, system stability, and the fate of folk ecologies. Certain analyses of population, food supplies, and famine in parts of the developing world incorporated some of the characteristics of this emerging approach. But political ecology as an essentially Marxist-inspired analysis of the human use of resources and its related impacts on environment is usually linked to two important works, by Blaikie (1985) and by Blaikie and Brookfield (1987).

Blaikie, addressing the problem of soil erosion in the less developed world, rejected the conventional explanation that referred to such factors as overpopulation, inappropriate use of resources, environmental difficulties, and collapse of markets. Instead, he emphasized social causes, specifically land managers, or households that were using local resources and that were obliged to extract surpluses thus leading to land degradation. Blaikie identified a chain of circumstances leading to soil erosion: first, the land managers have direct contact with the land; second, land managers relate to each other, to other land users, and to other groups in society; third, they are linked also to the state and to the world economy. The attractiveness of this political ecology needs to be understood within four larger contexts, in addition to those provided by the ecological traditions discussed in this chapter.

First, there is the context of political economy. For geographers, political economy is not easy to define, as it is really an umbrella term covering a variety of radical emphases mostly inspired by Marxism. It can be summarized as the idea that the political and the economic are irrevocably entwined, such that any study of human activities must incorporate both. Central to the political economic approach is the idea that much human activity is bound up with social and political struggles and with competing claims to such things as ownership and right of access. Political ecology is a radical perspective on the political economic approach, emphasizing social justice, equity issues, and unequal power relations between the people and groups that affect environments.

Second, there is the larger context of environmental and social concern that has been evident since about the late 1970s, especially in the context of the less developed world. Unlike the original environmental and social concern that was raised in the early 1960s, this later concern was bound up with larger questions of economic and cultural globalization and with increased awareness that humans were affecting a global environment. Political ecology is particularly critical of environmental policies that do not consider the impacts of global capitalism.

Third, political ecology needs to be considered in the context of its links with structuration theory as outlined by the sociologist Giddens (for example, 1984). Structuration is a complex set of ideas that seeks to integrate the work of a wide variety of thinkers, including Durkheim, Freud, Marx, and Weber. A key feature of this theory is its focus on links between human agents and the social structures within which they function, with individuals viewed as operating within both local social systems and larger social structures; indeed, there is a substantial and continuing debate in social science—the structure and agency debate—centered around this issue. Structuration theory views individuals as agents operating within both the contexts of local social systems (sometimes called locales) and the larger social structures of which they are a part. The most important characteristic of structuration is its identification of the dualities associated with social structure and human agency. Thus, social structure enables human behavior, while at the same time behavior can influence and reconstitute culture. Further, the rules of any social structure are both *constraining* and *enabling*: they are constraining because they limit the actions available to individuals, but they are also enabling as they do not determine behavior. Certainly, both political ecology and structuration theory consider the ways in which global, national, or regional processes connect over time with local-scale processes.

Related to structuration theory is the context provided by the socially informed new regional geography, which sees regions and regional change as bound up with social processes. Jarosz (1993:367) articulates this guiding principle of the new regional geography thus: 'Human activities do not cause regional change; rather human activities shape, and are shaped by, place and history. Human identities and activities constitute the economic, political, and ideological processes that form and transform regions. In turn, regions shape human activities due to particular contextual details of place.'

Fourth, political ecology can be considered in the context of a set of approaches, known as *liberation ecologies*, that explicitly incorporate ideas of entitlement, social justice, and livelihood. In a discussion of these Marxist inspired liberation ecologies, Peet and Watts (1996) incorporated aspects of development theory, various poststructural critiques and related discourse theory, and social movements and related political forms; these additional conceptual insights were noted in Chapter 3.

Political ecology is a multifaceted approach that shares conceptual and empirical concerns with several other interests. As with some of the other topics raised in this chapter—for example, the new ecology and ecofeminism—political ecology is an approach that continues to evolve as the underlying theoretical concepts are developed and clarified. This section now reviews three examples of political ecological research.

PEASANT–HERDER CONFLICTS

Since the early 1970s, large numbers of Fulani pastoralists in West Africa have moved south from semiarid environments in the Sahel, especially from Mali and Burkina Faso, into the northern savanna region of the Ivory Coast. Their migration has been welcomed by the Ivory Coast government because of the contribution the Fulani make to beef production, but one unfortunate consequence has been a series of conflicts between the incoming pastoralists and the local Senufo peasant farmers that they have encountered. The principal reason for conflict is the damage inflicted by Fulani cattle on crops grown by the Senufo, for which the Senufo have not been compensated. There are similar conflicts in other West African countries.

Traditional geographic explanations for such movements and conflicts rely on the circumstances of a declining resource base relative to population, while a political ecological focus stresses the need to consider human activities as responses in context, and looks at the chain of causality leading to conflict. As Bassett (1988:469) explains, 'It is at the intersection of Ivorian political economy and the human ecology of agricul-

It is not difficult understand how the movement of cattle through farming areas is likely to damage growing crops—usually cotton or sorghum—and to trample soil, making it more difficult for crops or grass to grow. (Jim Walls/www.fullpassport.com)

tural systems in the savanna region that one can begin to identify the key processes and decision-making conditions behind the current conflict! Peasant farmers are aggrieved at the consequences of pastoral intrusions into their cropping areas because of the impact of uncompensated crop damage on their already marginal standard of living. But although it might be suggested that this is the cause of conflict, it is not seen as a sufficient cause of the peasant uprising that included the murder of about eighty herders in 1986 and that is continuing today. Table 4.1 indicates the key determinants of conflict, distinguishing between ultimate causes and proximate causes, and identifying stressors and counter-risks.

As with any ecological approach, political ecology thus acknowledges complexity and interrelationships, looking beyond key factors that may be evident to understand an issue in context. In the case of peasant–herder conflicts in the Ivory Coast, three ultimate causes are noted, one of which is the Ivorian development model, which a traditional cultural ecological approach would not seriously consider. There are also more specific proximate causes that relate to the particular circumstances of the groups involved. Note that poverty, as expressed in terms of low incomes, is only a proximate and not an ultimate cause. It is necessary to know

how poverty comes about, rather than simply to use it as an explanatory factor.

BANANA EXPORTS AND LOCAL FOOD PRODUCTION

Historically, colonial powers encouraged banana production in colonies with the appropriate physical environment; thus, bananas were imported by France from Martinique and Guadeloupe, by Spain from the Canaries and the Azores, and by Britain from the Windward Islands in the eastern Caribbean. Indeed, for the islands of the eastern Caribbean, bananas have been the principal export crop since the early 1900s, with most production on small, usually less than 5-acre (2-ha), plots and by farmers who were often owner-occupiers.

In many islands of the eastern Caribbean during the early 1990s, there was evidence of increasing banana production accompanied by decreasing production of local foodstuffs and increasing food imports. The usual explanation for such circumstances assumes that it is the increase in export crop production that is the cause of the other changes. But in a political ecological analysis of this situation in the eastern Caribbean island of St Vincent, Grossman (1993:348) noted that such explanations do not 'specify the precise means by which banana production is supposed to interfere with local food

Table 4.1 Determinants of Peasant–Herder Conflicts in Northern Ivory Coast

Ultimate Causes	Proximate Causes	Stressors	Counter-risks
Ivorian development model: • surplus appropriation by – foreign agribusiness – the state • livestock development policies	Low incomes and beef consumption Insecure land rights	Uncompensated crop damage	Compensation Fulani expansion
Savanna ecology	Intersection of Senufo agriculture and Fulani semi-transhumant pastoralism	Political campaigns Theft of village cattle	Crop and cattle surveillance
Fulani immigration			Corralling animals at night

Source: T.J. Bassett, 'The Political Ecology of Peasant-Herder Conflicts in the Northern Ivory Coast', *Annals*, Association of American Geographers 78 (1988):456. (Reprinted by permission of Blackwell Publishing)

production', nor do they 'analyze the significance of the role of the state and capital in influencing these relationships'. Grossman (1993) cited, as an important political economic factor, the fact that marketing policies and state intervention result in banana production being advantaged over local food production, especially because of the guaranteed market and fixed prices for bananas; as an important local factor, he noted the reduced possibility of theft of bananas because of technological innovations in harvesting and packing that make it much more difficult for thieves to steal bananas at night and sell them in local markets the following day. He also considered the question of potential conflicts in the productive sphere by reference to any seasonal conflicts of time allocation, to spatial patterns of land use, and to issues of intercropping. Grossman concluded that such conflicts were not the principal cause of local food production decline. Rather, the explanation lies in the different political economic contexts of banana and local food production. The banana market is attractive for reasons related to the state and to foreign capital.

Interestingly, the value of this approach is further evidenced by changes in world trade regulations. In accord with protocols agreed to in 1957, European countries were able to afford preferential treatment to their banana-producing areas. In 1993, the European Union (EU) amalgamated the various protocols into a single preferential trading structure. This system of protectionism for specific banana-producing areas has, however, been condemned by the World Trade Organization (WTO) following a challenge initiated by the United States and the banana-producing countries of Ecuador, Mexico, Guatemala, and Honduras. This global economic change, only one aspect of the movement toward free trade, is a major blow to those banana-producing countries such as St Vincent that previously received preferential treatment. As another example of the complex relations between banana production and larger policies, the European Union issued a statement of standard in 1994 requiring bananas to be at least 1 inch (27 mm) wide, 5.5 inches (14 cm) long, and without abnormal curvature. The bananas

grown in the islands of the eastern Caribbean do not conform to such standards. Many of these details were resolved by the WTO in 2001.

VIOLENT ENVIRONMENTS

Some commentators from Malthus onwards have predicted that continuing population growth and related resource depletion will produce a human future fraught with problems. This **limits-to-growth thesis** was resurrected in the 1960s and has continued to dominate both scholarly and public opinion since then. The alternative **cornucopian thesis** sees our human future in rosier terms, with less human crowding, less pollution, and fewer ecological problems. Contemporary neo-Malthusian claims of impending anarchy related to the growing incidence of regional violence, such as those made by Kaplan (2000), have been criticized for their geopolitical underpinnings and their proposals for dealing with the perceived problem through a Western policy of containment and exclusion, which would effectively cut off parts of the less developed world. But a more important set of criticisms are those that emanate from a political ecological perspective.

In broad terms, political ecologists reject the argument that environmental scarcity and related problems of economic activity and migration are the cause of weakened states and subsequent violence. Rather, they see violence as 'a site-specific phenomenon rooted in local histories and social relations yet connected to larger processes of material transformation and power relations' (Peluso and Watts 2001:5). Any account of the political ecology of violence begins not with politics but with the fundamental geographic circumstance of human-and-nature relationships, with humans naturalized and nature humanized.

This debate about environmental security and violence is important, with Marxist thinkers stressing the historical and social appropriation of nature and favoring a political ecology approach, liberal thinkers being inclined to focus on environmental scarcity and favoring a limits-to-growth argument, and conservative thinkers emphasizing human ingenuity and technological advances and favoring a cornucopian view.

Environmental Ethics

How are we to determine what is ethically right when we consider our treatment of the natural world? Human attitudes toward environments and non-human animals have undergone significant changes since the dominant early nineteenth-century view that supported both the idea that it was right and proper for humans to use environments as they saw fit and also the idea that humans were different from all other animal species. Today, there is considerable philosophical debate concerning the proper assessment of humans as part of nature. This discussion involves consideration of a fundamental question: should we be behaving in some environmentally friendly manner because we need to for our own sake, or rather because it is ethically right for us to do so in terms of the natural world itself? The following sections outline seven different perspectives on this issue.

Religion and Environment

The debate about the relationship between Christian ideas and human impacts is but one aspect of the larger question of religion and environment. Most religions identify what is important to humans beyond everyday concerns and offer rules of behavior that are said to derive from some spiritual source. Inevitably, then, religion has something to say about human activities as these affect other humans and the environment that humans occupy. Overall, most religions provide guidelines for human activities and for an understanding of nature.

Many traditional religions were not as separated from the rest of human life as Christian religion has been. In traditional Hawaiian culture, god, humans, and nature were related to one another and formed a single interacting community, with all species—not just humans—considered sentient. Thus the entire world was seen as alive, capable of knowing and acting, in the sense that Western culture sees only humans as alive. Such beliefs allowed for a genuine kinship that is lacking in the Christian tradition; they also made for a quite different perception of the world.

These comments apply generally to a great many of the traditional belief systems that were replaced or substantially weakened as Europeans spread overseas, but they apply also to several of the major world religions practiced today.

In Islam, respect for the natural world is founded on the principle that Allah created all things. Islamic religion also acknowledges that humans are not the sole community living in the world. Use of the natural world to support humans is considered appropriate, but needless destruction is not sanctioned. Both Hinduism and Buddhism traditionally viewed humans as but one part of nature such that abuse of the natural world is also abuse of people. However, most observers suggest that the materialistic orientations of Western culture have affected these traditions with a corresponding decrease in concern for the natural world.

Extending the Principle of Equality

Environmental ethics breaks new ground from a philosophical perspective. All ethics is concerned with seeking an appropriate respect for life, but only environmental ethics asks whether non-human life should be included. Should we extend the principle of equality beyond our own species? If we do extend the principle of equality to include all sentient life, should we also extend it to insentient life? Note that asking these questions would not be necessary in some traditional cultures, as the discussion of religion and environment acknowledged.

Referring to the principle of equal consideration of interests, which allows us to claim that other humans should not be exploited because, for example, they have lesser intelligence or a different culture, it might be argued that it is similarly wrong for humans to exploit members of other species. Following this logic it is, for example, wrong to eat meat, wrong to use animals in experiments, and wrong to display animals in zoos. These are very complex ethical issues that are at the heart of our human relationship with parts of the natural world.

What are some of the basic arguments against keeping animals in zoos? Most obviously, the zoo environment is an alien one that restricts the animals' freedom of movement, limits their ability to seek out food, and, more generally, prevents them from behaving in ways that are natural to them. Further, it has been argued that animals often suffer in zoos, where many die young, suffer injury, or develop deformities. There is much evidence of animals engaging in disturbed, trance-like behavior: for example, elephants swaying back and forth, polar bears swimming in circles, parrots grooming themselves until they bleed, and gorillas regurgitating and then re-ingesting food. In response to ethical concerns, there has been considerable debate about the wisdom of keeping animals in zoos and whales and dolphins in aquariums, and several zoos have chosen to close their elephant exhibits in recent years.

But there is another rather more difficult argument against zoos that focuses on the question of why we wish to display animals in zoos.

In addition to conventional justifications concerning human entertainment, education, scientific analyses, and species preservation, it can be argued that they serve to reinforce our human sense of superiority. The fact that we are able to confine animals means that they are there for our use and pleasure. It is in this sense that zoos can be viewed as a reflection of an outmoded environmental ethic.

Ecology, Economics, Ethics

The growing field of environmental economics originated from the fundamental premise that a great deal of economic activity is unecological, and as a result there is a need to bring economic concepts of value more into line with ecological concepts of value. This may be a difficult task even though both concepts of value are concerned with the management of the household or home. Proponents of environmental eco-

Lulu the elephant stretches her trunk in her small enclosure at the San Francisco Zoo. The premature deaths of two other elephants raised concerns about the unhealthy zoo environment, and the city's board of supervisors passed a measure in December 2004 requiring the zoo to give elephants a habitat of at least 15 acres (6 ha). Unable to meet this requirement, the zoo moved Lulu to her new home, a sanctuary in the Sierra foothills, in March 2005. (Lou Dematteis/Reuters/CORBIS)

nomics recognize a need to adjust the traditional economic conception of a household as a given group of people to an ecological scale where that group is the entire world population and not some subset of that population. The issue is, then, one of social scale, as Hardin (1968) noted in the classic account of 'The Tragedy of the Commons', in which he asked whether we behave appropriately for ourselves only, or for ourselves and some others, or for ourselves and all others, with this final possibility implying also a respect for the natural world.

One major stumbling block to the widespread acceptance of environmental economics is that it does not accept the conventional economic wisdom that growth, measured in some monetary terms, is an end in itself. A possible resolution of this difficulty lies in the acceptance of the adaptive strategy of **sustainability**.

Behaving Ethically

It may be helpful to think in terms of environmental problems that result from the actions of individuals and organizations, and to try to understand, from a psychological perspective, why we behave as we do. There is a contradiction between the environmental behaviors that many of us practice and the environmental attitudes that we hold: many people may have pro-environmental attitudes and yet participate in environmentally destructive behaviors. How are we to understand and change this gap between attitude and behavior?

Rather than emphasizing additional environmental education, one approach to understanding the attitude–behavior gap might be to focus on the psychology behind the inconsistency. This kind of examination produces four general conclusions:

- First, environmentally damaging behavior is often not a result of indifference to the environment; rather, it occurs notwithstanding legitimate concern for the environment. Thus, it is possible that, for example, the decision not to recycle occurs despite both liking the idea and believing that it is good for the environment. It is also possible that requiring organi-

zations to adhere to certain environmental standards is counterproductive.
- Second, inappropriate behavior may result from faulty mental models of the world; in other words, we are not fully aware of the environmental problems that we are encountering and contributing to, and as a result we engage in behaviors that are less than ideal.
- Third, there are real difficulties in assessing the implications of environmental change for human life and welfare.
- Fourth, it is extremely difficult to place an economic value on environmental goods and outcomes.

Living with Discordant Harmonies

Can we achieve harmony with nature only by accepting the new ecological recognition that nature is ever changing and by using our technologies effectively to help us understand the natural world? Because humans are but one part of a larger, living, changing system that is global in scale, our aim should perhaps be to accept natural changes—that is, the discordant harmonies—and strive to make the earth as comfortable as possible for ourselves. But how can we expect to use the natural world wisely if we do not understand that natural world? For example, knowing that nature changes means that we need to rethink ideas about valuing wilderness for its intrinsic characteristics or conserving particular natural landscapes as we think they were before human impacts. If nature changes of its own accord, how can we possibly conserve it? Of course, these questions are not arguments against conservation and preservation, but rather arguments about the reasons for such actions.

Transforming Culture

One ecocentric perspective favors a deep ecology, or what is often called *sustainable development*. Naess (1989) proposed that humans need to develop a new worldview that sees nature as having value in its own right, that regards all humans as connected, that recognizes a need to work with rather than against nature, and that views ecosystem preservation as a primary goal.

In addition, such a deep ecology regards humans as part of nature, asserts that every life form has in principle a right to live, is concerned with the feelings of all living things, and is concerned about resources for all living species. Thus, deep ecology is both an environmental ethic and a spiritual basis for rethinking what it means to be human. Several of the basic concepts employed by Naess are derived from Hinduism.

In a related fashion, but pursuing an original argument, one writer has proposed that we need to move away from the oppression and emphasis on wealth that are characteristic of modern society through a reintroduction of the specific basic qualities that we once possessed, namely a connection with nature, a sense of belonging, and an egalitarian community (see Box 4.6).

The 'Golden Rule'

Another way to approach the question of environmental ethics is to think in terms of how we wish to be treated. We might approach this matter by questioning two of the frequently stated bases for developing an appropriate ethic, namely the related claims that humans are permanently damaging the earth and that we need to learn how to behave as stewards for the earth we are threatening. In one sense, both of these propositions reflect an exaggerated view of our human role because humans are but one of mil-

Box 4.6 Social Evolution and the Environmental Crisis

For Earley (1997), it is the continuation of modernity—the modern period—that poses the greatest threat to human survival, in the sense that our current environmental crisis results from the way our world is structured and our related view of nature and our place in nature. To move away from this modernity requires us to change our social structures and institutions. These points are clarified by means of a model of social evolution (Earley 1997:2–4). At the outset of social evolution, humans enjoyed three basic qualities:

- a connection with nature;
- a sense of belonging and richness of experience; and
- an egalitarian community.

Through time, populations grew, resources were limited, and conflict occurred, resulting in three emergent qualities that provided for more conscious choice, increased freedom from environment, additional population growth and resource exploitation, and a new understanding of ourselves and of nature. These are

- technology,
- reflexive consciousness, and
- social structure.

The difference between these two sets of qualities is striking. The initial ground qualities—what we were—reflect vitality and organic wholeness, while the later emergent qualities—what we have become—

reflect power and differentiated organization. Put simply, it is the emergent qualities, especially the resultant economic system emphasizing material growth, that are responsible for our **alienation** from nature, from others, and, indeed, from ourselves. The result is that we

- have gained technology, but destabilized the environment;
- understand more, but empathize less; and
- have developed social structure, but lost equality and community.

According to Earley, it is only in recent years that it has been possible to fully grasp the implications of this process of social evolution and to appreciate that it is insufficient and, of course, impossible, to return to the ground qualities. What humans need is a rediscovery of the ground qualities and an integration of ground and emergent qualities, resulting in

- ecological technology;
- integrated mind and heart; and
- a social structure that promotes equality and community.

In summary, Earley (1997:4) states: 'we must now consciously choose to regain our wholeness and vitality in conjunction with our complexity and autonomy, as individuals and communities, as organizations, as a world society, and as a living planet.'

This commercial poster highlights what is rightly regarded as a universal value: treating others as one wishes to be treated by others. This principle is also the first article in the 1948 United Nations Declaration of Human Rights: 'All human beings are born free and equal in dignity and rights. They are endowed with reason and conscience and should act towards one another in a spirit of brotherhood.' (© Paul McKenna 'The Golden Rule', reduced from 22×29, is reprinted by permission of Scarboro Missions/www.scarboromissions.ca)

lions of species and have, geologically speaking, appeared on the earth only recently. But to think in such grand terms is, of course, misleading.

To help develop an appropriate ethic, we need to think in more local terms, believing that we are capable of such devastating impacts as eliminating both the natural world upon which we depend and our very species. Following this logic, Gould (1990) proposed that an appropriate environmental ethic is one that can convince people to view clean air, clean water, reforestation, and so forth as being essential to the well-being of our environment and ourselves here and now, rather than one that stresses long-term consequences of neglecting such issues. The ethic in question is simply that of treating others as we wish to be treated by others.

Pause for Thought

Is it rather disturbing that so much popular and governmental attention is paid to the question of human impacts on environment when the problems of human impacts on other humans remain so distressing? It is commonplace today to claim that concern for environment is a fair measure of a society, but do we similarly assess ourselves according to our treatment of others? The answer seems uncertain. To paraphrase Wagner (1990:41), a principal goal of cultural geography should be to develop a theoretically well-grounded, intellectually vigorous, and practically effective discipline that might well assume, in time, a major role in guiding and guarding our attitudes toward and treatment of ourselves and others.

Global Landscape Change During the Past 12,000 Years

Any consideration of human-and-nature relationships through time addresses three critical and loosely related variables: resources, population, and technology. Expressed simply, more and varied resources, more people, and a higher level of technology mean increased landscape change. As this section makes clear, human-induced change over the long period of

human occupance of the earth does not occur separately and apart from natural change. Butlin and Roberts (1995:10) noted: 'Nature does not create landscapes, stop its work, and then hand over to human agency to complete the transformation.' Following a few definitional statements, this section outlines global landscape changes through time, especially over about the last 12,000 years, and further informs the account by referring to related changes in political and social systems.

Resources are cultural appraisals defined not only by their physical presence but also by human awareness, technological availability, economic feasibility, and human acceptability. The presence of a particular resource is no guarantee that it will be used, because human ability and need are prerequisites for use. Cultural change through time typically increases the resources available, especially through advances in technology and scientific knowledge. These simple ecological statements are, of course, a compelling argument against any simplistic environmental determinism. In short, natural resources are cultural appraisals.

Regardless of specific cultural interpretations, it is usual to divide resources into two discrete categories. **Stock resources**, such as land and minerals, are fixed in their supply as they cannot be replaced except over the long period of geological time. **Renewable resources** are replaced sufficiently rapidly after their use by humans so that they can effectively be regarded as always available. However, such a basic distinction is misleading, and it is better to think of a use renewability continuum (see Figure 4.4).

Resource use is related to **technology**—the ability to convert **energy** into forms useful to humans—and one of the most fundamental features of cultural evolution is the increasing capacity to use energy sources more effectively. Prior to the nineteenth century, most groups relied almost exclusively on solar power, but the relatively recent ability to tap the energy stored in fossil fuels combined with advances in scientific understanding have resulted in huge

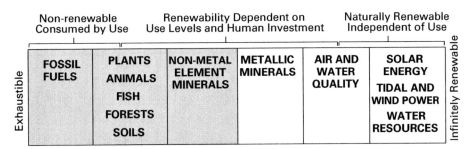

Figure 4.4 The Natural Resources Continuum. Rather than conceiving of resources as being either replaceable or irreplaceable, it is appropriate to recognize a continuum with all resources being replaceable and the difference between resources being the length of time needed for replacement to occur. At one end of the continuum are those resources that are renewable over a very short time period, the availability of which is therefore unrelated to use. Included in this group are solar energy, wind and tidal power, and water resources. At the other end of the continuum are those resources that are consumed when used and are effectively non-renewable. Fossil fuels are the principal example. Between these two extremes is a wide range of resources whose renewability is a function of the way in which humans use them.

Source: J. Rees, 'Natural Resources, Economy and Society', in *Horizons in Human Geography*, edited by D. Gregory and R. Walford (New York: Macmillan, 1989):369. (Reprinted by permission of Palgrave Macmillan)

increases in our ability to use resources. Also linked to the initial use of fossil fuels was an exponential increase in population numbers resulting from a decline in death rates without a corresponding decline in birth rates, a situation that characterized the industrial world of the nineteenth century. Basic links between technology and population are summarized in Figures 4.5 and 4.6.

Central to much ecological cultural geographic work is the concept of **carrying capacity**—the maximum human population that can be supported in an area given a particular level of technology. A useful amendment of this concept is the Boserupian thesis, which argues that increases in population serve to stimulate technological advances, hence increasing the carrying capacity of an area (Boserup 1965).

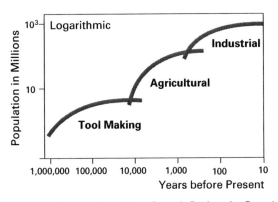

Figure 4.5 Human Population Growth During the Past 1 Million Years. The relationship between technological change and population growth is summarized in this figure. The use of a logarithmic scale enables each of three technological transformations, tool making, agriculture, and industry, to be placed in context.

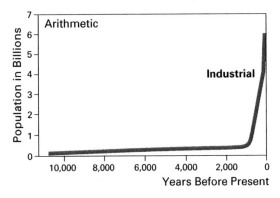

Figure 4.6 Human Population Growth During the Past 10,000 Years. The overwhelming impact of the transformations that occurred during the Industrial Revolution is evident in this overview of population change. Note that the scale employed in this figure is arithmetic, unlike that in Figure 4.5.

Foraging

Our understanding of human origins is always changing, but the dominant current view is that the human evolutionary line separated from that of the apes some 6 million years ago (mya), with the first appearance of modern humans before 100,000 ya, and with most of the earth colonized by 10,000 years ago (ya) (see Table 4.2 and Figure 4.7).

Most current evidence suggests that hominids were restricted to east and southern Africa until *Homo erectus*, which first appeared about 1.8 million ya, moved into warmer parts of Europe and Asia at the beginning of the **Pleistocene**, about 1.6 million ya. As they encountered different environments, these early humans devised cultural adaptations relating especially to clothing and shelter. There is evidence of permanent **pair bonding** and substantial advances in language about 400,000 ya. At this time, a new species evolved, archaic *Homo sapiens*, similar to modern humans but with some physical differences—Neanderthals are usually considered a subset of this species.

The appearance of modern humans, *Homo sapiens sapiens*, occurred before 100,000 ya. There is much debate concerning the specific location of origin, with some favoring multiple origins in Africa, Europe, and Asia from pre-existing *archaics*, or archaic groups (this is known as the multiregional hypothesis), and others favoring a single origin from archaics in eastern Africa only (this is known as the replacement hypothesis). Although this debate, like many other debates about human evolution, is not resolved, the weight of available paleontological, genetic, and archeological evidence favors the replacement hypothesis, with modern humans evolving from archaics in eastern Africa only, and then spreading across the earth, replacing earlier archaic groups because of advantages associated with cultural developments.

This process of human movement over most of the earth took place during a period of harsh climate, with the most recent glacial period— one of perhaps eighteen such periods in the past 1.6 million years—beginning some 80,000 ya and ending with the retreat of ice sheets from

Table 4.2 Basic Chronology of Human Evolution

6	mya	Probable earliest date for evolution of first hominids, the bipedal primate Australopithecus, in Africa
3.5	mya	First fossil evidence of Australopithecus afarensis in east Africa
3	mya	Probable appearance of Homo habilis in east Africa; evidence of first tools
1.8	mya	First appearance of Homo erectus in east Africa
1.5	mya	Evidence of Homo erectus in Europe; evidence of chipped stone instruments in Africa; beginning of Pleistocene
700,000	ya	First evidence of chipped stone instruments in Europe
400,000	ya	Homo erectus uses fire; develops strategies for hunting large game; first appearance of archaic Homo sapiens
100,000	ya	Appearance of Homo sapiens sapiens (in Africa only, according to favored replacement hypothesis)
80,000	ya	Beginnings of most recent glacial period
40,000	ya	Arrival of humans in Australia
25,000–15,000	ya	Arrival of humans in America
12,000	ya	First permanent settlements
10,000	ya	End of most recent glacial period; beginning of Holocene period

Note: mya = million years ago
ya = years ago

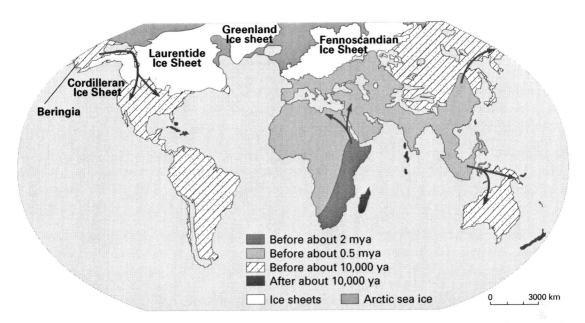

Figure 4.7 Human Colonization of the Ice Age Earth. This figure, and Table 4.2, suggest a chronology of human evolution and of human spread across the surface of earth during the ice age. Following a probable origin in east Africa about 6 mya, the genus *Homo* spread throughout most of the world by the end of the Pleistocene, 10,000 ya. The principal region unaffected by humans was Antarctica. The human movement shown in this figure was accomplished first by *Homo erectus* and later by archaic *Homo sapiens*. All this expansion was based on a foraging way of life and occurred in a physical environment that changed over time but that was typically different to that of today. The figure locates the principal ice sheets present during the Pleistocene glaciation.

According to the favored replacement hypothesis, modern humans—*Homo sapiens sapiens*—appeared about 100,000 ya only in Africa and then proceeded to settle much of Africa, Europe, and Asia by 40,000 ya, displacing earlier groups. Australia was reached about 40,000 ya, and the Americas perhaps as early as 25,000 ya but no later than 15,000 ya. Movement through northern Canada, Greenland, and some islands occurred during the last 10,000 years.

Sources: N. Roberts, *The Holocene: An Environmental History* (Cambridge: Blackwell, 1989):56; and, W. Norton, *Human Geography*, 5th edn (Toronto: Oxford University Press, 2004):97.

much of the northern hemisphere about 10,000 ya. This date, 10,000 ya, marks the end of the Pleistocene geological epoch and the beginning of the **Holocene**. The glacial period was characterized by some significant changes in average temperature, which suggests that humans were able to adjust to different physical geographic circumstances through some process of cultural adaptation.

Culturally, the late Pleistocene is referred to as the **Upper Paleolithic**, beginning about 40,000 ya and ending about 10,000 ya. During this cultural phase, humans depended on natural sources of food. Hunting, fishing, and gathering—described collectively as foraging—prevailed as the means of gaining subsistence.

In some parts of early Holocene Europe, **Mesolithic** cultures appeared in response to changes in the available animals and plants associated with the receding of ice sheets.

The widespread colonization of the earth occurred during the Pleistocene and was accomplished by cultures with subsistence economies. This fact, and evidence from the few remaining foraging groups, suggests that such cultures had an intimate knowledge of their natural environment, including the movement and behavior of game animals and the location and beneficial properties of different plants. The relative importance of animals and plants in overall diets varied from place to place, with animals more important in the cooler areas of Europe and Asia and plants more important in warmer African and Asian areas. All individuals were involved, with women and children as gatherers and men as hunters, although it is probable that in many societies women also hunted small game and men gathered food. It was possibly during this extended foraging phase, which lasted tens of thousands of years, that it became common to devalue women relative to men because of the perception that women were closer to nature whereas men were closer to culture.

It was customary to share food with other group members, and this activity encouraged social cohesion and stability. Groups were small, averaging perhaps about thirty, moved regularly, and required access to a sufficiently large area to support the group. Population densities were low, and a necessarily uncertain estimate of world population about 10,000 ya is 4 million.

Environmental Impacts

Few people in total, low population densities in any given area, and a nomadic lifestyle resulted in minimal impacts on the physical environment, especially long-term impacts. Early humans used fire to remove unwanted vegetation and encourage regrowth, and in animal kills. Although much of the evidence is circumstantial, humans may have caused animal extinctions either through overkill or, more probably, through human modifications of ecosystems. In Europe, several large mammals became extinct toward the end of the Pleistocene, while the Maori settlement of New Zealand has been linked to the extinction of about twenty species of flightless birds.

Pause for Thought

Traditionally, foraging has been seen as providing the basic necessities for survival, with a popular image of groups eking out a meager livelihood and being obliged to share in order to avoid starvation. However, this traditional understanding is now substantially revised, and the favored current view is of an affluent society with adequate food, a high degree of human security, and leisure time. Why the changing appraisal? It is possible that the earlier view was a consequence both of cultural relativism

and of the fact that observations of the few remaining foraging groups failed to acknowledge that such groups, which have been pushed into relatively inhospitable, marginal environments, might not be characteristic of earlier groups that depending on foraging.

Agriculture

Animal and plant domestication was a slow and gradual process, beginning perhaps 12,000 ya, toward the end of the Pleistocene period, and involving increasing numbers of people and expanding areas. By about 2,000 ya, most of the world population was dependent on agriculture. During the cultural phase of the **Neolithic**, agriculture spread throughout many parts of the world, replacing the earlier foraging activities of Upper Paleolithic and Mesolithic groups. What is often called the **agricultural revolution** is perhaps best described as the human use of **artificial selection** to modify animals and plants according to the needs of the human group.

Why the Transition from Foraging to Agriculture?

Domestication might be seen as the first major cultural, as opposed to biological, revolution. Certainly, ecological questions are at the heart of the agricultural origin problem, specifically the relative roles played by environment and culture and by individuals and groups. Questions of where, when, how, and why humans first domesticated plants and animals are crucial to an understanding of the history of humans and their relationships with nature. The introduction of agriculture as the dominant human means of subsistence revolutionized human impacts on environment and the human way of life. Questions about the origins of agricultural have been addressed by both physical and social scientists. Two pioneering contributions were made in 1926 by a botanist, Vavilov, who introduced the idea that there were centers of origin, and in 1928 by an archeologist, Childe, who popularized the idea that there was an agricultural revolution. Multidisciplinary research teams have used various methods and concepts, including the method of systems analysis and the concept of adaptation, to help answer questions about agricultural origins.

Accepting the new ecological argument that nature is inherently unstable requires us to rethink some of the basic concepts used to understand agricultural origins. Most notably, if, as proponents of the new ecology suggest, changes in nature are usual and not only the result of human activity and of extreme natural occurrences, then concepts such as those involving plant communities and climax vegetation are suspect. The new ecology implies that it was individual plant species that changed locations because of their particular requirements and adaptations rather than whole communities changing locations in response to climatic change. The idea that vegetational change proceeds through a succession to a climax state can also be questioned, on the grounds that it is a form of evolutionary thought, in the Spencerian tradition, presupposing that change is linear, developmental, and progressive.

A possible explanation for domestication is as follows. At the end of the Pleistocene, climatic change and associated receding of ice sheets prompted some groups to locate and settle in areas that permitted a more sedentary way of life—coastal and river areas for example. The establishment of more permanent settlements resulted in increasing population numbers, or at least denser populations, perhaps because of increasing fertility and a higher infant survival rate. In order to feed more people, these groups, which were already harvesting wild grass seeds such as wheat, rice, or corn, applied their knowledge of plant propagation to increase food supply by artificially selecting products useful to them.

Previously, it was widely assumed that there was one single origin area for plant domestication and that technologies diffused from that one hearth. The favored candidate was southwestern Asia, although Sauer (1952) argued for an origin in southeastern Asia. These diffusionist models were inherently ethnocentric, and the current consensus, supported by considerable archeological evidence, is that domestication occurred independently in several areas, namely southwestern Asia, Africa south of the Sahara, southeastern Asia, southern Europe, Central America, coastal South America, and central North America. Each of these areas had a different farming system; animals, for example, were a key part of the agriculture that evolved in southwestern Asia but were less important elsewhere (see Figure 4.8).

As noted, the transition from foraging to agriculture was related to the development of new technology, which gave humans the means by which they became aware of previously unused energy sources and acquired the ability to use them. Thus, for example, the domestication of plants involved humans acquiring control over a natural converter— plants convert solar energy into organic material via photosynthesis. Similarly, animal domestication involved control over another natural converter—animals change one form of chemical energy, usually inedible plants, into another form usable by humans, such as animal protein. In this sense, the domestication of plants and animals involved the use of new energy sources as a result of technological change. By about 2,000 ya, the earth included areas of relatively specialized agriculture, of peasant agriculture, and of nomadic pastoralism, in addition to areas that continued to emphasize foraging (see Figure 4.9).

ENVIRONMENTAL IMPACTS

The environmental consequences of agriculture were considerable, with the greatest impact on those plants and animals that were domesticated. Increasing human dependence on a few species initiated conflict with others. Because an agricultural landscape is an artificial environment, it favors some species at the expense of others, and animal and plant extinction resulted from loss of habitat as agriculture expanded. There were also impacts on soil as land was cleared for seeding, especially in areas of permanent agriculture. Notwithstanding these ecological changes, agricultural groups were tied to

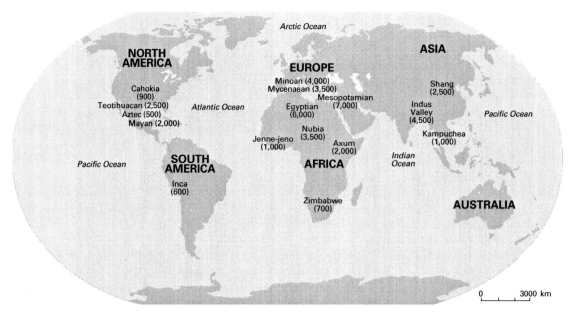

Figure 4.8 Global Distribution of Possible Hearths of Domestication. Although necessarily simplified, this figure provides an approximate chronological and spatial summary of animal and plant domestication. All dates are years before present.

Source: K.L. Feder and M.A. Park. *Human Antiquity: An Introduction to Physical Anthropology and Archaeology*, 3rd edn (Mountain View, CA: Mayfield, 1997):381–2.

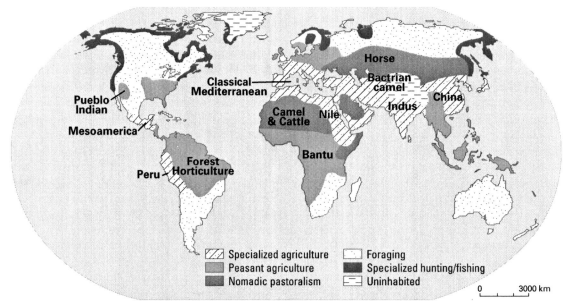

Figure 4.9 Humans and Nature During the Late Holocene. This figure provides a global summary of human ways of life about 2,000 ya. At this time there was a broad range of activities: sedentary specialized agriculture in such areas as the Mediterranean; specialized hunting/fishing in a few areas; peasant, often shifting, agriculture in more humid areas; nomadic pastoralism especially in semi-arid areas; and foraging in other large areas.

Source: N. Roberts, *The Holocene: An Environmental History* (Cambridge: Blackwell, 1989):121.

land, and their survival was dependent on maintaining a productive environment.

At first, agricultural groups grew crops in small areas close to their homes. In some cases, these groups found it necessary to move regularly because of declining yields associated with continuous cultivation of the same land. In other areas, groups were able to settle permanently, either because regular river flooding fertilized land naturally or because products such as human and animal dung and household waste were used to fertilize land. Gradually cultivation increased in many areas, and fire and axes were used to remove vegetation as necessary. Terracing was used on some steep slopes, and irrigation was sometimes used in areas of inadequate rainfall. During the first several thousand years, prior to the rise of a capitalist social and economic system, agriculture was typically a cause of ecological diversification that replaced, for

example, the limited variety of many forest ecosystems with a mosaic of habitats.

CULTURAL IMPACTS

The cultural consequences of agriculture were also considerable. Some areas produced food surpluses, sparking population growth and encouraging increased labor specialization. Population increased in the major agricultural areas more rapidly than at any previous time, and there were about 100 million people by 500 BCE, and about 170 million by 0 CE; growth rates then lessened until the sixteenth century. Agriculture, unlike foraging, requires a degree of permanence, and so a sedentary village lifestyle became increasingly common in some areas. This new permanence, along with food and labor surpluses, initiated a transition from essentially **egalitarian** social groups to situations characterized by social **stratification**. In a foraging society, human activities

Located about 9,000 feet (2,700 m) above sea level in the Peruvian rainforest, the Inca site of Machu Picchu remained unknown to Europeans until 1911, when Hiram Brigham, a Yale University professor, traveled there. He called it the 'Lost City of the Incas'. Built in the 1460s in difficult terrain high above the Urubamba River canyon, Machu Picchu may have served as a royal estate and religious retreat. The site is divided into two parts, one agricultural and one urban. As shown in the illustration, the agricultural part is a series of terraces that enabled corn and potatoes to be grown. Today, Machu Picchu is a fragile site because heavy rains erode the steep slopes. (Wolfgang Kaehler/CORBIS Canada)

varied primarily according to age and sex, and those in the same age and sex categories were essentially equal in terms of both power and wealth. But in agricultural societies these circumstances were dramatically altered. Agriculturalists produced food surpluses, freeing others to be involved in non–food-producing activities. Fixed layers of power and wealth became usual, with a few members of society exercising control over the majority and having access to a disproportionate share of the wealth; in many cases, the privileged few used their position to oblige others to work for them as personal servants or as laborers in construction projects.

It is these circumstances—the acquisition of wealth by some members of society, the rise of cities, and the eventual rise of states and empires—that are components of the **civilizations** that first appeared about 6,000 ya. Any small change—for example, a population increase or technological advance—that disturbed the delicate state of equilibrium of an agricultural group prompted a ripple effect, with one possible outcome being the emergence of a civilization. Agricultural groups in many parts of the world, notably Mesopotamia, the Nile valley, the Indus valley, northern China, the Mediterranean, southern Africa, lowland and highland areas of Central America, and mountain valleys in the Andes experienced these changes (see Figure 4.10).

Pause for Thought

The word civilization *is an uncomfortable one that many scholars studiously avoid as it can be interpreted to imply that 'I' am civilized and 'you' are not. Although, there is good reason to believe that many early 'civilizations', such as those in China, India, Greece, Rome, Persia, and the Middle East, saw themselves in precisely this way, it is important to recognize that the use of the word in this book does not have that implication. The word is used in this textbook because it is the basis for a substantial social science literature discussed especially in Chapter 6. It is understood as referring not to the idea that there is a simple dualism between civilization and other cultural groups but rather to the many complex societies that have risen and fallen during the long span of human history.*

FEUDALISM

The cultural changes brought about by the development of agriculture involved substantial social change, especially the already noted change from egalitarian to stratified societies, which have continued to be the norm through to the present. The earliest stratified societies sometimes incorporated forms of involuntary labor, notably slavery. Many included a ruling group based on noble birth, and in these cases a form of feudalism evolved, with several groups and estates distinguished. In the case of Europe, there was a land-owning aristocracy or gentry with considerable power and wealth; the clergy constituted a separate estate with some distinctive privileges and power; and the third estate (the great majority of the population) comprised serfs, free peasants, artisans, and merchants.

Unequal power, often involving exploitation of a majority by a minority, has been a key feature of most societies since the beginnings of agriculture. Feudalism transparently qualifies for this description, as a proportion of the production of serfs especially was often transferred directly to the aristocracy.

EUROPEAN OVERSEAS EXPANSION

The period between about 6,000 ya and about the fifteenth century involved the rise and decline of numerous civilizations. Several of these expanded to new areas affecting other populations, but none expanded globally. At the beginning of the fifteenth century three civilizations had the technological capacity to initiate large-scale migration : China, the Islamic world, and Europe. Of these, China and Europe were the most densely populated, with each having perhaps 25 per cent of the global population.

Both regions initiated overseas expansion, but it was Europe that capitalized on the early voyages to engage in an extended period of movement. Explanations of this phenomenon are varied, but most rely on such circumstances as a favorable physical geography, a broadly shared culture of Christianity, a fragmented economic and political framework, gradual advances over preceding centuries in agri-

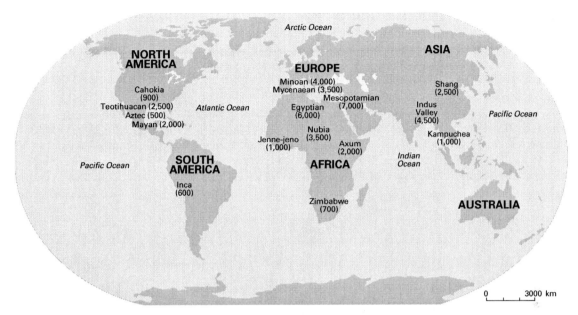

Figure 4.10 Global Distribution of Early Centers of Civilization. Although necessarily simplified, this figure provides an approximate chronological and spatial summary of the rise of civilizations. All dates are years before present.

Source: K.L. Feder and M.A. Park. *Human Antiquity: An Introduction to Physical Anthropology and Archaeology*, 3rd edn (Mountain View, CA: Mayfield, 1997):430 and 484.

culture and trade, and the emergence of **mercantilism**. A number of technological advances, notably papermaking, printing, the mariner's compass, and gunpowder, were also important. Although some of these advances had already been seen elsewhere, it was Europe particularly that used them to help begin a process of global movement and activity that continued until the late nineteenth century. Chapter 5 develops these ideas, and Chapter 7 includes a critical discussion.

Once begun, this process of European overseas expansion and related developments rapidly proceeded to dramatically affect the global population, estimated at about 350 million in the mid-fifteenth century This brought to an end the situation that had prevailed prior to the period of rapid expansion, in which numerous groups of people lived essentially separate, local rather than global lives. Since the fifteenth century, there has been unprecedented biological, cultural, and technological movement related to European expansion and colonization. The first

great European powers to enhance their wealth and status by this means were Spain and Portugal, but France, the Netherlands, and Britain quickly followed suit.

Industry

In addition to expanding throughout much of the world, diffusing Christianity and several European languages, Europe was also the site of a second major revolution (following the agricultural revolution), in the use and development of new energy sources and technological advances. But even before the industrial changes initiated in the eighteenth century, several changes in the practice of agriculture, including new animal and plant domesticates and new tools and techniques, permitted ever-increasing human control of energy sources. Following the onset of agriculture, three new energy converters were invented, namely the water mill, windmill, and sailing craft. But these were evolutionary rather than revolutionary technological advances, and it was not until the

Industrial Revolution that another dramatic set of changes occurred.

The Industrial Revolution involved the large-scale use of new energy sources via inanimate converters. The new energy sources were coal in the second half of the eighteenth century, oil and electricity in the second half of the nineteenth century, and nuclear power in the middle of the twentieth century. The critical initial technological change was the adoption of the steam engine in the late eighteenth century. The first developments occurred in England, spreading first to elsewhere in western Europe, second to the United States, and eventually globally.

ENVIRONMENTAL IMPACTS

The Industrial Revolution was much more than simply a series of changes in energy sources and technology. The use of machines involved a factory system that could support massive increases in output. Manufacturing centers

This photo of the Potteries landscape at Burslem, Staffordshire, appeared in Hoskins's classic book, *The Making of the English Landscape*, together with the following caption: 'This sad picture of an industrial landscape should be examined under a powerful reading-glass. There is the church, rebuilt in 1717, when the pottery manufacture was flourishing with 43 small manufactories, carried on by families until they were superseded in the Industrial Revolution. Next to the churchyard, children play around the Board School. All around, their homes lie intermingled with the filthy potbanks with their characteristic bottle shapes rising above the roof-tops. A blackened Anglican church, to cater for the enlarged Victorian population, stands in the middle of the view. Derelict ground occupies much of the scene. It is a formless landscape, usually thickly blackened with smoke. The photograph was taken on a favourable day or it could not have been taken at all. Imagine being born amid this ugliness: or worse still, buried among it.'

sprouted in locations where key resources were available, sparking the rapid growth of towns with localized areas for working-class residences. Transport links increased, improved, and diversified, with new canals, roads, and the invention of rail. Mechanization contributed to increases in agricultural productivity. Enhanced textile industries created a great demand for cotton and wool, both imported from outside of Europe.

CULTURAL IMPACTS

The Industrial Revolution produced numerous cultural changes, beginning with a rapid decline in the percentage of the total labor force engaged in agriculture. There was a dramatic decrease in death rates followed by decreases in birth rates, with the intervening period being one of rapid population growth. From about 500 million in 1650, the world population increased to 1.6 billion by 1900. In addition to considerable European overseas expansion there was much migration from rural areas to urban centers, creating new sites of high population density in industrial and urban areas. Feudal societies disappeared to be replaced by the social and political system of capitalism, labor was transformed into a commodity to be sold, and there was a related separation of the producer from the means of production. This was also the time during which the modern nation state assumed dominance, with several of these developing into world empires through processes of conquest and subjugation of other peoples and places. Global population has continued to increase rapidly into the twenty-first century, reaching a total of more than 6.4 billion in 2005 (see Figure 4.11).

More generally, building on Enlightenment assumptions concerning the practicality and desirability of scientific knowledge and the quest for social truths, this period witnessed the rise of modernism. The world in which we now live, notwithstanding some notable changes since the end of the Second World War, is very much a product of nineteenth-century industrialization and associated modernity (see Figure 4.12).

A Postmodern World?

According to Marx, societies are stable if there is a balance between economic structures, social relationships, and political systems; they become unstable when tensions or contradictions develop. The transition from feudalism to capitalism involved tensions related to the new manufacturing activities that then caused class conflict—most notably the French Revolution—such that the capitalist system assumed dominance. But, the capitalist system in turn involved tensions, specifically because of the increasing gap between rich capitalists and poor workers that would lead to a further transformation to socialism or communism. But the experience of capitalist societies has not followed the pattern anticipated by Marx, and it does not appear that industrial capitalism is heading in such a direction. Rather, current evidence points to a transition to a postmodern society and economy, also

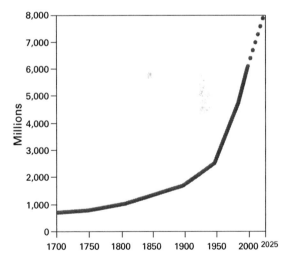

Figure 4.11 Human Population Growth, 1700–2050. Global population growth since 1700 relates to: (a) a decline in death rates followed by a decline in birth rates in what has become the more developed world, with the interval between these two declines being thus one of high rates of natural increase; and (b) a later decline in death rates followed by a decline in birth rates that is only now occurring in what has become the less developed world, with the still continuing interval between these two declines being a period of high rates of natural increase. As of 2005, a reasonable prediction for the year 2050 is 9.3 billion people.

Figure 4.12 Human-and-Nature Relationships Through Time. Different social and economic systems involve different human-and-nature relationships. Foragers were directly dependent on wild animals and plants, implying an intimate relationship. With domestication of animals and plants, humans began to assume some control over nature and to dramatically and permanently alter nature. The two remained integrated, but the bonds were weakened. It was with the onset of industry that humans and nature became separated. H refers to humans and N to nature.

Source: N. Roberts, *The Holocene: An Environmental History* (Cambridge: Blackwell, 1989):184.

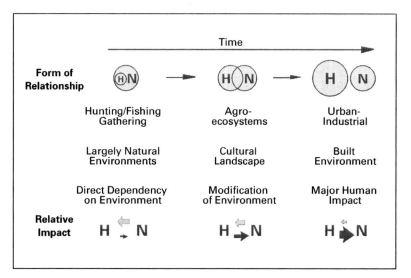

variously labeled the postindustrial, information, knowledge, or service society.

The rise of modernism is linked to the scientific revolution, the Enlightenment, and the Industrial Revolution, with their shared emphasis on the practicality and desirability of scientific knowledge. Indeed, the rise of an industrial way of life can be interpreted as a transition from tradition to modernity, or from *Gemeinschaft* to *Gesellschaft*. Modernism, the modern period, is based on the idea that the world is knowable and that there is a need to pursue some presumed knowable truth. Modernism assumes that social reality is logical, subject to laws waiting to be discovered. Explaining the human world will eventually lead to further technological advance, personal liberty, social equality, and the eradication of such things as poverty and oppression.

Certainly the contemporary world continues to show the effects of the dramatic changes initiated during the past 200 years, and modernity continues to be the goal for many countries and groups, especially in the less developed world. However, in the Western world there are reasons to suggest that a new set of processes is in evidence, namely postmodern forces that first began to appear after the Second World War and that gradually accelerated through the final decades of the twentieth century. This concept of

the postmodern is an especially difficult one, at least partly because it is always difficult to clearly appraise current circumstances.

Essentially, critics have suggested that contemporary societies are moving beyond industrial development—the period of modernism—to a new social form emphasizing the use of information or knowledge rather than machines and industrial energy sources. Employment rates in the service sector are increasing at the expense of those in the manufacturing sector. But there is much more than this social change associated with the concept of the postmodern. Rather, as the term implies, the postmodern is contrasted with the modern. Postmodernism rejects the grand claims associated with modernism because of a loss of faith in the modern world.

There are three postmodern claims worth noting here. First, there is no possibility of achieving objective, value-free knowledge. Second, knowledge is associated with power, and existing social institutions are not in any sense natural or right, but rather have been created to ensure the maintenance of power. Third, the self is fluid, continuously reinvented in discourses and invented anew in life; this postmodern claim replaces the modern Western conception of self as unified and given. Given these claims, postmodernism rejects the grand attempts at explanation involved in science and Marxism. These postmodern themes are particularly evident in the academic world, where they have led to a blurring of traditional—that is, modern—disciplinary boundaries and the creation of new interdisciplinary areas. This circumstance is clearly evident in the cultural geography reflected in this textbook.

The State of the Earth

Perhaps the most significant contemporary applications of ecology are those concerned with the global environment. Although human impacts on environment are studied in many disciplines, both physical and human, geography plays an important, often integrating, role. Within geography, contributions from cultural geographers are increasing in both number and significance. Particularly significant are a number of thoughtful proposals from Wagner, who has long demonstrated a concern for the study of human impacts both from a scholarly and from a practical and concerned perspective. In the 1960 volume *The Human Use of the Earth*, Wagner aimed to show how humans were continually remaking landscapes (Wagner 1960), but by 1972, in *Environments and Peoples*, Wagner realized that 'the cultural propensities of mankind in themselves tend to bring about a kind of "dislocation" of communities of men in respect of their habitats. That, I think, is the kernel of the modern ecological dilemma' (Wagner 1972:x).

Notwithstanding the recent rise of an ecocentric worldview, much human use of the earth continues to reflect an anthropocentric view. Briefly, our use of the earth involves pollution of land, water, and air, changes to global climate, loss of species, loss of land, loss of wilderness, loss of cultural diversity, and dramatically uneven patterns of consumption and quality of life. In this sense, many observers judge that there is a crisis of environment, with environment including humans and nature—effectively a crisis of contemporary culture. Discussions of the specific environmental issues are commonplace in both academic and other circles, and it is not appropriate to include any detailed account in this textbook.

There are many sources of information on the global environment that are geographic in character. Three important volumes merit mention. First, as noted earlier, *Man's Role in Changing the Face of the Earth* (Thomas, Sauer, Bates, and Mumford 1956) provides a substantive account of human impacts, but this is necessarily dated both conceptually and factually. Second, *The Earth as Transformed by Human Action: Global and Regional Changes in the Biosphere in the Past 300 Years* (Turner et al. 1990), a more recent volume in a similar vein, focuses on human impacts globally during the period since the onset of industrialization. This work discusses: changes in population, technology, culture, urbanization, and human perception of these and other changes; accounts of impacts on environment both generally and in specific regional studies, including such critical issues as **acid rain** and the human-induced **greenhouse effect**; and also thoughts about humans and nature and **maladaptation**, gender issues, and the relevance of ecological approaches. Third, *Regions at Risk: Comparisons of Threatened Environments*, by Kasperson, Kasperson, and Turner (1995), identifies causes and consequences of human-induced environmental change as unevenly distributed over the surface of the earth and locates critical environmental regions. Understanding critical regions requires acknowledgment of the following factors:

- Criticality is interactive, relating to the type and rate of environmental change, the characteristics of the ecosystem, and the culture in the area.
- Criticality needs to be studied in both temporal and spatial contexts.
- Criticality arises if the relations between culture and environment involve a situation susceptible to changes in economy or environment.
- Criticality refers to a reduction in environmental quality as this relates to human occupance.
- Future situations of criticality are often difficult to predict.
- Substantial human impacts usually result in economic gains.
- Current human activities have different degrees of sustainability, and there is usually a range of other options available.

• There are driving forces such as population growth, technological capacity, affluence, poverty, state policy, character of economy, and beliefs and attitudes.

In these and other works, cultural geographers contribute both conceptually and empirically to discussions of the state of the earth and especially to two important and unresolved debates. First, there is the debate between limits-to-growth theorists and cornucopians. Second, there is the debate between those who emphasize essentially environmental causes of problems and those who subscribe to a political ecological logic. With reference to such debates, Smil (1987:341) noted that the 'most fundamental difference is the initial mind-set'.

Pause for Thought

One striking fact that comes to light following any review of humans on earth is that it took an enormous length of time for major cultural and technological developments to occur. It is generally accepted that early humans were genetically very similar to humans today, with similar brains, emotions, hands, eyes, and so forth, but it has been only in about the past 10,000 years that major developments have occurred. Why? One answer is climate change: the Holocene saw the onset of significant global warming after more than 100,000 years of cooler temperatures. This warmer climate may have encouraged long-distance migrations, the domestication of plants and animals, and the related rise of urban centers.

Concluding Comments

'Human ecology means different things to scholars in different disciplines' (Porter 1987:414). Although, as this remark suggests, it is evident that there is no consensus in the social sciences as to what constitutes an ecological approach, it is equally evident that much useful research has been conducted under the general umbrella of ecology. Conceptual plurality is accompanied by substantial and quite varied research achievements.

Ecology is one means of thinking about human–human and human–nature relations. Accordingly, it is not central to any one social science, but rather it is of interest to geographers, sociologists, anthropologists, psychologists, historians, and others. Geography might appear to be the most natural home of ecology because of the physical and human content and the long-standing idea that the discipline can serve as some sort of bridge between physical and human sciences, but this has not proven to be the case, and the geographic use of the ecological approach has been erratic. There was limited acceptance of the pioneering call from Barrows, while later attempts to define the ecosystem as a geographical model have not won many converts. Stoddart (1967:538) proposed the idea of a geosystem as a replacement for ecosystem, arguing that systems analysis 'at last provides geography with a unifying methodology, and using it, geography no longer stands apart from the mainstream of scientific progress'—an appeal that has gone largely unanswered. Such a failure may reflect the fact that geographers do not seek a unifying methodology, especially given the diverse subject matter of the discipline. Perhaps the most successful uses of ecology in geography are the current linkage between ecology and environmental concerns, the development of a political ecological focus, and the uses of ecology in analyses of population change as it relates to technology.

Geographers have adopted and applied ecology as developed in other disciplines, especially in sociology and anthropology. For example, the Chicago school of sociology was a major inspiration for urban geography in the 1960s, despite such limitations as the omissions of a cross-cultural perspective and any reference to symbolism. In some cases, concentric patterns of urban social areas were identified and such processes as invasion and succession noted. Cultural geographers have also turned to anthropological cultural ecological analyses, especially to the adaptation concept developed in that discipline. More recently, geographers have been attracted to the new ecology in biology and to political economy as developed in politics and economics especially. In addition, in a discussion of Australian prehistoric cultural landscapes, Head (1993, 2000) employed some of the ideas associated with the new cultural geography in a study of landscape as a symbolic expression.

Ecological analyses by geographers have a rich and complex heritage. Criticisms of that heritage and of the many early studies are inevitable given recent advances in social theory, especially feminist concerns, but much current evidence strongly suggests that ecology is an increasingly popular and even more diverse tradition today than previously. There appears to be every indication that cultural geographers will continue their fascination with this approach and that conceptual revisions and additions will be accompanied by increasing numbers of empirical analyses. And it seems just as likely that ecological analyses, in all their variety, will continue to inform us about ourselves and the world in which we live.

Further Reading

The following are useful sources for further reading on specific issues.

The traditional sociological view was that geography was principally concerned with factual information about humans and their physical environments rather than with explanations of human distributions. Thus, Quinn (1950:339) suggested that geographers were concerned with direct human-and-land relations whereas ecologists were concerned with interrelationships. Hawley (1950:72) identified a similar distinction, but also stressed that geography, specifically regional geography, was atemporal while human ecology was not. Schnore (1961) conveyed a more positive view in an enlightened interpretation of then current human geography in ecological terms—work in economic, urban, and population geography was cited for ecological content.

In sociology, some initial human ecological ideas were outlined by Park (1915), with the first textbook account authored by Park and Burgess (1921).

The geographer Entrikin (1980) discusses human ecology in early sociology.

Hawley (1950, 1968) built on the human ecological work of Park to propose closer links with biology and a de-emphasis of humans, while Firey (1945, 1947) argued, with specific reference to Boston, for the importance of symbolic variables rather than physical science-based human ecological variables in any account of urban areas.

Kroeber (1928) notes a neglect of ecological considerations in then current ethnological studies, but chose not to pursue the theme any further. Steward (1936, 1938) relates group characteristics to ecological circumstances. Bennett (1976:48) and Vayda and Rappaport (1968:492) outline criticisms of the cultural ecology proposed by Steward.

The plethora of ecological approaches in anthropology prompted Meggers (1954) to discuss the limits set on cultural evolution by environments, even to the extent of arguing for cases of determinism, and resulted in Netting (1977) suggesting that any theoretical or methodological overview of the approaches was not practicable.

Headland (1997) proposes that anthropology adopt a historical ecology and identifies a number of myths that such an approach could address.

Reviews of and proposals for ecological history are provided by Worster (1988, 1990), Merchant (1990), White (1990), and Cronon (1992). Some examples of ecological studies by historians include those by Cronon (1983) on environmental changes caused by different cultural groups in New England, by Crosby (1986) on the biological expansion of Europe, by Merchant (1989) on changing ideas of nature, gender, and science in New England, and by Arnold (1996) on environment, culture, and European expansion. A succinct review of the American West environment is by Jacobs (1997). Somewhat separately, explanations of the rise of the Western world also focus on the relevance of physical geography as it relates to human activities (Landes 1998, Diamond 1997).

For a historical geographer's perspective on environmental history, see Baker (2003). Articles on the study of past environments by historical geographers and environmental historians are included in a Special Issue of the *Geographical Review* (April 1999).

Crosby (1986) and Griffiths (1997) stress links between

use of environment and political power in an historical context.

Crosby (1995:1186) chooses 'for the sake of convenience' to identify the work of Sauer and the 1955 symposium, which produced the seminal volume *Man's Role in Changing the Face of the Earth* (Thomas, Sauer, Bates, and Mumford 1956), as the scientific debut of the ecological approach in history.

The ecological psychology outlined by Lewin is more fully developed into a substantive research area by Barker (1968), with the term 'behavior settings' used to describe the larger context within which behavior occurs. The relative significance of the individual and the environment to an understanding of behavior is related to the character of the environment, with the individual being more important if the environment is relatively stable and the environment being more important if it is varied and changing. A related advance is that of environmental psychology (Ittelson, Proshansky, Rivlin, and Winkel 1974).

Boulding (1950) outlines some links between economics and ecology.

Stoddart (1965, 1966) discusses geography and ecology in general terms, and Martin (1987b) details the ecological tradition in geography.

Schnore (1961:209) describes the article by Barrows as 'little more than a piece of intellectual history', while Koelsch (1969) notes Barrows's impact on historical geography.

Sauer's interest in human ecology is noted by Leighly (1987).

Clarkson (1970) and Grossman (1977) outline geographic versions of human ecology.

Chapman (1977), Huggett (1980), and Wilson (1981) provide broad-based geographic accounts of the systems approach. The most substantial applications of a systems approach by geographers are studies in the natural hazards tradition (see Kates 1971), while there are also studies of subsistence (Nietschmann 1973) and of civilizations as ecosystems (Butzer 1980).

Using a cultural ecology framework, the anthropologist Geertz (1963) studies land use change in Indonesia in generic systems terms, while Rappaport (1963) analyzes human involvement in island ecosystems. A key assertion evident in much of the geographic and anthropological

work is that human populations can be treated in much the same way as any other population. Rappaport (1963:70) notes that the 'study of man, the culture bearer, cannot be separated from the study of man, a species among other species'.

Concerning terminology employed in adaptation studies, Goldschmidt (1965:402) notes: 'Our investigation is a study in cultural adaptation, in ecological analysis, in the character of economic influence on culture and behavior in social micro-evolution—depending upon which of the currently fashionable terminologies one prefers.'

Berry (1984) outlines ecocultural psychology; Berry (1997) uses the approach in a study of immigration and subsequent acculturation; Bennett (1993) considers adaptation and social scale in psychology; Cronk, Chagnon, and Irons (2000) review adaptation in anthropology. Porter (1965) is an early statement on adaptation and cultural geography.

Some ecological analyses employ the term 'cultural landscape' to refer to altered nature (Rowntree 1996:129)—a not uncommon use of the term in British and European literature (see also Birks et al. 1988; Simmons 1988; Svobodová 1990; Dieterich and van der Straaten 2004). A larger context for what is sometimes called historical ecology is contained in Crumley (1994).

Hawley (1998) comments on developments in human ecology.

There are many geographic and other commentaries on nature as socially constructed and on the related idea of hybridity; these include Gerber (1997), Coates (1998), Proctor (1998), Whatmore (1999, 2002), Eden (2001), Franklin (2002), Huckle (2002), and Castree (2003). Von Maltzahn (1994) focuses on the social construction of nature from a humanistic perspective; Evernden (1992) outlines the need to rethink our culture of nature by refocusing on the wildness that is inherent in nature; Eder (1996) considers the need to understand why humans continue to abuse the environment, proposing that the real culprit is our continuing taken-for-granted idea that nature is there to be dominated; Oelschlaeger (1991) focuses on the cultural idea of wilderness from a historical perspective, beginning with the experience of preagricultural societies to show that there has long been an idea of wilderness as other. Wilson (1999) discusses the place of nature in the context of La Sierra, Colorado. On a rather different theme, Salerno (2003) discusses human estrangement from nature.

Current geographic interests in ecology are outlined by Fitzsimmons (2004) and Robbins (2004).

Lovelock (1982) details the Gaia concept.

Wolch and Emel (1995) consider the claim that social theories are becoming increasingly anthropocentric.

Murdoch (1997) outlines actor network theory.

Wolch, Emel, and Wilbert (2003) describe animal geographies.

Botkin (1990) notes some psychological difficulties involved in the idea that nature is constantly changing.

The origins of Western environmentalism are discussed by Grove (1992); Paehlke (1995) provides an encyclopedia of conservation and environmentalism; de Steiguer (1997) discusses the age of environmentalism; Inglehart (1990) identifies changing attitudes to environment. The life and work of George Perkins Marsh is covered by Lowenthal (2000).

According to Plummer (1993), nature, women, and other subordinated groups have been manipulated as a result especially of the two processes of backgrounding (that is, denying value) and instrumentalism (that is, service without recognition).

Shiva (1988), Buckingham-Hatfield (2000), Warren (2000), and Domosh and Seager (2001) examine women and environment.

The Chipko movement is discussed by Dwivedi (1990), Merchant (1995), Mellor (1997), Rose, Kinnaird, Morris, and Nash (1997), and Warren (1997).

Dickens (1996) notes links between the division of labor and our understanding of nature.

Wisner (1978) developed an early argument for a socialist human ecology (see also Bennett 1976; Ellen 1982).

Baker (1997) identifies the Landcare movement, a community-based approach to the land degradation crisis in Australia that has the potential to tackle larger issues of ecological sustainability.

Kjekshus (1977), Porter (1979), Watts (1983), and Grossman (1984) are early studies in the political ecology tradition. More recent studies include those by Zimmerer (1996b) on peasant life in the Peruvian Andes, Zimmerer and Young (1998) on the complex character of environmental change in the less developed world, and Simmons (2004) on land conflict in the eastern Brazilian Amazon. Also see Batterbury (2001), Bebbington and Batterbury (2001), Zimmerer and Bassett (2003), Forsyth (2003), Paulson and Gezon (2005), and the various articles in a Special Issue of *Economic Geography* (Vol. 69, no. 4, 1993).

Fairhead and Peach (1996:10–15), Mayer (1996), Bryant (1997, 2001), Keil, Bell, Penz, and Fawcett (1998), Low and Gleeson (1998), and O'Connor (1998) provide accounts of the larger intellectual context supporting political ecology. Walker (2005) provides a critical review.

With reference to the rural American West, Walker (2003) discusses a political ecological approach to environmental conflicts in advanced capitalist societies.

Kaplan (1996, 2000) offers interesting global arguments for adopting a limits-to-growth perspective.

Simon and Kahn (1984) present a forceful argument in favor of the cornucopian thesis, and Lomborg (2001) gives a thoughtful and provocative global overview.

Des Jardins (2001) provides an introduction to environmental philosophy. Smith (2004) discusses ethics and the human environment.

Links between the Christian tradition and environmental abuse are noted by White (1967); Gore (1992) considers problems with this argument; Kinsley (1994) reflects on aspects of both points of view. Some accounts of religion and environment are Spring and Spring (1974), Gottlieb (1996), Chapple and Tucker (2000), Redekop (2000), and Jamieson (2001). Hunter and Toney (2005) examine the distinctiveness of contemporary Mormon environmental perspectives as contrasted with the larger United States population.

Geographic discussions about humans and animals include Anderson (2000) and Philo and Wilbert (2000), while Singer (1993) and Jamieson (1985, 1997) raise philosophical questions relating to our human treatment of animals, including animals in zoos.

Diesendorf and Hamilton (1997) outline environmental economics.

Tenbrunsel, Wade-Benzoni, Messick, and Bazerman (1997:2) note discrepancies between environmental attitudes and behaviors.

Botkin (1990) discusses the concept of living with discordant harmonies.

Carneiro (1960) looks at the concept of carrying capacity. There are numerous studies, especially of agricultural areas in the less developed world, that employ this concept (see, for example, Bernard and Thom 1981).

There are many recent publications that focus on humanity and environment at a global scale and over an extended time period. These include Butlin and Roberts (1995), Southwick (1996), Fernandez-Armesto (2001), Hughes (2001), Mithen (2003), Richards (2003), Simmons (2003), and Cook (2004).

Flannery (2001) provides an ecological history of North America.

Atkins, Simmons, and Roberts (1998) examine the history of relations between landscape, culture, and environment.

Sahlins (1974) addresses the argument that foragers are the original affluent society.

Feder and Park (1997:384-426) summarize and evaluate eight proposed explanations for the origin of agriculture, concluding that five of the proposals are complementary, not competing, and together provide a plausible explanation.

Smith (1995), Harris (1996), Anderson (1997), and Mathewson (2000) look at the agricultural revolution

Feder and Park (1997:438-43) identify and evaluate seven proposals for the rise of civilization, concluding that no one explanation fits all cases and that myriad factors were involved (see also Scarre and Fagan 1997:1-20).

Mungall and McLaren (1990) and Simmons (1997) are two of many studies that detail human impacts on environment.

5

Landscape Evolution

Looking back on a long and distinguished career in human geography, Morrill (2002:34) observed: 'I define geography as the study of the evolution of landscape—the physical and human forces that shape the earth's surface.' Turning his focus to the human forces, Morrill (2002:34) continued: 'The elements of study of geography as a social science are the *behavior* of persons, households, groups, institutions, and social systems that define and change the landscape. Geography as a social science tries to explain this behavior and analyze processes of change. . . .' These statements about human geography are generally in close accord with the principles of the cultural geography subdiscipline introduced by Sauer. Although Sauer's early methodological statements emphasized the behavior of cultures as the cause of landscape change, much subsequent empirical work used the other social scales noted by Morrill.

One challenge laid down by Sauer was to study the evolution of cultural landscapes. This was a challenge because the orthodox view of geography in the 1920s was that it was an atemporal regional discipline; any historical geographic studies were limited to the study of regions in past times. What Sauer proposed was, within the North American tradition, unorthodox because the study of change was a challenge to the hegemony of regional geography. But the response to his proposal was both appreciative and considerable, resulting not only in the rise of a cultural geography studying landscape evolution but also in the rise of a relatively distinct historical geography separate from the tradition

of studying the geography of past times. That this historical geography of changing landscapes was accepted into mainstream North American geography was evidenced by the inclusion of a chapter on historical geography in the seminal 1954 volume *American Geography: Inventory and Prospect*. Written by one of Sauer's students, this chapter spells out a historical geography concerned with both changing landscapes and past times (Clark 1954). Thus, in addition to studies of cultural landscape evolution, it is appropriate to see Sauer as a leading figure in the rise of a more general historical geography.

This chapter discusses studies of cultural landscape evolution as well as related work in historical geography. The idea that informs most of this work is the recognition that humans, often through their cultural identities, transform physical landscapes through time to create human landscapes. Thus, it is usual to identify close links between the characteristics of the culture and the landscapes resulting from cultural occupance. This chapter contains

- a discussion of the landscape consequences of cultural change as these relate to processes of diffusion, with reference to three distinct approaches, concerned respectively with culture traits, spatial processes, and political economy;
- a discussion of the landscape consequences of cultural change as these relate to contact between different groups, with emphasis on the movement of Europeans overseas and resultant encounters between Aboriginal and European cultures. There is also reference to

our changing understandings of the character of these encounters;

- a review of the tradition of historical geography, with reference to a number of approaches and to examples of empirical work and with emphasis on the shaping or making of landscape. This review examines approaches to the analysis of change through time, frontier studies, evolutionary regional landscape studies closely associated with the Sauer tradition, an essentially British local-history approach to the study of landscape change, and the tradition of reading landscape as exemplified in much of the work published in the magazine *Landscape*;
- an account of *imagined landscapes*, subjective environments or images that must be considered in almost any discussion of landscape evolution as they may be significantly different to the objective environment. The three regional examples detailed are those of colonial western Australia, the American Great Plains, and prairie Canada.

Throughout there are close links with the discussions in Chapter 6 of cultural regions, which can be considered as dynamic outcomes of evolutionary processes. Several evolutionary regional concepts are also discussed in Chapter 6.

Cultural Diffusion

Given Carter's (1978:56) claim that 'In the broadest sense diffusion is the master process in human culture change,' it is not surprising that diffusion studies have been a central feature of work in cultural geography since the beginnings of the landscape school. It is notable, however, that the basic thrust of these studies has undergone two significant changes, with the result that it is appropriate to identify three principal approaches in geographic diffusion research. Interestingly, these three approaches mirror the larger twentieth-century history of human geography. The first, the traditional approach, which focuses on the spread of a specific **cultural trait**, is identified with the landscape school. The second, the spatial analytic approach, which is designed to uncover empirical regularities in the diffusion

process, evolved from the traditional approach in the 1960s, but is best seen as a part of the theoretical and quantitative movement in human geography at that time. The third, the political economy approach, which is concerned with links between diffusion, culture, and power, began as a reaction against the perceived dehumanization of the spatial analytic approach. Regardless of the specific approach, in most studies the concern is with the diffusion of an **innovation**.

The Spread of Culture Traits

The traditional approach to studying cultural diffusion has close ties with early twentieth-century anthropology, especially with the Boasian school, and is an explicit component of the landscape school of cultural geography. Accordingly, the concern is with diffusion that is largely controlled by culture but that does not exclude physical environment. For the landscape school, culture history is the core of human geography, and diffusion is the basis for understanding cultural origins, cultural landscape evolution, and the creation of cultural regions. Not surprisingly, then, there is concern with those elements of material culture that are visible in landscape. The pioneering exponent of this type of study was Kniffen, who employed this focus largely to demonstrate certain generalizations involving questions of regionalization and diffusion: 'The material forms constituting the landscape are the geographer's basic lore. The cultural geographer deals primarily with the occupance pattern, the marks of man's living on the land. He finds his data, his evidence, in buildings, fields, towns, communication systems and concomitant features' (Kniffen 1974:254). Most analyses in this tradition are empiricist and inductive, and focus on a single culture trait—agricultural fairs, covered bridges, and house types, for example—identifying its origins, its routes, and its present distributions.

THE EXAMPLE OF COVERED BRIDGES

Culture trait studies were justified on the grounds that an understanding of the origins and diffusion of a particular trait aids understanding of spatial variations in cultural landscapes. A study of covered bridges in the United States stressed the feature's European antecedents, with covered wooden bridges especially evident in the Rhine valley region of Switzerland. The original reason for covering is uncertain, although it does provide shelter and a location for market stalls, offers defensive advantages, and significantly extends the longevity of a wooden bridge. Once covering became common, the practice probably continued for reasons of tradition. In North America, covered bridges became popular after about 1810, with the area from New England to New Jersey proposed as the **cultural hearth**. Subsequent diffusion of this landscape feature is illustrated in Figure 5.1.

Kniffen explained the diffusion of covered bridges first in terms of physical geographic considerations, specifically the availability of suitable streams and the necessary timber for building, second in terms of cultural considerations, with covered bridges seen as just one component of the larger New England culture, and third in terms of technological considerations, with new iron structures appearing after about 1860. This analysis of covered bridges included several generalizations about cultural differences in North America: for example, Kniffen suggested that within Canada, Quebec was more receptive than Ontario to cultural innovations from the United States.

Linking Trait Studies to Culture

Culture trait research typically focused on describing and mapping the diffusion of a single landscape feature, with emphasis on particular cultural identities and descriptions of related impacts on landscape. But although these analyses tended to study single traits, they often viewed such traits as representative of the larger culture, a tendency that is most apparent in studies of agricultural diffusion.

Of the nineteen covered bridges built in Madison County, Iowa, only five remain. Built in 1880, Holliwell Bridge is the longest covered bridge in Iowa at 122 feet and remains in its original site over the Middle River southeast of Winterset. Renovated in 1995 at a cost of $225,000, the bridge is one of several featured in the movie, *The Bridges of Madison County*. The movie was adapted from the best seller by Robert James Waller and tells the story of a four-day affair between an Iowa housewife and a traveling National Geographic photographer who is in Madison County to photograph the picturesque covered bridges. Today the covered bridges of Madison County are a popular tourist attraction. (William Norton)

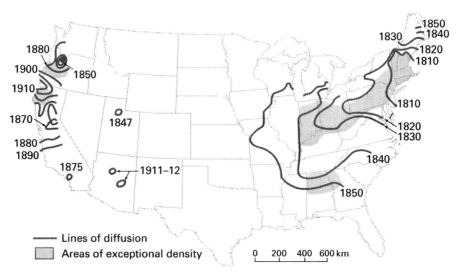

Figure 5.1 Diffusion of the Covered Road Bridge in the United States. This is an example of the type of diffusion map produced by researchers working in the landscape school tradition. Areas of high density are marked, and there are isolines indicating the date of acceptance of the landscape feature. In explaining this map, Kniffen (1951b:123) stated: 'The covered wooden bridge tended to spread freely among those possessing kindred patterns of thought and action.' More specifically, the feature spread unevenly west with migrants reaching southeastern Ohio by 1820, southern Indiana by 1840, and attaining a maximum expansion in the eastern region by 1850. After this date, the number of covered bridges within the eastern region increased substantially and includes many examples in the Canadian province of New Brunswick. The principal area in the western United States extends from southwestern Washington to central California, a largely humid and wooded area west of the Sierra Nevada Mountains. The isolated instances of covered bridges in Utah and Arizona are seen as special cases, and not as parts of larger regional patterns.

Source: F. Kniffen, 'The American Covered Bridge', *Geographical Review* 41 (1951):119. (© The American Geographical Society; reprinted with permission)

For example, an analysis of the diffusion of cigar tobacco production in the northern United States, published in 1973, demonstrated links between the diffusion of this cultural trait and certain cultural—and, in particular, religious—identities. The author of the study, Raitz, found a twofold significance of cultural identity: first, he uncovered a relationship between culture (notably religion) and the interest in cultivating a crop that is labor-intensive; second, Raitz established that networks of local groups promoted the cultivation of cigar tobacco by spreading necessary knowledge. The diffusion pattern is shown in Figure 5.2. The introduction of cultivation was associated in Pennsylvania with Germans, especially with Old Order Amish; in the Connecticut valley with Polish Catholics; in Ohio with German Lutherans, Baptists, and Old Order Brethren; and in Wisconsin and Minnesota with Norwegian Lutherans. In all major

production areas today, there continues to be an association between cigar tobacco production and a specific cultural group.

A second example of this development in culture trait research involves an ambitious study by Sauer (1952), *Agricultural Origins and Dispersals*, in which he expanded earlier local studies of plant diffusion in Mexico into a thesis about agricultural origins and diffusion in a global context. Central to this argument was his claim that cultural changes such as the development of agriculture occurred in areas that lacked environmental stress. In other words, according to Sauer, necessity did not prompt invention; rather, it was a sedentary lifestyle and food surpluses that provided the necessary leisure time for such activities as agricultural experimentation. It was with this argument in mind that Sauer proposed two areas of agricultural origin—Southeast Asia and the lands around the

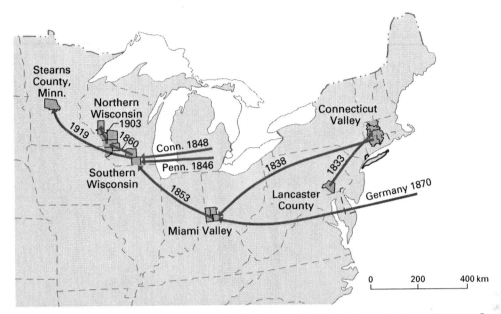

Figure 5.2 Diffusion of Cigar Tobacco Production in the Northern United States. Beginning in southeastern Pennsylvania, cultivation diffused to the Connecticut valley of Massachusetts and Connecticut in 1833; a single migrant then diffused cultivation from the Connecticut valley to southwestern Ohio in 1838, with German immigrants later reinforcing the tradition. Diffusion to southern Wisconsin occurred independently between 1846 and 1843 from three sources, namely Pennsylvania, Connecticut, and Ohio (an earlier 1838 diffusion, not shown on the map, occurred from a tobacco area in New York). Finally, there were later diffusions west and north from southern Wisconsin towards northern Wisconsin in 1860 and in 1903, and also to central Minnesota in 1919.

Source: Adapted from K.B. Raitz, 'Ethnicity and the Diffusion and Distribution of Cigar Tobacco Production in Wisconsin and Ohio', *Tijdschrifte voor Economische en Sociale Geografie* 64 (1973):296.

Caribbean, both areas offering food from water and land. Sauer postulated that vegetative planting cultures developed in and diffused outwards from these hearths. Although most subsequent work on the origin of agriculture does not provide compelling support for these ideas, they remain as a stimulating contribution to an important debate, and the 1952 book is widely acknowledged as a classic.

MIGRATION AND DIFFUSION

Cultural geographers have long been interested in the movement of cultural groups and how the cultural diffusion associated with this movement affects the related landscape. The distinction between migration (the movement of people) and diffusion (the movement of a culture trait that may or may not involve the movement of people) is rarely made in these geographic analyses, a reflection of the fact that the central interest is

the effects rather than the process of diffusion. Salter (1971a:3–4) summarized the links between migration and diffusion thus: 'The cultural geographer views man's mobility with a tripartite perspective: the catalyst for movement, the effect of movement on trait or people in motion, and the consequences of such movement.'

A study of Amish populations conducted in the 1970s addressed aspects of all three of these concerns (Crowley 1978). The Amish are members of a strict Protestant Anabaptist sect that, led by Jakob Amman, broke from the Mennonite Church in the 1690s. Amish populations migrated throughout much of western and central Europe by the early nineteenth century, largely as a consequence of religious persecution. About 500 Amish moved to North America between 1717 and 1750 with a further 1,500 moving between 1817 and 1861. The first group settled in southeastern Pennsylvania, while members of the second group settled in Ohio, Illinois, Iowa, and Ontario. From these early North American locations, Amish settlements have diffused throughout a much larger area of North America and also into Latin America, often moving long distances, often settling on the frontier, sometimes experiencing failures, and further subdividing into different groups. Despite this diffusion process, most settlements continue to be in the original hearth areas; thus, of 98 Old Order Amish colonies in the United States in 1976, 29 were in Pennsylvania, eleven in Indiana, ten in Missouri, nine in Ohio, and nine in Wisconsin. This type of diffusion process is especially difficult to explain in general terms because many of the movement decisions reflected particular circumstances.

A Spatial Analytic Emphasis

The second approach to studying cultural diffusion is the spatial analytic approach. Although this method, as one component of the larger theoretical and quantitative movement in geography, was not popular until the 1960s, its seeds were sown much earlier in the pioneering work of the Swedish geographer Hägerstrand.

THE WORK OF HÄGERSTRAND

Hägerstrand studied the diffusion of new agricultural practices—innovations—by mapping spatial patterns through time and trying to understand the process of diffusion that was responsible for the changing patterns. In doing this, he introduced some original concepts and procedures of analysis. Moving beyond the primarily descriptive analyses being undertaken in the landscape school tradition, Hägerstrand developed a series of 'Monte Carlo' **simulation** models. He chose the term 'Monte Carlo' to acknowledge the fact that

these models incorporate an element of chance. The basic model, which he derived from empirical work, assumed that the probability of an innovation being adopted in a particular location is related to distance from the location of first adoption. Hägerstrand thus described a probabil-

The diffusion of both people and culture traits has made a significant and lasting contribution to the visible and material landscape of much of North America. Although there are numerous regional variations in the landscapes of Amish settlement, members of the group typically live without many of the appurtenances of modern technology, including electricity, running water, telephone service, and automobiles. Reflecting their philosophy of humility, Amish people favor simple, conservative clothing. They grow most of what they need to eat, and harvest crops using horse-drawn equipment. As shown in this photograph, the horse-and-buggy is the primary source of transportation, sharing country roads, main highways, suburban streets, and parking lots with the more ubiquitous automobile. (Robert Grim/Photo Network Stock)

ity surface—essentially a map showing the probability of the occurrence of adoption as one of decreasing probabilities with increasing distance from the site of first adoption—and then simulated the spatial diffusion process. The procedures for conducting a Hägerstrand-type simulation are straightforward:

- The diffusion of an innovation, such as a new agricultural practice, and the changing of the cultural landscape that results are explicitly interpreted as influenced both by the prevailing pattern of communication in a region and by chance.
- A probability surface is constructed on the basis of assumptions about the process of communication flow. The simplest kind of probability surface assumes that information about an innovation is passed from one person to another. It is a probability surface precisely because chance is incorporated in the process.
- **Surrogate** data, such as information on telephone calls, are used to calculate specific probabilities. The probability surface is a distance–decay surface, meaning that it shows a decreasing probability of diffusion with increasing distance.
- Once probabilities are mapped, usually in a square grid format, the simulation process is conducted and a surface of projected adoption dates is created.
- The result of the simulation is compared to the known real-world pattern of diffusion, and a close correspondence between the two is interpreted as an indication that the probability model—necessarily a simplification of reality—is a reasonable approximation of the diffusion process.

Spatial diffusion analyses typically identified empirical regularities of the diffusion process; three examples are noted in Figure 5.3, namely the neighborhood effect, the hierarchical effect, and the s-shaped curve.

There were close links between this approach and work in rural sociology, from which geographers borrowed two ideas. The first was the recognition that adopting an innovation is not immediate upon receipt of the relevant information; rather, there is an adoption process that proceeds through stages—awareness, interest, evaluation, trial, and, finally, adoption. The second idea borrowed from sociology was the recognition that individuals varied in terms of innovativeness, that is, in their willingness to adopt an innovation. The standard assumption is that individual innovativeness follows a normal statistical distribution, with a few innovators who are often opinion leaders accepting the innovation quickly, a few laggards resisting the innovation for some length of time, and the vast majority of the population placed between these two extremes.

SPACE NOT CULTURE?

Studies employing the methods pioneered by Hägerstrand were quite different to culture trait studies, favoring such topics as urban settlement and disease rather than specific landscape features—a classic example is Morrill's (1965) work on Swedish urban settlement. The most substantive example of a cultural geographic theme being analyzed in this manner is probably a study of settlement diffusion in the Polynesian region, which involved the use of a comprehensive simulation model to describe and explain a regional history of cultural movement; however, this work failed to encourage cultural geographers to follow suit (Levison, Ward, and Webb 1973).

The spatial analytic approach to studying cultural diffusion incorporates processes aimed at understanding the creation of regional cultural and landscape differences This reflects the fact that Hägerstrand was a cultural geographer in the Swedish tradition and primarily interested in landscape change. Nevertheless, because Hägerstrand's work emphasized quantitative and modeling strategies, it was seen as lacking explicit cultural content and proved more attractive to North American quantitative geographers than to North American cultural geographers. Specifically, studies that employed the procedures introduced by Hägerstrand largely ignored the cultural content, focusing instead on the mechanics of the diffusion

Figure 5.3 Three Empirical Regularities of the Innovation Diffusion Process. A principal concern of the spatial analytic approach to diffusion research was identifying regularities in the process of diffusion that were evident regardless of the trait studied.

In Figure 5.3a, the *neighborhood effect* refers to a situation where acceptance of an innovation is distance-biased, with adoption first achieved by a group of individuals living in close proximity to one another. Subsequent expansion occurs such that the probability of new adoptions is higher among those who live nearer the early adopters than among those who live further away. This regularity reflects a particular pattern of communications, most characteristic of rural or pre-industrial situations. It is useful to think of this regularity as similar to the result of throwing a pebble in a pond with ripples moving outwards and losing intensity with increasing distance from where the pebble fell.

In Figure 5.3b, the *hierarchical effect* refers to a different interpretation of distance, with initial adoption occurring in the largest urban center and subsequent diffusion occurring both spatially and vertically down the size ladder of urban centers. Knowledge and adoption of the innovation leaps from center to center rather than spreading in a wave fashion as with the neighborhood effect.

There are two key considerations affecting which of the two distance effects is more likely to occur. The first is the character of the innovation itself—a new variety of corn seed will diffuse throughout an agricultural landscape, whereas a new style of fashion will diffuse through the urban system. The second is the larger cultural landscape in which the process is occurring—a face-to-face communication system favors a neighborhood effect, while a technologically more advanced system favors a hierarchical effect.

Figures 5.3a and 5.3b describe alternatives for the spread of an innovation in geographic space. Figure 5.3c, the *s-shaped curve*, describes the expected growth through time of innovation adoption. The curve summarizes a process that begins slowly, then picks up pace, only to slow again in the final stage. There is a close relationship between the normal distribution of innovativeness in a population and this s-shaped curve—a cumulated normal distribution produces an s-shaped curve.

Note that the three empirical regularities are not restricted to studies of diffusion processes; indeed, they are evident in various other areas of geographic analysis.

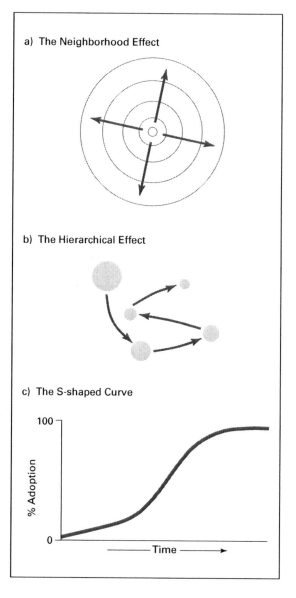

a) The Neighborhood Effect

b) The Hierarchical Effect

c) The S-shaped Curve

process. Thus, the spatial analytic approach to diffusion, although explicitly derived from traditional interests, was concerned mostly with process and model construction and less with culture and cultural landscape. This preference was in accord with prevailing interests and preferred approaches in the human geography of the 1960s.

Interest in spatial analytic diffusion research declined with the decline of interest in spatial analysis generally after about 1970. A growing number of critics of this approach to diffusion studies identified several weaknesses, the most notable of which was the exclusion of culture, indeed of any recognition of human distinctiveness. Consequently, and as one part of the changes taking place in cultural geography in the early 1970s, a rather different approach to diffusion that reintroduced an explicit cultural emphasis came to the fore.

Pause for Thought

On the basis of this account, does the spatial analytic approach to diffusion, with its emphasis on process, appear to be dehumanizing—excluding humans and their cultural identities? Certainly, most cultural geographers appear to have reached this conclusion, for they have largely ignored this work, preferring instead either to continue in the established Sauerian culture trait tradition or to seek alternative approaches. As noted, the spatial approach views diffusion as a communication process, transforming a landscape that is empty of a particular feature to one that includes that feature. Accordingly, there is limited concern with cultural origins and even less concern with cultural identities. Perhaps most critically, the simulation process effectively treats all individuals as identical to one another. On balance, it is not surprising that the approach never became popular among cultural geographers.

Diffusion, Culture, and Power

Although cultural geographers and others had long acknowledged that all members of a population are not equally receptive to an innovation—recognizing, for example, a distinction between more innovative urban dwellers and more conservative rural dwellers—the spatial analytic tradition often assumed uniform behavior for reasons of simplicity. Stimulated principally by ideas from political economy, the third approach to diffusion explicitly acknowledges that some groups, such as the poor, the less educated, the aged, and the unemployed, may be disadvantaged in having limited access to innovations. Accordingly, studies in this tradition stress the social, economic, and political conditions over which most individuals have little control. Because different diffusion processes operate in different contexts and have different causes, it is important to understand the political state and institutions. In some instances, those who are in a position to affect the innovative behavior of others may preempt valuable innovations.

The political economy approach recognizes that a process of innovation diffusion not only results in the presence of that innovation among a cultural group but also affects use of resources in space and over time. Thus, some innovations

are time-saving, causing substantial shifts in the daily time budgets of household members, while other innovations, such as a village school in an agrarian society, are time-demanding. Analyses that emphasize these aspects are oriented less around the diffusion process and more around cultural change. This is not a return to the traditional approach, but it does represent a renewed concern with culture. Much work in the political economy tradition is in the context of the diffusion of agricultural innovations in the less developed world.

AGRICULTURAL DIFFUSION IN KENYA

Two analyses of agricultural diffusion in Kenya illustrate the difference between the spatial analytic and the more politically based approaches. First, Garst (1974) described the diffusion of six agricultural products in one district of Kenya, indicating the way in which both spatial diffusion and the characteristic S-shaped adoption curves were affected by a wide range of physical, cultural, and infrastructural factors, including the availability of information, the location of processing plants, and various attempts by authorities to restrict some practices, such as tea growing, to particular areas (see Figure 5.4).

In the second analysis, Freeman (1985) discussed agricultural change throughout Kenya, focusing on the diffusion of coffee, pyrethrum, and processed dairy products, and concluded that preemption by early adopters was a critical part of the process, explaining both spatial spread and temporal growth (see Figure 5.5). Preemption—the process whereby those who are relatively privileged are able to adopt innovations early while effectively prohibiting others from adopting—may often be a critical factor permitting the rise of landed elites and rich peasants within a larger landscape of poverty. In many parts of the less developed world especially, those who adopt new innovations at an early stage may be entrenched elites who are able to transform initial profits related to early adoption into permanent profits. Impacts on the cultural landscape include a strengthening of the elites within a larger pattern of rural poverty and a dramatic reduction in the number of adopters of innovations.

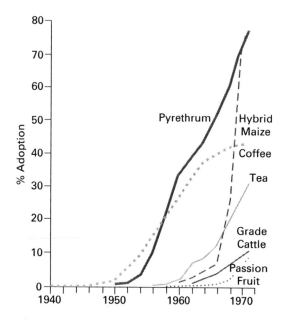

Figure 5.4 Agricultural Diffusion in Kisii District, Kenya, 1940–1971. The Kisii district of western Kenya, occupied primarily by the Gusii cultural group, is a densely settled agricultural area that has experienced significant landscape change since about 1940 as a result of the introduction of new agricultural practices. For six of these, Garst (1974) discussed the diffusion process by identifying the location of initial adoption, mapping the spatial spread, and plotting temporal growth. Garst (1974:311) here summarizes the results of the figure, which shows the adoption curves for coffee, pyrethrum, tea, passion fruit, grade cattle, and hybrid maize:

> Innovation diffusion among the Gusii follows a classic pattern. It is generally characterized by an initial rapid outward spread of adoption at low intensity with relatively little contrast between adopting and non-adopting areas. Later, new diffusion nodes appear at scattered locations while other nodes develop into peaks of higher adoption above the general low adoption level. As the peaks of higher adoption approach 100 percent the adoption surface becomes highly irregular. Finally, the peaks spread out forming plateau-like surfaces of saturation acceptance that soon coalesce, producing saturated regions.

Source: Adapted from R.D. Garst, 'Innovation Diffusion Among the Gusii of Kenya', *Economic Geography* 50 (1974):303.

Pause for Thought

You will have noticed that this section on diffusion has not, as you might have expected, provided a substantive account of regional cultural landscape evolution. This is

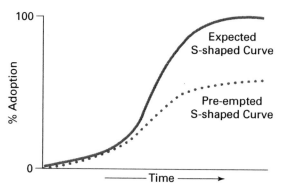

Figure 5.5 The Preempted S-shaped Curve. In a discussion of the importance of being first, Freeman (1985:17) contended that many diffusion analyses overlook 'the frequent cases of preemption of valuable innovations by early adopters'. With reference to the diffusion of coffee, tea, and processed dairy products in Kenya, preempted curves were explained by the presence of entrenched elites able to adopt early and then to restrict later adoption by others. Restriction was typically accomplished by means of production quotas, limiting the number of processing plants, and manipulating credit, farm extension, and social overhead capital.

Source: D.B. Freeman, 'The Importance of Being First: Preemption by Early Adopters of Farming Innovations in Kenya', *Annals of the Association of American Geographers* 75 (1985):18. (Reprinted by permission of Blackwell Publishing)

because most cultural geographers concerned with diffusion, regardless of the particular approach favored, have focused on specific questions rather than larger regional issues. This is an important point because it suggests that, notwithstanding the claim by Carter that opened this section, there is much more that cultural geographers might accomplish in their studies of diffusion and landscape. Issues relating to cultural diffusion and related regional change are discussed in the following chapter.

Cultural Contact and Transfer

Much cultural and related landscape change occurred as a consequence of population movement and related contact with others. Indeed, the expansion of European groups overseas between about 1450 and 1900 has directly or indirectly affected the lives and landscapes of peoples throughout much of the world. Accordingly, this discussion includes a general account of European overseas movement and a consideration of related Aboriginal–European contacts. A fuller consideration of links between European overseas movement and global regional concerns as developed in world systems theory is left to the following chapter. One feature of the material included in this section is the emphasis on the United States and Canada, along with some references to central and southern parts of the Americas, South Africa, Australia, and New Zealand. In this respect, the discussion is a reasonable reflection of the interests of cultural geographers.

Europe Overseas

> How strange it is to find Englishmen, Germans, Frenchmen, Italians and Spaniards comfortably ensconced in places with names like Wollongong, (Australia), Rotorua (New Zealand), and Saskatoon (Canada), where obviously other peoples should dominate, as they must have at one time. (Crosby 1978:10)

Indeed, before about 1450, Europe, Asia, and Africa were linked only by a few overland trade routes, while the Americas and Oceania were separate and isolated regions; only the Islamic civilization had spread significantly by sea, with expansion east of the Arabian hearth to the island region of southeastern Asian. However, after about 1450, five European countries—Spain, Portugal, France, the Netherlands, and Britain—embarked on overseas movement to areas outside of Europe previously unknown to them.

WHY EUROPE?

More than curiosity as to what lay beyond the horizon motivated European overseas movement; the colonization was not only scholarly, it was also political and economic—in short, it was about power and wealth. The desire to gain land and resources was probably the most compelling motive.

Two fundamental issues, addressed primarily by historians, concern why exploration took place in the second half of the fifteenth century, and why this overseas movement started in Europe. After all, many of the advances in late medieval European culture, such as the invention of the compass and the astrolabe, were diffused from China, often via the Islamic world. Basic answers refer to a number of factors, including the advances in navigation and the increased understanding of trade winds and ocean currents already made by Portugal in the first half of the fifteenth century, the presence of government or merchant company support, the desire to expand trading activities and acquire wealth beyond the confines of Europe, and the missionary zeal of Christianity, which was especially apparent in the context of the ongoing conflict with Islam. It can further be argued that Europe was able to move overseas because of the region's distinctive physical geography and culture. The physical factors include Europe's climatic and topographic variety, its related spatial diversity of natural resources, and its extensive system of natural waterways that together prompted both co-operation through trade and territorial conflicts that, in turn, contributed to the rise of several centers of power. The cultural factors include Europe's political fragmentation, which created a number of ambitious and competitive states, its systematized investigations in science, which led to a number of technological advances, and its conception of individual property rights. An alternative argument might stress the proximity of Europe to North America, compared to the distance that Chinese or Islamic explorers needed to travel to reach the American continent.

WHY NOT CHINA?

Any discussion about why Europe initiated overseas expansion in the fifteenth century cannot help but reflect on why China did not. China was a great naval power in the early fifteenth century, and during his 21-year reign (1403–24) Emperor Zhu Di sponsored several voyages for the purposes of exploration, mapping, and collecting tribute from other peoples. Chinese navigators under the leadership of Admiral Zheng He reached the eastern coast of Africa and, according to Menzies (2002), may have circumnavigated the globe. But, with the emperor's death, the country entered a long period of isolation, and these expansion initiatives were not pursued further.

Considering the above circumstances, a major difference between Europe and China was that the latter was a centralized empire whereas Europe comprised a number of competing political units. In China, one decision by one emperor was sufficient to terminate the exploratory process; in Europe, it would have been necessary for several countries to make such a decision at the same time—a highly unlikely circumstance—in order for European overseas expansion to end. In turn, explanations of this political difference between China and Europe often refer to basic physical geography, with China a compact landmass with a relatively short coastline in contrast to Europe's many peninsulas and islands and relatively long coastline According to Landes (1998), the physical geography of China facilitated the spread of a single dominant cultural group and thus the creation of a centralized state.

Cultural differences between China and Europe are also often cited to explain the different experiences of overseas voyaging. Chinese Confucianism, with a key ethic of harmonious living, encouraged the belief that its adherents are living at the center of the world. This represents a significant difference from the Christianity of fifteenth-century Europe, with its more aggressive monotheistic and universalizing character that encouraged looking outwards rather than inwards.

ECOLOGICAL IMPERIALISM

European explorers travelling overseas brought with them their conflicts, languages, religions, customs, and economic systems. At first, the Europe that was carried abroad was one characterized by feudal economic and social systems that privileged a few—many segments of the European population, including women, were seen as inferior beings—and by an attitude to nature that was essentially one of exploiting at will. Most of those who moved overseas, at least prior to the late eighteenth century, were male. Most of the colonial administrators were members of the privileged class, while many of the colonists were religious dissidents persecuted in their home areas. Their attitudes towards both the lands and the peoples they encountered were in accord with the European **norms** and **values** of the time, and the behaviors precipitated by those attitudes in many ways mirrored what was happening in Europe—a not unimportant point. Land was to be used as desired, and weaker peoples were to be used as labor—even as slaves—and moved as desired. If the indigenous population was judged unsuitable as a labor supply, slaves might be imported. In brief, Europeans, or more correctly those Europeans who moved overseas, saw themselves as superior to the peoples they met. The transition from feudalism to capitalism did little to change these general characteristics.

Of course, it was not only Europeans themselves who moved overseas but also European plants, animals, and diseases. Europeans were accompanied by what Crosby (1986:194) calls 'a grunting, lowing, neighing, crowing, chirping, snarling, buzzing, self-replicating and world-altering avalanche'. Indeed, many of the most aggressive plants in the overseas temperate humid regions today are of European descent, while horses, cattle, sheep, goats, and pigs were all introduced to overseas areas by Europeans. But it was the introduction of disease, notably smallpox, that was to prove devastating to Aboriginal ways of life. The effects of European diseases were overwhelming, especially in the Americas, which had been isolated from the larger world since the land bridge between Asia and North America was reduced to the Aleutian Islands as a result of a rise in sea level following the end of the most recent ice age. Aboriginal population estimates for the Americas prior to European contact are uncertain, but Denevan (1992) estimates that the population decreased by 89 per cent between 1492 and 1650 (from 53.9 million to 5.6 million).

The effects of disease varied spatially. Demographic collapse was not common in either Africa or Asia, whose populations, being less isolated from Europe than those of the Americas, had built up an immunity to some of the diseases carried by Europeans. Nor did it occur in those more isolated areas, such as New Zealand, that

In 1519, the Spanish explorer Hernando Cortez landed in Mexico with about 400 men. Conflict with the Aztec Indians began in 1520. Despite their experience in warfare and numerical advantage, about 4,000 Aztecs were soon defeated, and Cortez established Mexico City on the site of their major city. But the Aztecs were not really defeated by the Spanish but by small-pox, a highly contagious disease that the Spanish had unknowingly brought with them and against which the Aztecs, like other North American Aboriginal populations, had no immunity. This illustration depicts the progression of the disease. Initial symptoms include fever, head and body aches, and nausea. A rash that develops on the tongue becomes sores that break open, releasing the virus inside the mouth. A rash starts on the face, and within about a day it spreads to the arms and legs, then to the hands and feet, and then to all other parts of the body. The rash turns into bumps, and scabs and pustules form. The disease is often fatal, with death occurring within about two weeks. (The Granger Collection)

were settled by Europeans after the introduction of vaccines. Even in the Americas, where some groups such as the Yahi of California and the Beothuk of Newfoundland disappeared altogether, there were others such as some Maya groups in Guatemala that survived relatively unscathed. Not surprisingly, the debates about population numbers in the Americas prior to 1492 and about the role played by disease in subsequent declines are far from resolved, with much of the debate linked to the divergent views of history that a choice of numbers implies.

Aboriginal–European Contacts: Changing Understandings

The long history of human movement has included numerous instances of different cultural groups coming into contact with one another. Contacts between foraging groups or shifting agriculturists were often an incidental consequence of the movement that was an integral part of the lifestyle of these groups. Most contacts of this kind were between groups with broadly similar cultures, including comparable levels of technology, and one result of contact was a gradual agreement, through experience, on territorial limits. Another result was the transfer of ideas and goods, a process of **transculturation**. In this way, a series of linked cultural regions came into being.

However, the context for European movement and the resulting contact with Aboriginal groups was quite different. As noted, the principal motivations for movement included the acquisition of power and wealth, while the expansion was carried out by cultures that typically viewed themselves as superior to those they contacted. Further, European newcomers were not simply fishermen or traders; they were representatives of Europe. The commercial system of early capitalism was both a cause and a component of contact. Aboriginal–European contacts were, then, of a different order from previous contacts between cultural groups: they involved very different cultures that had very different motivations for their behavior at the time of contact. The dissimilar cultural views of Europeans and Maoris are diagrammed in Figure 5.6.

Meinig (1986:205–13) supplied an overall context to clarify the specific details of cultural contact situations. He proposed a usual sequence of contact, followed by Aboriginal population loss and changes to social and ecological patterns, followed by the appearance of some form of dominance–dependence relationship, followed by some population recovery and further changes to social and ecological patterns, followed finally by some cultural stabilization. At any given time, different areas contacted by Europeans were in different stages in this process of cultural change.

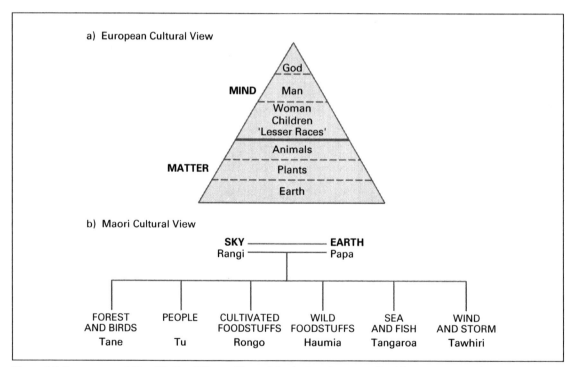

Figure 5.6 European and Maori Cultural Views. Figure 5.6a depicts the generalized European view of the world during the extended period of overseas expansion. There is a clear separation of humans and nature. As an indigenous group, Maoris are included in the 'lesser race' category. Unlike the European view, the Maori view (5.6b) is premised on a close relationship between humans and nature, with nature providing humans with both material and spiritual sustenance.

Source: E. Pawson, 'Postcolonial New Zealand'. In *Cultural Geographies*, 2nd edn, edited by K. Anderson and F. Gale (Harlow: Longman, 1999):29.

DIFFERENCE AND INFERIORITY

There is another aspect to this discussion of European contact. The accounts written by Europeans did not present a fair and accurate picture of Aboriginal populations, who were not the unchanging peoples they were once thought to be. The obvious but typically neglected point is that the European view was necessarily flavored by ideas and assumptions that were a part of Europeans' own background and experience. It is because their cultural baggage shaped their perceptions that Europeans frequently lacked respect for Aboriginal cultures and values. Eliades (1987:33) states: 'From the beginning of white–Indian contact, the Europeans, with only a few exceptions, refused to see the richness and diversity of Indian cultures. Instead of perceiving and accepting the native cultures as different, the Europeans saw them as inferior.'

Sauer (1971:303) arrived at a rather different conclusion: 'Throughout the eastern woodlands, the observers were impressed by the numbers, size, and good order of the settlements and by the appearance and civility of the people. The Indians were not seen as untutored savages, but as people living in a society of appreciated values.' Similarly, with reference to the Canadian prairies, Friesen (1987:18) observed: 'when contact was made, Europeans were quite prepared to recognize both the legal existence and the military power of the native peoples, to accept native possession of the land, and to negotiate with them for privileges of use and occupancy by the customary tools of diplomacy, including, of course, war.' Nevertheless, the prevailing view in the scholarly literature continues to emphasize the negative assessments made by European newcomers. Simply expressed, Europeans equated difference with inferiority. A logical extension of this argument is that Aboriginal voices have been largely unheard because of the unequal power relations inherent in the process of contact and in subsequent events.

The conceptual framework for much current research in this vein is based on contemporary social theories, including postmodernism. Many Europeans overseas commented on what they observed, and these commentaries were informed by their ideological baggage. How, then, are contemporary readers to understand these commentaries? It is helpful to approach this challenge by identifying a dominant and culturally specific complex of assumptions and attitudes—usually White, male, middle-class, and with a specific national identity. Perceptions and descriptions of Aboriginals are to be understood by reference to these assumptions and attitudes, to what Geertz (1973:3) describes as a 'web of significance'. In other words, the accounts provided by Europeans are not to be treated as objective, factual, and truthful because such qualities cannot be achieved; rather, the accounts are mere representations, a handful among many possible understandings of reality. Work in the postmodern tradition often claims that knowledge is necessarily uncertain and that there is no one truth waiting to be discovered by the contemporary scholar; this idea is central to much current work on cultural contact scenarios. It is now usual to emphasize how the voices of the subaltern were usually omitted or misrepresented in official documents. Box 5.1 offers a discussion of changes in the scholarly understanding of cultural change.

Aboriginal–European Contact: Cultural Change

The contact between incoming European populations and indigenous Aboriginal populations resulted in both groups experiencing cultural change, and in many cases it is not possible to clearly separate the roles played by the two groups in details of changes that followed contact. Certainly, the cultural change that took place involved both groups and did not simply involve the less dominant group almost unwittingly changing to adapt to the more dominant group. Thus, Europeans often adopted selected Aboriginal cultural traits judged to be advantageous in the new environmental setting. In early French Canada, for example, the birchbark canoe was an important borrowing, as it was made from local materials, was able to carry heavy loads, and yet was light enough to be carried when necessary. More generally, much European exploration relied extensively on the knowledge provided by Aboriginal groups; in Australia, for example, inland movement by Europeans was dependent on Aboriginal knowledge of environments and of peoples encountered.

RESISTING CHANGE, ACCEPTING CHANGE

Although contact typically prompted some cultural change by the Aboriginal group, the specifics of such change were often the choice of the receiving group rather than an imposition by the incoming group. Raby (1973:36) offers the example of Native groups on the Canadian prairie: 'The Indian bands of southern Saskatchewan . . . had by no means been converted into the competitive agrarian individualists sought by their white guardians. Much had been said to them of the virtues of agriculture. Farming, and not the supposedly demoralizing

pursuits of hunting and fishing, was identified as work.' Indeed, as this quote suggests, there was often resistance to Europeans' efforts to impose radical cultural change.

Aboriginal populations not only resisted many of the European attempts to impose change but in fact often accepted only those aspects of European culture they considered to be to their advantage and to involve the least disruption to their existing cultural context. For example, in the Queen Charlotte Islands of British Columbia between 1774 and the 1860s, the Haida experienced substantial population losses (because of disease), dramatic changes in their settlement pattern, and detailed changes to their seasonal cycle of activities. Changes in the Haida's settlement pattern stemmed both from the abandonment and fusion of villages as a result of population loss and from the desire to

Box 5.1 Cultural Change

It should not be surprising that scholarly understanding of Aboriginal–European contacts has changed a great deal since the first accounts appeared. These accounts were produced by Europeans and were necessarily from a European perspective. The characteristic view of Europeans at the time, shared by historians and anthropologists, was that civilization had encountered savagery. According to this view, Aboriginals had no history because their cultures were unchanged for long periods prior to contact—in other words, their history began with contact. Static cultures and their related unchanging cultural regions were often assumed to be simple responses to physical environment. This view was gradually replaced by the idea that a more complex civilization had encountered a less complex civilization, a distinction that was explicit in a 1930 study of the Canadian fur trade (Innis 1930).

Of course, the suggestion that there was Aboriginal cultural change prior to contact meant that an understanding of Aboriginals before contact could not be based on the observations of Europeans, for those Europeans were describing a dynamic, not a static, culture. Further, a traditional **ethnography** necessarily described cultures after, rather than before,

contact, and was therefore describing a culture affected by contact. Archeological data are needed for descriptions of Aboriginal groups prior to contact, and there is today a large amount of archeological evidence to suggest that many of the apparent results of contacts were not really dramatic revisions of earlier culture, but rather logical extensions of change that was occurring before contact. For Huron groups in Ontario, for example, there was 'a significant revival of intentional trade in the late prehistoric period and it was along the networks that supplied traditional prestige goods that European materials first seem to have reached the interior of eastern North America' (Trigger 1985:162). Of course, European movement and diffusion of their goods generated changes, but many of these changes need to be understood in the wider context of Aboriginal cultures that were already evolving. Further, an increasing number of scholars today recognize that it is not appropriate to assess the consequences of contact for Aboriginals in terms of such values of European culture as progress and change. Certainly, arguments based on the traditional cultural enrichment hypothesis need to be approached with caution as they may be both simplistic and Eurocentric.

enhance their participation within the fur trade. Impacts on the seasonal cycle involved changes in detail that did not affect the broader principle of seasonality (see Figure 5.7). This was because the cycle served important cultural functions that proved resistant to substantive change. The introduction of a sedentary way of life was forcefully advocated by missionaries but was rejected because it was not in accord with Haida values.

Similar conclusions can be drawn concerning the impacts of fur trading on the Canadian interior, where the continuity of traditional ways was a dominant feature of the early fur trade period. The European trade was conveniently fitted between established annual migrations

that followed changes in animal location and plant resources. It is increasingly evident that Aboriginals were often very careful in their interactions with Europeans, adapting to the new challenges posed and often regarding the newcomers with disdain.

IMPOSED CHANGE

In some areas there were active attempts to alter Aboriginal lifestyles. In Guatemala, for example, the military contact between 1524 and 1541 resulted in a Spanish takeover of both land and people followed by conscious efforts to impose a Spanish way of life on the conquered groups. This forced cultural change involved both a mandatory resettlement—a process that invariably sev-

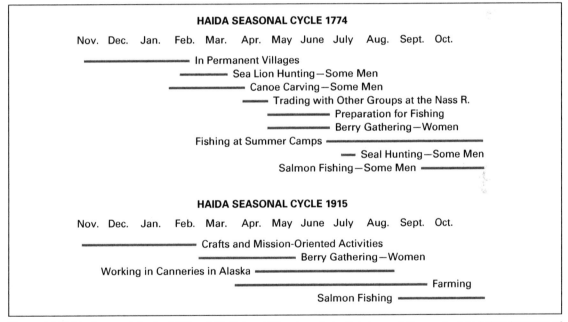

Figure 5.7 Haida Seasonal Cycle, 1774 and 1915. In addition to substantial depopulation and related abandonment of villages that followed the first European contact in 1774, the Haida of the Queen Charlotte Islands experienced some detailed changes to the activities of their seasonal cycle. The cycle provided a framework for social and economic life, with the regular movements supporting the basic rhythm of life and confirming human continuity with nature.

Although the seasonal cycle depicted for the pre-contact situation in 1774 and that depicted for the post-contact situation in 1915 are different in detail, the basic logic of the cycle, namely spatial movement on a seasonal basis, is unchanged. Fur trade contacts after 1774 until the 1830s did little to change the basic seasonal pattern, with the fur trade incorporated into the established framework of life. By the 1830s, a decline in the availability of sea otter for fur meant reduced Haida–European contact. Renewed contact was initiated in the 1850s with the discovery of gold, copper, and coal, but again these mining activities impacted little on the seasonal cycle. Even the gardening introduced by European missionaries towards the end of the nineteenth century was incorporated into the cycle.

Source: J.R. Henderson, 'Spatial Reorganization: A Geographical Dimension in Acculturation', *Canadian Geographer* 12 (1978):11, 16. (Reprinted by permission of Blackwell Publishing)

ered ties between the Aboriginal people and their ancestral lands—and heavy demands on native labor. The relationship between Spanish and Aboriginals became one of oppressor and oppressed, and institutionalized exploitation was characteristic of the colonial period; the hacienda system, for example, forced many Aboriginals into servitude on large plantations and introduced new tools, crops, and animals to the region. But such attempts to impose change were not always successful: for instance, Spanish control of Aztec populations in Mexico did not prevent those native groups from successfully retaining many of their traditional cultural values.

DEPENDENCE

One controversial issue that has emerged in recent scholarship on European contact in the western hemisphere concerns the view that the overexploitation of both fur-bearing animals and game by North American Aboriginal populations was caused not by a developing dependence on trade for European goods but rather by a growing belief among Aboriginal peoples that animal spirits were responsible for the ravages of disease. Some scholars argue that this new belief led Native groups not only to reject their traditional beliefs in favor of new Christian values but also to launch a deliberate extermination of animals (Martin 1978). This issue is part of two larger debates.

First, there is a debate about the characteristic Aboriginal human-and-nature relationship prior to European arrival. The conventional wisdom is that Aboriginal peoples were typically in harmony with nature, although there is evidence to suggest that some groups abused their environments and that Aboriginal peoples in general are not best described as ecological custodians. Second, there is a debate about the extent to which Aboriginal groups became dependent on the newcomers. Certainly, European culture often failed to satisfy Native aspirations. Referring to Australian Aboriginals, Reynolds (1982:129) stated that 'young blacks who went willingly towards the Europeans fully expected to be able to participate in their obvious material abundance. Reciprocity and sharing were so

fundamental in their own society that they probably expected to meet similar behavior when they crossed the racial frontier.' This kind of expectation was rarely fulfilled.

Cultural change, **acculturation** but not necessarily **assimilation**, was an inevitable consequence of Aboriginal population losses, European attitudes of disdain and imperialism, and differences in technological sophistication, such that some version of a dominance–dependence relationship often arose. Although specific details vary from place to place, it was not unusual for Aboriginal peoples to become economically and/or politically dependent on European newcomers. Sometimes, as in many parts of Spanish America, this was effectively a forced dependence, while on other occasions, as in many North American fur trade areas, it was an unintentional effect of contact. Sometimes it occurred quickly, sometimes more slowly. Regardless, some form of dependence is a not uncommon consequence of contact that, in many instances, has carried through in some form to the present.

In the case of the North American fur trade, there is debate between those who emphasize that Aboriginal groups retained the ability to make choices about their economic lifestyles and those who see overriding forms of economic and political dependence. Certainly, in general terms, the ever-increasing exploitation of a limited resource base meant that those Aboriginal groups that became reliant on the trade often suffered once the trade declined. European cultural traits, such as guns and steel traps, facilitated the exploitation of resources that eventually resulted in the termination of the trade. It is possible that involvement in the trade may have prompted the rejection of traditional alternative ways of making a living. Further, before they began trading furs for European goods, Aboriginal peoples exchanged among themselves only commodities that they produced, and this was a more equitable form of commerce. In the case of northern Manitoba, it is clear that the fur traders became wealthy while the Native peoples received only a marginal return for their labor (see Figure 5.8).

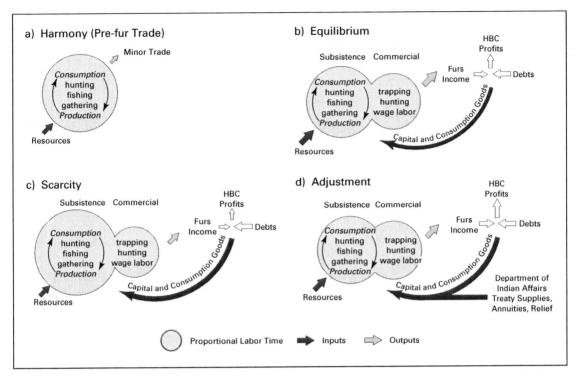

Figure 5.8 Changes in the Aboriginal Economy of Northern Manitoba. Links between the two sectors of the Aboriginal economy—subsistence and commercial—changed over time. Before the fur trade a harmonious economy prevailed, with production and consumption in balance, as shown in 5.7a. After about 1670, with establishment of the Hudson Bay Company, a commercial economic sector developed and a delicate balance was in place (5.7b). Subsequent scarcity of subsistence resources meant that less time was available for the commercial sector, prompting import of food products (5.7c); for the Company, it was cheaper to import food than to experience a drop in fur production. In the final phase, the Canadian government assumed responsibility for the import of food, thereby subsidizing the Aboriginal economy while supporting fur production (5.7d). This sequence of stages supports the idea that the 'concept of paternalism, and not partnership, seems to capture the historic relationship' between Aboriginals and traders (Tough 1996:9).

Source: F. Tough, *As Their Natural Resources Fail: Native Peoples and the Economic History of Northern Manitoba, 1870–1930* (Vancouver: University of British Columbia Press, 1996):16.

Pause for Thought

Understanding the details of culture contact is far from easy. Facts are uncertain, and different ideologies promote different evaluations of available evidence. It is common to suggest that Aboriginals and Europeans entered into an uneven relationship characterized by misunderstandings about such matters as access to land and resources. The European desire to expand territory and to settle, in what were for them, new areas, necessarily clashed with Aboriginal understandings. In some areas forced movement occurred, while in other areas treaties were made that, from a European perspective, provided them with exclusive access to land and resources but that, from an Aboriginal

perspective, merely allowed Europeans to share. Treaties are a good example of the different understandings that were part of the contact process, and the current disputes over Aboriginal land claims are one outcome.

Shaping Landscapes

Several German geographers, including Schlüter and Ratzel, advocated analyses of landscape evolution at the turn of the nineteenth century, making the concern with landscape evolution a central feature of the German cultural geography tradition. Similar ideas were evident in France with the writings of Vidal and his followers. Schlüter defined geography as the study of the visible landscape as it changed through time and 'was the first to raise the landscape forming activity of man to a methodological principle' (Waibel, quoted in Dickinson 1969:132). Ratzel, meanwhile, 'did not exaggerate the potency of the physical environment. . . . What saves him from such naiveté is the recognition of the time factor. . . . No one could emphasize more than Ratzel the force of past history' (Lowie 1937:120). In similar vein, Vidal introduced the term 'personality' into the geographic literature, noting that 'geographic personality is something that grows through time' to create a distinct regional landscape (Dunbar 1974:28). Further, the French *Annales* school of historical research, initiated by Febvre and Bloch in 1929, has close intellectual ties with *la tradition Vidalienne*.

In the United States, Sauer built on these German and French ideas to provide an English-language conceptual basis for evolutionary analyses. For Sauer, natural landscapes when modified by humans become cultural landscapes. With this basic idea in mind, proponents of the evolutionary approach aim to understand contemporary landscapes as the outcome of long-term processes of changing relations between humans and land. The literature of cultural geography includes many studies of areas where the primary concern is the changing landscape, with one goal being to demonstrate that an understanding of past circumstances is a prerequisite to understanding the present. The following statement from Sauer, recorded by one of his students in 1936, emphasizes this focus: 'We are what we are and do what we do and live as we live largely because of tradition and experience, not because of political and economic theory' (Sauer 1985:1). Interestingly, Sauer's ideas failed to attract much interest in British geography, where the early work of the leading British historical geographer, Darby, lacked both visible landscape and evolutionary content, and focused on past—rather than changing—landscapes.

Approaches to Historical Geography

How best to organise historical studies of landscape has been a question addressed by geogra-phers for decades and there is clearly not a single, optimal solution: there is no methodological Holy Grail in historical geography, however diligently and ingeniously it is sought. (Baker 2003:133)

Historical geographers have identified and tested various approaches to the study of landscape change through time, often recognizing, however, that the differences among approaches are rarely substantive. Box 5.2 outlines relevant approaches as noted by three reviews of work in historical geography. Three of the more popular of these are briefly discussed here, namely *narratives*, *cross-sections*, and *sequent occupance*.

The use of narrative in both history and historical geography is a well-established way to describe and even explain change through time. Indeed, for some historians, all history is narrative, since a narrative both describes and explains change. Although other historians disagree, seeing the narrative as failing to explain why things happen, historical and cultural geographers have chosen not to enter this debate, preferring instead to integrate a narrative approach with the cross-sectional approach.

The aim of the cross-sectional approach is to select a particular moment in time, describe the geography of that time, and then devise a narrative connecting this to another moment in time. Studies such as these do not necessarily convey details of change through time, but rather focus on amounts of change between selected times; these studies are concerned with changing geographies rather than geographical change. Some cross-sectional studies are better characterized as descriptions of past geographies in that, for reasons of data availability, they are often limited to one moment in time. The classic work in this vein was conducted by the leading British historical geographer, Darby, in several book-length studies of England in 1086 that used data from the Domesday Book to create a series of more than 800 distribution maps (see, for example, Darby 1977).

A study of Scandinavian historical geography employed a genetic and evolutionary approach to the cross-sectional method. Mead identified five stages, or cross-sections, and prepared a series of connecting narratives to link successive stages (see Figure 5.9). 'The object is to pause at certain stepping stones in the stream and to look at the surrounding stones. They will be scenes of the same place, but they will present patterns and distributions particular to different times. The changes between the scenes are explained by the processes—physical and human—that have characterized the intervening years' (Mead 1981:1).

Following Sauer, most cultural geographers who have employed cross-sections have done so in a rather specific format, namely the approach known as **sequent occupance**. Studies in this vein recognize that a landscape remains relatively stable for a period of time, but then

Box 5.2 Changing Geographies—Some Historical Geographic Approaches

Historical geographers have embraced a wide range of approaches to their studies, and there are many methodological writings that attempt to classify these approaches. This box summarizes three of these attempts. The approaches identified refer to the two principal historical geographic concerns, namely study of the past and study of change.

First, Smith (1965:120) notes four approaches, two of which (those indicated with an *) refer to the study of change:

- operation of the geographic factor in history
- reconstruction of past geographies
- study of geographic change through time*
- evolution of the cultural landscape*

Second, Newcomb (1969) noted twelve approaches, six of which (those indicated with an *) refer to the study of change:

- historical regional geography
- areal differentiation of remnants of the historic past
- genre de vie
- theoretical model
- pragmatic preservation of landscape legacies
- past perceptual lenses
- temporal cross-section*
- vertical theme*
- cross-section–vertical blend*
- retrogressive*
- dynamic culture history*
- humans as agents of landscape change*

Third, Prince (1971) noted three themes, namely the study of real worlds, perceived worlds, and theoretical worlds. With regard to real worlds, Prince identified studies of past geographies, of geographical change, and of processes of change and listed a total of fourteen approaches, nine of which (those indicated with an *) refer to the study of change:

- past geographies
  - géohistoire (reconstructing past landscapes in order to aid understanding of everyday life)
  - urlandschaften (reconstructing the prehuman landscape)
  - static cross-sections
  - sources and reconstructions
  - narratives of changes*

- geographical change
  - sequent occupance*
  - evolutionary succession*
  - episodic change*
  - frontier hypothesis*
  - morphogenesis of cultural landscapes*
  - agency of humans*
  - rates of change*

- processes of change
  - dynamics of change*
  - inadequacy of inductivism* (an appreciation that facts do not speak for themselves)

Although historical geographers have found it possible to classify their preferred approaches in a variety of ways, it is clear that many of the approaches refer to change through time, including one that Prince chooses to include under the heading of studies of the past. This preference is notwithstanding the opposition that this approach faced from the then dominant regional approach to geography before about the mid-1950s. Certainly, the popularity of studies of change owes a great deal to the advocacy of Sauer and his student Clark.

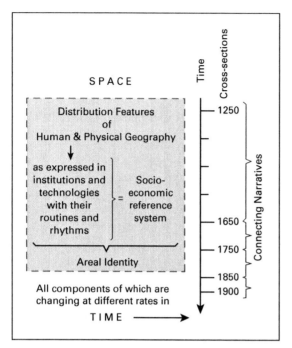

Figure 5.9 Cross-sectional Historical Geographies of Scandinavia. The study consists of five cross-sectional analyses describing geographic patterns for the years 1250, 1650, 1750, 1850, and 1900. The interval between cross-sections decreases because the pace of change increases through time, with, for example, more change occurring during the final 100 years studied than occurred during the preceding 500 years. Each of the cross-sections is discussed in a systematic manner, with accounts of political, population, agricultural, resource, and communication geographies. 'At any given cross-section, a number of people with particular sources of energy at their disposal will be viewing the options open to them and shifting the directions of Scandinavian development responsively. Simultaneously, they will bequeath a legacy of evidence for the geographical understanding of their times which will reflect their degree of literacy and numeracy' (Mead 1981:3). Connecting narratives that describe processes of change explain the transition from one cross-section to the next, and there is an additional narrative for the period since 1900. Each cross-section and each linking phase is discussed in a separate chapter. This approach is based on that developed by Darby (1973) in a study of the historical geography of England.

Source: W.R. Mead, *An Historical Geography of Scandinavia* (London: Academic Press, 1981):2. (© 1981 Reprinted by permission of Elsevier)

undergoes change that is both rapid and substantial, leading into another period of stability. Thus, a succession of cultures is associated with a succession of cultural landscapes. Each period

of stability reflects a particular cultural occupance, but is transformed into a subsequent period by some evolutionary or diffusionist process. One stage may be transformed into another because of the in-movement of a different cultural group or because of some substantial change in, for example, technology or economic system. In this sense, any landscape is a **palimpsest** in that it comprises the consequences of a series of different occupations. Whittlesey (1929:162) stated: 'The view of geography as a succession of stages of human occupance establishes the genetics of each stage of human occupance in terms of its predecessor.'

A classic sequent occupance analysis in the Sauerian tradition is Broek's (1932) study of the Santa Clara valley, which was designed to understand landscape change as a result of a succession of different cultural occupances within a 200-year period. The first cultural occupance was Aboriginal; the second stage was Spanish occupation, which was characterized by missions and cattle ranches; the third stage was an early American economy of wheat and cattle farming; and the fourth stage was dominated by horticultural activity. Looking back at this work in 1965, Broek (1965:29) noted that it was appropriate to add a fifth stage dominated by urbanization.

More generally, many world histories that consider human activity during the past 10,000 years or so stress the transitions from foraging to agriculture and from agriculture to industry as relatively brief periods of substantial change separated by lengthy periods of relative stability. Indeed, this was the approach taken in the Chapter 4 discussion of global landscape change.

The differences between a historical geographic cross-sectional analysis and a cultural geographic sequent occupant analysis are worth noting. In the cross-sectional analysis, there is no suggestion that each cross-section is a stage separated from the previous and the following stages by some significant change of occupance. Rather, the cross-sections selected are intended to be representative of some larger period; relatively speaking, they are arbitrarily chosen. Thus, in a cross-sectional analysis, the cross-sections are particular moments in time and are separated by much longer periods of change. In a sequent occupance analysis, the stages described are lengthy and separated by briefer periods of rapid and significant change. Expressed another way, a cross-sectional approach emphasizes patterns at particular moments in time, albeit with process-oriented linking narratives, whereas sequent occupance places a higher priority on the processes operating to establish and maintain a particular stage of occupance.

Frontier Experiences

As part of the attempt to explain changing geographies in areas of European overseas expansion, especially the United States, cultural and historical geographers have focused on the frontier concept through analyses of the cultural and landscape changes that occurred as European groups embarked on new lives in what were, for them, new places.

The American historian Frederick Jackson Turner (1861–1932) introduced the influential **frontier thesis** in the late nineteenth century. The American **frontier** was understood as both place and process. As a place, it was defined in terms of population density, with a figure of two people per square mile suggested as marking the end of the frontier phase. In terms of process, Turner, employing terms popular in anthropology at the time, proposed that the American frontier was the meeting point between savagery and civilization, with the European newcomers initially borrowing many cultural traits from Aboriginal groups before developing their own distinctive cultural identity. Thus, the frontier was where Europeans became Americans: 'American development has exhibited not merely advance along a straight line, but a return to primitive conditions on a continually advancing frontier line. . . . American social development has been continually beginning over again on the frontier' (Turner 1961:38). Within history, these ideas have been further developed in major studies of the Great Plains and other world frontiers. Despite being typically ill defined, the frontier concept has been seen as key to understanding the larger European experience overseas.

Meinig (1993:258–64) proposed an alternative to Turnerian logic, namely a six-stage transition from a North American traditional system, through a modern phase, to a world system. The six stages are titled *Indian society*, *imperial frontier, mercantile frontier, speculative frontier, shakeout and selective growth*, and *toward consolidation*. Each of these stages is characterized by some particular cultural, economic, and political circumstances. Figure 5.10 provides a schematic summary of the Turner and Meinig conceptions of changing frontiers.

Transferring Cultural Baggage

'It is a striking insight', comments Harris (1977: 469), 'that Europeans established overseas drastically simplified versions of European society'. One possible explanation for such **simplification** is that only a fragment of larger European culture moved, and this movement was followed by an often lengthy period of isolation in a new and different environment. This combination of a particular cultural fragment and subsequent isolation resulted in a new and simpler identity. Cultural change in Australia might be appropriately interpreted in terms of this fragment concept. An alternative explanation centers on the particular conditions, specifically the confrontation with a new land, that Europeans experienced. In early French Canada, South Africa, and New Zealand, for example, Europeans with a strong sense of family and a desire for private ownership of land encountered areas of inexpensive land and limited markets for agricultural products, with the result that a homogeneous society based on the nuclear family dominated; in other words, a process of simplification occurred. The initial transplanted population might have been culturally differentiated, but the resulting settler society was culturally uniform.

Rejecting the Turner thesis because of its explicit environmental determinism, Sauer (1963:49) observed that the European frontier experience overseas was affected by 'the physical character of the country, by the civilization that was brought in, and by the moment of history that was involved'. This observation

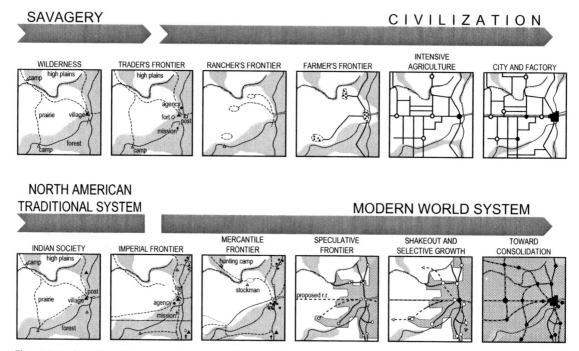

Figure 5.10 American Cultural and Landscape Change. Following a critique of the Turner thesis (*above*), which empha-sized 'the simplistic, optimistic view of society as an organism evolving inexorably from simple to complex', Meinig pro-posed an alternative framework (*below*) that places greater emphasis on imperial, capitalist, and cultural features and that 'offers a better sense of the swirl and complexity of the American scene' (Meinig 1993:259).

Source: D.W. Meinig, *The Shaping of America: A Geographical Perspective on 500 Years of History. Volume 2: Continental America, 1800–1867* (New Haven: Yale University Press, 1993):260–3. (Reprinted by permission of Yale University Press)

raised three questions: What cultural baggage did an incoming population bring to an area? What happened to that baggage? To what extent did the shaping of landscape reflect cul-tural characteristics?

The simplification thesis suggests a loss of baggage, perhaps implying that settlers from different backgrounds in the same area did not shape markedly different landscapes. A loss of cultural traits—and, therefore, of identity—is indeed characteristic of many cultural groups after movement to a different area. One expla-nation for such developments relates to the fact that the motivation behind most settler activ-ity was economic, and adopting appropriate behavior given the market economy tended to be more important than preserving some cul-tural traits. Thus, Swedes and Norwegians in the Upper Midwest grew wheat, while Germans in Texas adopted cotton and tobacco cultivation.

Evidence from eighteenth–century southeastern Pennsylvania shows that differences in land use could not really be attributed to differences in settler source area, while evidence from Kansas points to an even more rapid loss of cultural identity than occurred in eastern North Amer-ican areas settled earlier and under different circumstances. Evidence from Scandinavian converts in the Mormon region of the inter-montane West suggests that the role played by the new religious identity was so important that culture traits associated with area of origin were quickly lost.

But there is also a considerable body of lit-erature demonstrating the effects of different groups in areas of European overseas settlement. For example, North American studies have often examined the role played by the ethnic back-ground of settlers. For many cultural groups, the institution that played the largest role in minimizing trait loss was the church; indeed, in cultures primarily united by religion, the rate of trait loss was often very low. In the case of Swedish settlement in Minnesota, group membership was able to influence economic decisions, but not sufficiently to prompt farmers to behave in an inappropriate economic manner; the community that evolved from cultural background rather than cultural identity itself was seen as influencing behavior.

An illuminating example of factors affecting retention or loss of cultural baggage is that of Irish settlers in nineteenth-century eastern Canada. Taking into account material folk culture and settlement morphology, the movement of Irish groups to Canada typically involved a loss of cultural traits, but the rate of loss varied substantially between three eastern Canadian locations. Retention of cultural baggage was greatest in the Avalon Peninsula of Newfoundland, and loss was greatest in the Peterborough area of Ontario; the Miramichi River area of New Brunswick represents a middle ground. The different experiences in these areas reflected two principal considerations, namely physical environment and proximity to other incoming groups. Thus, the Avalon peninsula encouraged retention of traits because the environment was similar to that of the source area and because there were few other newcomers with whom to interact. The environment of the Peterborough area, on the other hand, was unlike that of the source area and included a number of other cultural groups. The loss of traits in the Peterborough area did not, therefore, necessarily imply a simplification of culture as there was trait borrowing from other groups to compensate for the lost traits.

Known as the White House, this home in the residential community of Portugal Cove, on the south side of Conception Bay, close to St John's, was built in the early nineteenth century and is a good example of Irish construction techniques. Most notably, there is the traditional Irish fireplace, large, open, and central, made of local stone, with two benches seating up to ten people. In addition to being used for cooking and heating, large open fireplaces were centers of family life. The house also features a hipped rather than a gabled roof, characteristic of Newfoundland-Irish homes. This building was designated a Registered Heritage Structure in October 1992. (Heritage Foundation of Newfoundland and Labrador)

In the case of German settlers in Texas, imported traits were an important factor related to agricultural landscapes, despite the fact that some of these traits were lost soon after arrival, while some new traits were adopted from other groups. Certainly, there are numerous examples of particular agricultural practices associated with certain ethnic groups and with distinctive cultural landscapes. Trait loss may indeed have been usual in response to economic circumstances, but it was certainly possible for some of the new traits acquired to subsequently become associated with a particular group.

In summary, it appears that trait retention by a cultural group moving to a new environment was most likely if

- the physical environments of the source and destination areas were similar;
- a dominant goal of the group was to seek a location separate from others;
- contact with other groups was limited;
- the group was socially cohesive, possessing shared values;
- the impact of external institutions was limited;
- the number of immigrants was substantial; and
- settlement was contiguous.

Pause for Thought

Although the simplification of some European cultures overseas is well documented, especially prior to the large-scale movements of the nineteenth century, the significance of such simplifications is debatable. Indeed, it is possible that the frontier experience played only a minor role in shaping immigrant cultures in areas of European overseas expansion. This is because many frontiers were settled by few people and soon overwhelmed by the spread of a commercial economy. Prolonged frontier circumstances were common only in a few relatively undesirable areas, such as Appalachia or much of South Africa. More characteristic was the Shenandoah valley in eighteenth-century Virginia, where the frontier population, which included many opportunistic individuals, moved quickly into the new economy of commercial agriculture, shedding unnecessary cultural traits and adopting necessary new traits.

Evolutionary Regional Studies

Landscapes, cultural regions, or geographic personalities develop in response to cultural occupance through time, becoming 'as it were, a medal struck in the likeness of a people' (Vidal, quoted in Broek and Webb 1978:32). With this evocative phrase, Vidal captured both the concept of *genre de vie* and the Sauerian idea of humans occupying physical landscapes to create cultural landscapes through time.

Writing in 1930 with reference to the American West, Sauer (1963:45) offered a clear statement of the historical geographic or evolutionary approach to landscape study: 'The three major questions in historical geography are: (1) What was the physical character of the country, especially as to vegetation, before the intrusion of man? (2) Where and how were the nuclei of settlement established, and what was the character of this frontier economy? (3) What successions of settlement and land utilization have taken place?' Further: 'In order to evaluate the sites that were occupied, it becomes necessary to know them as to their condition at the time of occupation. Only thus do we get the necessary datum line to measure the amount and character of transformations induced by culture' (Sauer 1963:46). With these arguments in mind, Sauer discussed the physical landscape, the Aboriginal landscape, and the several European frontiers that moved west, resulting in a series of cultural successions on the landscape. He regarded frontiers as secondary cultural hearths.

In an earlier discussion of Mexico, Sauer (1941b) had used the term *personality* to refer to the relationship between human life and land. Because the roots of Mexican life lie in a distant past, Sauer looked to transformations that had occurred in that past in order to gain an understanding of contemporary Mexico. Thus, he identified a pre-Spanish and a Spanish past as the principal bases for the present cultural geography. He also made a critical distinction between the two cultural regions of northern and southern Mexico that were evident before, during, and after Spanish occupation (see Figure 5.11). Sauer described the north as less advanced and the south as more advanced, and concluded: 'The old line between the civilized south and the Chichimeca has been blurred somewhat, but it still stands. In that antithesis, which at times means conflict and at others a complementing of qualities, lie the strength and weakness, the tension and harmony that make the personality of Mexico' (Sauer 1941b:364).

A work such as this, full of generalizations that were supported with factual detail, is characteristic of much early landscape school work—ambitious, thoughtful, and provocative. Box 5.3

notes a few of the many other examples of work in this 'making landscape' tradition.

Reading the Landscape

Observing and reading a landscape is far from easy, as McIlwraith (1997:377) attests when he comments: 'It has been said that sight is a faculty but that seeing is an art that must be cultivated and kept finely honed through persistent field observation' (McIlwraith 1997:377). There are two important traditions concerned with seeing and reading landscapes, and though they appear to

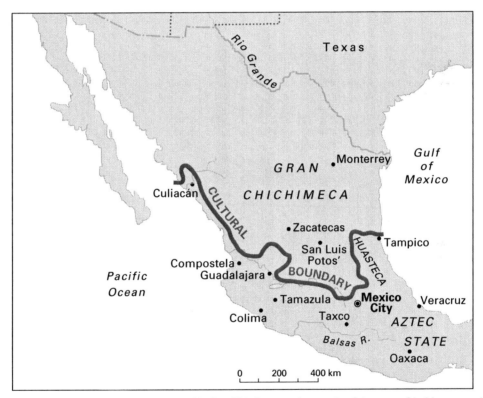

Figure 5.11 A North–South Cultural Division in Mexico. This is a good example of the type of bold map produced by Sauer and other cultural geographers in the early landscape school tradition. The boundary drawn is explained in cultural, not environmental, terms, although there is also a climatic distinction in that the north is mostly arid or semi-arid but includes some alluvial valleys and good upland soils, while the south is generally better watered upland that includes some rainforest. The Spanish distinguished between Aboriginals of the more cultured south, the Aztec region, and those of the less cultured north, the Chichimeca region. Indeed the southern region is one of the great cultural hearths of the world (see Figure 4.10). Further, the Spanish recognized the south as a region of peace and the north as a region of conflict. At the time of Spanish in-movement, the south was largely agricultural, while the north was a mix of agricultural and foraging activities.

Source: C.O. Sauer, 'The Personality of Mexico', *Geographical Review* 31 (1941):355. (©The American Geographical Society; reprinted with permission)

be quite different, they are in fact not so dissimilar. The first is an explicitly historical approach, essentially a form of narrative, as described above, that is concerned with detailed field investigations of local landscapes. It is primarily the product of British researchers and is closely associated with the historian W.G. Hoskins. The second is a more broadly conceived endeavor, with strong historical content certainly but with some additional defining characteristics, such as a concern with landscape symbolism and with the vernacular in landscape. This is the *Landscape* magazine tradition and is closely associated with J.B. Jackson, a private scholar who described himself as a lay geographer.

THE LOCAL AND REGIONAL NARRATIVE HISTORY TRADITION

A classic study concerned with the making of landscape appeared in 1955. *The Making of the English Landscape* (Hoskins 1955) was a trailblazing book that, although initially received without any real acclaim, led to a series of local county landscape histories and, indeed, to a television series in the 1970s. Meinig (1979b:209) quoted Hoskins as saying: 'I once wrote a book with the simple title of *The Making of the English Landscape*, but I ought to have called it *The Morphogenesis of the Cultural Environment* to make the fullest impact.' Certainly, the book is now considered a pioneer work in the field of local history.

Hoskins wrote the book to provide a detailed analysis of the landscape as it was made over time by humans. He achieved this with a series of chronologically arranged chapters, each dealing with a particular period of landscape-forming activity. As Hoskins makes clear, landscape evolution is his primary concern and the motivation for his work:

Despite the multitude of books about English landscape and scenery, and the flood of topographical books in general, there is not one book which deals with the historical evolution of landscape as we know it. . . . What I have done is to take the landscape of England as it appears today, and to explain as far as I am able how it came to assume its present form, how the details came to be inserted, and when. At all points I have tried to relate my explanation to the things that can be seen today by any curious and intelligent traveller going around his native land. There is no part of

---

Box 5.3 **Making the Regional Cultural Landscape**

There are several book-length works in historical geography that are closely related to the evolutionary landscape school approach; four examples are noted briefly in this box.

First, Davidson (1974) studied a small group of islands, the Bay Islands, off Honduras. Focusing on a description of the physical landscape and on a reconstruction of past landscapes, he explicitly identified culture as a cause of change. He incorporated a sequent occupance focus, noting eight cultural stages, each affecting the material landscape. He identified as the principal cause of change over the long term an ongoing conflict between English and Spanish.

Second, Williams (1974) discussed the making of the South Australian landscape, stressing the details and consequences of land clearing, such as swamp draining and woodland removal, as well as the consequences of changing technologies and incoming cultural groups.

Third, Lambert (1985), in a historical geography of the Netherlands, described the physical landscape and analyzed the impacts on that landscape, focusing on the resulting material landscape as the key concern. Notwithstanding this focus, Lambert's study displayed relatively little concern with culture as a cause of landscape change, concentrating instead on economics.

Fourth, a study of the Cistercian monastic order in medieval England is in a rather different vein, being described as an attempt to 'recreate the changing landscape of a culture rather than a cultural landscape' (Donkin 1997:248).

Several other types of study concerned with regional landscape creation are considered in Chapter 6.

England, however unpromising it may appear at first sight, that is not full of questions for those who have a sense of the past. (Hoskins 1955:13–14)

The originality of this work as a piece of historical scholarship lay in the use of the landscape itself as a source of information so much more valuable, given the aims of the study, than documents and books on local history. Hoskins wanted to read the landscape, not simply to read about it. His work demonstrates a recurring concern with the local rather than the larger regional or national scene. Further, Hoskins (1955:231) shows a concern about what he perceived to be some of the more unfortunate, even painful, recent developments in landscape: 'especially since the year 1914, every single change in the English landscape has either uglified it, or destroyed its meaning, or both.'

THE *LANDSCAPE* MAGAZINE TRADITION

The most substantive contribution to the literature concerned with the making of landscape, other than that associated with the landscape school itself, is without a doubt the tradition initiated and sustained by Jackson that began with the publication of the magazine *Landscape* in 1950. Over the next 17 years, Jackson financed, published, edited, and contributed much material to the periodical. Initially subtitled *Human Geography of the Southwest*, the magazine soon outgrew its regional focus as well as the link with a single discipline to develop into an acclaimed outlet for articles adding to understanding of landscape. The distinctive, thought-provoking contents of this magazine struck a chord with many academics and others interested in landscape topics, successfully filling a void in the literature that Jackson had identified.

Six principal characteristics of the tradition initiated by Jackson combined to capture the interest of readers. First, Jackson's magazine recognized that landscapes are evolving, such that there is no such thing as a static or finished landscape, a suggestion that was in accord with the ideas of cultural geographers who were steeped in the landscape school tradition. Second, Jackson stressed that to change

a landscape in a desirable direction necessarily implied changing the culture of the occupants, an idea that is always a controversial matter—recall the reference to social engineering in Box 2.3, for example. For Jackson, as with Hoskins, this topic was an important one because of the sometimes undesirable tendencies of contemporary landscape-forming activity. Third, *Landscape* emphasized the vernacular, the everyday commonplace features in landscape such as houses, and it is in this context that Lewis (1975:1) used the phrase 'Common houses, cultural spoor'. Fourth, the tradition viewed landscapes as related to prevailing ideologies and interpreted them in relation to the values of their occupants. Landscapes were, according to this view, composed of material things but also had symbolic character. Fifth, recognizing the symbolic quality of landscape introduced the need to read the landscape in order to learn both about that landscape and about those who authored it. For Lewis (1979), a landscape was seen as an 'unwitting autobiography'; more accurately, it is an autobiography that is forever being revised. Sixth, Jackson was concerned with the idea that landscape was both a reflection of the unity of humans and nature and a means by which it was possible to understand those who helped to create the landscape.

The tradition initiated by Jackson is original, but it would be misleading to give the impression that it is quite different from landscape school writing. There are clear links with the ideas of Sauer, and not surprisingly, Sauer wrote for *Landscape* magazine on several occasions. With regard to the study of vernacular landscapes, recall that Bowman (1934:149–50), in an early review of human geography as a social science, explicitly identified the landscape school concern as follows: 'Wherever man enters the scene he immediately alters the natural landscape, not in a haphazard way but according to the culture system which he brings with him, his house groupings, tools, and ways of satisfying needs.' Further, the original landscape school diagram prepared by Sauer referred to landscape features including housing plan and housing structure (Figure 2.2). Sauer was by no means unconcerned with the vernacular landscape. Nevertheless, the *Landscape* magazine tradition has inspired many cultural geographers with such original ideas as its concerns with symbolism and landscape reading, ideas that clearly anticipate some of the work on symbolic landscapes discussed later in this book. Box 5.4 suggests that much of the work accomplished in this tradition might be best thought of as the geography of the everyday landscape.

MAKING THE AMERICAN LANDSCAPE

In introducing *The Making of the American Landscape*, Conzen (1990a:vii) noted: 'This book has

its roots in the fertile bicontinental traditions of landscape study nurtured by William G. Hoskins and John Brinckerhoff Jackson, and it was written in the belief that nothing quite like it exists in the American literature, and that there is a place for it.' The organization of the book reflects the complex processes that were understood to have contributed to the building of the American landscape. Thus, there is an overview of the physical environment, an account of the approximately 15,000–year Aboriginal presence, and discussions of the principal colonizing cultures, namely Spanish, French, and British. Conzen argues that relatively speaking, the Spanish and French cultural occupa-

---

Box 5.4 Everyday Landscapes

The concern in much of the writing associated with the *Landscape* magazine tradition is a concern with the everyday landscape, with the majority of people in a landscape and not with those few who are wealthy and exercise power. These everyday landscapes come into being and continually change because they reflect the changing ideas, values, and needs of ordinary people. For geographers such as Aschmann, landscapes were studied 'as places where we could live and work and celebrate together' (Jackson 1997b:vii), while Clay (for example, 1980, 1987, 1994) has written a series of interpretive essays on the landscapes and character of urban areas. 'Ready and waiting to be found out there in our visible landscape, there exists an observable and universal order that speaks eloquently to us, that will tell us far more of ourselves and our future than we can see at any one moment' (Clay 1987:1). From this perspective, we only have to see what is right in front of our eyes. This interpretation sees an orderly, indeed grammatical, landscape. In a similar vein, Hough (1990) focused on the way that traditional vernacular landscapes reflected the character of an area and the identity of people, thus emphasizing differences between places, while designed landscapes tended to negate the differences between places as a result of the fact that local populations were not responsible for their creation.

There are many other books and articles that reflect this *Landscape* magazine interest in the every-

day world. A few examples of titles, or phrases taken from titles, are the following:

- *House Form and Culture* (Rapoport 1969)
- *The Accessible Landscape* (Jackson 1974)
- *The Look of the Land* (Hart 1975)
- *Ordinary Landscapes* (Meinig 1979a; Groth and Bressi 1997)
- *Common Landscapes* (Stilgoe 1982)
- *Learning from Looking* (Lewis 1983)
- *Discovering the Vernacular Landscape* (Jackson 1984)
- *Wood, Brick, and Stone* (Noble 1984)
- *Right Before Your Eyes* (Clay 1987)
- *The Visual Elements of Landscape* (Jakle 1987)
- *Real Places* (Clay 1994)
- *The Gas Station in America* (Jakle 1994)
- *People and Buildings* (Marshall 1995)
- *Landscape in Sight* (Jackson 1997a)
- *Fast Food: Roadside Restaurants in the Automobile Age* (Jakle and Sculle 1999)

Capsule phrases such as these reflect this tradition of reading landscapes created by the majority of people to serve the needs of their daily lives rather than focusing on the behavior and landscape consequences of elite populations. An interesting corollary of this focus is that unusual landscape features were necessarily of lesser concern, and *Landscape* magazine occasionally published pieces that were critical of some conservationist groups that emphasized the unusual at the expense of the everyday.

tions were spatially restricted, leading to the formation of cultural regions that are losing their distinctiveness over time. The British cultural occupation involved a series of traditions representing various population groups and the variety of physical environments they encountered. The most influential British tradition originated in New England and southeastern Pennsylvania and spread west over much of the larger American area, while the British plantation tradition expanded west throughout much of the American South.

These sections on principal colonizing cultures are followed in the book by accounts of the national settlement policy that was initiated after political independence, and of the effects of that policy on three principal physical environments: forest, grassland, and desert. The book identifies and discusses two landscape preferences: the first involved shedding cultural baggage, typically to create a landscape of single-family farms and homes; the second involved retention of cultural roots. Conzen then evaluates industrial, urban, and transport processes as they have involved changes in landscape. The book concludes with sections on the role of central authority, the importance of power and wealth, and the vernacular landscape.

The multilayered organization of *The Making of the American Landscape* is thus an accurate reflection of the complexity of the subject matter, an acknowledgment of the fact that landscapes are shaped through time by a multitude of processes that are not always easily or readily distinguishable.

MAKING THE ONTARIO LANDSCAPE

A second valuable addition to the body of literature on the reading of landscape is the study by McIlwraith (1997) on two centuries of landscape change in Ontario. McIlwraith's stated concern is regional, not local, and his approach is very much in the *Landscape* magazine tradition. His aim is to read landscape as a reflection of the way in which ordinary people lived their lives, and hence he shows an interest in both material and symbolic land-

scape. McIlwraith achieves this aim through two sets of discussions. First, he establishes a context through accounts of attitudes to land, of historical matters, of the land survey system, of the process of determining place names, and of the technologies and resources employed in the making of landscape. Second, he proceeds from these contextual issues with a consideration of particular landscape features and of the clustering of features. The book includes discussions of traditional house architecture as it reflected the consistency of vernacular behavior, as well as accounts of churches, schools, barns, fences, industrial buildings, and gravestones. Clusters of features that McIlwraith considers include farms, roadsides, transport systems, and town streetscapes.

MAKING THE IRISH LANDSCAPE

There is a tradition of historical cultural geography in Ireland that stands somewhat apart from the various approaches identified so far, but that is explicitly concerned with understanding both how the Irish landscape was shaped through time and how landscape reflects the culture of those who did the shaping. Indeed, one reason this tradition is different is that it is so eclectic in character, being based on an integration of ideas from geography, anthropology, and history. In this respect it stands apart especially from the English historical geography tradition, within which, until recently, very few attempted to integrate geography, anthropology, and history—the principal exception being Fleure (1951).

The key figure in the Irish tradition was E. Estyn Evans (1905–89), whose influential university career extended from 1928 to 1968. For Evans, an understanding of Ireland and the Irish landscape involved a wealth of interests and approaches. There were concerns shared with *la tradition Vidalienne* and the related *Annales* historical school, especially regarding *genre de vie* and *milieu*. Such interests are not surprising in light of the plurality of regional cultural identities evident throughout Ireland. Evans was also concerned with the evolution of the material cultural landscape in response to cultural occupance, in close accord with the landscape school ideas of Sauer. As noted, Evans's willingness to incorporate these ideas was in marked distinction to the general neglect of these ideas in the more dominant English historical geography tradition. In accord with both Vidal and Sauer, Evans employed the term 'personality' to refer to the idea that through a long history of human occupance, a distinctive character was impressed on the Irish landscape.

Evans maintained a focus also on fieldwork and related observation of both people and landscape. Such approaches were advocated by Sauer and practiced in the British local history tradition initiated by Hoskins. The concern with fieldwork also involved interests in common people and their way of life, and therefore in the related vernacular landscape. Further, the explicit linking of culture and landscape included recognition of the symbolic content of landscape, although this was expressly limited to the rural scene. The interests in vernacular landscapes and in landscape as symbol are, of course, central to the *Landscape* magazine tradition.

Although this eclectic approach to historical cultural geography is consistent with the related ideas of both Vidal and Sauer, the additional interests result in it being a distinctive tradition. It was practiced by Evans and others throughout much of the twentieth century with minimal formal interaction with other scholars and with generally little recognition outside of Ireland.

Pause for Thought

Because landscapes are indeed shaped through time, is it possible and appropriate to observe a landscape and to seek to read the detailed history of that landscape—to regard the landscape as a text as suggested in Box 3.5? Rather than emphasizing the cultural causes and the landscape consequences in the Sauerian tradition, why not investigate the contemporary landscape in terms of how each of the features came into being, asking questions such as When was the feature added? Under what circumstances was it added? Why was it added? Who was responsible?

Imagining Past Landscapes

In order to comprehend nature in all its vast sublimity, it would be necessary to present it under a twofold aspect: first objectively, as an actual phenomenon, and next subjectively, as it is reflected in the feelings of mankind. (von Humboldt, in Saarinen 1974:255–6)

Despite this pioneering statement, and the fact that some of the writings of Vidal and Sauer included strong visual content, the idea that both objective and subjective worlds need to be considered in studies of landscape evolution was not central either to the Sauerian school or to historical geography more generally. Even two specific proposals for a geographic study of subjective worlds, published respectively in 1947 and 1951, had minimal impact at the time, and their influence continues to be debated today.

The first of these was not clearly related to earlier or contemporary geographic literature, nor did it adequately signpost future research directions, but Wright did explicitly distinguish between the subjective and the objective:

Objectivity . . . is a mental disposition to conceive of things realistically. . . . The opposite of objectivity would, then, be a predisposition to conceive of things unrealistically; but, clearly, this is not an adequate definition of subjectivity. As generally understood, subjectivity implies, rather, a mental disposition to conceive of things with reference to oneself. . . . While such a disposition often does, in fact, lead to error, illusion, or deliberate deception, it is entirely possible to conceive of things not only with reference to oneself but also realistically. (Wright 1947:5)

Thus, a subjective view of the world might range anywhere along a continuum from complete error to complete accord with reality, and Wright proposed a new discipline—'geosophy'—to examine these true and false geographical ideas. However, these original ideas were related neither to other disciplines, such as psychology, nor to other developments in geography, and

A traditional building style in many of the more remote areas of Europe, including much of the Celtic fringe, was the long-house. The main entrances, front and back, were in the middle, as was the fireplace. Traditionally, on one side was one large room where all family members and any visitors ate, worked, and slept. On the other side were the animals. Sometimes the floor on the human side was elevated, or if the house was built on a slope then the animal part was located down slope. Windows were rare, as warmth was more important than light. The example in this photo retains the vernacular tradition but without the inconveniences of earlier times. It is not difficult to understand why Evans was prompted to use the term 'personality' when discussing the distinctive character of the Irish material cultural landscape as it reflected the long history of human occupance. (Elan Valley Ranger Services/www.elanvalley.org.uk)

there was no substantive empirical attempt to pursue the argument.

The second contribution to the geographic study of subjective worlds, by Kirk (1951), was similarly uninfluential at the time of publication, at least partly because it was published in a journal that was relatively inaccessible to North Americans, so that the work was not generally known until the 1960s. Kirk's was a more explicit attempt to incorporate behavior into geographic analyses, making a distinction between the world of facts—both physical and human—and the environment in which these facts were culturally structured and in which they acquired cultural values. Kirk labeled the world of facts the *Phenomenal Environment* and the world in which they received meaning the *Behavioral Environment*. Twelve years later,

Kirk (1963) developed these ideas to include the suggestion that human behavior in the Behavioral Environment was rational. These original ideas, for geography, have not become central to subsequent geographic research.

Wright and Kirk identified the presence and importance of subjective worlds, as Humboldt had a century earlier, but these contributions did not form a continuous development of ideas, nor did they stimulate research into subjective worlds. It was not until the 1960s and the rise of a behavioral geography initially associated with spatial analysis and subsequently inspired by humanistic concepts that cultural and historical geographic studies began to focus explicitly on the role played by subjective worlds in landscape evolution.

Creating Images

Much of the individual and group behavior that was critical for landscape formation occurred in situations of uncertainty, and what happened in these uncertain environments may have been more closely related to views developed from myth than to the realities of the environment itself. Indeed, as Salter (1971b:18) points out, myth may be a 'prime generating factor in the decision to migrate. Though the mythical goal is seldom realized, the processes of mobility and discovery are just as real as if the myth had been

---

Box 5.5 Learning in Landscape

Learning often occurs in environments of uncertainty, where a discrepancy between objective and subjective worlds forces people to make adjustments to their behavior. Colonization invariably imported cultural systems in confrontation with new, strange, and often inhospitable environments. Rarely were the particulars of these environments consistent with colonists' perceptions of them, and even more rarely were they totally amenable to the resolution of colonists' aims (Figure 5.12). Perceptions and aims therefore had to be modified, and this initiated a process of active, conscious adjustment of learning that continued until perceptions were consistent and aims became realistic and attainable (Cameron 1977:1).

Further developing this idea, Cameron (1977) devised a model of adaptive learning that provides a framework for analysis of settlement in pre-1850 western Australia (see Figure 5.13). The colonization process involves the establishment of behavior pat-terns in a new environment, and these patterns are a consequence of a complex interplay of factors. The model is useful, although it does not focus explicitly on such factors as prior experience of the colonizer, including any preadaptive traits, and the attitudes held by individuals or groups. Clearly, much behavior in landscape is affected by prior experience and the relative importance of individual identity and group membership, as in the example of Irish settlers in eastern Canada mentioned earlier in this chapter.

Prior experience may be a key consideration if the goals established in the new area are similar to those pursued previously, if the new environment is similar to the old, if there are limited links with other groups, if the image held of the new environment is such that repetition of prior behavior is judged appropriate, if institutional considerations do not play a major role, and if group membership is more important than individual characteristics. The char-

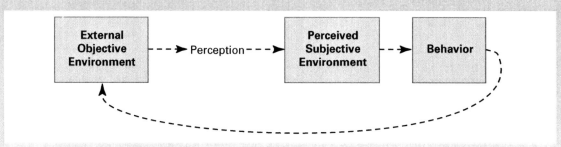

Figure 5.12 Perception, Behavior, and Landscape. This is a schematic diagram outlining the essential logic underlying many studies of perception in cultural geography (see also, Haynes 1980:2). There is an external objective environment that is perceived by humans. The environment that is perceived—the subjective environment—is necessarily different to the objective environment. Further, the perceived environment is different for each individual, although individuals who belong to a group, such as a culture, are liable to have perceptions that are comparable. This is because perception is affected by both individual characteristics and group characteristics. Behavior in landscape is a response to the environment as perceived, not a response to the objective environment. Behavior changes landscape and thus changes both the objective and subjective environments through time.

Several attempts have been made to conceptualize these ideas more explicitly, and Kitchin (1996) reproduced nine rather more sophisticated formulations centered on the theme of cognitive mapping. There has been emphasis on trying to understand what is going on in the human mind as images are constructed, with most of this work being built on the ideas of cognitive psychology and including perception as one component of cognition.

realized.' Although inappropriate behavior often resulted from perceptions that were significant distortions of reality, errors were normally not repeated (see Box 5.5).

Since the 1960s especially, historical and cultural geographers have shown much interest in understanding the differences between objective and subjective environments—particularly in areas of European overseas expansion during the periods of exploration and early European settlement—because of the realization that any comprehensive understanding of the evolution

---

acteristics of each individual are likely to be important factors if group membership patterns are weak, if institutional considerations are lacking, if goals are unclear, and if the environment is relatively benign. Group membership will be important if individual characteristics are invariable, if institutions do not dominate, if links with other groups are limited, and if there is a shared environmental image.

As one example of the importance of perception in this larger context, consider the reaction of European settlers to North American forests. In southern Ontario, civilization and progress were deemed incompatible with the continuing presence of forest, and so colonizers attacked forests 'with a savagery

greater than that justified by the need to clear the land for cultivation' (Kelly 1974:64). This was despite the fact that the agricultural value of retaining some forest was understood. A lack of prior experience was also important in this situation because most incoming settlers were ill prepared for the vastness of North American forested areas. The goal of establishing commercial farming required some land clearing, but not on the scale undertaken. In this example, the behavior of settlers largely resulted from their lack of prior experience with the type of environment that they encountered, from the behavior of others, from the perception of the forest as some sort of threat to advancement, and from related attitudes.

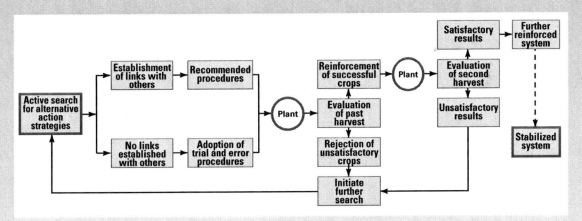

Figure 5.13 A Simplified Model of Colonizer Behavior.
- First, movement into a new environment prompts a search for those actions that will satisfy preconceived goals. Hopefully, a colonizer can simply imitate others who are already achieving the equivalent goals, but if this is not possible, then the colonizer must take actions determined subjectively on the basis of the image held of the area.
- Second, actions are taken.
- Third, when the consequences of actions are known, the colonizer undertakes a process of evaluation. Actions that were successful in achieving the desired goals are repeated, and those that were unsuccessful are rejected and a new search procedure begun.

This sequence—search, operate, evaluate—is repeated as many times as needed until the desired goals are achieved. Through this process, the colonizer learns a set of actions that become repeated on a regular basis.

Source: J.M.R. Cameron, *Coming to Terms: The Development of Agriculture in Pre-Convict Western Australia* (Perth: University of Western Australia, Department of Geography, Geowest 11, 1977):3. (Reprinted by permission of the author)

of settlement and the transformation of land-scapes over large areas of the globe requires such understanding. The key idea supporting this type of research is that humans do not respond directly to the environment but rather to their mental image of the environment, an image that is the product of perception. Thus, human land-scape-making activity is related to the human image of the environment. Although it is possi-ble to conceive of more complex relationships and interactions, the basic logic of this approach is straightforward. Certainly, the idea that people represent environmental information in the form of an image in their heads has proven very attractive to cultural geographers. It is also well established that these images err in certain con-sistent ways. For example, local areas are usually known in relative detail while more distant areas are less known. Three principal issues have been addressed in the historical context:

- How were subjective worlds—images—that were different from reality created?
- What was the extent of the difference between images and reality?
- Is it possible to understand behavior and the changing landscape by reference to images?

These are not easy questions to answer either conceptually or empirically, especially given the difficulties of collecting information on imagined worlds in a historical context. Attempts to answer the first of these three questions have stressed the role played by those who provided descriptions. It is often appropriate to distinguish between the accounts of those who saw the area for them-selves and those who produced second-hand descriptions, and it is also recognized that the impression conveyed relates to the motivations of those who are involved. In the context of much European overseas activity, it may be use-ful to identify the different motivation of such individuals and groups as promoters, officials, settlers, travelers, and natural historians: each brought their particular biases to their written assessments, their various images of reality.

Promoters, such as leaders of immigration schemes and shipowners, were especially pro-lific writers and typically exaggerated the ben-eficial features of the environment in order to encourage settlement. The expectations of immigrant settlers were largely based on the writings of such promoters. Literature produced by officials was more variable, sometimes mis-leading and sometimes relatively objective. The usually optimistic—especially during the initial phase of settlement—settler literature was taken seriously by potential settlers, but was limited in quantity. Travelers and natural historians produced the least useful accounts because they often focused on topics of very limited interest to prospective settlers.

One of the more notorious instances of a subjective environment concerned the supposed existence of a navigable North West Passage across the northern part of North America, link-ing the Atlantic and Pacific oceans. Proposed both by explorers and by armchair geographers in Britain, the misguided image based on this fabled waterway had a profound influence on the history of the area:

The whole enterprise was founded on a misappre-hension, a geographical fiction, a fairy tale, spring-ing out of the kind of stories sailors tell to amaze landsmen or to delude other sailors, to which were soon added the inferences, speculations and downright inventions that scholars manufactured to amaze themselves. (Thomson 1975:1)

The three examples that follow are not quite so dramatic and colorful as the example of the North West Passage, although they certainly illus-trate the important role played by the distortion of reality, a distortion that was often intentional.

THE SWAN RIVER COLONY

An excellent example how substantially incor-rect images were created concerns the Swan River Colony of western Australia during the period from 1827 to 1830. An 1827 exploratory voyage by Captain Stirling and Fraser (a botanist) was the sole substantive source of information about the Swan River area during this period. Although their visit to the area was brief, the subsequent report to the British Colonial Office supported the idea that Swan River was a suitable location for a colony and settlement because of fertile soils, a suitable climate, and adequate water.

What is remarkable about this example is the way the media presented the contents of the report to the public. As Heathcote (1972:81) remarks, 'Between the "place" and the "people" themselves, however, lay the "media" by which the so-called facts of the one were made available to the audience of the other'. Stirling and Fraser's report was made available to the public only in the form of an article published in a journal *Quarterly Review* (1829) but this article served as the basis for later journal articles in the *Mirror* (1829), *New Monthly Magazine* (1829), and *Westminster Review* (1830). These publications were all directed at the middle class and entrepreneurs.

The first article, in the *Quarterly Review*, was written by Barrow, the acknowledged expert on the southern hemisphere, and was aimed at an audience of shipowners, merchants, and wealthy potential settlers. Stirling and Fraser's generally favorable report was the basis for the article, but Barrow employed various strategies, such as deleting unfavorable comments and reporting

This 1827 painting shows Stirling and a small party of men exploring the Swan River about fifty miles upstream. Aboriginals observe the newcomers with little apparent interest. The vegetation along the river is luxuriant, and the black swans would have been exotic by European standards. Note that the mountains are not too far away. (National Library of Australia, nla.pic-an2260474)

optimistic conjecture as fact, to better promote the area. The three later articles continued this process of image distortion to such a degree that the *Westminster Review* article showed only superficial resemblance to the original report.

Two examples of the extent of the distortion are shown in Table 5.1: the two most important features of the area for potential settlers, namely the size of the cultivable area and the quality of the climate, were distorted almost beyond recognition. The conditions that favored distortion in this example include the short time period involved, the existence of only one basic source of information, and the pro-colony mood of the British public at the time, but the sequence by which the information was disseminated differed but slightly from the way news was communicated about other colonial areas.

Undoubtedly, Swan River is a compelling instance of the misleading details that were made available to those considering investing in or moving to an overseas area. Consider the following two statements. First, the *New Monthly Magazine* reported that 'the soil is a fine brown loam, alternating with broad valleys of the finest alluvial soil; the hills appeared finely timbered; the valleys produced an immense luxuriance' (in Cameron 1974:65). Second, one early settler complained: 'Not a blade of grass to be seen—

Table 5.1 Image Distortion in Swan River Colony

	Stirling and Fraser 1827	*New Monthly Magazine* 1829
Cultivable area (sq. mi)	100	22,000
Conjecture, as % of total description of climate	9.7	51.1

Source: J.M.R. Cameron, 'Information Distortion in Colonial Promotion: The Case of Swan River Colony', Australian Geographical Studies 12 (1974):66, 68. (Reprinted by permission of Blackwell Publishing)

nothing but sand, scrub, shrubs, and stunted trees, from the verge of the river to the top of the hills. . . . I may say with certainty, that the soil is such, on which no human being can possibly exist' (in Cameron 1974:74). The subjective world presented to settlers by the media and the reality—or, more correctly, the perception of reality—that the settlers encountered were worlds apart.

GREAT PLAINS OR GREAT AMERICAN DESERT?

Although the Great Plains is probably the most studied example of a subjective landscape in the historical cultural geographic literature, it is also one of the most confusing. It has often been claimed that the popular image of the plains in the period from about 1820 to about 1870 was that the area was a desert; indeed, the area was known as the Great American Desert. But this scholarly perception—held by both historians and geographers—that the area was recognized as a desert has itself been challenged as an incorrect perception.

The dominant scholarly view, based largely on a study of the reports of explorers and educational texts, is that the area was perceived as desert, an idea that was initiated by early nineteenth-century European explorers. Scholars supporting this view have offered three principal reasons to explain the emergence of this perception and its continuance for about fifty years. First, individuals tend to record what impresses them most and, for explorers from the eastern United States, the small areas of desert in the plains merited substantial description. Second, relative to the eastern region, the treeless plains could fairly be described as desert. Third, there was a powerful political impetus to move west to the Pacific coast as quickly as possible, and promoting a view of the extensive plains region as desert was conducive to that goal. In these three ways, the plains were being defined in eastern terms. Corrections to the myth of the Great American Desert came around 1870 with improved knowledge of the area.

But a major challenge to this dominant scholarly view claims that 'the desert belief was far from universally held in America in the mid-nineteenth century' (Bowden 1976:119–20).

According to this argument, the incorrect idea that the plains were perceived as desert can be understood in terms of the variable climate of the area, with some years being very dry and others much wetter: travelers passing through the region during a dry year were likely to interpret the region as a desert. Further, several influential historians, particularly, Turner, Webb, and Malin, stressed differences between the humid east and the drier plains, thus contributing to the desert image.

'A New and Naked Land'

Settlement of the Canadian prairies by Europeans between about 1895 and 1914 was not simply the next logical progression in the process of overseas movement and a response to the demand for wheat; it was also prompted and sustained by an organized campaign of intentional distortion of the geography of the region. As Rees (1988:4) comments, 'The more distant and inhospitable the land, the greater the blandishments necessary to attract settlers to it. Cold, dry, treeless plains half a continent and—for Europeans—an ocean away were hardly alluring.'

In the mid-nineteenth century, the region was viewed as either a northern continuation of the Great American Desert or a southern continuation of the Arctic—the area was indeed *The Great Lone Land*, as Butler (1872) called it. But this was an unsatisfactory state of affairs for both British and Canadian officials who were concerned about the lack of settlement between Ontario and the Pacific coast, especially given the possibility of American northward expansion. Accordingly, in 1857 there was a British expedition to the area under the leadership of Captain John Palliser and a Canadian expedition the following year under the leadership of Henry Youle Hind. The reports of these expeditions introduced the idea that the Prairies had agricultural potential, with the exception of the grasslands area in the southwestern portion, which became known as Palliser's Triangle. This was the start of a process of image change.

This illustration, a wood engraving, is based on a watercolour painted by the artist–engineer John A. Fleming, who was a member of the 1858 Hind exploratory expedition to the Canadian West. It shows wagons being used to transport canoes, with the Qu'Appelle River in the background. Although the painting does not particularly suggest that the area has agricultural potential, neither does it reinforce the *Great Lone Land* image, considering its foreground that includes several figures and evidence of human activity. (National Archives of Canada, C-003692)

During the 1860s, 1870s, and 1880s, numerous favorable reports of the Canadian prairies appeared, often written with a limited factual basis. Thus, A.J. Russell, the inspector of Crown agencies in Canada, stated in 1869 (without having visited the Prairies) that both of the expedition reports had probably underestimated the agricultural potential of the grass-land region, while John Macoun, a government botanist and author of *Manitoba and the Great North-West*, declared in 1881 that the area was 'all equally good land' (in Rees 1988:7). These and other largely unjustified claims about the physical environment encouraged massive campaigns by the two principal promoters of the area, the Canadian Pacific Railway and the government. But despite this ongoing process of image change, the results were disappointing, with little immigration from the United States, Britain, or western Europe. Something else was needed to attract immigrants to the Prairies, a fact that was recognized by Clifford Sifton, who was placed in charge of the Department of the Interior in 1896.

Sifton initiated three principal policy changes. First, he extended the range of advertising into non-traditional immigrant source areas, notably into eastern and southeastern Europe and Ukraine. Second, he substantially increased the quantity of the promotional literature. Third, he began to make major changes to the content of the promotional literature. For example, Sifton disallowed overseas publication of Manitoba temperatures. In addition, he required that the words 'snow' and 'cold' not be used in government publications, that snow not be included in any illustrations, that the often inadequate rainfall in the growing season be hidden through the use of annual averages, and that the great open expanses of the prairies be softened through descriptions that focused on enclosures. In this example, image distortion was intentional and substantial.

The Department of the Interior produced this 1911 poster to encourage prospective immigrants from the northern United States to settle in the Canadian prairies. Aimed at bringing 40,000 workers into southern Saskatchewan, the poster advertises $12 fares, its highly stylized representation meant to suggest what an easy move it is to make. Although the appeal is directed primarily at agricultural workers, part of the image is of a settled urban landscape far removed from frontier circumstances. (National Archives of Canada, C-056088)

It is clear that the promoters of settlement were operating in a context of high emotion and high expectations. There was a perceived need to settle the West as quickly as possible, and it seems that this justified the use of dubious strategies to entice immigrants to the area. For many settlers, the reality they encountered on arrival was quite different from the expectations generated by the image makers: an unliterary child from Ontario described the prairies as 'a new and naked land' (in Rees 1988:36).

The conclusion reached by Christopher with reference to southern Africa applies equally well to the Canadian prairies and to many other areas of European overseas movement:

To a large extent men believed what they wanted to believe or that which they were given to believe, rather than the truth; and there can be little doubt that large tracts of southern Africa would never have been settled by Europeans or would have been settled in different circumstances had the true state of affairs been appreciated. Immigrants' illusions were sometimes shattered soon after they reached southern Africa, but illusion had brought them, and they had to make the best of it. (Christopher 1973:20)

The discussion of subjective landscapes demonstrates that the subjective environment is often a significant distortion of the objective environment and that, as a consequence, much behavior occurs in situations of uncertainty. Certainly, humans know little of the consequence of behaviors that are not repetitive or that are not being imitated. Given an initial uncertainty, it is reasonable to suggest that a learning process usually proceeds and involves increasingly more appropriate behaviors through time (as described in Figure 5.13).

Pause for Thought

This is an opportune time to raise a very thought-provoking idea: If it is reasonable to suggest that much human behavior during the early stages of settlement in a new environment had a rather flimsy rationale, does this mean that alternative behaviors might well have occurred if better information had been available? The suggestion that different decisions might easily have been taken and then might have had different implications for landscape does seem reasonable. This is a stimulating and seductive idea for cultural geographers to consider and is explicitly addressed in Box 5.6 and Figure 5.14.

Concluding Comments

The evolution of the cultural landscape. For many cultural geographers, this one phrase identifies the central concern of the landscape school tradition as initiated by Sauer in the 1920s. As evident from discussions in this chapter, many contemporary cultural geographers continue to work in the landscape school tradition but while referring to a more specific evolutionary concern. Indeed, a historical orientation was the first of the seven 'persistent preferences' of cultural geography noted by Mikesell (1978:4), while Williams (1983:3) notes that in 1936 Sauer described historical geography as 'the apple of my eye'. Similarly, Donkin (1997:248) writes: 'a fascination with "origins" is both primitive and widely shared—the product, perhaps, of "the essential time bond of culture rather than its looser place bond", some rather uncomfortable words of John Leighly, which I've often pondered.' It bears repeating that this emphasis on time and change was in marked contrast to the regional geographic approach (as favored by Hartshorne and others) that dominated North American geography between about 1900 and the mid-1950s.

Writing about the physical landscape, Sauer (1925:36) noted that it was to be understood in terms of both space and time, since it was continually changing in response to geomorphologic and climatic processes. He also noted that the presence of humans and the consequent introduction of cultural processes resulted in landscapes that were no longer only physical but that also reflected cultural occupance. Physical geographers studied the physical processes generating change, while cultural geographers studied the cultural processes generating change. Although Sauer himself apparently showed little interest in such methodological statements after they were written, their impact on others seems undeniable. Indeed, notwithstanding the various earlier European traditions that Sauer acknowledged, it is not an exaggeration to claim that he was responsible for prompting the English-language interest in both the evolutionary study of cultural landscapes and the historical geographic tradition of studying change through time. Mikesell (1978:3) asserted that 'Sauer acted as a catalyst for cultural geography, as well as an initiator of specific trends.'

Cultural geographers continue to be concerned with studies of changing landscapes and with the somewhat loosely defined historical geography tradition. This is evidenced by a considerable body of literature focusing on the evolution of cultural regions, especially in the United

Box 5.6 What Might Have Been? Part I

One of the most interesting yet rarely pursued questions that historical cultural geographers might consider raising is '*What might have been?*' Given that much landscape-making behavior occurred (and indeed continues to occur) in situations of uncertainty, it is clear that behaviors different from those that took place might have taken place, and that any such different behaviors might have resulted in different landscapes. In this sense, the basic logic behind **counterfactual** queries is simple, and their use can be justified as follows: 'Every statement of causation implies the counterfactual proposition that in the absence of the causative factor the event would not have occurred' (North 1977:189). A counterfactual is any statement that is untrue in that it typically queries, 'What would have happened if . . . ?'

Because the usual object of scholarly enquiry is what is real, there is an implicit assumption that this reality is in some sense a necessary consequence of what went before; this can be described as the bias of hindsight. To think counterfactually is to acknowledge that things might have happened differently. If things had occurred differently and a different outcome had resulted, then this would suggest that the presumed causes were indeed important. If, however, things had occurred differently and the outcome was similar to the reality, then this would suggest that the presumed causes might not be critical to the outcome. One of the dangers of not thinking counterfactually is that we tend to adjust our thoughts so that the expected probability of something occurring that has in fact occurred is increased, while the expected probability of something occurring that has not occurred is decreased. Contingency may be ignored if counterfactual logic is not employed.

The principal debate about causation and counterfactuals takes place in philosophy, although historians have long taken an interest in such matters. Typically, historical analyses in this vein belong to two types. First, some counterfactual histories may be largely speculative, involving reasoned assumptions about what might have happened had some part of the past been different; military historians have been attracted to this type of argument because they have recognized that many critical events depended on individual decision-making or on circumstances that might very well have been different, such as weather conditions. Second, other counter-

factual histories are designed to specifically test ideas about the past, for example, the idea that North American westward expansion was closely related to railroads; most of these studies are in the tradition of the new economic history that developed in the 1960s and involve formal procedures of hypothesis testing by means of quantitative methods.

The following are three examples of historical cultural analyses using counterfactual logic. First, an Australian historian whose many works have inspired geographers because of their recurring concern with spatial issues, made a compelling argument for counterfactual analyses:

> We easily forget that every statement we make of why events happened is in part speculation. In looking at history we have to ask: what might have happened? Every time we affirm the profound importance of a particular event—whether the finding of gold in 1851 or the failure of the North Queensland separation movement in the 1880s—we unmistakably imply that such an event was a turning point and that society—but for that event—would not have changed direction. There can be no discussion of a powerful event without realizing that it is like a traffic juncture where a society is capable suddenly of changing direction. In writing history we concentrate more on what did happen, but many of the crucial events are those which almost happened. The born and the unborn may seem completely different but essentially they have to be analyzed and discussed in the same way. (Blainey 1983:202–3)

Second, a leading North American historical geographer thoughtfully described some of the consequences for the emerging political unit of Canada that might have resulted from minor changes in the exploratory activities of Champlain during the years 1605–8 (Clark 1975). Third, as part of a larger account of the North American landscape, Meinig (1993:215) created two thought-provoking alternative political landscapes (see Figure 5.14).

Given circumstances of uncertainty and the role of chance, it is possible that cultural geographers might pursue counterfactual analyses if it seems reasonable to suggest that the details of the process that is presumed to have affected landscape creation might have been different.

A Greater United States

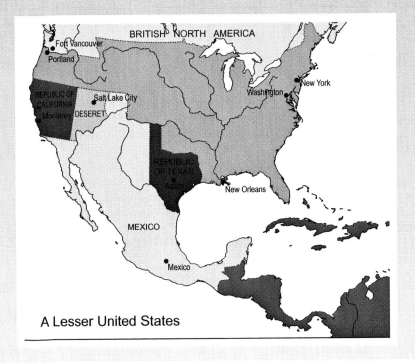

A Lesser United States

Figure 5.14 A Greater and a Lesser United States. Meinig is renowned for his evolutionary analyses of Australian and North American regional landscapes, which combine incisive description and original explanations. Especially compelling are the many bold maps concerned with regional evolution. This figure suggests two North American political landscapes as alternatives to the one that actually came into being. Both were possible realities and counterfactuals such as these are a 'fresh and forceful way of reminding ourselves that there was no destiny manifest. . . . ' (Meinig 1993:217).

Source: D.W. Meinig, *The Shaping of America: A Geographical Perspective on 500 Years of History. Volume 2: Continental America, 1800–1867* (New Haven: Yale University Press, 1993):215. (Reprinted by permission of Yale University Press)

States, and on the evolution of groups of similar cultural regions—known as cultural realms—that are evident at a global scale. These contributions are discussed in the following chapter.

But the general Sauerian influence is also accompanied today by a series of other concerns that are associated with developments in geography since the late 1970s. For example, the evolutionary study of cultural regions is enhanced by a consideration of the way in which humans relate to each other and thus identify themselves as members of groups. Similarly, the evolutionary study of cultural realms has been enhanced both by the arguments of world systems analysis and by the cultural studies tradition referred to in Chapter 3, especially by postcolonial theory. These contributions are discussed in subsequent chapters.

Further Reading

The following are useful sources for further reading on specific issues.

Jones (2004) notes a disciplinary trend favoring analyses of shorter time periods and argues that a lenghthening of timeframes enables different stories to be told.

Hugill and Dickson (1988) discuss diffusion as a multidisciplinary concern, and Entrikin (1988) provides an overview of diffusion in the context of the debate about naturalism.

Analyses of culture trait diffusion include Kniffen (1949, 1951a) on agricultural fairs; Kniffen (1936, 1965), Kniffen and Glassie (1966), Walker and Detro (1990), and Jordan (1983, 1985) on house types and log construction; Stanislawski (1946) on the grid-pattern town; Leighly (1978) on place names; Seig (1963), Brunn (1963), and Carter (1977, 1980, 1988) on agricultural practices; and Arkell (1991) and Carney (1994a, 1996) on musical styles.

In addition to the 1952 book, which is a seminal work (see Conzen 1993), several innovative arguments about the origins and diffusion of agriculture were proposed by Sauer (1968, 1969, 1970).

Leighly (1954), a colleague of Sauer, discusses both the traditional cultural and the spatial analytic approaches to diffusion.

Hägerstrand (1951, 1952, 1967) is the principal figure in the spatial analytic approach to diffusion.

Gould (1969), Abler, Adams, and Gould (1971), and Haggett (2001) include comprehensive accounts of the simulation procedure developed by Hägerstrand.

Rogers (1962) outlines the rural sociological approach to diffusion.

Pyle (1969) analyzes the diffusion of cholera during three pandemics in nineteenth-century North America, and Haggett (2000) provides a review of the geographical structure of epidemics.

Blaut (1977) argues (1) that the traditional approach to diffusion provided the needed foundations for further theory development, precisely because of the breadth of the approach, and (2) that the spatial approach was too narrow because of the tendency to view innovations in isolation and to ignore the consequences of diffusion.

Carlstein (1982) discusses diffusion and use of resources.

Yapa (1977, 1996), Blaikie (1978, 1985), and Browett (1980) analyze agricultural diffusion in the developing world.

Hauptman and Knapp (1977) contrast Dutch movement and activity in the two areas of Formosa (Taiwan) and New Netherland in North America.

Fox (1991) discusses the rise of the modern European world; Jones (1987) outlines the European 'miracle' argument, and Blaut (1993b) critiques that argument.

Sauer (1935), Veblen (1977), Lovell (1985, 1992), and Cook and Lovell (1992) have all contributed to the debate on the impact of disease on American population numbers.

Trigger (1982, 1985) and Ray (1996) are examples of studies of cultural change in North American Aboriginal groups prior to European contact.

A stimulating overview of the history of the Americas is provided by Fernandez-Armesto (2003). Studies focusing on Aboriginal identities and landscapes include Denevan (2001), Doolittle (2001), and Mann (2002). The October 1991 issue of *National Geographic* reviews the Americas before Columbus.

Vibert (1997:3–23) and Radding (1997) discuss how the voices of the subaltern were usually omitted or misrepresented in official documents.

Bauer (2003) reviews the cultural geography of colonial American literatures.

Ray (1974), Judd and Ray (1980), and Peterson and Anfison (1985) study Aboriginals in the fur trade.

Unstead (1922) and Smith (1965:128) discuss historical geography as the study of the past.

Williams (1983, 1987) and Donkin (1997) describe the approach to historical geography initiated by Sauer.

Norton (1984), Butlin (1993), Conzen, Rumney, and Wynn (1993), Hornbeck, Earle, and Rodrigue (1996), and Baker (2003) detail the varied methodologies practiced in historical geography. These works aid considerably in placing the landscape school emphasis in a larger historical perspective, making it very clear that Sauer initiated a 'sweeping and majestic brand of culture history' (Earle 1995:457). Grey (1994) continues this tradition in a study of New Zealand.

Examples of the use of cross-sections include Cumberland (1949) on New Zealand, Clark (1959) on Prince Edward Island, and Darby (1973) on England.

Mikesell (1976) and Conzen (1993:33–7) discuss the sequent occupance approach.

Webb (1931, 1964) studied frontiers in the Great Plains and in a world context.

Mikesell (1960), Wishart, Warren, and Stoddard (1969), Eigenheer (1973–4), Hudson (1977), Christopher (1982), and Guelke (1987) consider spatial aspects of the frontier and definitional issues concerning new land settlement and cultural interaction.

Mitchell (1977) studies the Shenandoah Valley frontier.

Clark (1959), Jordan (1966), Lemon (1972), Mannion (1974), Rice (1977), McQuillan (1978, 1990, 1993), Ostergren (1988), and Conzen (1990a) examine the loss or retention of cultural traits, especially agricultural practices, by immigrant ethnic groups in North America.

Norton (1988) and Conzen (1990b) outline general considerations relating to cultural trait loss or retention.

Early examples of evolutionary regional studies include those by Kniffen (1932), who focused on physical and later cultural landscapes of the Colorado delta area and recognized three stages of occupation and related landscape, and by Meigs (1935), who focused on the Dominican mission frontier of lower California, also identifying physical and cultural landscapes and the processes of change. Studies of Mexican and American landscapes are included in two valuable collections of the writings of Sauer, namely *Land and Life* (Leighly 1963), and *Selected Essays, 1963–1975* (Sauer 1981).

Other British work comparable to the tradition initiated by Hoskins includes the historical writing of Beresford (1957), the historical geographic writings of Darby (1940, 1956) on the drainage of the fens and the clearing of the woodland, and numerous historical geographic analyses of farms, fields, fences, and other features of the cultural landscape (see, for example, Baker and Butlin 1973; Slater and Jarvis 1982). A popular paperback book in this broad tradition is *The Penguin Guide to the Landscape of England and Wales* (Coones and Patten 1986).

Wilson and Groth (2003) pay homage to J.B. Jackson in an edited collection of landscape studies.

Lowenthal (1968) focuses on the perceptions of the American landscape that emphasized the overwhelming character, the great size, the wild and unfinished feel, and the formless and confused visual impression, especially in comparison to the landscape of England (see also Lowenthal and Prince 1964, 1965).

Mills (1997) studies American physical and cultural landscapes and the way these have been interpreted in popular culture.

Other examples of landscape studies are Widdis (1993) on the cultural landscape of Saskatchewan; Ennals and Holdsworth (1981) on the vernacular architecture of the maritime province area of Canada; and Francaviglia (1991) on mining landscapes in the United States.

There are many studies of regional landscape evolution, including Wyckoff (1999) on Colorado, Meyer (2000) on Illinois, Lewis (2002) on Michigan, Schroeder (2002) on Missouri, Jordan-Bychkov (2003) on the Upland South, Rehder (2004) on Appalachia, and Straw and Blethen (2004) also on Appalchia. Other studies that focus on the evolution of regional landscapes are discussed in Chapter 6.

The life and work of E. Estyn Evans is discussed by Buchanan, Jones, and McCourt (1971), Glasscock

(1991), and Graham and Proudfoot (1993); the classic work is Evans (1973), and for a collection of relevant essays see Evans (1996).

Powell (1977) discusses the role of images in the transformation of global environments.

Merrens (1969), Heathcote (1972), Christopher (1973), and Johnston (1979, 1981) discuss images and image-makers during European overseas expansion.

Francis (1988) describes the image of the Canadian West during the settlement period.

Dicken and Dicken (1979) analyze historical perceptions of Oregon, and Peters (1972) studies surveyor perceptions in Michigan in the early nineteenth century.

There are discussions of counterfactuals in historical geography (Norton 1984:14–15; also see Norton 1995 for an example), in world politics (Tetlock and Belkin 1996), in history (Ferguson 1997), and in mostly British politics (Brack and Dale 2003). Fogel (1964) is a classic statement on counterfactuals in the new economic history.

6

Regional Landscapes

The study of the earth as the home of humans. For many human geographers this simple phrase captures the spirit and intent of the discipline, acknowledging the importance of physical geography but stressing the role played by humans changing that geography. Beginning about 12,000 years ago and with increasing intensity as cultures and technologies changed and as population numbers increased, our tenure on the earth has been characterized by our practice of transforming physical landscapes into cultural landscapes. We build ever-changing homes designed to satisfy our needs and wants, and one outcome of this process of landscape evolution, an outcome that reflects both physical geography and the spatial clustering of cultures through time, is the formation and ongoing reformation of regional landscapes.

Much of the content of preceding chapters anticipated this discussion of regional landscapes in three ways. First, Sauer's methodological principles incorporate regional concepts. Second, ideas about humans and nature see regions as one consequence of the human-and-nature relationship. Third, studies of landscape evolution recognize that cultural occupance of an area and related landscape change often involve the creation of a **cultural region**. The underlying theme of this chapter is that groups of people who share some cultural characteristics and live in close proximity are likely to produce a landscape that reflects the shared characteristics—that is to say, a cultural region evolves.

Cultural regions develop at a variety of scales, from local to global, in response to the many different criteria that serve as the bases for human identification with others. Notwithstanding the presence of such regions at a variety of scales, it is worth noting that the relatively small, local, ethnically defined region and the much larger, global **cultural realm** are parts of a single dialectic. Regardless of scale, the presence of a cultural region demonstrates the critical role in landscape creation played by individuals in their capacity as members of groups, as well as the critical role that may be played by those members of a group who are able, for whatever reason, to affect the behavior of others to ensure some consistency in behavior among group members. In this chapter you will find the following:

- an opening account of the regional theme, which includes an overview and evaluation of the approach as used by cultural geographers. The several types of region—formal, functional, and vernacular—are noted, and some problems and limitations of regionalization are identified. Also included are references to criticisms of what is sometimes described, rather inaccurately, as the traditional approach to region identification;
- an outline of a number of concepts devised to understand region creation at the subnational scale. These concepts build upon discussions of shaping landscapes in the previous chapter;
- an overview, building on the idea that human geography is concerned with human behavior and on a body of social science concepts, of an objectivist approach to the analysis of landscape evolution, with reference to a specific example;

- a discussion of principal examples of regional cultural geography at the subnational scale, with emphasis on the concept of regions as homelands and on the importance of a socially cohesive group in region formation. Shared religious beliefs and values are the factors that most commonly link such groups, which are often described as ethnic groups. Principal examples discussed are those of the Mormon and Hispano regions in the American West. There is also an account of ethnicity and related features of material landscapes;
- an account of the regional concept as applied at the global scale, including a discussion of the way in which the modern world has evolved. Central to this discussion are several major analyses of civilizations and also the concept known as world systems theory, which aims to explain the evolution of the modern political, economic, and cultural world. This interest is clearly different from the traditional aim of regional cultural geography, which is to describe and explain particular visible and material landscapes;
- a discussion of the contemporary outcome of global evolutionary processes. The pioneering 1951 global regionalization developed by Russell and Kniffen is presented, as are two other more recent attempts at regionalization;
- brief concluding comments that reintroduce the question of how best to delimit cultural groups and, accordingly, how best to delimit the regions that those groups occupy.

What Are Cultural Regions?

Geographers have always been concerned with describing parts of the surface of the earth. In the mid-seventeenth century, Varenius formally identified a regional approach, which was proclaimed the geographic method by Kant in the late eighteenth century and then proposed as the basis for the newly institutionalized discipline by Richthofen and others in the late nineteenth century. One consequence of this concentration of activity was that regional geography became the dominant **paradigm** in the discipline for most of the first half of the twentieth century.

Indeed, it is not unusual to regard the landscape school of cultural geography as an alternative view of the discipline. Notwithstanding the possibly competing concerns of regional geography and the landscape school, cultural geographers fully appreciated the need to structure their studies of cultural landscapes in a regional framework. Delimiting regions was not the goal of cultural geography but rather a logical outcome of the primary interest in landscape.

Culture Areas in Anthropology

In stressing the value of a focus on regional landscapes, cultural geographers turned not to the leading regional geographers, such as Richthofen, Hettner, and Hartshorne, but rather to Ratzel and to related anthropological applications of the culture area concept. Culture area analysis, although it was proposed by Ratzel, was first explicitly used in anthropology to delimit North American ethnic regions. For anthropologists, the concept was employed initially as a means to establish some semblance of order on seemingly diverse cultural phenomena. In this sense, identifying culture areas was one way to classify data. There were overtones of environmental determinism in some of this work, which identified close links between natural and cultural areas.

A culture area was defined specifically as 'a part of the world where inhabitants tend to share most of the elements of culture, such as related languages, similar ecological conditions, economic systems, social systems and ideological systems. The separate groups within the system may or may not all be members of the same breeding population' (Foley 1976:104). In most of the anthropological studies the culture area concept was a tool used to recognize *cultural wholes*; a cultural whole—the term is virtually synonymous with *cultural region*—was an area dominated by a culture diffused from a 'cultural core' that appeared in the most favorable part of the area.

Regions in Geography

The landscape school incorporated ecological, evolutionary, and regional concerns. Of these three, the regional concern is the most confusing methodologically because it has been common to conceive of Sauer, the pioneering cultural geographer, and Hartshorne, the foremost advocate of geography as the study of regions, as being pitted against each other. The quarrel between the two was not, however, based on substantive differences about the role of regions, but rather concerned the legitimacy of incorporating time in geographic studies (the rift between the two was complete in 1941 at the time of Sauer's major address on historical geography) and also the Sauerian position that studies of landscape should focus on visible material features. Although Sauer did not accept the regional approach advocated by Hartshorne, he did not question the need for regions in geographic study; indeed, Sauer employed the culture area concept, usually favoring the term 'region' in his lecture courses. The difference of opinion on the matter of regions was essentially one of terminology, emphasis, and degree, not one of substance. Indeed, there is some confusion regarding the use of four terms to refer to those parts of the surface of the earth that can be regarded as internally homogeneous and as different from the surrounding parts:

- **Landscape**. Sauer favored this term to characterize the geographic association of visible facts, though he also used the terms 'area' and 'region'.
- **Area**. Anthropologists favored this term.
- **Region**. Following Hartshorne, and as used by many geographers outside of the landscape tradition, this term was seen as the basic concept of the discipline of geography until the 1950s.
- **Place**. This fourth term, though it was also used by Hartshorne, is now associated primarily with some more recent approaches to cultural geography.

Terminological confusion aside, Sauer was explicit concerning the definition of a geographic culture area:

The geographic culture area is taken to consist only of the expression of man's tenure of the land, the culture assemblage which records the full measure of man's utilization of the surface—or,

one may agree with Schlüter, the visible, areally extensive and expressive features of man's presence. These the geographer maps as to distribution, groups as to genetic association, traces as to origin and synthesizes into a comparative system of culture areas (Sauer 1931:623).

Pause for Thought

Terminological uncertainties are one outcome of the often deliberate practice of scholars using different terms in order to distance themselves from other scholars expressing similar ideas. As evident from the account of key concepts in Chapter 1, many instances of terminological uncertainty cannot be avoided. Landscape *and* region *are two terms used throughout the current discussion, with* landscape *referring to the visual scene and* region *to a part of the earth's surface that has a similar landscape. From our cultural geographic perspective, the two terms are not interchangeable, but they are closely related.*

Delimiting Regions

Having identified what is meant by the term *region*, we run into the question of how regions are delimited. Alternatively phrased: *What is the regional method?* The basis for the answer that follows is the idea that delimiting formal regions (the most usual type) is essentially a process of classification, with classification being a procedure designed to impose some order on complex realities. Box 6.1 distinguishes between three types of region, namely formal, functional, and vernacular.

CLASSIFICATION

From a conventional scientific perspective, there are two approaches used to classify formal regions. One approach is to classify all possible locations individually according to a certain set of criteria, and then, where appropriate, to draw boundaries around groups of locations that are deemed to be similar, thus delimiting regions. This is described as grouping, or classification from below, and it is a procedure that has often been informally applied in attempts to regionalize a national area, or, more simply, to identify a single region. An alternative approach is to treat all the possible locations as a single set and then draw boundaries between smaller groups of locations by means of some predetermined criteria. This is described as logical division, or classification from above. Classifying in this manner requires extensive knowledge of the locations so that sensible distinguishing criteria can be used. Most of the traditional world regional classifications conducted by cultural geographers and other scholars are of this second type.

Interestingly, some efforts at regionalization made by cultural geographers following Sauer rely on a procedure that is different from both of those associated with the exercise of classification. Rather than beginning with the finished product (that is, the region or set of regions), an evolutionary approach has been used. Specifically, the aim is to identify a cultural hearth and then trace the diffusion of the culture outwards over some larger area until a region has been identified.

FOUR DIFFICULTIES

There are four general issues that need to be addressed in any attempt to demarcate a cultural region or a set of cultural regions. First, regions are continuously evolving, and this makes the process of regionalizing difficult, since a set of regions once delimited can change. This issue follows closely from the contents of the previous chapter, especially from the idea that regions are ever-changing outcomes of cultural occupance and transformation of landscape, but it also relates to the Sauerian interest in identifying cultural hearths and tracing diffusion. This chapter includes two sections that focus on the evolution of cultural regions, one referring to the subnational scale, the other to the global scale.

Second, region identification must consider questions of spatial scale and cultural scale. Linking these two scales stresses the fact that cultural regions are, logically, regions that are usually occupied by distinguishable cultural groups, and are not just evident because of different visible landscapes. Indeed, most attempts at regionalizing the world are classifications of people rather than classifications of visible landscape. The significance of different scales, spatial and cultural, is that the tasks of regionalizing the world,

Box 6.1 Three Types of Region

First, a **formal region** is characterized by uniformity of a given trait or traits; it can be of any size. The culture area concept used by anthropologists and the regional concept used more broadly by geographers apply to this type. There are numerous criteria that might be used to identify formal regions. Characteristics of language, religion, and ethnicity may be used because they are reflected in the landscape. Alternatively, the concern may be more directly with the landscape, as in the recognition of a corn belt or prairie landscape. Cultural geographers typically use multiple traits in identifying formal regions. Lewis (1991:606) notes that formal regions are created through human relatedness, that is, 'the notion that important commonalities unite certain groups of individuals to varying degrees, while separating them from those in other, similarly defined communities'.

Second, a **functional region** is one ranging in scale from a single home to the entire world that in some way operates as a unit. Unlike formal regions, functional regions result from human connectedness rather than human relatedness—in other words from some process of interaction between locations rather than from some shared values or characteristics. Clearly, a part of the surface of the earth that is integrated in a functional sense, such as a political unit or a religious settlement, is quite likely to exhibit some landscape similarities as well, since landscape may be affected by the functional factors unifying the region. Accordingly, functional regions often overlap with formal regions in cultural analysis.

A third type of region, the **vernacular region**, is a locally perceived regional identity and name. Although a vernacular region may exhibit some distinct visible material landscape, its distinguishing characteristics relate more to a perceived sense of identity than to a visible sense of identity. There is considerable overlap between formal and vernacular regions, with many cultural regions being both formal and vernacular; there may also be overlap with a functional region. Many small, relatively local tourist areas, such as Niagara Falls, qualify as vernacular regions.

regionalizing a country, and regionalizing a sub-national area necessarily require use of different criteria and involve different levels of expectation concerning the internal homogeneity of the various cultures and landscapes delimited.

A third difficulty, apparent in most regionalization exercises, concerns the precise location of boundaries. Most functional regions have relatively precise boundaries, but both formal and vernacular regions are usually much more susceptible to particular interpretations made by the researcher. The reality is, of course, that most cultural regions do not have sharp boundaries separating them from other regions. Rather, the boundaries are zones of transition with a gradual merging into neighboring regions. Although cultural geographers are well aware of this situation, there is often an understandable temptation to map regions with boundaries that imply a line of separation.

The final difficulty is the most problematic: on what basis are regions delimited? What criterion is—or what criteria are—to be used? How is this choice to be justified? Of course, there are no simple responses to questions such as these. The chosen criteria typically reflect the specific aims of the regionalization. For example, if a set of religious regions is the objective, then some religious criteria are employed. But statements such as these do not really address this issue. The key point is that multiple criteria might be legitimately employed in the act of demarcating cultural regions, and any interpretation of the resulting regionalization must consider the criteria used. There is always a danger of extending a regionalization beyond the intended purposes. What this point tells us is that there is no such thing as a correct system of regions.

These four difficulties often result in a regionalization being the personal product of an individual researcher, such that in many cases the boundaries of regions are located in a relatively arbitrary manner. Hart (1982:21–2), a leading advocate of regional geography, has stated: 'Regions are subjective artistic devices, and they must be shaped to fit the hand of the individual user. There can be no standard definition of a region, and there can be no universal rules for

recognizing, delimiting, and describing regions.' The extent to which this state of affairs is a problem is, of course, debatable, as it is dependent on philosophical persuasion. Humanist geographers are comfortable with individual researchers playing key roles, while more scientific geographers are much less comfortable.

Pause for Thought

Given these four difficulties, is it worthwhile to try to delimit regions? For most cultural geographers, the answer is yes, for at least three reasons. First, regions are an integral part of geographic research generally. Second, regionalization is one means of imposing some order on a complex reality. Third, the favored Sauerian procedure of identifying hearths and diffusion routes does minimize these four difficulties. Nevertheless, it is important to ask whether maps of regions assume a life of their own, implying order and structure where these do not really prevail. Any regional exercise needs to acknowledge, as Hart suggests, that regions are not real things with unique characteristics.

THE DECLINE OF TRADITIONAL REGIONAL GEOGRAPHY?

Two quotations provide us with two very different opinions about the need for a regional approach to geography. First, Hart (1982:1) argued for some type of traditional regional geography, asserting that such work was 'the highest form of the geographer's art'. Second, Thrift (1994:200) expressed the opinion that the regional approach required 'exhumation rather than resuscitation', and bemoaned the fact that it appeared to have become 'an acceptable form of professional nostalgia, conjuring up memories of a golden age, now (thankfully) defunct'. A reasonable assessment of the regional approach lies somewhere between these two statements.

It is commonplace in much recent human geography to make two related claims: first, that traditional regional geography has declined, and second, that a revised regional geography, one informed by recent advances in social theory, is at the forefront of much geographic research. The basic logic behind the first claim is clear: the regional approach, as advocated by Hartshorne, effectively dominated English-language geography from about the 1920s until about the mid-1950s, at which time it was challenged by the scientifically inspired spatial analysis. Since then it has not succeeded in re-establishing itself as the geographic paradigm.

But although it is correct to say that the regional approach declined in popularity after the mid-1950s, it is also true that regional geography has remained central to much geographic research and continues to be advocated and practiced by many geographers. More important, for our purposes, the discussion of regions continues to be a popular and worthwhile approach for the study of cultural landscapes. Indeed, some of the very best cultural geography of recent years is explicitly concerned with identifying and understanding regions at a variety of scales.

Principal criticisms of traditional regional geography are that the approach is of limited value because it is concerned primarily with observing and recording facts and because most studies of regional geography implicitly assume that human geographic regions are organized according to differences in physical geography. These criticisms relate to the empiricism and naturalism evident in some regional geography, but do not really apply to the regional approach as it is used by cultural geographers who have always been concerned with a level of generalization, notably in the identification of cultural hearths and diffusion routes, and who have always rejected simplistic physical and human associations. Indeed, the more socially informed new regional geography is built partly on the ideas of cultural geography, specifically the idea that regions are related to spatial variations in culture and not to physical geography. The principal difference between traditional and new approaches, so far as cultural geography is concerned, relates to the traditional interest in material landscape and the newer interest in culture as socially constructed.

Although traditional regional geography is out of favor, relatively speaking, having been replaced by a more socially informed regional geography, there seems no doubt that it continues to thrive because, as part of the landscape school, it did not incorporate some of the principal limitations of the larger traditional concern. Certainly, some of the criticisms of the traditional regional approach neglect to consider the substantive achievements of cultural regional geography, achievements that are discussed throughout the remainder of this chapter.

Forming Cultural Regions

The processes by which distinctive areas of cultural landscape—cultural regions—come into being are complex. For the United States, circumstances that might encourage retention of cultural traits in turn might contribute to region formation. These circumstances include a substantial number of immigrants, a pattern of contiguous settlement, a set of shared values, and a lack of contact with other groups at the time of settlement.

This section details nine useful concepts developed by cultural geographers and used in the process of region identification and delimitation. The discussion of these nine concepts is illustrated with several examples of cultural regions that are explicitly related to the concepts described, while the succeeding section reviews a number of other major studies that focus on the idea of regions as homelands.

Cultural Hearths

Geographers working in the landscape school tradition and inspired by its particular interest in diffusion have paid considerable attention to the concept of the cultural hearth as an original source area that possesses some distinctive cultural attributes and from which these attributes have been diffused. Sauer stressed the importance of locating hearths from which settlement expanded as a basis for region identification, and the concept has been used in numerous cultural geographic analyses. Most of this work in region identification has been carried out on American and, to a lesser extent, Canadian landscapes. This preference is related to the fact that the European settlement of North America involved distinct cultural groups settling separate areas, thus contributing to the creation of cultural hearths and making North America particularly well suited to this kind of region identification; it also reflects the unique attachment of North American cultural geographers to the landscape school. Examples of hearth identification are discussed later in this section together with examples of some of the more elaborate conceptual formulations.

Core, Domain, and Sphere

A stimulating conceptual contribution to the literature on landscape evolution and related region formation is the *core-domain-sphere* model as developed and used by Meinig. The model formulation was introduced in an exemplary analysis of the emergence of the Mormon cultural region. Following accounts of the cultural group, of the movements of the group, and of the evolution of the region, Meinig outlined this generic model as a basis for both delimiting a region and identifying variations within the region. There are three components to the model (see Figure 6.1).

The first component of Meinig's model is a cultural core. This is not quite the same as the hearth area noted above. Rather, the core is the center of cultural control and the zone of most intense activity, but is not necessarily the original cultural source area; in other words, the core may or may not be the hearth. In the Mormon case, the core was founded following movement of the cultural group away from the area of cultural origin and hence is not the hearth. Surrounding the core in Meinig's model is a domain, the area over which the culture diffused and became dominant. Although the Mormon domain lacks the intensity of occupance and complexity of development that are evident in the core, it is the area that most obviously displays the culture in the visible landscape. Beyond the domain is the sphere, an area that belongs only partly to the cultural region in question because other cultural influences from adjacent cultural regions are also present.

Certainly, the Mormon region for which this model was constructed is an unusual one

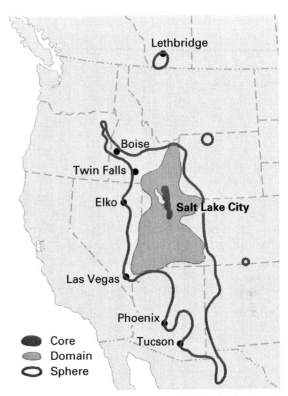

Figure 6.1 Core, Domain, and Sphere. This is the classic application of the core-domain-sphere model that Meinig developed in order to accommodate a common problem, namely mapping cultural regions that contain internal variations. Meinig's innovative solution was to use generic concepts—core, domain, and sphere—that could reflect such variations. Because cultures often spread away from a core, the transition from core to domain to sphere is likely to suggest spatial and temporal patterns. The success of the model in the Mormon context seems undeniable, but it is also clear that the Mormon case is far from typical, especially when the isolation of the region and the distinctiveness of the group are considered.

Source: D.W. Meinig, 'The Mormon Culture Region: Strategies and Patterns in the Geography of the American West, 1847–1964', *Annals of the Association of American Geographers* 55 (1965):214. (Reprinted by permission of Blackwell Publishing)

because of the initial isolation of the group in the area occupied and the especially distinctive character of the Mormon cultural group. This combination of isolation and a distinctive culture is probably a prerequisite for the evolution of a core (or hearth), domain, sphere regional identity that is as readily recognizable as it is in the Mormon case. Other applications of this model are therefore more limited, precisely because most groups are in close contact with other groups in adjacent areas, and because most groups do not actively seek to emphasize their differences from wider society, as the Mormon group has. The principal value of the model, aside from the specific Mormon application, is the explicit focus on evolution and the acknowledgment, in the reference to a sphere, that the boundaries of cultural regions are zones of transition and not lines.

As one part of a study concerned with urban street patterns in eighteenth-century Pennsylvania, the Pennsylvanian cultural hearth, domain, and sphere were mapped (see Figure 6.2). A hearth rather than a core was identified, as the center of cultural activity was also the area of cultural genesis. The Pennsylvania case was probably more typical of the North American situation than was the Mormon region in that several cultural groups, rather than just one group, were present during the period of expansion. Accordingly, the cultural region developed quite differently. This argument was also extended to the larger region of the northeastern United States. Thus, the model was tentatively revised to include, for each of several cultural groups, a hearth, a domain, and a sphere (see Figure 6.3).

First Effective Settlement

'Whenever an empty territory undergoes settlement, or an earlier population is dislodged by invaders, the specific characteristics of the first group able to effect a viable, self perpetuating society are of crucial significance for the later

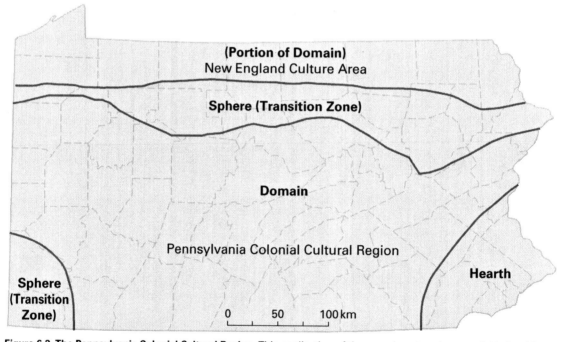

Figure 6.2 The Pennsylvania Colonial Cultural Region. This application of the core-domain-sphere model is in a historical, regional, and cultural context that is quite different to that of the original Mormon application, although the model is used by Pillsbury for essentially the same reason as it was developed by Meinig, namely to accommodate the fact that the cultural region studied displayed internal variations. The most obvious differences between the two cases are (1) that the Pennsylvania colonial cultural region is not isolated from other cultural influences, and (2) that the core is also the genesis area, the hearth. This figure derives from a study of urban street patterns in Pennsylvania, hence the region is mapped only as it is located within Pennsylvania.

Source: R. Pillsbury, 'The Urban Street Pattern as a Cultural Indicator', *Annals of the Association of American Geographers* 60 (1970):445. (Reprinted by permission of Blackwell Publishing)

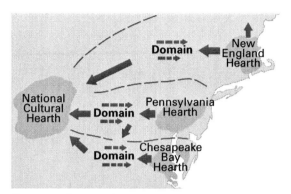

Figure 6.3 Colonial Cultural Regions in Eastern North America. This suggestive diagram locates three colonial cultural hearths in the eastern United States, namely New England, Pennsylvania, and Chesapeake Bay; a fourth North American cultural hearth, French Canada, was also identified. The three American hearths were previously proposed by Kniffen (1965) in a study of the diffusion of folk housing. One way to interpret this suggestion that eighteenth-century North America comprised four regional cultural hearths is by reference to the simplification thesis discussed in Chapter 5. According to this thesis, the large number of local cultural regions, or *pays*, evident in Europe were simplified after being transferred to an overseas area, with one result being a smaller number of hearths. One implication of this proposal for four hearths is that some cultural groups, such as Swedes in Delaware and Dutch groups on the Hudson, became assimilated shortly after arrival. This figure also suggests that the three American hearths merged as they diffused west, thereby contributing to the formation of a national American culture.

Source: R. Pillsbury, 'The Urban Street Pattern as a Cultural Indicator', *Annals of the Association of American Geographers* 60 (1970):446. (Reprinted by permission of Blackwell Publishing)

social and cultural geography of the area, no matter how tiny the initial band of settlers may have been' (Zelinsky 1973:13). This concept of **first effective settlement** is similar to the initial occupance proposal suggested by Kniffen (1965:551) in a diffusion analysis of folk housing. Zelinsky used the concept to map modern cultural regions, and the resulting regionalization is depicted in Figure 6.4. Zelinsky designated five regions as 'first-order' settlements, namely New England, the Midland, the South, the Middle West, and the West.

The significant variations in regional size are related to differences in physical geography and in settlement experiences. For each of these five,

Zelinsky (1973:119) indicated the date of first effective settlement and the date of related cultural formation, as well as the major sources of culture for region formation. Each of the first-order regions is subdivided into secondary regions, while three areas (Texas, Oklahoma, and peninsular Florida) are considered too difficult to classify given the criterion employed. All but one of the first-order regions have more than one major source of culture, the exception being New England, which has England as the sole source area. Because of this, only New England, among the first-order regions, has a date of first effective settlement noted; for the other four Zelinksy deemed it appropriate to indicate first effective settlement dates and sources of culture for secondary and, in three instances, tertiary regions. Further, in most of the cases, Zelinsky identified multiple sources. The concept of first effective settlement enables this regionalization, but it cannot be applied on a simple one date–one cultural source–one region basis. Certainly, the regionalization is a valuable application of the concept, but not surprisingly, the regional scene is much too complex to be represented by a single concept.

The basic logic of first effective settlement is often used in accounts of cultural landscapes. In the case of the Musconetcong valley landscape of New Jersey, for example, many elements reflect a continuity from the eighteenth-century pioneer period to the present; thus, houses and barns along with many auxiliary structures reflect types established in the area during the eighteenth century.

Again, as with the core–domain–sphere model, the first effective settlement model has seldom been applied to areas outside of the United States. Certainly, in Europe and in many areas of European overseas expansion where the experience was not as numerically overwhelming as it was in North America, the concept is less meaningful. This is both because the first effective settlement is much older than in the American examples, and because contemporary cultural identity is often the product of a series of contributions rather than a relatively clear reflection of a single dominant contribution.

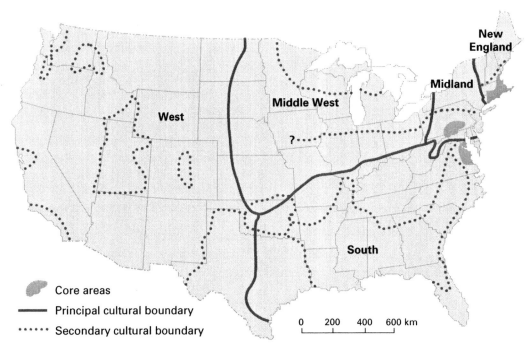

Figure 6.4 First Effective Settlement in the United States. This is the classic regionalization of the United States based on the concept of first effective settlement. Zelinsky preferred this concept to the core-domain-sphere model as a basis for regionalization because most of the core areas identified lack the isolation required to create a recognizable core-domain-sphere cultural region; this same point is made by Pillsbury (see Figure 6.3).

Source: W. Zelinsky, *The Cultural Geography of the United States* (Englewood Cliffs, NJ: Prentice Hall, 1973):118. (© 1973, reprinted by permission of Pearson Education, Inc.)

Duplication, Deviation, and Fusion

The Turner frontier thesis explained the formation of American cultural regions in terms of the westward moving frontier, but there are several problems with this frontier argument. These include: the assumption of an initial subsistence phase of economic activity, which was in fact not typically present, since many settlers moved quickly into commercial activity; the assumption of an initial regional isolation, which was not typically the case, since most areas were closely linked to other areas from the outset of settlement; and the assumption that American culture was derived from the frontier experience west of the Appalachians

It is possible that the cores of the three American colonial cultural regions—those located in southern New England, southeastern Pennsylvania, and the Chesapeake tidewater—acted as dynamic cultural regions, contributing first to the creation of intermediate regions, and later to the creation of regions west of the Appalachians. Each of these regions included internal differences such that hearths and domains can be distinguished.

Three mechanisms might have been responsible for the creation of regions west of the colonial hearths: duplication of the traits evident in the hearth; deviation from the traits evident in the hearth because of different local circumstances; and fusion of traits from two or more hearths with the resulting formation of a different cultural form. Of these three, 'it is the fusion process, aided by the appearance of symbols of national unity, that seems to have been most important in the cultural formation of the early trans–Appalachian west' (Mitchell 1978:67). This proposal, a development of both the core–domain–sphere model and the first effective settlement concept, allows for a variety of formative processes (see Figure 6.5).

Stages of Regional Evolution

In a penetrating analysis of the American West, Meinig (1972) identified a set of six dynamic regions. For each of these regions, Meinig then identified four general stages of development in each of four categories of regional features, namely population, circulation, political area, and culture. The sequence of stages for culture (see Figure 6.6) was as follows:

- Stage 1: *transplant* – a selected and experimental adjustment to the new environment
- Stage 2: *regional culture* – the formation of a cohesive regional society

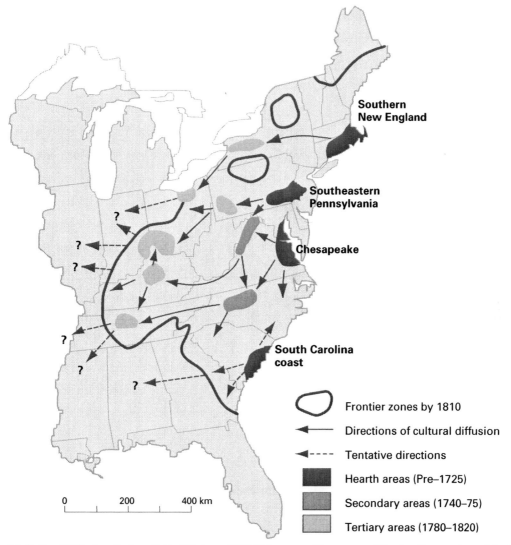

Figure 6.5 Cultural Diffusion, Eastern United States, *c.*1810. This map outlines some of the directions and possible directions of cultural diffusion from four colonial cultural hearths between about 1725 and 1820. Diffusion of traits, involving duplication, deviation, and fusion, contributed to the formation of other cultural regions.

Source: R.D. Mitchell, 'The Formation of Early American Cultural Regions'. In *European Settlement and Development in North America*, edited by J.R. Gibson (Toronto: University of Toronto Press, 1978):75. (Reprinted by permission of the publisher)

- Stage 3: *impact of national culture* – the consequences of improved communication and marketing
- Stage 4: *dissolution of historic regional culture* – a new awareness of regional values

The distinction between the first two and later two stages is particularly important, as the beginning of the third stage, impact of national culture, marks the end of insularity and local cultural identity and the onset of a rapid increase in the power of those forces encouraging national cultural uniformity. Much of the early regional distinctiveness has been eroded because the twentieth-century experience has typically been one of cultural integration related to the impact of a national American culture.

Figure 6.6 Four Stages of Cultural Region Development. This diagram shows one of four related schema, for *culture*, with the other three referring to different categories of regional features (*circulation*, *population*, and *political area*). Each of the six Western regions shown in Figure 6.7 is seen as passing through these four cultural stages.

Source: Adapted from D.W. Meinig, 'American Wests: Preface to a Geographical Introduction', *Annals of the Association of American Geographers* 62 (1972):163. (Reprinted by permission of Blackwell Publishing)

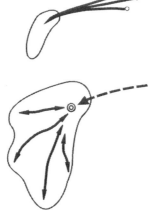

Selected **transplant** from one or more source regions; never a complete cross-section of the older society; experimental adaptation of imported cultural traits to new environment.

Regional culture; new amalgam of people forming cohesive society and adjusting to insularity and new environment; high potential for cultural lag and divergence.

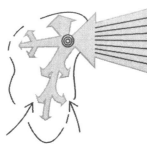

Strong **impact of national culture**; nationwide communications, marketing networks, and control of facilities diffuse national culture through central place network. Only subcultures with tenacious social patterns (religion, language, race) can persist as distinct.

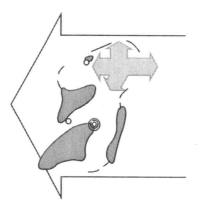

Dissolution of historic regional culture; all areas directly exposed to national culture; emergence of ethnic mosaic and new innovative centers; new consciousness of local environmental and cultural values.

SIX REGIONS IN THE AMERICAN WEST

This approach to regionalization is developmental, synthetic, and generic. Applied to the American West, the approach resulted in the mapping of six major nuclei, or cores (the first stage) and six recognizably distinct regional cultures (the second stage), along with a number of secondary nuclei, some of which were inside and some outside the six major regions (see Figure 6.7). Because settlers in the American West came mostly from areas in the East and brought with them certain cultural traits, their behavior was more imitative than innovative. It was this behavior, combined with environmental differences, that contributed to the formation of the six regions. The six major regions identified as being present by the end of the second stage, defined as that of regional culture, are as follows.

Hispano New Mexico was the first region to form, following Spanish settlement of the upper Rio Grande valley in the late sixteenth century. This was an isolated region of Hispanic European culture for over 200 years until in-movement by Anglos initiated change. By the late nineteenth century, it was a complex cultural area with relatively few links to the larger national system. The Mormon region formed beginning in 1847, with the initial in-movement of a distinctive religious group seeking isolation from larger American influences; a cultural region was created during about the next fifty years. Both the Hispano and Mormon regions are discussed in greater detail in the following section.

The region of the Oregon country was formed beginning in the 1840s, primarily by settlers who were part of the larger westward movement, with no one cultural group dominating the in-movement. The character of the region diverged from that of the migrant source areas in the East because of differences in location and environment. Northern California had limited settlement until the discovery of gold in 1848, an event that precipitated rapid and dramatic population growth accompanied by related cultural and economic change. The incoming population was especially heterogeneous. The region of southern California was unaffected by the gold rush and remained as a dual Mexican and Anglo-

American culture until a real estate boom in the 1880s initiated a process of rapid growth. Regional development began in eastern Colorado in the late 1850s following the discovery of gold, and the area developed as a cattle and irrigated farming area. None of these final four regions is as readily identifiable as a specifically cultural region as are the Hispano or Mormon areas; their identification as regions within the West is based more on economic and other criteria than on cultural criteria.

Texas can be interpreted as a distinct cultural region that has evolved through four stages, each of which has left a mark on the present scene. The four stages are implantation, which reflects

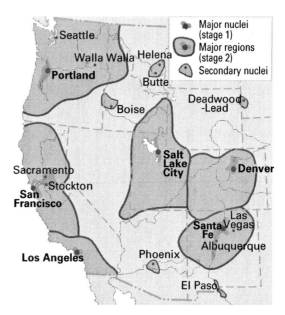

Figure 6.7 Major Nuclei and Regions in the American West. Meinig proposed six regions for the late nineteenth century. Hispano New Mexico has a major nucleus in the upper Rio Grande valley centered on Santa Fe. The Mormon region has a major nucleus at the base of the Wasatch Mountains centered on Salt Lake City. The Oregon country has a major nucleus in the Willamette valley centered on Portland. In all three cases, the nucleus is a prime agricultural area. Northern California is centered on San Francisco, Southern California on Los Angeles, and Colorado on Denver. In addition, Meinig located several minor nuclei, each of which showed some degree of local autonomy.

Source: D.W. Meinig, 'American Wests: Preface to a Geographical Introduction', *Annals of the Association of American Geographers* 62 (1972):169. (Reprinted by permission of Blackwell Publishing)

Box 6.2 The Shaping of North America

The fundamental logic behind both the core-domain-sphere model and the stages of regional evolution approach has been further pursued by Meinig who argued the need to interpret larger American cultural and economic development from a series of initial units:

> . . . the most important task in the historical geographic study of colonial America is to define as clearly as possible this sequence of territorial formation from points to nuclei to regions on the North American seaboard and to describe the changing geography of each in terms of spatial systems, cultural landscape, and social geography. (Meinig 1978:1191)

Our understanding of North American cultural region evolution is enhanced significantly by the three (as of 2005) published volumes of the proposed four-volume set *The Shaping of America: A Geographical Perspective on 500 Years of History* (Meinig 1986, 1993, 1998). These volumes elaborate on the earlier work on core, domain, and sphere, and on the related stages of regional evolution framework. Thus, in a study of the United States at the end of the colonial period, Meinig charted American national cultural, political, and economic expansion using a core-domain-sphere model. Meinig situated the core on the eastern seaboard, especially in New York and Philadelphia; the domain is the area settled by European Americans, which was divided into a northern free domain and a southern slave domain; and the sphere is the area claimed but not settled by European Americans (see Figure 6.8). Beyond the sphere in 1800 were areas of foreign territory, namely British North America, Louisiana, and Florida.

For the late 1850s, Meinig developed these ideas at a transcontinental scale. The New York City–Philadelphia axis, with extensions north to Boston and south to Baltimore, remained the core area, with two important extensions west along the Hudson–Erie Canal and through central Pennsylvania (see Figure 6.9). The domain is the area of contiguous European

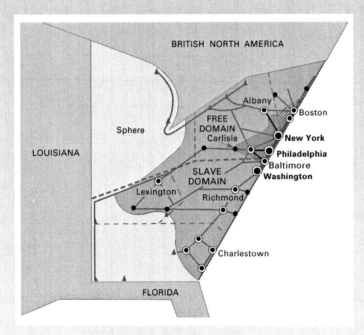

Figure 6.8 Core, Domain, and Sphere in the United States, *c.*1800. This schematic diagram reflects an innovative extension of the core-domain-sphere model to the emerging political unit of the United States. The concern here is with nationalism and the new nation rather than with culture and landscape. Philadelphia and New York comprised the true core, the area around which the new United States was beginning to form. Even in 1800, however, this was only an incipient core. Aspects of a new national lifestyle began in the core with little concern for what was happening elsewhere. Boston was not strictly a part of the core because it reflected a regional New England interest, while Baltimore looked south rather than north. The domain was only loosely linked to the core. Within the domain, there was a distinction between those states with large numbers of slaves and other states. Beyond the domain was the sphere, an area available for future settlement. The outer boundaries of the sphere were determined by the presence of colonial powers.

Note that the terms, *core, domain,* and *sphere,* do not have quite the same meanings as they did in the original Mormon application of the model. A revised definition of the term *domain* is noted in the comments accompanying Figure 6.9. Meinig (1986:404) stated: 'The tentative character of all these regional patterns was evidence that the United States was no more than a nation abuilding and it was being shaped not after the design of a simple model but uniquely, experimentally, within the framework of a federation.'

Source: D.W. Meinig, *The Shaping of America: A Geographical Perspective on 500 Years of History. Volume 1: Atlantic America, 1492–1800* (New Haven: Yale University Press, 1986):402. (Reprinted by permission of Yale University Press)

American settlement, an area that was much more extensive in the 1850s than it had been in 1800. At this national scale, it is clear that the domain comprised several parts, each of which exhibited some variation of larger national cultural characteristics. Thus, the domain in the late 1850s included some backwoods areas adjacent—and clearly subordinate—to the core, as well as some major areas (notably the South and West) that were a part of the nation but that contained their own variants of the national culture and their own regional landscapes.

The remarks accompanying figures 6.8 and 6.9 include an important distinction between use of the term 'domain' in the Mormon study referred to earlier and use of the same term in these more ambitious analyses. More generally, this application of core, domain, and sphere is rather different from the earlier Mormon application. In the Mormon case, the concern was with a cultural region; in this more ambitious application, the concern is with a developing national culture and with those areas that, although linked to the national culture, have some distinctive characteristics of their own. This is simply a way of saying that the spatial and social scales are changed and, therefore, the criteria for identification and the expectations for regional homogeneity are also changed. Later in this chapter the spatial and social scales change again, and in the discussion of regions at a global scale the criteria and expectations for regional homogeneity are adjusted accordingly.

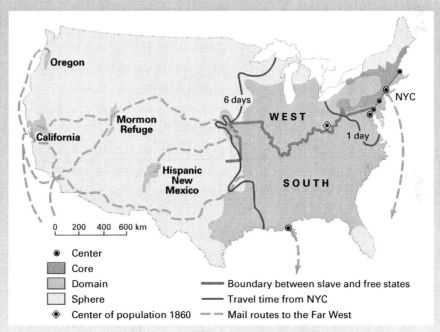

Figure 6.9 Core, Domain, and Sphere in the United States, late 1850s. For the late 1850s, Meinig mapped the core, domain, and sphere in a more conventional manner than he had used for his study of the United States at the end of the colonial period, though he continued to be concerned primarily with the idea of nation building. The core is the heart of the nation and the source of most of the innovations that diffused throughout the country. The core is also 'multicentered', with New York City as the financial and commercial capital, Philadelphia as a prestigious city, Boston as a center that sought to shape moral character, and Washington as the political capital.

The domain is the remainder of the contiguously settled area, but it is far from culturally uniform. Within the domain, for example, the South exhibited many characteristics that differed from those found in the core. Indeed, Meinig (1993:423) stated: 'Domains are realms in which distinct regional variants arising from different physical environments, resources, local economies, and mixes of people are likely to show through the veneer of national culture.' Expressed rather differently, the domains in this and the previous figure can be conceived of as incipient in comparison to the Mormon domain shown in Figure 6.1. The West was more closely linked to the core with principal differences relating more to low population densities and to how recently the area was settled. By the late 1850s, the sphere extended continuously to the Pacific.

Source: D.W. Meinig, *The Shaping of America: A Geographical Perspective on 500 Years of History. Volume 2: Continental America, 1800–1867* (New Haven: Yale University Press, 1993):424. (Reprinted by permission of Yale University Press)

both Spanish and Mexican influences; assertion, which includes the periods of republic and early statehood; expansion, following the Civil War; and elaboration, which involves more recent developments. Box 6.2 reviews some of the more ambitious uses of the core–domain–sphere model combined with the synthetic logic of the stages of evolution approach.

Culturally Habituated Predisposition

A pioneering analysis of agricultural—not specifically cultural—region evolution interpreted change in three regions as a result of cultural rather than environmental or economic processes. The three regions that were examined were the American corn belt, the Philippine coconut landscape, and the Malayan rubber landscape. Although each case was different in detail, a cultural (specifically psychological) process was evident in all three cases. The American corn belt, for example, was seen as the 'landscape expression of . . . the totality of the beliefs of the farmers over a region regarding the most suitable use of land in an area' (Spencer and Horvath 1963:81). Thus, formation of the corn belt followed a decline in sheep populations, which resulted from a general lack of interest in sheep rather than any crucial environmental or economic factors. The relevant cultural process was that of innovation diffusion. Once a process of change began, the mechanics of local communication networks ensured that the change became widespread; these networks were particularly effective in an area undergoing a process of first effective settlement as they were not complicated by the presence of earlier networks

In a related fashion, the coconut landscape of the southern Philippines resulted from a particular farming mentality that was favorably predisposed to coconut planting. This psychological mindset, which was shared by most farmers in the region, was a strong force in the evolution of the coconut landscape. Expressed rather differently, it was a culturally habituated predisposition toward a particular crop capable of providing a stable return that initiated the formation of a particular agricultural region. The

origin of this region was similar to, although different in detail from, that of the corn belt, particularly because all farmers in the region were familiar with the coconut and with appropriate techniques for harvesting the crop. The specific stimulus for the development of a commercial coconut landscape was a change in demand that farmers were able to respond to accordingly.

A third example of cultural causation was that of the Malayan rubber landscape. This landscape developed as a result of psychological change among the local population. Malayan farmers initially viewed rubber as an alien system of agriculture but changed their attitude toward this activity and adjusted their behavior accordingly. Of particular importance in this change was the realization that the daily work schedule for rubber, with most work done during the cooler hours, was preferable to that of rice field labour, which required a large amount of hard work during the heat of the day.

Preadaptation

Persistent disagreement over the nature of American frontier settlement has led to the replacement of the principal theories by a cultural theory. The American frontier of the eighteenth and early nineteenth centuries is best understood not as a time, place, or process, but as the preemption of a vast domain by one preadapted, syncretic American culture. The new culture was the Upland South of Turner and Kniffen; the processes were diffusion and migration; and the mechanism was cultural preadaptation (Newton 1974:143).

The concept of **preadaptation** was introduced specifically to explain the diffusion of culture from the American Upland South, an area stretching from southeastern Pennsylvania to eastern Georgia, covering much of the United States as far west as the plains. Preadaptation is the condition of a culture that already possesses the cultural traits required to allow it to successfully occupy a previously unencountered environment prior to moving to that environment; preadapted groups have a competitive advantage over other groups moving to the same environment. Eleven traits that enabled the culture of the Upland South to expand have been identified (see Box 6.3).

Settlers in the Upland South possessed a set of cultural preadaptations that enabled them to spread their Upland South culture beyond the Appalachians and into the plains. This essentially ecological concept was later developed to argue that seventeenth-century Finnish settlers of the lower Delaware valley were preadapted to life on the American frontier because of their previous European experience. As such, this group can be seen as the single most important contributor to the culture of the American backwoods frontier. Although this identification of a specific, highly localized source area and of a particular ethnic group is debatable, the basic argument is clear: preadapted settlers were necessarily effective settlers, and hence, if they were the first to arrive in an area, they were the first to establish an effective settlement.

Cultural Islands

Expressions of group identity in landscape result from three factors in particular: the tendency of members to settle in close proximity to one another; the importation of distinctive culture traits; and the presence of particular attitudes, perceptions, and behaviors. Immigrant groups are sometimes described as **ethnic** to indicate that they share such important characteristics as a place of origin, a sense of tradition, a language, and a religion. For many immigrant groups, cultural inertia played a role in the creation of the landscape they first occupied, such that the presence of a regional landscape is often some measure of their group conservatism. Certainly, most groups faced two opposing forces when settling in a new environment: there was, on the one hand, a powerful sentiment in favor of maintaining shared traits and identity, while on the other hand, there was a recognition among settlers that many advantages accrued to those willing to break with their traditions. Distinctive landscapes are most likely to develop and be maintained where the group abides by certain common rules of organization. This claim is supported by numerous examples of rural landscapes of religious and other ethnic

groups that display a preferred architectural style, which may be reflected in house and farm building and in church construction. In some of these landscapes, the ethnic groups themselves reflect a particular lifestyle in their choice of clothing and food.

Cultural islands are areas that are occupied by ethnic groups and that exhibit some landscape features that reflect these occupying ethnic groups. Cultural islands associated with Amish and Mennonite populations—two closely related groups—are among the most frequently studied because they tend to reflect aspects of a particular religious worldview, notably the commitment to an agricultural way of life and the preference for a closed and introverted society. The initial Amish and Mennonite groups that settled in North America favored dispersed settlement landscapes comprising farmsteads located apart from each other, while later groups that settled in southern Manitoba from Russia favored settlement in street villages with homes that had large rooms and a centrally located kitchen.

Some other well-researched and rather more widespread examples of cultural islands are certain Ukrainian landscapes evident in a broad belt across the Canadian Prairies from Manitoba to Alberta; several Finnish landscapes in the upper Great Lakes, evidenced by the many Finnish place names (see Box 9.5) and by the sauna that is a typical addition to farms; and certain German landscapes in the Texas hill country, where resistance to acculturation is evident in church affiliation and reflected in a landscape of distinctive ecclesiastical architecture, cemeteries, and graveyards. There are also many cultural islands outside of North America, and cultural geographers have identified and described many of these, including Japanese areas on the Brazilian frontier and German towns in southern Brazil.

In some areas, the development of a cultural island is dependent on the ability of the ethnic group to impose their own form of land survey rather than having their landscape surveyed in common with other areas. This is because land surveys often have lasting impressions on the boundaries of property, on road networks, and on other aspects of cultural and economic occupance. Long lot systems in the Hispano area, along the lower St Lawrence, and in Louisiana were a marked contrast to prevailing land survey systems and are vivid indicators of a specific cultural impress. Land subdivision imposed upon an area may have the effect of minimizing potential impacts of local cultures.

The Authority of Tradition

In an account of the development of the American corn belt from cultural hearths in the East—specifically in the Upland South—Hudson (1994:3) referred to the idea of the 'authority of tradition'. This phrase is a useful way to summarize a key interest of cultural geographers and one that has been implicit in much of the preceding discussion. Recall the statement from Sauer, cited in Chapter 5: 'We are what we are and do what we do and live as we live largely because of tradition and experience, not because of political and economic theory' (Sauer 1985:1). The authority of tradition is clearly evident in landscapes, and folk components in a landscape are especially useful to the cultural geographer attempting to understand that landscape's regional distinctiveness; on the other hand, cultural landscapes reveal little meaning unless each component is observed and understood both for what it is and for how it fits into the larger context.

Box 6.3 Cultural Preadaptation

There were eleven preadaptive traits that, according to Newton (1974:152), made it possible for the culture of the Upland South to expand west to the edge of the Great Plains quickly and successfully. The eleven traits are as follows:

- *dispersed settlement*, which allowed a relatively small number of people to lay claim to an extensive area;
- *kin-based dispersed hamlets*, which was a form that could be readily replicated and adapted to varying physical and economic circumstances;
- *dispersed services*, such as mills, stores, and churches, which offered numerous focuses for settlement;
- *a combined stockman-farmer-hunter economy*, which encouraged diverse economic activity;
- *log construction*, which allowed forests to be exploited;
- *universal construction techniques*, which allowed people from diverse ethnic backgrounds to share building tasks;
- *a productive and adaptable food-and-feed complex* that included cattle, hogs, corn, and vegetables and did not depend on restrictive tree or shrub cultivation;
- *extreme adaptability concerning the choice of commercial crop to generate income*;
- *an evangelical Protestantism combined with antifederalism*, which together encouraged settlements to control their own affairs;
- *an open class system*, which allowed White settlers to advance socially;

- *a courthouse-town system* that provided a focus on civil order and emphasized skills of the elite over others.

For Newton, it was these eleven preadaptive cultural traits characteristic of the Upland South culture that allowed the culture to advance westward. The culture itself was formed east of the Appalachians in the early to mid-seventeenth century, and moved west to the Great Plains, where the semiarid and treeless environments required a different set of preadaptive traits.

Jordan and Kaups (1989) addressed essentially the same problem, but chose to identify the preadapted culture by using the term 'backwoods.' Arguing that the culture was formed in the mid- to late seventeenth century in eastern Pennsylvania, they also emphasized the role that culture played in the settlement of the western United States following a brief pause at the eastern edge of the Great Plains (see Figure 6.10). Jordan and Kaups added the following traits to the list of eleven proposed by Newton:

- *considerable mixing with Aboriginals*;
- *an almost compulsive mobility*, which contributed to some individuals and families moving on as many as five occasions during their lifetime;
- *expansion that often leaped ahead of the frontier to create islands of settlement*;
- *no interest in conservation*.

Pause for Thought

These nine ideas about the evolution of cultural regions reflect a substantial body of scholarship aimed at understanding the cultural regionalization of the United States. There have been analyses of landscape evolution focused on other parts of the world, but they have been less fruitful. One reason that other areas of European overseas settlement might be less suited to such analyses is that there was less variety in the first effective settlement; Australia, for example, received most of its early European settlers from Britain. Many other parts of the world have such longstanding complex cultural geographies that these nine concepts are simply not relevant.

Following Rules?

A principal debate in social science concerns the issue of naturalism, the view that the social sciences—including cultural geography—can be modeled on the physical or natural sciences and can therefore apply objectivist approaches to the study of humans. While few cultural geographers explicitly argue the naturalist case, it is clear that several of the concepts discussed above have naturalist overtones. In particular, first effective settlement, culturally habituated predisposition, preadaptation, cultural islands, and the authority of tradition all suggest that people engage in repetitive behavior; unless there are compelling reasons to do otherwise, humans do what they

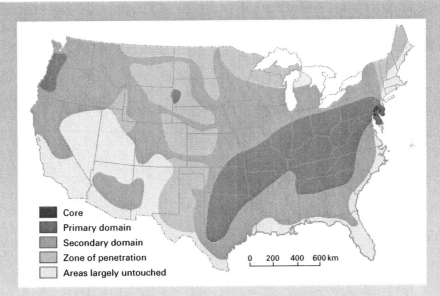

Figure 6.10 Diffusion of the Preadapted Backwoods Pioneer Culture. This map indicates a proposed diffusion of the preadapted backwoods culture, using the concepts of core, domain, and sphere in a modified format. There is a core area, where the culture first formed among Finnish settlers between about 1640 and 1680 (the precise date and specific ethnic source continue to be debated). The proposed core area, the lower Delaware valley south of Philadelphia, is much more localized than that suggested by Newton (1974); it is also smaller than and north of the cultural core proposed by Mitchell (1978), and different in detail to that described by Meinig (1993). Necessarily, then, the ethnic composition proposed by Jordan and Kaups is also different to that proposed by the several other authors, who typically identify a mix of backgrounds, usually including English, Scots-Irish, and German. The primary domain is the area of most intense cultural transplant, which began about 1700 and ended about 1850. The secondary domain was settled between about 1725 and 1875, and required considerable cultural adjustment. The zone of penetration is comparable to the sphere, showing even less evidence of the original culture.

Source: T.G. Jordan and M. Kaups, *The American Backwoods Frontier: An Ethnic and Ecological Interpretation* (Baltimore: The Johns Hopkins University Press, 1989):8–9. (© 1992 The Johns Hopkins University Press; reprinted by permission of the publisher)

During the 1870s, nearly 7,000 Russian Mennonites settled in southern Manitoba in townships set aside specifically for Mennonite development. The town of New Bergthal, Manitoba, located on the West Reserve (west of the Red River), is typical of the Mennonite street village, with lots fronting on the main street and houses set back from the road behind trees and gardens. Central lots were reserved for the church and school, and smaller lots towards the end of the village were reserved for non-farmers, such as tradesmen and retired farmers. The arable land surrounding the village was divided up and allocated for individual use, while the remaining land was used for common pasture (Kalman:347–9). (National Air Photo Library, *A-16615-23: 17-7-59* © 2005. Produced under licence from Her Majesty the Queen in Right of Canada, with permission of Natural Resources Canada)

have done before. This same logic was evident in the model of colonizer behavior noted in Chapter 5 (see Figure 5.13). As members of cultural groups, individuals often behave in ways that are consistent with established group practice, given that previous behavior has proven successful, and thus their behavior is predictable. In particular, many cultural geographic studies of groups in landscape implicitly recognize the role played by adherence to cultural norms, or what might be called rule-governed behavior.

Consider the implications of the following comment about members of the Dutch-Reformed Church in southwestern Michigan:

The basic principles followed by the adherents to Dutch-Reformed ideology can be stated simply as follows: (1) there are particular rules governing the conduct of life which must be obeyed literally; (2) man is obliged by these rules to perform both physical and spiritual work; and (3) opposition or intrusion of conflicting rules of conduct cannot be tolerated, because life after death depends upon the literal conduct of life on earth on a principled basis and is not subject to individual interpretation. (Bjorklund 1964:228)

Given the importance of imitative behavior, tradition, and adherence to established group practice in many cultural region studies, and building on claims that human geography is first and foremost about human behavior (Morrill 2002:34), it is appropriate to consider whether or not cultural geographers might make use of the concepts and principles of **behavior analysis** as developed in psychology and employed also in other social sciences. Informed by the tradition of behaviorist psychology introduced by Skinner, these concepts and principles are based on the claim that good things produced by a certain behavior make it probable that the behavior will be repeated, while bad things produced by a certain behavior make repetition of that behavior unlikely. With this simple idea in mind, behavior analysts identify antecedent conditions, operants, and consequences. Combined, these three factors constitute what is known as a contingency statement.

Accounts of a physical environment (its climate, landforms, soils, and vegetation), of a cultural environment (its religion, language, or ethnicity), of an economic environment (such as capitalism or socialism), and of a technological environment (such as agriculture or industry) are accounts of what behavior analysts label *antecedent conditions*. For many groups two particular antecedent conditions are especially significant. First, there may be an explicit or implicit hierarchical administrative structure, in which members typically respond to requests from leaders. Second, group members may practice co-operative effort and offer support to other members.

The behaviors that people engage in are *operants* and result in *consequences*. If the consequences are culturally satisfying and economically successful, they reinforce the behavior. However, in the cultural geographic context these reinforcing consequences are not usually direct-acting because they are typically delayed; consider, for example, the time span between planting and harvesting. In such circumstances, there will be cultural rules that specify both the behavior and the consequences of that behavior, so that much of the behavior responsible for transforming landscapes is a form of what behavior analysts call rule-governed behavior. Box 6.4 summarizes an application of these behavior-analytic concepts in the context of regional landscape change in nineteenth-century southeastern Australia.

Regions as Homelands

The Homeland Concept

The preceding section discussed several conceptual formulations relating to the formation of cultural regions in areas of European overseas expansion, and noted some applications of these concepts, primarily in an American context. This section builds on that discussion in the context of what is proving to be a particularly interesting regional concept—that of **homeland**. As detailed by Nostrand and Estaville (2001a:xv–xvii), this concept was used as early as 1971 by Carlson in a discussion of the Hispano region of the United States, but it is only in recent years that it has been widely applied to North American cultural regions. Nostrand and Estaville (1993a:1) defined homelands as 'places that people identify with and have strong feelings about'; they identified five ingredients common to all homelands (see also Nostrand and Estaville 2001a:xviii–xxi):

- a distinctive, self-consciously aware group, typically ethnic in character;
- a distinctive cultural regional landscape;
- an emotional bonding of the group with the region;
- a degree of institutional control of the region;
- sufficient time for these four conditions to develop.

This argument that there are distinctive, mostly ethnically based, homelands within North America also acknowledges that there are at least two other types of homeland. Most notably, for most Americans and also for residents of most other nation states, including Canada, the national unit is itself a homeland. Further, in many parts of the world there are subnational homelands whose residents are focused primarily on achieving some degree of independence from the national unit. Nostrand and Estaville (2001a) do not suggest that homelands within North America typically have such political aspirations—the principal exception to this generalization is the French Canadian cul-

Box 6.4 Explaining the Pastoral-to-Arable Transition in Nineteenth-Century Southeastern Australia

Why was there a transition from a pastoral landscape to an arable landscape in nineteenth-century southeastern Australia? From a behavior-analytic perspective, a preliminary answer is that the antecedent conditions changed to produce a new contingency with new economic, social, and political environments that encouraged wheat farming. Thus:

- there were changes in the British economy that decreased the demand for wool and increased the demand for wheat;
- there were changes in world prices for the two products to the advantage of wheat;
- there were increases in the number of local wheat markets;
- some of the pastoral area was conducive to wheat cultivation, especially given developments in technology and railroad construction;
- there was a need to feed the rapidly increasing mining population;
- there was a powerful political impetus to provide land for the poor by taking land out of the hands of a few wealthy pastoralists;
- there were neighbors cultivating wheat.

The physical environment also served as an antecedent condition, a cue informing farmers what to do in order to achieve the economic success that they desired.

Wheat farming is an operant that was reinforced and hence maintained by its consequences. Although the key reinforcer of economic success was not direct-acting in that there is a long interval between sowing seed and receiving income for the crop, there were direct-acting contingencies related to the statement of rules that specify the behavior of wheat farming and the consequences of wheat farming. Examples of rules include the prevailing view held by most members of the group that wheat farming was a profitable activity because of such factors as the railroad network and the availability of markets. The principal delayed consequence was economic success. There were also direct-acting contingencies related to both the statement of rules and to the way these rules were followed, and these contingencies served to support wheat farming until that activity paid off with the consequences noted. For example, the relevance of rules favoring wheat cultivation was evident to a new farmer who was able to observe and imitate existing farmers already experiencing the reinforcing consequences of economic success and social stability. These direct-acting contingencies encouraged wheat farmers to continue arable behavior until the reinforcing consequences specified by the rules came into play.

A general regional context for this example of land use change is included in Box 6.5.

tural region. Rather, the many examples of homelands they discuss in two edited collections (Nostrand and Estaville 1993b, 2001b) are mostly ethnically based regions with distinctive cultural landscapes that are best understood as dynamic outcomes of particular human and environment relations. In this sense, the homeland is a version of the Vidalian concept of *pays*.

In the United States and Canada, the presence of such homelands is explained especially in terms of the history of immigration and the ingrained cultural pluralism. These homelands demonstrate six human values that are rooted in place: love for one's birthplace and home, emotional attachment to the land of one's people, sense of belonging to a special area, loyalty defined by geographical parameters, strength that comes from territoriality, and the feeling of wholeness and restoration when returning to one's homeland (Nostrand and Estaville 2001a:xxiii).

CONTRARY VIEWS

The homeland concept as detailed above is not without critics. Some cultural geographers consider it inappropriate to investigate the landscapes of North America in terms of different ethnic groups and related identities and homelands. Two substantive criticisms are worth noting.

First, Hart criticized regional cultural studies based on ethnic identity as lacking sophistication:

Cultural geographers, who should have been able to make a special contribution to our understanding and appreciation of the importance of noneconomic values, have let us down rather badly, at least in the United States, because their approach to values has been so unimaginative. They have focused on those groups that 'have worn name tags', at the expense of groups that have not been easily identifiable by such obvious distinctions as country of birth, language, and religious affiliation, and they have largely ignored the importance of region of birth, social class, and other more subtle distinctions. (Hart 1982:26)

A second criticism of the homeland concept favors use of the term 'ethnic' to refer only to national identities and not to more localized identities: 'there is really no serious challenge to a pervasive, if largely subconscious, code governing the proper ways in which to arrange human affairs over American space' (Zelinsky 1997:158). Such an interpretation may involve either a real or an imagined community, and is consistent with a view of ethnic groups as named human populations with shared ancestry myths, histories, and cultures, features that promote a sense of togetherness as well as an association with a place. Conceiving of 'ethnic' in this fashion suggests that the cultural landscape of the United States is best understood in terms of a single dominant culture, namely the Anglo-American ethnic group. From this essentially superorganic perspective, ethnically based regional variations are acknowledged but should not be exaggerated.

The idea that homelands are best understood as applying to national groups is further developed by Conzen (2001:251), who argues that the key components of homelands involve three dimensions and nine criteria, which he outlines as follows:

- identity
  - ethnogenesis: a sense of peoplehood
  - indigenization: time to develop in place over multiple generations
  - exclusivity: promoted through geographical isolation
- territoriality
  - control of land and resources
  - dedicated political institutions
  - coherently manageable spatial unit
- loyalty
  - defense of the homeland against 'alien' intrusions
  - compulsion to live within the homeland
  - production and veneration of 'nationalistic' landmarks

Based on these criteria, Conzen (2001) argues that in the case of the United States, two categories of homeland predominate, namely the national homeland and a number of subnational Aboriginal homelands, although a case can be made for Mormon, Hispano, and Cajun ethnic homelands.

To clarify the meaning of the term 'homeland', the following sections present the examples of French Canada and French Louisiana. Following these examples, two of the most easily recognized and well-researched cultural regions in North America, namely the Mormon intermontane region and the Hispano Southwest, are discussed in greater detail.

Pause for Thought

Is it helpful for cultural geographers to discuss cultural regions in terms of the admittedly emotive concept of homeland? One implication of the homeland concept might be that if a group is sufficiently distinctive and occupies a particular territory, then that group deserves to be labeled as

different and perhaps even as a nation with legitimate arguments for separate statehood. As discussed later in this section, this is precisely the view that was adopted by Mormon leaders in the mid-nineteenth century. Certainly, it is not difficult to see how the homeland concept can assume powerful political overtones. On the other hand, it does appear that some regions aspire to separate political status without an explicit recourse to ethnicity or to a homeland concept. Box 6.5, which outlines the example of the Riverina region in southeastern Australia, notes that although this region lacked cultural distinctiveness as this is usually understood, there were attempts in the mid-nineteenth century to achieve some sort of separate status for the region.

FRENCH CANADA

No sooner permanently settled in the St Lawrence Valley at the beginning of the 17th century than

---

Box 6.5 The Riverina Region

Australia became part of the European overseas world in the late eighteenth century, a relatively late involvement that is most easily explained in terms of the distance between Australia and Europe. Movement of goods within Australia was also delayed, again because of the distances involved or, perhaps more correctly, because of the goods being produced, transport costs, and market availability.

The early British settlement of southeastern Australia centered around three locations: Sydney, Melbourne, and Adelaide (see Figure 6.11). In all three cases the immediate hinterland was not well suited to agricultural production, largely for topographic reasons. The search for new and better pastures for cattle and sheep was the stimulus for inland expansion following exploration in the 1820s. European penetration and settlement was little affected by the presence of Aboriginal populations, although some friction occurred. From their perspective, these settlers, known as squatters, were moving into a new and empty land and were entitled to lay claim to vast areas. Squatters moving from Sydney reached the eastern Murray Valley by the mid-1830s. Pastoral occupation continued west and was reinforced in the early 1840s by movement north from Melbourne, so that the region was fully claimed by squatters by

about 1850. The pastoral landscape that emerged included both cattle and sheep until sheep assumed prominence by the 1860s.

The landscape associated with this squatter settlement was not one that demonstrated substantial economic progress. Squatting involved dispersed and low-density settlement, and for many, the insecurity of land tenure limited substantial capital investment in the landscape. Neither poor nor wealthy had any great incentive to change the landscape, although regulations passed in the 1840s for New South Wales provided some security of tenure, thus encouraging investment, especially in fencing. Squatters had little need for towns, so urban centers were few and small in size.

As immigration to Australia continued and population numbers increased, there was growing objection to the fact that so much land was in the hands of so few, and a series of land acts gradually opened the squatter lands to closer settlement by wheat farmers. The transition from a landscape dominated by squatters to one dominated by wheat farmers was not an easy one. By about 1860, squatters were a powerful force throughout much of southeastern Australia and proved very resistant to the intrusion of wheat farmers into their lands. Successful arable

the French spread beyond the great river in quest of continental adventure. (Louder, Morissonneau, and Waddell 1983:44)

Unlike most of the American regional examples discussed in this chapter, this French Canadian hearth was quickly deserted by those who first settled there: the ambitious *coureurs du bois*, French and Métis fur traders who spread throughout much of North America from the St Lawrence to Louisiana and from Acadia to the Rocky Mountains. This expansion was doomed to failure, as the territory occupied was extensive but with few permanent locations established. Following the French defeat in North America, which was formalized by the 1763 Treaty of Paris, a gradual process of retreat to the St Lawrence occurred. Quebec was confirmed as

the French homeland in North America, an insular reserve surrounded by Anglos. But the idea of expansion remained, and by about 1870 three different ideologies prevailed among French leaders in Quebec:

- A continental ideology resurfaced, prompted by substantial movement of French Canadians into New England to work in textile mills.
- A Canadian ideology was supported by some who saw the opening of the Canadian West as an opportunity to expand the French area.
- The third, northern, ideology was the one that came to fruition. This was really a retreat into Quebec, but into all of Quebec, including the extensive northern area. This commitment to Quebec ensured the survival of French culture such that Quebec today might be considered

settlement required revised land laws and was associated with a rapidly developing economic infrastructure of towns and communication lines. The expansion of the arable landscape was encouraged by a variety of factors, including demand for wheat and the need to settle surplus populations on the land. These two factors combined to ensure the success of wheat farming in the long run.

It is in this context that arguments favoring the separation of the Riverina area can be understood. Wheat farmers, beginning in the 1850s, had argued for the transfer of the Riverina from New South Wales to Victoria as a means of destroying the security of tenure that Riverina squatters had gained. For precisely this reason, annexation was not favored by the squatters, who had begun to concentrate in the Riverina to seek relief from what to them were unfavorable Victorian land laws. Separation arguments appeared in the 1860s, with squatters leading the campaign for independence because of the perceived threat of land acts passed in New South Wales. This second development involved a much larger area than the Riverina proper, essentially all of western New South Wales.

In the end, both of these movements were unsuccessful. The group that favored the annexation of Reverina to New South Wales was never sufficiently powerful, while the squatter claims for independ-

ence were short-lived as the land acts were not as detrimental to their interests as many of them had anticipated. Perhaps the most compelling reason for both failures was that, by the 1860s, Britain was uninterested in creating additional colonies in eastern Australia.

Figure 6.11 The Riverina Region in Southeastern Australia. The Riverina, although located in New South Wales, is generally closer to Melbourne than to Sydney. The precise boundaries are difficult to locate: 'Generally speaking, in the nineteenth century, every man defined his own Riverina to suit his own purpose' (Buxton 1967:3). Twentieth-century delimitations tend to focus on the area enclosed by the Murray and Murrumbidgee rivers eastward from their junction to a line joining Albury and Wagga Wagga.

both part of and apart from the rest of the North American continent.

FRENCH LOUISIANA

The French Louisiana homeland has been described as a relic folk landscape 'faintly scattered here and there' that, because of the Anglo domination of the larger region, is today 'more myth than reality' (Estaville 2001:100). The development of this homeland can be seen as comprising seven stages. The first three stages were French and the final four English.

Transplantation occurred between 1700 and 1760, creating a cultural hearth along the Mississippi between Baton Rouge and New Orleans; this Lafourche hearth was characterized by the French language, Catholicism, and various French food and clothing preferences; it also included a large slave population and some German settlers. Between 1760 and 1800 hearth expansion occurred and was reinforced by the in-movement of French settlers deported from Acadia by the English in 1755. (Acadia, on the shores of the Bay of Fundy, was the smaller of the two French colonies in Canada; the larger was on the St Lawrence.) These Acadian settlers formed a secondary French hearth—the Teche hearth—west of the initial location and adjusted to their new environment in several ways. In particular, they abandoned wheat, barley, and other crops that had been grown in Acadia, switching to corn, rice, and other crops suited to the subtropical environment. Thus, by 1800, there were two rather different hearths, the original Lafourche French Creole hearth and the secondary Teche Acadian or Cajun hearth. Large numbers of Anglos moved into the area between 1800 and 1840, especially following the Louisiana Purchase of 1803, and this initiated a period of competition, especially in the Lafourche hearth, which lost its earlier isolation and purity, though the French managed to retain significant political control until 1845.

Between about 1840 and 1880, a process of accommodation and French cultural loss occurred, mostly as a result of increasing numbers of Anglos and the expansion of the rail network. These changes continued in the attraction period (1880 to 1920) such that although the French were able to cling to a sense of place, their cultural intensity was significantly diminished. The period between 1920 and 1960 was one of assimilation, involving continued dissolution of the original hearth and further Anglo intrusions into the secondary hearth. Many of these changes were linked to economic developments, such as the rise of the oil industry. Finally, since 1960 there has been evidence of a resurgence of French—especially Cajun—identity, bringing about a process of revitalization. This final phase has involved the beautification of the Cajun identity and the Cajunization of French Louisiana.

The Mormon Homeland

In order to understand the Mormon cultural region it is important to consider both the religious background and the physical environment of the Mormons, what Francaviglia (1978:3) describes as the 'bold, semi-arid backdrop which frames all cultural elements'.

PEOPLE AND REGION

As mentioned in Chapter 1, Mormons are members of the Church of Jesus Christ of Latter-day Saints, a Protestant group formed in Fayette, New York, in 1830. In 1847, following a series of movements in Ohio, Illinois, and Missouri, members of the church began a movement to an area well beyond the frontier, the Great Salt Lake area of the intermontane West. The Mormon Church is characterized by a strong sense of community related to an emphasis on co-operative endeavors, and also by a powerful administrative hierarchy; church members typically respond positively to the requests of their church leaders. The combination of these two characteristics contributed to a relative uniformity of behavior in landscape after 1847, and this facilitated the development of a formal cultural region and a homeland for the Mormon people.

On first seeing the Salt Lake valley in 1847, Mormon leader Brigham Young declared that this was the place, a powerful symbolic assertion that set the scene for subsequent expansion of settlement, region formation, and homeland creation. The physical landscape of the region is semiarid and comprises extensive mountain areas with numerous valleys, meaning that agricultural settlement had to take place in a few selected locations. The first areas settled—valleys close to the initial Salt Lake City location—included areas of good soil well suited to irrigated agriculture. Subsequent expansion of settlement, diffusion from the core, focussed on some other attractive valley locations in central Utah and southern Idaho. By the 1880s, settlement was attempted, with varying degrees of success, in some isolated intermontane locations and in arid areas of southern Utah and northern Arizona not well suited to agriculture. Political difficulties also prompted movements into northern Mexico and western Canada.

Despite the very real physical challenges to settlement, colonization was both rapid and successful, with almost 400 communities pioneered by 1877. This remarkable effort was largely under the personal direction of Young himself, variously called the 'American Moses', the 'Colonizer', and the 'Lion of the Lord'. The extent of the cultural region is mapped in figures 6.1 and 6.7.

LANDSCAPE

Speaking in 1874, church president George Smith observed:

> The first thing, in locating a town, was to build a dam and make a water ditch; the next thing to build a schoolhouse, and these schoolhouses generally answered the purpose of meeting houses. You may pass through all the settlements from north to south, and you will find the history of them to be just about the same. (quoted in Church News 1979:2)

Given this relatively uniform behavior, it is not surprising that a cultural region characterized by a distinctive visible landscape was formed.

One factor that contributed to the shaping of a distinctive Mormon homeland was the preference for towns—where all the local population, including farmers, resided—over more dispersed agricultural settlements. It is likely that in encouraging town settlement, Joseph Smith Jr, the first president of the church, was both reflecting his own eastern American background and anticipating the difficulty, encountered elsewhere in America, of maintaining a community in areas of dispersed farmsteads. Factors that influenced nucleated settlement include the favorable physical geography of the region, the desire to maintain the solidarity of the group, and Smith's original plan, detailed in 1833, of the 'City of Zion'. Mormon towns were laid out according to a regular grid pattern and oriented to the cardinal compass points. There were square blocks, wide streets, lots as large as 2.5 acres (1 ha) with backyard gardens, buildings of brick or stone construction, and central areas for church and educational buildings. Farm dwellings were inside the town, giving a cluttered rural appearance. Mormon towns thus both enhanced and reflected the community focus.

Within towns, the most distinctive house type was the central-hall house, a house plan that was well established in the eastern United States but one that was being modified and even replaced by the time settlement reached the plains. The plan was transferred to the West by Mormon pioneers and became a distinctive landscape feature. The two principal versions of this type are the 'I' house, one room deep and either one-and-a-half or two stories high, and the 'four over four' plan, which is two rooms deep. Both are symmetrical, usually with a chimney at each end.

Arable agriculture dominated the rural landscape and was supported by a network of irrigation ditches. Irrigation was necessary because the area being settled was typically semiarid, unable to support arable agriculture without irrigation, and also because the desired Mormon way of life required village settlement and agriculture together. Other distinctive features of the early Mormon landscape include the widely cultivated lombardy poplar which is not native to the region, the hay derrick, and the large number of unpainted fences and barns. The combination of these features clearly distinguished the Mormon landscape.

Interestingly, the Mormon landscape achieved a high level of uniformity in spite of

Photo Essay 6.1 The Mormon Landscape. The Mormon rural and small-town landscape is a distinctive one characterized by a number of specifically Mormon features. The exceptionally wide main street of the agricultural small town remains evident in the contemporary landscape; this photograph (*top*) is of Manassa in southern Colorado. Less evident in the present landscape (as it is no longer used for its original purpose of stacking hay) is the Mormon hay derrick; this example (*center*) is from Cove Fort, central Utah. In the summer of 1878, a group of Mormon pioneers, including Erastus Snow and William J. Flake, were sent on colonization missions by Brigham Young to the Lower Little Colorado River region in Arizona. They settled on the Stinson Ranch and named the townsite Snowflake. With a population of 4,230 today, the town has a Mormon temple and nine Mormon churches. The pioneer monument shown in this photo (*bottom*) was dedicated in 2000. The first Mormon pioneers to cross the American border into the North-West Territories settled in Cardston, in present-day Alberta, in 1887. The town was named for Charles Ora Card, a son-in-law of Brigham Young. The Mormon temple in Cardston (*opposite top*), like similar temples in many of the larger Mormon settlements, provides a powerful statement of the importance of the Mormon Church in the local community. Dedicated in 1972, the Provo Temple in Salt Lake City (*opposite bottom*) is located on a 7-acre site. Modern and functional in design, the exterior finish includes white cast stone, gold anodized aluminum grills, and bronze glass panels. There is a single painted spire. Inside there are six ordinance and twelve sealing rooms. It is located close to Brigham Young University and adjacent to the church Missionary Training Center. (William Norton)

the diverse origin areas of the Mormons, many of whom were converted to Mormonism in Europe before traveling to the Salt Lake area. The willingness of these immigrants to jettison ethnic cultural baggage is an indication of the overriding importance of their religious identity and of the role of co-operative effort in the creation of their homeland, the 'Great Basin Kingdom'. More so than many other immigrants, Mormons really were making a fresh start in terms of both place and identity and the new Mormon identity was more important than the authority of any earlier tradition.

Early experiences of the Mormon group suggested a need for political autonomy, and so church leaders, on arrival in the intermontane West, initiated the process of seeking statehood. A constitution was written, ambitious boundaries proclaimed, a symbolic name (Deseret) proposed, and a provisional government established. This tentative state was dissolved in 1851, following the creation of the smaller territory of Utah, with Brigham Young as the first governor. Further unsuccessful attempts to achieve self-government were made in 1856, 1862, 1867, 1872, and 1882. The boundaries of the hoped-for state of Deseret included all of present Utah, much of Nevada and Arizona, about one-third of California, and parts of five other states.

The Hispano Homeland

Is there a Hispano homeland region in the American Southwest? Two cultural geographers hint at an answer in the titles of their books: *The Spanish-American Homeland* (Carlson 1990) and *The Hispano Homeland* (Nostrand 1992).

PEOPLE AND REGION

Hispanos are those people of Spanish-American ancestry who settled in the upper Rio Grande valley of New Mexico. The physical environment of the Hispano region comprises arable floodplains and an extensive semiarid highland with a sparse vegetation cover. The core of this Spanish-American, or Hispano, region is the Río Arriba including Santa Fe, and the Río Abaja including Albuquerque. Prior to the late sixteenth century, about 45,000 sedentary and peaceful Pueblos—

members of various American Indian peoples—were settled in perhaps eighty villages in this core area, while the surrounding plateaus and mountains were occupied by foraging groups, Apaches to the west and east and Navajos to the north. Spanish colonization from Mexico began in 1598, and this outpost of the Spanish empire remained largely isolated from other influences for over 200 years.

The relationship between Pueblos and the Spanish was one of Spanish domination and passive resistance by Pueblos punctuated by revolts. Following an uprising in 1680 the Spanish in 1693 regained control of the region, which subsequently grew because of increases in the Hispano population, which spread in all directions. Expansion was especially marked between 1790 and 1900. By the late 1840s, the larger region contained about 70,000 Hispanos and fewer than 10,000 Pueblos. Together these peoples effectively represented an island in a sea of nomadic Indians, linked only tenuously to European–American people to the east. In 1848 the territory became a part of the United States and was immediately converted, politically speaking, from a region in the far north of Mexico to a region in the southwestern United States. Intrusion of Anglos (a term that included French trappers and traders) began around 1821, when Mexican independence sparked changes in frontier policies, but was greatest beginning in the 1860s. These movements added another dimension—the third of three peoples—to the cultural mix of the region.

The details of regional morphology reflect both Hispano expansion and contact with other groups. The Hispano region was in its heyday in 1900, when it extended over parts of five states, with Santa Fe at its center and a population that was two-thirds Hispano (see Figure 6.12).

LANDSCAPE

Within the larger Spanish Southwest the Hispano region of the upper Rio Grande valley is distinctive for two principal reasons: first, the Spanish settlers came earlier and tended to come directly from Spain, and second, the area was isolated for an extensive period following Spanish colonization. These two details are similar to factors that influenced the shape of the Mormon region, and Hispanos, like the Mormons, created a distinct region both by adjusting to their natural environment and by impressing their cultural identity on their surroundings. The principal distinction between Mormons and Hispanos is that, while Mormon behavior was largely community-focused and dictated by the wishes of church leaders, the behavior of Hispanos was more individual and family-focused, although various institutions, especially the Catholic Church, also played a role.

During the period of formative colonization, prior to substantial Anglo in-movement, a network of missions, *presidios* (military settlements), towns, farm villages, and ranches began to dominate the landscape. For the Spanish and, later, Mexican governments, this was an important frontier area where land policies were implemented to attract settlers and to ensure orderly settlement. The most distinctive landscape feature was the long lot, a system of land subdivision that involved long, narrow, ribbon-like lots, often fronting on a river. This system provided settlers equal access to irrigation water and permitted them to live close to each other in village settlements, a desirable state of affairs for both defensive and social reasons. Beyond the long lots were large pastoral areas, and transhumance (seasonal movement of animals) was practiced.

The landscape of this region is distinctive, especially because of the long lots, irrigation systems, adobe buildings and outdoor ovens, village settlement with plazas, and profusion of Spanish and religious place names. Many of these cultural traits were evident earlier in Spain and New Spain. The early Spanish colonists built buildings with sun-dried clay bricks, or adobes, reflecting the shortage of timber for building purposes in the favored floodplain areas. Houses were frequently added to over time to accommodate additional family members, resulting in L-shaped or even U-shaped forms, and they were often equipped with an outdoor adobe oven. There were also homes built of stone, while log buildings were usual only in the upland areas. Early villages comprised either houses on small lots around a plaza or rows of houses on lots along a riverfront. The latter form was especially popular after about 1850, when defense was less of a consideration. The impress of religion is also striking, with Catholic churches and religious meeting houses being characteristic features of the village landscape. The religious meeting houses (*moradas*) of Penitente chapters are of special symbolic importance. Spanish (often religious) place names are usual throughout the landscape and provide an especially compelling aspect of the regional identity.

In a discussion of regional morphology as evident in the Hispano landscape, Nostrand (1992:226–30) used the terms *stronghold*, *inland*, and *outland*, to refer to stages of change in geographical processes. The stronghold, defined as the area that is more than 90 per cent Hispano,

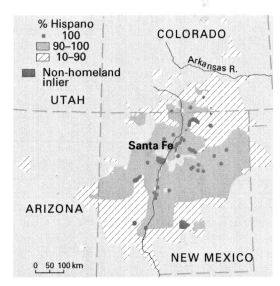

Figure 6.12 The Hispano Homeland, 1900. This map shows the Hispano homeland during its heyday. The demographic center is Santa Fe in the upper Rio Grande valley, the largest Hispano center and the political and religious capital. In 1900, 90 per cent of all Hispanos lived within 150 miles of Santa Fe. 'Thus, like an island on the land, much of the Homeland in 1900 stood tall and flat, so uniformly high were its Hispano proportions. Its edges were like steep cliffs, so few Hispanos lived off the island' (Nostrand 1992:232). Since 1900, regional identity has waned somewhat, especially because of significant intrusions of Anglo populations. It remains, however, as a distinctive region today.

Source: R.L. Nostrand, *The Hispano Homeland* (Norman: Oklahoma Press, 1992):231. (© 1992 University of Oklahoma Press; reprinted with permission of the publisher)

Photo Essay 6.2 The Hispano Landscape. The impress of the Catholic religion is very evident in the Hispano landscape, as the buildings shown in these photographs suggest. Built in the classic New Mexico Spanish mission style in 1780 and situated on the original town plaza, this small church (*top*) was originally named Santo Tomás, but the name was changed in 1881 to San José de Gracia. It is located in Las Trampas, northern new Mexico, a Spanish colonial village established in a mountain valley. It is an active parish church and was fully restored in the 1970s. The adobe church (*center*) is in Abiqui, also in northern New Mexico, and is located at one side of the old plaza. The *morada* (*bottom*), a much less ornate construction than either church above, is one of several in the southern Colorado–northern New Mexico region of Hispanic settlement; the example in this photograph is on the outskirts of the small settlement of San Pablo in southern Colorado. Located south of Tucson, Arizona, the mission of San Xavier del Bac (*opposite top*), known as the 'White Dove of the Desert', was built in 1783 after an earlier structure was destroyed. The interior is filled with brightly painted carvings of apostles and saints and ornate decor statues that are draped in real clothing. It is an important Catholic pilgrimage site. Another important pilgrimage site is El Santuario de Chimayo (*opposite bottom*) in northern New Mexico. This shrine, which has been described in some tourist literature as the 'Lourdes of America', was constructed in 1816 and remains an important symbolic location today. (William Norton)

existed by 1790 and peaked in 1900, but had largely disappeared by 1980. The inland, 50–90 per cent Hispano, was a series of separate islands that enlarged notably between 1850 and 1900 and coalesced to replace the stronghold by 1980. The outland, 10–50 per cent Hispano, was present only between 1870 and 1930.

There is, of course, a similarity here with the core–domain–sphere model as it was developed to explain the development of the Mormon cultural region. The intent is the same, namely to acknowledge variations within the region rather than simply presenting the region as internally homogeneous and sharply differentiated from surrounding areas.

Pause for Thought

Homelands are cultural regions that are especially closely linked to a distinctive group. But this seemingly straight-forward idea is not without critics. For example, some cultural geographers have called into question both the identification of the Hispano group and the reference to a Hispano homeland, citing two 'myths' involved in these ideas, namely the myth that there is a distinctive Hispano group and the myth that there is a real and discrete region occupied by the supposed group, a homeland. 'It is hard to imagine anything more powerful by way of an existence-statement about a culture than to depict its map location, its boundaries, and its internal subregions' (Blaut 1984:159). This is an important criticism that merits consideration.

Shaping the Contemporary World

Cultural geographic interest in the evolution of regions has focused largely on the subnational scale, and there has been relatively little concern with the question of global divisions. This seems natural given the intellectual antecedents of the landscape school in the work of such European geographers as Schlüter and Vidal and in the culture area concept in anthropology. Indeed, there has also been little interest in world regions in the larger discipline of geography except as basic teaching aids, with world regions seen as outcomes of human–and–land relations, usually with an emphasis on physical geography

as cause. A principal exception to this focus was the identification of human regions as outcomes of human energy as well as of physical geography, an approach that permitted the recognition of regions of hunger, debilitation, increment, effort, industrialization, lasting effort, and wandering (Fleure 1919).

The pioneering cultural geographic contribution to the study and identification of evolving major world regions is the textbook *Culture Worlds* (Russell and Kniffen 1951), while a more recent effort in this direction is a work by Lewis and Wigen (1997). Other than these important studies of world regions, cultural geographers have paid little attention to this topic. However, there is a substantial literature in comparative history, anthropology, and sociology on the larger topic of the rise of civilizations, as well as a concern with the evolution of the modern world as a system of related areas. Both concerns are discussed in this section.

Civilizations as Global Regions

Civilizations, or 'giant cultures', have been identified by scholars, often called civilizationists, such as Toynbee, Spengler, Sorokin, Quigley, and Kroeber. Each of these writers proposed different processes, stages, and civilizations, often in significant disagreement with others, but most agreed in noting cyclical patterns in history. For most writers in this tradition, the distinction between a culture and a civilization is one of scale and level of achievement, with civilizations being larger and more complex. Indeed, civilizations usually involve an integration of many languages and religions. Civilizationists have made a major contribution to the study of world history and world geography because of their explicit concern with civilizations globally rather than with the European civilization as it formed in association with other areas. As such, they can be described as scholars of world history.

The best known example of this genre of world history is probably the 12-volume work of Toynbee, *A Study of History* (1934–61). Although civilizations were not explicitly defined, they were implicitly regarded as large cultural and social groups that evolved from earlier primitive societies. The process by which a primitive society was transformed into a civilization was one of challenge and response, a form of adaptation. Toynbee used an organismic analogy, with civilizations passing through a life cycle of four stages: genesis, growth, breakdown, and disintegration.

Toynbee proposed five types of challenge or stimulus that were required for genesis to occur, namely a difficult physical environment, the availability of new territory, a military defeat, regular aggression from outside, and discrimination by others. These challenges needed to be mounted at an appropriate level—that is, sufficient to encourage a creative response without being too devastating. Following genesis, a group would grow if it continued to respond to subsequent challenges. In both the genesis and growth stages, a creative minority of the population took the lead in initiating responses to challenges. Continued growth eventually led to a situation where growth was no longer possible, and the third stage, breakdown, began. This occurred when there was a loss of unity and direction related to the demise of the creative minority. New institutions were needed but not created, such that responding to a new challenge consumed large amounts of energy, making it difficult to respond to subsequent challenges. Finally, disintegration occurred, often as a result of major social schisms.

Following these criteria, Toynbee identified 23 civilizations in world history, subsequently increasing the total to 26, of which he defined 10 as living. The 10 living civilizations were Western Christendom, Orthodox Christendom, the Russian offshoot of Orthodox Christendom, Islamic culture, Hindu culture, Chinese culture, the Japanese offshoot of Chinese culture, Polynesia, Nomads, and the Inuit, with the last three defined as living but arrested. As these regional names suggest, the basic criterion used was religion. Toynbee excluded Africa from this classification on the grounds that it was primitive and not culturally committed.

The typical reaction to this corpus of work has been predominantly negative, at least until fairly recently. This is not surprising given the difficul-

ties of working at a global scale, especially given the generalizations required. Toynbee's basic naturalist idea that civilizations can be treated as organisms has been criticized by those who would prefer to see civilizations described as adaptive systems or ecosystems that can be understood as responses to ecological opportunities.

Notwithstanding these problems, it is increasingly recognized that there is some value in viewing civilizations at a grand global scale, since this may involve a useful integration of economics, politics, society, and culture, and may offer insights into both past and present that more local scale analyses are not able to achieve. Indeed, McNeill (1995) has argued for and applied an approach to world history that acknowledges circumstances of culture contact and the role of communications and that incorporates accounts of biological and ecological encounters. The idea of some dogmatic sequence of stages is not popular, but the idea that civilizations are meaningful units for analysis remains. Civilizations may be more akin to functional than to formal regions in that they are linked by human connectedness rather than by human relatedness or uniformity—recall the important distinction noted in Box 6.1.

The Evolving World System

World system ideas are a more recent alternative to approach long-term historical change. Whereas civilizationists often stress the isolation and distinctiveness of regional civilizations, the world system concept focuses on integration and interaction. A world system is a large social system that possesses three principal distinguishing characteristics. First, it is autonomous, meaning it can survive independently of other systems. Second, it has a complex division of labor, both economically and spatially. Third, it includes multiple cultures and societies.

Historically, world systems covered only some part of the entire world, but the current capitalist world system, which began about 1450 with the onset of European overseas expansion, has broadened (that is, expanded spatially) to cover the entire world. This broadening has involved a division into three unequal but inter-dependent regions: a dominant core, a semiperiphery, and a subordinate periphery. In order to maintain a position of privilege, the core needs to maintain the underdevelopment of the periphery. The current capitalist world system has also deepened, or evolved into a more complex system; this deepening involves a set of processes that are more fully outlined in Chapter 7. Further, both the broadening and deepening of the system can be understood in terms of the Marxist concept of ceaseless capital accumulation. Though the world system approach has focused primarily on the modern world system, it is possible to extend the ideas to a much longer time scale comparable to the 5,000-year time scale employed by civilizationists.

CIVILIZATIONS AS WORLD SYSTEMS

Expressed rather simply, the greatest single difference between the civilizationist and world system approaches to the study of long-term historical change and the shaping of the modern world is the civilizationist interest in culture and the world system interest in politics, economy, and connections between regions. This distinction, which was never very clear to begin with, has become increasingly blurred, as evidenced by the discussion above as well as by recent attempts to integrate the two approaches.

One extension of civilizationist ideas views civilizations as multiple cultures or polycultures, which are best seen as sociopolitical units comprising multiple states linked either in alliance or conflict. Extending this general idea, world history can be interpreted in terms of a process of integrating civilizations that has resulted in a single dominant central civilization. The single global civilization is the lineal descendant of, or the current manifestation of, a civilization that emerged in the Near East around 1500 BCE, when Egyptian and Mesopotamian civilizations collided and fused. This new fusional entity has since then expanded over the entire planet and absorbed, on unequal terms, all other previously independent civilizations (Wilkinson 1987:46). Table 6.1 lists these various civiliza-

tions, showing central civilization as the forerunner of the current world system. Central civilization formed about 1500 BCE, grew spatially and demographically, and is now global in scale. As such, it can be regarded as the key entity in any consideration of world systems and world economies. The central civilization is the same entity as the modern world system as defined by Wallerstein.

'The Greatest Topic in Historical Geography'

Cultural and historical geographers have chosen not to address the challenging question of the formation of global regions as one of their principal research topics. Certainly, no equivalent to either the civilizationist or the world system approaches has been developed within the discipline. This failing exists in spite of Meinig's assertion (1976:35) that this is 'the greatest topic in historical geography'. Given the presumed cultural importance of central civilization or the capitalist world system to any understanding of global cultural issues, cultural geographic interest in this topic might increase significantly.

There are three aspects to the cultural geographic interest in civilizationist and world system concepts, each of which is noted briefly here and discussed more fully later, with the first topic as the focus of the next section and the second and third topics being reintroduced and further discussed in Chapter 7. First, civilizationist concepts link directly with cultural geographic attempts to delimit regions at the world scale. Second, according to the basic logic of both central civilization concepts and world system theory, the significance of the fact that Europe has played the critical role in the last 500 years of world history is that Europeans not only have effectively taken over the world but have reconstructed the world in their image. This reconstruction has been political, economic, and cultural, and is variously called *westernization* or *modernization*. The significance of this situation is difficult to exaggerate. Third, the growth of a capitalist world economy can be seen as a major global transformation in cultural as well as in economic terms; indeed, it is often asserted that the world has changed from

Table 6.1 Incorporation of Civilizations into Central Civilization

Civilization	Duration	Terminus
Mesopotamian	before 3000 BCE–1500 BCE	Coupled with Egyptian to form Central
Egyptian	before 3100 BCE–1500 BCE	Coupled with Mesopotamian to form Central
Aegean	2700 BCE–560 BCE	Engulfed by Central
Indic	2300 BCE–1000	Engulfed by Central
Irish	450–1050	Engulfed by Central
Mexican	before 1100 BCE–1520	Engulfed by Central
Peruvian	before 200 BCE–1530	Engulfed by Central
Chibchan	? –1530	Engulfed by Central
Indonesian	Before 700–1590	Engulfed by Central
West African	330–1550	Engulfed by Central
Mississippian	700–1590	Destroyed (pestilence?)
Far Eastern	before 1500 BCE–1850	Engulfed by Central
Japanese	650–1850	Engulfed by Central
Central	1500 BCE–present	?

Note: All dates are approximate.
Source: D. Wilkinson, 'Central Civilization', *Comparative Civilizations Review* 17 (1987):31–59.

a mosaic of groups with different cultures to a more culturally unified world. This change is related to the increasing importance during the past 500 years of the nation state and to the roles played by the three hegemons, namely the seventeenth-century Dutch, the nineteenth-century British, and the twentieth-century Americans. Thus, the transition is from a world differentiated in cultural terms to a world differentiated in political terms.

Global Regions

Lewis and Wigen (1997:14) argued: 'we would insist that world regions—more or less boundable areas united by broad social and cultural features—do exist and that their recognition and delineation are essential for geographical understanding.' This section outlines three proposed regionalizations.

The first regionalization is the classic cultural geographic regionalization derived from the Sauerian tradition that recognized a number of *Culture Worlds* on the grounds that the 'logical approach to regional geography is one that is based on cultural outlines' (Russell and Kniffen 1951:viii). This work had a considerable impact on human geography generally, as it helped replace environmental determinism with a form of cultural determinism. The second regionalization is based on the work of a political scientist who identified cultural regions as the appropriate basis for understanding the contemporary world. He justifies this approach to regionalization on the grounds that conflicts are no longer based on politics and ideologies but rather on differences in culture, and as a result the modern world is seeing *The Clash of Civilizations* (Huntington 1996). The third regionalization is an original and innovative geographic contribution that critiqued earlier work, stressed the heuristic value of global regions, and explicitly identified *The Myth of Continents* (Lewis and Wigen 1997). In addition, cultural geographers have traditionally emphasized the roles of language and religion as indicators of various cultural differences from place to place, and Box 6.6 provides a brief overview of these two variables in the global context.

Culture Worlds

As noted earlier, landscape and cultural region concepts were not devised, methodologically speaking, with a global scale in mind. Nevertheless, there have been attempts to divide the world into cultural regions as a means of studying general world culture. The pioneering textbook in world cultural regional geography grouped people according to culture, related culture to area in order to derive culture worlds as of the mid-twentieth century, and showed that such culture worlds had formed over a long time period. The problems of regional classification at the global scale were recognized, and various subregions were identified. The seven culture worlds, shown in Figure 6.13, were summarized as follows.

The European world was described as aggressive in the sense that Europeans typically assumed that most of their everyday cultural matters were universal, even though they may have been foreign to those outside of Europe. Although contemporary discussions in cultural geography are more theoretically and socially informed than they were in the past, it is clear that this point was not ignored in earlier discussions of cultural geography. Key European cultural traits were identified as those of field agriculture, industrialization, urbanization, and labor specialization.

Two separate cultural hearths were noted for the Oriental world, namely northern China and the Indus valley. Although the Indus valley hearth showed some similarities to the European world, it was argued that it was closer culturally to northern China. Between the European and Oriental worlds, and also extending through northern Africa, is an area of semiarid and arid

climate that separated and discouraged interaction between these two worlds. In this area, the Dry world, nomadic peoples developed ways of life quite different from those in the more humid neighboring areas.

The Dry world also served to separate the African world from the European. In the African world, people solved problems in a different manner than elsewhere and therefore developed different cultures. The isolated Polar world, characterized by a severe climate, extends across northern Eurasia and northern North America. Both the American and Pacific culture worlds developed largely in isolation from other areas until the beginnings of European overseas movement. Although both of these areas displayed significant internal cultural differences, sufficient similarities were noted to allow them to be recognized as culture worlds.

This regionalization, and many subsequent regionalizations included in introductory human and regional geography textbooks, were essentially attempts to classify the world into a fairly small number of discrete areas that demonstrated a degree of internal homogeneity and that were different from other areas. As such, they served as heuristic devices to facilitate larger textbook accounts of world regional geography. As attempts at delimiting meaningful cultural regions, such regionalizations are easily criticized, principally because the larger the area to be divided into regions, the more superficial or more numerous the regions become. As previously noted, world regional classifications typically result from a process of logical division of a large area. Such classifications require that a great deal be known about the locations beforehand so that sensible distinguishing criteria can be used.

The Clash of Civilizations

The second world regionalization discussed here grew out of concerns quite different from those of Russell and Kniffen. In an attempt to understand the principal schisms in the world following the end of the Cold War in the early 1990s, Huntington suggested that these schisms would be located along the peripheries of the major world civilizations or cultures. The central argument he put forward was that cultures and cultural identities that can be broadly conceived as civilizations are shaping relations in the contemporary world in the sense that groups sharing cultural values are increasingly co-operating with one another. Huntington further suggested that the world since about 1990 can be seen as multicivilizational—this is an important claim that contrasts with the ideas of central civilization and of world systems, both of which see a single large area, the West, as dominant globally.

The most intriguing aspect of this argument for cultural geographers is Huntington's claim that, post–Cold War conflicts, unlike most nineteenth- and twentieth-century conflicts based on economic and ideological antagonisms, are related to differences in culture. As a result, it is critically important to map cultures and thereby identify those areas (called *fault lines*) susceptible to conflict. This regionalization contains no great surprises given the earlier work of civilizationists as discussed in the preceding section, and given the standard regionalizations proposed by cultural geographers, as exemplified by Russell

Box 6.6 Language and Religion— Classification and Distribution

There are many human languages, each subject to changes in rules, content, and dialect. Interestingly, there is much debate concerning the number of languages spoken today (with the usual estimate around 6,000), although one recent major report identified about 10,000 (see Carvel 1997). To help understand the origin, diffusion, and current distribution of these languages, cultural geographers use the concept of a **language family**—a group of closely related languages that show evidence of having a common origin—although it is now widely accepted that the various language families are themselves related. Specific classifications vary in detail, but most recognize at least fourteen principal families. Within language families there are usually several subfamilies. Different languages are by definition mutually unintelligible, but each individual language may comprise several dialects, or mutually intelligible varieties of that language.

The numerically largest and spatially most widespread language family is the Indo-European, which includes most European languages and also most of those of the Indian subcontinent. The dominance of this family in the world context is related to the process of European colonial expansion. Principal subfamilies within Indo-European include Romance, which evolved from Latin and includes Portuguese, Spanish, French, and Italian; Germanic, which includes English, Norwegian, Swedish, and German; Slavic, which includes Polish, Ukrainian, and Russian; and Indic, which includes Hindi and Bengali.

There are smaller Celtic and Baltic subfamilies and also some individual languages, such as Greek and Albanian, that do not clearly belong to a subfamily. The common bond between all of these languages is their origin in eastern Europe perhaps some 8,000 years ago in the form of proto Indo-European, itself but one branch of a much larger linguistic tree.

Classifying religions has proven more straightforward, with the two major types recognized by cultural geographers being ethnic and universalizing. An **ethnic religion** is closely identified with a particular group and does not actively seek to convert others. Numerically, the principal ethnic religion is Hinduism, the oldest of the major religions, originating in the Indus valley about 4,000 years ago. Other ethnic religions include Judaism, Confucianism, Taoism, and Shinto.

A **universalizing religion** actively seeks converts because of the claim that the religion is proper for all people. The three principal examples are Buddhism, Christianity, and Islam, each of which can be further subdivided. Buddhism, an offshoot of Hinduism, developed about 500 BCE; Christianity developed from Judaism at the beginning of the Common Era, and Islam developed from both Judaism and Christianity about 600 CE. Historically, Christianity has been the most concerned with seeking converts, and because it was associated with Europe during the period of overseas expansion from the fifteenth century onwards, it has spread over much of the globe along with some of the Indo-European languages, notably Spanish, Portuguese, French, and English.

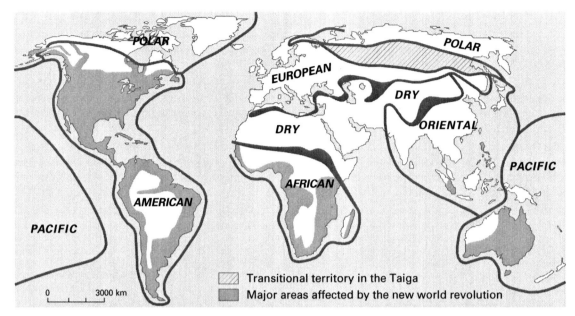

Figure 6.13 World Cultural Regions I. 'Through the device of recognizing seven culture worlds, the geography of the inhabited parts of the earth is presented in an orderly manner. Each culture world is a reasonably unified subdivision of the earth's surface occupied by peoples who are strikingly alien to inhabitants of other culture worlds' (Russell and Kniffen 1951:viii). The global regions mapped are the European World, the Oriental World, the Dry World, the African World, the Polar World, the American World, and the Pacific World.

A transitional area is located between the European and Polar Worlds, and large areas outside of Europe are seen as affected by a New World Revolution, notably the American, African, and Pacific Worlds. Each region is discussed in considerable physical, historical, and cultural detail. This pioneering use of the cultural region concept at the world scale is a major development within the Sauerian tradition. A principal question about this contribution concerns the use of the label *Dry World*, as the use of a climatic term does not accord with the idea of a cultural region. There are many other schemes for world regionalization employed by textbook authors for essentially heuristic reasons, but most are similar, although different in detail, to this pioneering effort.

Source: R.J. Russell and F.B. Kniffen, *Culture Worlds* (New York: MacMillan, 1951).

and Kniffen. The key criterion used to identify civilizations is that of religion, as several of the regional names indicate (Figure 6.14).

This argument has generated substantial discussion in political science, as it represents a marked departure from the conventional concern with states as the key units in a conflict situation. There has been some broad agreement concerning the idea that future conflicts are likely to occur in boundary areas, but not necessarily the boundaries between world civilizations. The argument is particularly open to criticism for attempting to build the West around conservative values and for stressing the dangers presented to the West by others, such as Islamic fundamentals and immigrants.

Perhaps the most credible alternative suggestion (again of great interest to cultural geographers) concerns the prospect of future conflicts related to discord within political units that are prompted by massive variations in quality of life, which are in turn often related to ethnic and cultural differences, or, more directly, to resurgent ethnicities.

Although cultural geographers have not traditionally concerned themselves with conflict issues, these two suggestions about likely future conflict emphasize the close links between political, economic, and cultural themes. Good cultural geography, like good political and economic geography, needs, at times, to be broadly concerned with all other aspects of larger human geography.

The Myth of Continents

Attempts to regionalize the world using cultural criteria are problematic. It is not an easy task to delimit large areas that can be meaningfully argued to be in some sense internally homogeneous. Of course, the conventional approach, both within and outside of geography, is to use continental divisions. Most of us tend to think of Europe, Asia, Africa, North America, and Latin America as regions, at least in some informal sense. The principal alternative approach is based on literate civilizations, as discussed earlier and as used by Huntington. It is not difficult to see that the attempts at world regionalization

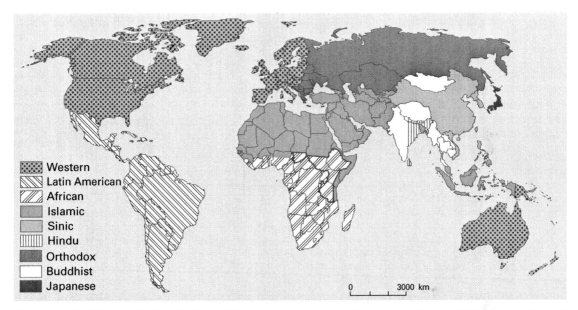

Figure 6.14 World Cultural Regions II. Although Huntington insisted on the importance of regional civilizations as the basis for understanding contemporary world conflict, the task of identifying those regions was not straightforward. Huntington (1993:25) named, but did not map, 'seven or eight major civilizations. . . . Western, Confucian, Japanese, Islamic, Hindu, Slavic-Orthodox, Latin American, and possibly African civilization'. Africa was listed as 'possible' because of the divisions into Islamic and non-Islamic and into Saharan and sub-Saharan.

This figure shows the nine civilizations mapped by Huntington (1996). But even this second regionalization contains some uncertainties and apparent confusions. The map and the discussion of the map are not in complete accord. Thus, in referring to the map, Huntington (1996:21) noted that there are 'seven or eight major civilizations', although nine are included on the map. What is the justification for these nine mapped civilizations?

- Five of the nine were accepted on the basis of earlier civilizationist ideas; these are Sinic (also called Chinese and previously called Confucian, as noted above), Japanese, Hindu, Islamic, and Western.
- Some civilizationists also add Orthodox Russian civilization.
- Latin American and possibly African were considered useful.

The inclusion of the Buddhist civilization was not explained.

But following the statement, 'The major civilizations are thus as follows,' Huntington (1996:45) listed and briefly discussed only seven civilization—Sinic, Japanese, Hindu, Islamic, Western, Latin American, and African (possibly); neither Orthodox nor Buddhist were identified as major civilizations even both were included on the map. Huntington noted separately from the list that there is a Theravada Buddhist civilization in Sri Lanka, Burma, Thailand, Laos, and Cambodia.

These problems evident in the initial identification and mapping of civilizations are confusing from a cultural geographic point of view. However, it should be noted that the extensive discussions of political and related issues that constitute the bulk of the book focus on all nine civilizations.

Source: S.P. Huntington, *The Clash of Civilizations and the Remaking of World Order* (New York: Simon and Schuster, 1996):26–7. (Reprinted by permission of Simon & Schuster Adult Learning Group; base map © Hammond World Atlas Corp.)

discussed so far can be criticized for occasionally flawed logic. It may be that these essentially traditional approaches to regionalization, which are often based on presumed continental divisions, are at best misleading and at worst wrong.

One response to this problem is based on the following argument: 'World regions are multicountry agglomerations, defined not by their supposed physical separation from one another (as are continents), but rather (in theory) on the basis of important historical and cultural bonds' (Lewis and Wigen 1997:13). This does not mean civilizations as discussed by civilizationists, which typically assume literacy; rather, global regions, large sociospatial groupings, are identified on the basis of a shared history and culture. The resulting regionalization (see Figure 6.15) is based on processes and on traits, ignores political and economic boundaries, and employs cultural boundaries.

First, it is helpful to think of Africa, Europe, and Asia as comprising the supercontinent of Afro–Eurasia. Effectively, Europe can be thought of as Western Eurasia or as Northwest Afro–Eurasia. Second, note that denying that Europe is a continent does not prevent it from being treated as a cultural region because, of course, an area does not need to be a continent in order to merit cultural region status. Third, the regionalization of Asia is reconceived, being divided into five regions, an explicit acknowledgment of both size and regional complexity. The inclusion of a Central Asian region is a distinctive addition to most regionalizations, as is the reference to Lamaism (that is, Mahayana Buddhism). Fourth, the regionalization of the Americas is

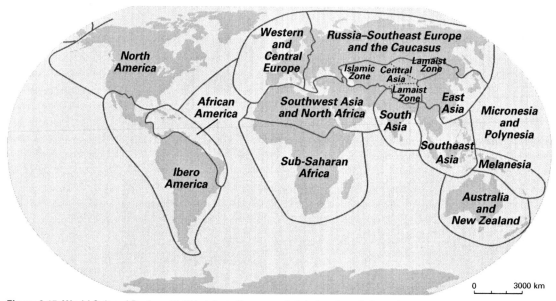

Figure 6.15 World Cultural Regions III. This interesting map of global regions followed an extended account that stressed some problems of regionalizing at this scale, but that also identified the value of regionalizing at this scale. It was acknowledged that 'all human geographical entities are conventional rather than natural,' and that there is accordingly a need to use regional names that are generally understood rather than names that have different meanings in different contexts (Lewis and Wigen 1997:196).

The fourteen regions recognized are those of East Asia, Southeast Asia, South Asia, Central Asia (divided into Islamic and Lamaist zones), Southwest Asia and North Africa, Sub-Saharan Africa, Ibero America, African America, North America, Western and Central Europe, Russia–Southeast Europe and the Caucasus, Australia and New Zealand, Melanesia, and Micronesia and Polynesia.

Source: M.W. Lewis and K.E. Wigen, *The Myth of Continents: A Critique of Metageography* (Berkeley: University of California Press, 1997):187. (Reprinted by permission of University of California Press)

similarly reconceived. Latin America is rejected as a regional name, and the Central American area is named African America.

Pause for Thought

One continental division that may be particularly unjustifiable is that between Europe and Asia, given that Europe, as it is not a distinctive and separate landmass, is essentially a western peninsula of Asia. Is it reasonable to regard Europe as a continent and India as a subcontinent? The use of continents, or presumed continents, as a basis for regionalizing is certainly flawed reflecting, as it does, a particularly Eurocentric view of the world—precisely the point made by Russell and Kniffen. Indeed, it is increasingly clear that the particular views of the world that a group holds are a reflection both of reality and of their perceived position in the world. The cultural studies tradition, especially in the form of postcolonial theory, addresses these ideas more fully. Overall, does it appear that the three world maps of world regions are meaningful contributions, or are they oversimplifications of complex realities?

Concluding Comments

A regional approach to cultural geography derives both from the established tradition of geographic regional study and from the specific incorporation of regions in the landscape school. Notwithstanding some terminological confusion regarding the words *region*, *area*, *landscape*, and *place*, cultural geographers have embraced the approach. *Landscape* and *region* are the terms preferred in this chapter, with *landscape* as a general term referring essentially to the visible scene, and *region* as the more specific term referring to a part of the surface of the earth that has a similar landscape. Applying the concept at the subnational scale is the favored approach, with most studies centering on American regions. Applying the regional concept at a global scale in order to aid in identifying and mapping cultural realms, has a venerable and controversial tradition and is currently experiencing something of a resurgence. There are two different, perhaps contradictory, approaches to

global regionalization, namely the civilizational and the world system approaches. The resurgence of interest in global issues is related to the dramatic changes occurring in the contemporary world, especially concerning globalization.

There is a key outstanding issue that has been raised on more than one occasion in this chapter but that has not been addressed directly. Recall that some regions can be described as homelands, a term that is used to reflect an especially close link between people and both the physical and built landscape of their region. This idea has sparked some controversy and is a useful entry point into a larger debate about the legitimacy of distinguishing distinct groups of people and related distinct regions.

Notwithstanding some methodological concerns, and as the discussion of regional studies in this chapter makes clear, many cultural geographers both favor and continue to work in a regional tradition that does not place priority on matters of social production. The success of this continuing and continually changing regional perspective, as demonstrated in this chapter, is undeniable.

Pause for Thought

There is little doubt that questions of group identity are complex, elusive, and certainly changing—are we really able to divide people into neat parcels such as cultures or ethnic groups, or are these groupings really arbitrary creations? There is no clearcut response to this question, and it is sufficient to raise these concerns in the conclusion to this chapter without attempting any resolution. Indeed, in addition to the elusiveness of group identity, it is now increasingly acknowledged that both human identity and human landscapes, regions, and homelands, are socially constructed or produced. Although there is debate on the merits of this perspective, the popularity of the idea in much contemporary cultural geography is undeniable. This chapter on cultural regions has largely ignored both this perspective and the debate surrounding it, as such matters are more appropriately introduced in later chapters as part of a more socially informed cultural geography.

Further Reading

The following are useful sources for further reading on specific issues.

Mason (1895) and Wissler (1917, 1923) are early anthropological applications of the regional concept, a tradition that culminated with the publication of a major work on the cultural and natural areas of North America (Kroeber 1939).

According to Carter (1948) the anthropological contributions attracted much geographic attention. A survey of the culture area concept as used in anthropology and a comparison with geographic applications is found in Mikesell (1967:621–4, 626–8).

Hartshorne (1939, 1959) authored the classic works on geography as the study of regions.

Butzer (1989a) discusses the methodological conflict between Sauer and Hartshorne.

Harvey (1969) details regions as an example of classification procedures.

Gilbert (1988), Pudup (1988), Johnston (1990), and Lee (1990) are discussions of the new regional geography.

Wishart (2004) discusses the concepts of region and period as similar types of generalization.

There are few examples of cultural region formation and transformation outside of North America that have employed core, domain, sphere concepts, notwithstanding the generic flavor of the model. For Wales, two principal cultural regions can be delimited according to the relative strength of the English and Welsh languages, and a bilingual zone between the two can be equated with the Meinig domain (see Carter and Thomas 1969; Pryce 1975). Also see Withers (1988) for an account of the Scottish Highlands cultural region.

Harris (1978) addresses the historical geography of North American cultural regions.

Uses of the first effective settlement concept include the following: Gastil (1975), in an account of areas of cultural homogeneity; Elazar (1984, 1994), on the United States political scene with reference to the westward migration of cultural traits but without any explicit use of the work of cultural geographers; Wacker (1968), on the Musconetcong valley of New Jersey; and Wacker

(1975) and Wacker and Clemens (1995), on the larger area of New Jersey.

Aschmann (1965) used the preadaptation concept, without using the particular term, in a discussion of five Apache cultural traits that made their occupation of the American Southwest possible.

Jordan and Kaups (1989) offer a discussion on the linking of specific ethnic groups with specific environments, and Garrison (1990) provides a critical commentary on this idea.

West and Augelli (1966:11–16) identify cultural regions and related landscapes in Central America, emphasizing the interplay of physical geography and history as reinforced by isolation. Jordan-Bychkov and Bychkova Jordan (2002) map regions in Europe based on particular variables, such as language and religion, while attempting a broadly defined cultural regionalization of the continent in terms of North and South, East and West, and core and periphery. Ostergren and Rice (2004) provide an overview of the cultural geography of Europe.

Luebke (1984) and Shortridge (1995) explain many developments in the Great Plains in terms of distinct incoming cultures.

Meinig (1969) examines Texas as a cultural region; Hilliard (1972) and Nostrand and Hilliard (1988) treat the American South; Cobb (1992) looks at the Mississippi Delta; Wyckoff (1999) focuses on Colorado; Hausladen (2003) includes accounts of the American West as place and myth.

Jordan, Kilpinen, and Gritzner (1997) discuss the folk landscape of the North American mountain West.

Rose (1972) and Jeans (1981) address the possibility of delimiting cultural regions in Australia.

Christopher (1984:193) notes that, throughout much of Africa, 'New European societies were formed which were able to impress their ideas upon the landscape and create an image of France or England overseas'; nevertheless, there has been little interest in identifying any resultant cultural regions.

Accounts of behavior-analytic concepts and principles are included in standard introductory psychology textbooks; Martin and Pear (1996) provide clear and sympathetic details. These concepts and principles are applied to the southeastern Australian landscape by Norton

(1997b) and to the Mormon landscape of the intermontane West by Norton (2001, 2003a).

Kollmorgen (1941, 1943), Augelli (1958), Lehr (1973), Raitz (1973b), Gerlach (1976), and Gade (1997) discuss aspects of ethnicity and cultural islands.

A Special Issue of the *Journal of Cultural Geography* edited by Nostrand and Estaville (1993b) outlines the homeland concept and includes several regional studies. The concept is further detailed in *Homelands: A Geography of Culture and Place Across America* (Nostrand and Estaville 2001b), and additional regional examples discussed. See also Hurt (2003) for a further refinement of the homeland concept and for the example of the Creek nation.

Trépanier (1991) and Estaville (1993, 2001) discuss French Louisiana.

Meinig (1965) is the pioneering cultural geographic analysis of the Mormon region, although the presence of the region was generally acknowledged prior to that analysis (Zelinsky 1961); Francaviglia (1978) provided the seminal landscape analysis (see also Jackson and Layton 1976, Jackson 1978, and Bennion 2001); the symbolic importance of the region to Mormons has been described by Jackson and Henrie (1983). Smith and White (2004) discuss the case of Mormons located in Chihuahua, Mexico, in terms of their detachment from the homeland. Toney, Keller, and Hunter (2003) and Yorgason (2003) analyse aspects of twentieth-century change in the Mormon homeland.

Meinig (1971), Carlson (1990), and Nostrand (1992, 2001) examine aspects of the formation of the Hispano cultural region and the characteristic landscape; Smith (1999) focuses on cultural change occurring at the northern edge of the Hispano homeland. Nostrand (2003) provides an unusually detailed and insightful account of a single village.

Francaviglia (1994) discusses the understanding of the American Southwest as it has been continuously redefined through a series of cultural occupances.

Melko (1969), Sanderson (1995), and Wilkinson (1987) discuss the writings of civilizationists.

Wolf (1982), Gills (1995), and Sanderson and Hall (1995) address aspects of world systems.

The principal contributions to world systems theory are by Wallerstein (1974a, 1974b, 1980, 1989).

Hugill (1997:348) notes that the 'the most vital debate in social science' is the attempt to develop a theory of the history of the capitalist world, and possibly even a theory for all human history. Political and economic geographers have made major contributions to this debate (Dicken 1992; Taylor 1996).

Mosely and Asher (1994) is a comprehensive atlas of language; Renfrew (1988) stresses the relationships between language families; the role of language as a basis for group delimitation and communication between group members has been noted especially by Wagner (1958b, 1974, 1975); cultural geographic overviews of religion include those by Sopher (1967, 1981), Gay (1971), and Park (1994).

Sauer (1940) outlined a global regionalization that stressed the historical origins of regions in early civilizations.

Ó Tuathail (1996:240–9) notes possible failings of the Huntington world regionalization, and Kaplan (1994, 1996, 2000) suggests that conflicts might be based on differences in qualities of life and access to resources.

7

Power, Identity, Global Landscapes

There is a recurring interest in much current cultural geography with questions of human **identity**, especially with how identities relate to the many human and landscape inequalities that prevail in the contemporary world. The link between identity and inequality is often explained by the fact that a principal basis for distinguishing one group identity from another is in terms of relative **power**, an idea that is central to Marxist, feminist, and postmodernist ideas, as discussed in Chapter 3. The interest in identities is not new, as cultural geographers have longstanding interests in such characteristics as way of life, language, religion, and ethnicity. Nor is the interest in inequalities new, as many of the concepts devised by cultural geographers to explain the rise of regional landscapes and homelands—concepts such as first effective settlement—explicitly acknowledge that some groups are able to impose their identity on landscape while others are not. But the current concerns with identity and inequality are distinctive for at least two reasons.

First, cultural geographers are now making use of more sophisticated theoretical ideas than those belonging to the landscape school tradition, ideas that explicitly consider identity formation and inequalities. Second, there has been a broadening of the culture concept, from the possibly superorganic entity used in the landscape school tradition to one that encompasses a more meaningful view of human identity. Ideas about culture emerging from the new cultural geography acknowledge that many aspects of identity cannot be subsumed under a single general concept of culture but are rather socially constructed and continuously changing. Most notably, there is a recognition that dominant groups are typically able to assert their chosen identity while imposing identities on subordinate groups, thus creating a cultural world that is structured on their terms. This is, of course, precisely the argument that Marxist geographers use in their critiques of capitalist society and economy, that feminist geographers use in their critiques of a male–dominated world, and that postmodern geographers use in their studies of oppressed and excluded groups. More generally, another component of this broadening of the culture concept is contemporary cultural geographers' willingness to incorporate traditional social geographic subject matter, such as accounts of class, into their analyses, a circumstance that has at least partly removed the rather artificial distinction between the two subdisciplines.

The principal concern of this chapter is how identities relate to the myth of races within the human species and to related ideas about ethnic identity. The principal empirical accounts of inequalities are at global and national scales. In the following chapter the discussion of identity is further developed to include such topics as gender, sexuality, and disadvantaged groups generally, and these are discussed mostly with reference to inequalities at community and local scales. In this chapter you will find

- a discussion of the idea that we live in one world, but one world that is significantly divided by cultural variations;

- a historical account of the illusion of races within the human species, which concludes that race is not a valid means of classifying individuals into groups and that the term is therefore not an appropriate one to use;
- a consideration of the apartheid era, from 1948 to 1994, in South Africa, and several instances of genocide, which illustrate that although race is a myth, racist thought and practice are an everyday reality;
- an account of ethnicity and ethnic identity that considers the roles played by language, religion, and national identities;
- an assessment of globalization from a political, an economic, and most significantly, a cultural point of view. Attempts to explain the current division into more and less developed worlds are reviewed, and the possibility that political states might play a lesser role in the future is discussed;
- an evaluation of the claim that we are moving towards a world dominated by a single cultural identity;
- a brief concluding section.

One World Divided

A pioneering world regional textbook of human geography that was first published in 1964 highlighted a concern that is at the forefront of much of the current work in cultural geography, namely that human groups and their landscapes exhibit the characteristics of *One World Divided* (James 1964). The text was written with the premise that the great problems requiring attention by geographers were those related to inequalities, as these are evident in patterns of poverty, hunger, overpopulation, relations between peoples, environmental destruction, and violence. James argued that the global distribution of these problems could be understood in the context of two larger economic and social transformations. The logic behind his argument was as follows.

The world is divided both by differences in physical geography and the availability of related resources and by differences in human geography. Differences in physical geography, such as differences in climate, landforms, and

soil, are significant but essentially fixed. Differences in human geography, such as differences in language, religion, and way of life, are, on the other hand, constantly changing. As examples, James cited two revolutionary processes of change, those of industrialization and democracy, that were initiated in western Europe and proceeded to reshape global human geographies. One outcome of the spread and growth of these processes was the evolution of twelve global regions: Northern American, Southern American, European, North African–Southwest Asian, Middle African, Southern African, Soviet, South Asian, East Asian, Australia–New Zealand, and Pacific. These regions are not dissimilar to those typically produced by other geographers as evident in the discussion in Chapter 6.

But James (1964) proceeded further. In reviewing global changes, especially during the past 200 years, he asserted that there was one particularly tragic circumstance:

> Man himself has created divisions and barriers far more difficult to cross than those caused by nature. Man has brought into being mountains of hate, rivers of inflexible tradition, oceans of ignorance. And as technology brings men closer together the man-made barriers loom larger and more forbidding. (James 1964:3)

Since James invoked these powerful metaphors in 1964, most evidence substantiates his claim that cultural divisions are prominent. The long list of regions involved in recent or ongoing conflicts that at least partly reflect cultural divisions includes Northern Ireland, Quebec, the Middle East, Cyprus, the Kurdish region, Uganda and other countries in central Africa, Nigeria and other countries in West Africa, and the former Yugoslavia. Terrorist activities occurring in many different locations also reflect these cultural divisions. All of these conflicts occur in spite of improved communication technologies that James could not, of course, have anticipated in 1964, and at this time there seems little reason to believe that ongoing globalization processes will reduce the significance of human–made divisions.

Much of the content of both the present and the following chapter highlights this divided world. Although James was concerned with divisions at the global scale, it is clear that divisions are apparent at all spatial and social scales, and most current cultural geography favors more local scales of analysis. Similarly, although James was concerned mostly with highlighting such fundamental cultural variables as language and religion, it is clear that there are many other bases for division, including gender, sexuality, and disadvantage generally.

The Mistaken Idea of Race

Race is a biological term that, when applied to humans, is a myth. This section uncovers ways in which the term is incorrectly used to describe human groups; anticipating our conclusion, not only is the term meaningless but, further, it is so laden with tragic baggage that it is extraordinarily unhelpful to use it today. Nevertheless, although race is a myth, so many people today harbor negative views of others based on presumed racial identity that it is important to know what is meant by such terms as *racism*, *prejudice*, and *stereotyping*.

The Unity of the Human Species

A single and simple fact to begin: there is only one species of humans in the world, namely *Homo sapiens sapiens*. It is correct to say that this human species displays variation among its members, but it is biologically incorrect to say that these differences are sufficient to merit using race—that is, subspecies—to make distinctions.

The biological differences that do exist among humans result principally from chance and from the fact that human groups migrated globally and were isolated from each other for an extended period of time, specifically between about 100,000 years ago and about 10,000 years ago. It was this separation that allowed some selective breeding and some adaptation to physical environments to occur. Together these two biological processes facilitated the evolution of some biological differences. What is important, however, is that the period during which groups were separated was not long enough to allow races, in the biological sense of genetically distinct subspecies, to form.

The most obvious adaptation to physical environment—obvious in the sense that its results are clearly visible—was the evolution of several different skin colors, with lighter skins associated with temperate environments and darker skins associated with tropical environments. This adaptation has to do with the pigment melanin, which develops in the hair, skin, and iris of people (and animals) as a defense against the ultraviolet radiation that can cause cancer; humans and animals in tropical environments produce more melanin than those in temperate regions. But neither this nor any other difference means that humans can be neatly classified into racial groups. Indeed, there is no one biological trait that can be used to classify people into such groups. If we take three of the ways in which humans vary biologically—blood group, hair texture, and skin color—we discover that each is influenced by different genes and is thus inherited independently of other characteristics. What this means is that the identification of such typically defined groups as Negroid, Mongoloid, Australoid, and Caucasoid is fundamentally flawed and therefore meaningless. Indeed—and this is an important observation—there is typically more genetic variation within one supposed race than there is between supposed races.

In summary, humans are a biologically variable species, subject to the same processes that have produced variation in other living things, namely chance and natural selection in different physical environments, but humans cannot be divided into biological subspecies or races. The statement that all humans are members of the same species is biologically confirmed by the fact that any human male is able to copulate and procreate with any human female. In the context of humans, race is not an objective fact.

The Illusion

There are no races within the human species, and yet until quite recently it was widely accepted that races exist. How and why did this myth arise? What purpose did it serve? Why is it necessary to discuss this myth in the present context? Answering these questions and understanding the context for the emergence of the race myth is simple in principle, but difficult in practice.

Simply expressed, the notion of race, misguided though it was, became an important component of nineteenth-century scientific and political discourse in part because it seemed logical to many observers trying to understand the apparent diversity of humans. If we step back from this statement, it is not too difficult to appreciate that emphasizing difference in this way was but one of two directions these observers could have taken. The other direction, the one they did not take, was to realize that humans have many more characteristics that unify than that separate them.

But the misconception of race arose also because it suited the aims of many nineteenth-century Europeans. The principal purpose was a scientific one: race enabled physical anthropologists to devise formal classifications using what appeared at first sight to be sound scientific grounds. But there was another important and much more dangerous purpose, namely that of using presumed physical differences to explain cultural differences. It is therefore necessary to discuss this myth of race because it has been routinely used (perhaps a better word is *abused*) to justify rankings of groups on the basis of perceived abilities and to justify discriminatory thoughts and behaviors. In other words, notwithstanding the fact that there are not distinct human races, the supposed existence of such races is an idea that became influential in the nineteenth century and remains important in the contemporary world, and it is this idea, this socially constructed myth, that is the subject of the present discussion.

It is often claimed that the idea of race has a long history, and that many groups in the past perceived themselves and others as having racial status. This claim lacks credibility, since most recent evidence favors the idea that the myth of race was essentially a nineteenth-century creation. There is no word corresponding to the English word *race* in Jewish, Greek, or Roman writings, although there is much evidence of divisions between groups, especially the political and cultural divisions between citizens and barbarians. For Hannaford (1996:12), the Greek

'political idea involved a disposition to see people not in terms of where they came from and what they looked like but in terms of membership of a public arena', and it therefore 'inhibited the holding of racial or ethnic categories as we have come to understand them in the modern world'. Historically, both Islam and Christianity, in their capacity as universal religions, welcomed converts. They saw themselves as communities of believers, not as racial groups.

The word *race* dates only from the fourteenth century, and it was rarely used before the sixteenth century. It seems likely that there was no significant racial thought in the European world prior to the late seventeenth century and that the claim that there was such thought is itself largely a product of the nineteenth century. The myth of race is not some timeless concept; rather, in historical terms, it is a fairly recent invention.

CREATING THE ILLUSION

Seven principal developments contributed to the rise of the European idea of race. First, following the Reformation, polygenetic theories of human origin appeared; these theories contradicted the previously dominant Augustinian view of a single human origin and became a key element of the myth of race. Second, during the extended process of European overseas expansion, beginning about 1450, the idea that races existed seemed logical to Europeans because some of the people they encountered really did look different. It requires little imagination to appreciate how perplexed both sides of a cultural encounter might be on experiencing a first sight of the other. Third, with the rise of a global system of commercial capitalism related to colonial expansion, Europeans in some overseas areas, notably the southern United States, began to rely on slave labor; their growing dependence on slave labor fueled an extremely lucrative industry for slave traders and effectively institutionalized slavery as an accepted social relation, thereby fixing Africans in a position of permanent inferiority.

A fourth development that contributed to the rise of the race myth involved the creation of republics following revolutions in late-eighteenth century France and the United States, which served to integrate the concepts of **nation** and people. This raised new ideas about nations existing as racial communities. A fifth development had to do with major advances in the scholarly classification of humans—such as those by Carolus Linnaeus (1707–78), Johann Friedrich Blumenbach (1752–1840), and Buffon—which were one part of larger scientific advances. It should be pointed out that these understandable, indeed laudable, attempts to understand human variability and what it meant to be human did not attach any values to the classes identified. Sixth, Enlightenment philosophers such as Rousseau and Montesquieu paved the way for environmental racial theories with their ideas about civilization and savagery. Overall, the period from the late seventeenth- to the early nineteenth-century witnessed a rise in popularity of the idea that civilization advanced not through the 'public debate of speech-gifted men and the reconciliation of differing claims and interests in law but through the genius and character of the *Völker* naturally and biologically working as an energetic and formative force in the blood of races and expressing themselves as Kultur' (Hannaford 1996:233). The seventh development contributing to the European idea of race was the emerging field of ethnology, which developed as both a natural and human science concerned with identifying the biological and historical origins of particular groups. It was in this context that the myth of an Aryan race emerged.

Thus, it was primarily nineteenth-century Europeans who formulated the myth of race and created a history of racist thought through a reinterpretation of the past using the new language—discourse—of national identity, progress, and social Darwinism. Barthold Georg Niebuhr (1776–1831) rewrote the history of Greece and Rome using racial rather than political classes, while Augustin Thierry (1795–1856) performed a similar task for French history. Together, these works prompted the idea of an original European Aryan culture superior to those of both Greece and Rome. More generally, the tendency for most cultures at most times to dislike and even fear strangers and those who are in some way different was conflated with **racism**. Critically, the new doctrine of **nationalism** was supported by the idea of race, with national group membership contingent upon such characteristics as language and skin color, and with nation—whether real or imagined—replacing religion as a primary basis for group identity. It was in this context of the rise of the myth of race that racial classifications, now including hierarchies, were proposed by many scholars.

CLASSIFYING PEOPLE

Western academics and intellectuals of the nineteenth and early twentieth centuries generally accepted that it was appropriate to classify people into racial categories and, further, that the classes identified reflected not only physical criteria but also intellectual ability and cultural characteristics generally (see Figure 7.1). For example, Count Arthur de Gobineau (1816–82), in his *Essay on the Inequality of the Human Races* (1853–5), classified Europe into races with Germanic people as the superior group. Gobineau doubted the ability of Africans and Native Americans to become civilized, and generally saw race as the sole explanation for progress. For Gobineau, the existence of discrete races, with definable physical, intellectual, and moral characteristics, was not just a fact but the most important fact for any understanding of humans. And, given the existence of fixed races, he expressed considerable concern about mixing races and the loss of racial purity that would ensue.

Probably the most influential and widely accepted work on races was the 1899 study of civilization *The Foundations of the Nineteenth Century*, by Houston Stewart Chamberlain (1855–1927). Chamberlain argued that race explains everything, and he claimed civilization as the sole prerogative of the German race with its Aryan roots. Jesus, he argued, was a member of the Aryan race. Overall, Chamberlain was much less scientific than Gobineau, referring frequently to race souls, to the power of pure race, and to the importance of remaining racially pure.

A typical minutely calibrated sliding scale of human groups equating ability and culture with physical characteristics placed Nordic or Teutonic peoples at the top, followed by Alpine, Mediterranean, Slavic, Asiatic, and African groups. One influential racial classification was developed as recently as the second half of the twentieth century (Coon (1962). This classification distinguishing five principal presumed groups: Australoid, Mongoloid, Caucasoid, Capoids, and Congoid (or Negroid). In principle, each group was distinct from each other group, and Coon asserted his belief in the reality of racial purity.

It is important to acknowledge that many scholars accepted the idea that humans could be divided into distinct and pure racial categories, and even the idea that races could be ranked was widely viewed as correct. Certainly, by the early twentieth century, race was a dominant organizing idea, and it was generally believed that humans could be properly classified and ranked according to appearance. In other words, the idea of race as both a scientific category and a value judgment about people was part of the discourse of the time.

In some instances, racial classifications were taken to especially unpalatable extremes. For example, one regular feature of racial classifications and rankings was the explicit distinction made between German Christians and Jews. Jews in the West were a well-defined group traditionally excluded from the state, but political anti-Semitism developed only in the late nineteenth century, in relation to race studies. In Germany, ideas about the racial inferiority of Jews contributed to the policies of the National Socialist Party, which declared that only members of the supposed German race could be citizens of the German state; these policies were used to justify the mass murder of Jewish and some other groups during the Second World War.

Of course, racist ideas, with roots in Enlightenment and nineteenth-century thought, were not influential only in Germany. Rather, they were widely accepted and implemented in var-

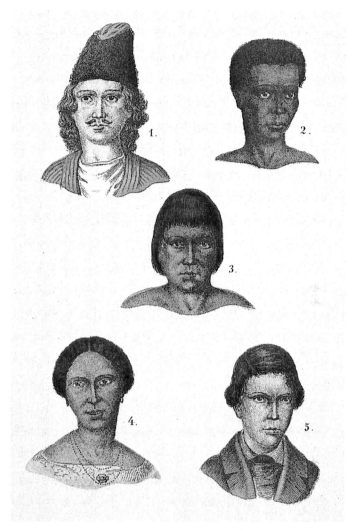

Figure 7.1 Classifying Humans in the Nineteenth Century. The idea of race was routinely accepted by scientists in nineteenth-century Europe, and this figure shows the frontispiece of one typical volume on natural history. Five principal races are identified in the book along with several mixed races. The frontispiece shows three of the principal races and two of the mixed races as follows: 1. The White or Caucasian Race; 2. The Black or Negro Race; 3. The American or Red Race; 4. Mulatto; 5. Mestizza. The other two principal races identified in the text are the Yellow or Mongolian Race and the Brown or Malayan Race.

Source: S.A. Myers, trans., *Martin's Natural History*, First Series (New York: Blakeman & Mason, 1862):frontispiece.

ious ways, both practical and scholarly, throughout much of the world, especially in Europe and areas of European overseas expansion. Indeed, nationalism, **colonialism**, and **imperialism** were all supported by the idea of race, as **neocolonialism** continues to be at the present time. In the late nineteenth and early twentieth centuries, racist immigration policies designed primarily to limit or totally exclude immigrants from Asia were introduced in such countries as the United States, Canada, Australia, and New Zealand.

It is significant that the various social science disciplines were institutionalized at a time when race was an important organizing idea. Certainly geography, anthropology, sociology, and psychology were all caught up in the idea of race. In geography, Ratzel was a founder of a German journal devoted to the study of race. He was also responsible for popularizing the idea of **lebensraum**, a related concept referring to the living space required by a political state. Several works by Taylor, beginning with *Environment and Race* (first published in 1927), applied what was understood as the fundamental biological and physical anthropological fact of race to the study of larger human geographies. But Taylor did not proceed to rank races, noting that: 'The science of ethnology is so largely a product of Western European peoples that it is natural for all classifications of the races of man to accept as a fundamental principle that this group is the highest type in human evolution. In this study we are concerned with ethnical not ethical status' (Taylor 1937:43). Other early twentieth-century geographers concerned with grand global-scale issues acknowledged race but did not attach cultural importance to what they saw as a biological fact. In outlining the principles of human geography from two quite different perspectives, both Huntington (1920) and Vidal (1926) saw race as largely irrelevant. But in many respects, Thurston in 1926 effectively stated the prevailing view of race in the early twentieth century, in a reference to China:

Although yellow men have been leading settled civilised lives in the valleys of China quite as long as the white men elsewhere, they have not made the great strides that the white man has made in modern times, so that although they are almost as numerous as the White race they have remained in their original homelands, leaving the New World to be developed by the white people. (quoted in Lambert 2002:297)

Race was also an important area of analysis in the discipline of statistics. In particular, the study of eugenics, a term referring to the use of selective breeding designed to promote certain characteristics while breeding out others, was popular in the early twentieth century. The term was coined by the British statistician Francis Galton in 1881, but it was promoted especially by Karl Pearson, who argued that eugenics was essential if human degeneration was to be avoided and human progress maintained:

How many centuries, how many thousands of years, have the Kaffir or the negro held large districts in Africa undisturbed by the white man? Yet their intertribal struggles have not yet produced a civilization in the least comparable with the Aryan. Educate and nurture them as you will, I do not believe that you will succeed in modifying the stock. History shows me one way, and one way only, in which a high state of civilization has been produced, namely the struggle of race with race, and the survival of the physically mentally fitter race. (Pearson, quoted in Weitz 2003:40)

However breathtakingly arrogant such a statement may seem today, it is critical to understand that such logic was part of the discourse of the time. The illusion of race was so powerful that it was a part of the taken-for-granted world both in the scholarly context and in popular understanding.

Demolishing the Illusion

The first explicit suggestion that this myth of race was unscientific came from the anthropologist Boas (1928). His claim was more fully detailed by Huxley and Haddon (1936), who argued that the term was fundamentally flawed and should not be used. But the most important statement was Montagu's treatise *Man's Most Dangerous Myth: The Fallacy of Race*, first published in 1942. In this landmark study, written at the height of racist policy and practice in Nazi Germany and also at a time when racial segregation was the norm in the American South, Montagu (1942) explained that, notwithstanding almost universal acceptance of the reality of race, the term was without scientific credibility and thus invalid. He also noted that any myth is most effective when it is not recognized as such. Montagu understood not just that race was a myth but that the word itself was racist. In the introduction to the sixth edition of *Man's Most Dangerous Myth* (1997) he wrote as follows:

> I have never before put what is wrong with the idea of 'race' in the form of a formula. Let me do so here. What the formula shows, in simplified form, is what racists, and others who are not necessarily 'conscious' racists, believe to be the three genetically inseparable links that constitute 'race': The first is the phenotype or physical appearance of the individual, the second is the intelligence of the individual, and the third is the ability of the group to which the individual belongs to achieve a high civilization. Together these three ideas constitute the concept of 'race'. This is the structure of the current concept of race to which most people subscribe. *Nothing could be more unsound, for there is no genetic linkage whatsoever between these three variables*. And that is what this book is designed to discuss and make clear. (Montagu 1997:31)

It is an astounding commentary on late twentieth-century society that Montagu found it necessary to state this some forty-five years after first publication of the book, at a time when the myth was exposed by irrefutable biological facts and following so much evidence of political and social misuse. Such is the seductive quality of the myth.

As stated by Stanfield (1997:27), 'Race as a myth is a distorting variable that convolutes and in other ways distracts attention from the variables that really matter in understanding how and why human beings think, act, and develop as they do. . . . If we can learn race, we can unlearn race.' This is precisely the approach adopted in this text.

Perpetuating the Illusion?

The great irony, indeed tragedy, surrounding the race myth is that studies exposing the falsity of the idea of race did not prevent the rise of scholarly race relation studies that typically were critical of racism but that at the same time accepted the idea of race as a legitimate basis for analysis.

Thus, sociological analyses of race initiated in the 1920s by Park were continued by the Swedish economist Gunnar Myrdal (1898–1987), who produced a major study of race relations in the United States (Myrdal 1944). In this way, the study of race relations became established as a legitimate component of the social studies curriculum, which included human geography, and it has remained so through to the present. Social science studies also influenced the development and implementation of the race relations legislation in many Western countries and the racial labels routinely used for various official purposes.

Thus, although it is consistently acknowledged today that race is a fallacy, the term continues to be used in mainstream social science, albeit sometimes framed by quotation marks. The term is also commonly used in the mass media, as any casual review of newspapers or similar sources indicates. And in sociology especially it continues to be quite usual to conflate the terms *race* and *ethnicity*. A recently published text states: 'In Canada, ethnic and racial diversity is a fact of life', with *race* referring to a 'group that is defined on the basis of perceived physical differences such as skin colour' (Rosenberg 2004:403, 405). Thus, the term is used even though, at the same time, the author stresses that 'the existence of distinct and definite biological races is largely a myth' (Rosenberg 2004:405). Similarly, the term is regularly used in geography, although not, of course, in the context of meaningful biological and cultural differences. Rather, geographers tend to treat the term the way Jackson (1989:177) does when, in a discussion of racism, he defines *race* as a 'social relation across deeply entrenched lines of inequality'.

Concern with 'race and racism' remains a principal research tradition of both cultural and social geography, dating from the 1960s and linked to sociological analyses of race relations. The approach includes descriptions of immigration, residential segregation, **ghetto** formation, and assimilation, especially with reference to North American and British cities. Certainly, notwithstanding the inherent flaws, the myth of race linked with the concept of ethnicity has proven to be a difficult idea for cultural geographers to ignore. Probably the most substantial area of work linking the two relates to the concept of a **plural society**. Initially developed for the study of colonial societies in southeast Asia, this concept distinguishes between societies that are culturally homogeneous with normative integration and societies that are deeply divided on an ethnic basis and held together only through strong state control.

Why is it so usual to uphold a term that is both unsound as a fact and that has been used to justify so many recent atrocities? As Gates (1986) observes, scores of people are killed on a daily basis in the name of differences ascribed to race. The answer appears to be that some social scientists consider it helpful to use the word *race* in discussions of racism. While this makes some superficial sense, it has been persuasively argued that the dangers of perpetuating the myth are simply too great. As explained as early as 1942 by Montagu, the term is so powerful and dangerous that it needs to be relegated to the garbage can of history. This section has described the history of the myth, while also acknowledging that race continues to be a taken-for-granted fact by far too many people. This text includes a definition in the glossary but only as the term is correctly used in a biological sense.

Pause for Thought

It is all too clear that a principal difficulty with many of our current well-meaning attempts to combat racism is the simple fact that, at the same time, we persist in using the term race *as though it had some meaning. Even if we understand that it is a myth, the simple act of using the term can be interpreted as implying that it is an objective reality. Surely it is sensible to abandon this dangerous word altogether. It is not enough to note that it is 'a social construction rather than a natural division of humankind' (Jackson 1987:6). Rather we should accept the fact that* race *applied to humans is a myth, a fallacy, an illusion, and thus abandon use of the word. Use of the term is especially prevalent among, but by no means restricted to, more conservative scholars (see, for example, Sowell 1994). The more the term is used, the more credibility it gains and the more likely it is that people will see themselves and understand their own identity through such a warped lens.*

The Reality of Racism

Even today, racism remains a fact of life for many people. It assumes a variety of forms, from mild silent disapproval of another person to mass murder. Race may not be a fact, but racism most certainly is. Racism is a particular form of prejudice that typically incorporates prejudgment of others. **Prejudice** refers to the holding of negative opinions about another group, negative opinions that are based on inaccurate information and that are extremely resistant to change. Racism similarly incorporates a schema or blueprint about others, and a **stereotype** is a set of beliefs and expectations about another group that is difficult to change even where there is clear evidence that these beliefs and expectations are incorrect. Box 7.1 outlines the divergent political attitudes towards both past and present racisms that were highlighted at a 2001 conference sponsored by the United Nations.

Our concern with racism in this section is with some tragic twentieth-century examples that were applied at national scales, that affected large numbers of people, and that have had overwhelming cultural geographic relevance. The first example is **apartheid**, a set of racist policies introduced and practiced in South Africa by Europeans to ensure that they would be able to retain power in a country in which, unlike most other mid-latitude settler colonies, they were a minority of the total population. Although there has been considerable debate about the reasons for the emergence of apartheid, especially between liberal and Marxist scholars, the present account situates the issue in the larger context of the discussion of the myth of race as presented above.

Apartheid in South Africa, 1948–1994

During the extended period of European overseas expansion between the fifteenth and nineteenth centuries, there were numerous encounters between incoming European groups and indigenous populations. The indigenous population of South Africa included several Khoisan groups located primarily in the western part of the country as well as Bantu groups in the east and north. The Dutch East India Company established a settlement at the southern tip of South Africa, now Cape Town, in 1652, but there was little subsequent immigration. The area came under British administration in 1806, and although there was some British immigration to the colony, the area never proved to be as attractive a destination as the various other British mid-latitude possessions. In the late seventeenth century, Dutch settlers had begun to move inland across the western area occupied by the Khoisan group, and this expansion and the related introduction of diseases prompted a decline in the indigenous population that was similar to declines in other mid-latitude areas such as the United States. But the Bantu groups in the north and east were less susceptible to imported diseases, and a series of conflicts occurred between the Bantu and both the migrating Dutch (known as Afrikaners) and the incoming British. There were also ongoing disputes between Afrikaners and British colonial

authorities, especially related to the Afrikaner republics in the north, which had formed during the nineteenth century. Further, following a practice carried out in several other British colonial areas, indentured laborers were brought from India. Most of these laborers worked in the sugar cane plantations in Natal colony in the northeastern part of South Africa.

Thus, South Africa comprised several groups involved in a series of hostilities, and the creation of the Union of South Africa from four British colonies in 1910 was really a compromise between two powerful but quite different European groups. In 1910 the population was about 5 million, of whom about 1 million were European, although the proportion of Europeans steadily declined to about 12 per cent by the end of the apartheid era in 1994. The apartheid era began in 1948 with the election of the Afrikaner-dominated Nationalist political party, and at that time about 55 per cent of the Europeans were Afrikaners.

Box 7.1 **Debating Racism Today**

The United Nations–sponsored conference against racism, racial discrimination, xenophobia, and related intolerance, held in Durban, South Africa, in 2001, provided some insights into current attitudes about racism, both past and present. Conceived as an opportunity for creating a new world vision at the beginning of the twenty-first century, the conference focused especially on the causes and forms of racism and on the provision of redress in the form of reparations for persecuted groups. Unfortunately, different countries interpreted these two matters differently.

Before the conference began, the American and Israeli delegations withdrew on the grounds that the intent of the conference was itself racist. The principal concerns of the two delegations related to two issues. First, the agenda included language, inserted at the insistence of some Arab countries, that described Israeli treatment of Palestinians as genocide. Second, the agenda called for a full apology for trans-Atlantic slavery to be issued. Many European countries as well as the United States objected to this demand, which was inserted at the insistence of some African coun-

tries. Their objections were based on the claim that slavery was legal at the time it was practiced, but also reflected their concerns that a flood of lawsuits might ensue if an apology was issued. From the perspective of the US and Britain especially, the emphasis on the ongoing conflict between Israel and the Palestinians and on past circumstances of slavery threatened to overshadow such current instances of racial discrimination as China's treatment of Tibetans, Zimbabwe's actions in Matabeleland, Russia's war in Chechnya, and the caste system in India.

Following extended and often bitter negotiations, the two controversial positions were rejected. The final declaration acknowledged that the trans-Atlantic slave trade was an appalling tragedy, especially in light of the way it dehumanized victims, but did not include an apology. There were no direct criticisms of Israel, although the Palestinians' right to self-determination was included. Perhaps more than anything else, this conference succeeded in exposing some bitter differences of opinion within the world community. On balance, these differences prevented the much anticipated coming together from being achieved.

GROUP IDENTITY IN SOUTH AFRICA

A key consideration in any account of racist geographies is the self-conception of the dominant group and the identities assigned by that group to subordinate groups. In this instance, it is important to ask: Who were the Afrikaners, the self-declared 'White Tribe' who assumed control of the country in 1948?

As noted, Dutch settlers beginning in the late seventeenth century slowly moved inland over considerable distances, establishing an extensive farming economy, such that a distinctive Afrikaner identity is generally considered to have been initiated during an almost 250-year period of movement. Factors that helped to shape this identity include the process of land acquisition and settlement, the opposition to both Dutch and British colonial authorities, isolation from others, a Calvinist religious tradition, the Dutch-based Afrikaans language, conflict with Bantu groups, the formation of republics, and war with the British (1899–1902). Lester equated the rise of

Encounters such as the one shown here must have taken place on many occasions during the extended process of European overseas expansion. This representation from the 1830s reflects a European perspective of an April 1652 meeting between Jan van Riebeeck and a few members of his party and some of the local indigenous population. Armed with instructions to establish a base that would provide Dutch East India Company ships en route between Europe and Asia with fresh meat and vegetables, van Riebeeck was little concerned with inland movement, preferring to establish cordial relations and trade with local populations. The painting, by Charles Bell—who later became the surveyor general of Cape Colony—suggests the grandeur of the natural setting, but more significantly highlights the clear superiority of the incoming Dutch with their fine clothing, impressive weaponry, flag, evident activity, and dominant placing in the painting. In contrast, the local population is depicted as peripheral, subordinate, and vulnerable. (The Granger Collection)

Afrikaner identity with nineteenth-century nationalist movements, noting that there was an

. . . implicit externally derived, but internally manifested, threat to the culture, especially religion and/or language of a body of people; a constructed but widely perceived history, the interpretation of which rests on the shared experience of 'the group' as opposed to alien 'others'; a threat to the economic fortune of most members of this historically defined group; and the creation of a set of symbols by which the group is demarcated and exhorted to act. (Lester 1996:60)

It is important to stress that for Afrikaners the self-conception included both a distinctive culture and a pure race, although the introduction of apartheid after 1948 necessarily involved grouping all White people—Afrikaners, British, and other Europeans—together.

Quotes from two ministers in the new Nationalist government serve to present the Afrikaner rationale for implementing apartheid. First, in 1953 the Minister of Justice stated:

If a European has to sit next to a non-European at school, if on the railway station they are to use the same waiting rooms, if they are continually to travel together on the trains and sleep in the same hotels, it is evident that eventually we would have racial admixture, with the result that on the one hand one would no longer find a purely European population and on the other hand a non-European population. (quoted in Christopher 1994:4)

Second, in 1954 the Minister of Labour stated:

I want to say that if we reach the stage where the Native can climb to the highest rung in our economic ladder and be appointed in a supervisory capacity over Europeans, then the other equality; namely political equality, must inevitably follow and that will mean the end of the European race. (quoted in Christopher 1994:2)

Both of these powerful political statements are premised on the completely incorrect notion

that there are such things as races within the human species. As discussed in the previous section, there are not. The two principal intents of apartheid were, first, to separate groups seen to be different and, second, to allow Whites to continue dominating the country.

DIVIDING PEOPLE, DIVIDING SPACE

It is not easy to convey the enormity of the apartheid endeavor, this 'uniquely selfish and degrading concept', as Christopher (1994:8) calls it. Box 7.2 outlines the key legislation designed to ensure complete social and spatial separation so that the politically and economically dominant White group could retain power over the country in spite of increasing demands for political rights by the Black majority. The basic mechanisms devised to guarantee the survival of the White

A beach sign during the apartheid era made it plain that non-Whites would not be welcomed. Such signs were commonplace throughout the landscape of apartheid South Africa, being found on washroom doors, park benches, many entrances to buildings, and so forth. It is difficult to exaggerate the humiliation suffered by those who were restricted to inferior beaches, facilities, and services. Often described as 'petty apartheid', this segregation of landscape was in fact far from petty, as it was both insulting and dehumanizing. (William Norton)

group were the removal of anyone not considered to be White from those areas designated as being for the occupation or enjoyment of the White population and the elimination, so far as was feasible, of all contact between Whites and others.

The key first step in the implementation of apartheid was to formally classify all individuals according to their supposed racial type, hence the 1950 Population Registration Act. Inevitably, the task of assigning a racial identity was con-

troversial in many specific cases, and Figure 7.2 shows the changes in racial description that occurred between 1983 and 1990 as a result of appeals. As the legislation listed in Box 7.2 indicates, there were three levels or spatial scales.

First, there was an ambitious attempt to redraw the political map by removing all–Black areas from the state and creating independent Black homelands. Figure 7.3 gives an indication of the population movements involved in this

Box 7.2 Segregation in South Africa—Legal Milestones

In South Africa, as in most other colonial areas, forms of spatial and social segregation were evident in both rural and urban areas from the early days of European involvement. As a result, it is fair to say that apartheid was rooted in the colonial era. This colonial segregation was intensified and formally institutionalized by the dominant European group beginning in 1910, after the colonial period, when the four former colonies united as the Union of South Africa. The principal acts were the following:

- **Natives Land Act, 1913**: This act, which was regarded as the first stage in drawing a permanent line between Europeans and the indigenous population, identified those rural areas assigned to the indigenous population.
- **Natives (Urban Areas) Act, 1923**: This act required local authorities to establish separate locations for the indigenous population.
- **Natives Administration Act, 1927**: This act effectively made those areas designated for the indigenous population subject to a political regime that was separate from that governing the rest of the country.
- **Slums Act, 1934**: This allowed inner-city suburbs to be demolished and any residents of indigenous ancestry to be removed.
- **Native Trust and Land Act, 1936**: This act extended the rural area assigned to the indigenous population.

In addition to these acts and numerous amendments, there were other acts for specific regions designed to limit the locational choices open to Asian and Colored populations.

Beginning in 1948 with the election of two allied parties that united to form the National Party in 1951,

the policy of apartheid—total segregation—was introduced, building on the segregation of the colonial period and the Union period. The principal acts were the following:

- **Prohibition of Mixed Marriages Act, 1949**: This act prohibited marriages between Whites and others.
- **Immorality Amendment Act, 1950**: This banned extramarital relations between Whites and others.
- **Population Registration Act, 1950**: The cornerstone of apartheid, this act provided for compulsory classification of all individuals into what the government considered to be distinct racial categories. Initially three groups were identified: White, Black, and Colored, with the Colored group subdivided into Cape Malay, Griqua, Indian, Chinese, and Cape Colored.
- **Group Areas Act, 1950**: This was designed to effect the complete spatial segregation of the designated population groups within cities, with respect to both residences and businesses.
- **Reservation of Separate Amenities Act, 1953**: This act was designed to create separate social environments for the various population groups—the so-called petty apartheid.
- **Promotion of Bantu Self-Government Act, 1959**: This created a hierarchy of local governments for rural reserve areas.

The end of apartheid came as a complete surprise to most observers. During the 1980s, there was a gradual dismantling of petty apartheid. A series of political developments led in 1992 to a referendum, in which over two-thirds of voters (all designated as Whites) accepted the principle of an undivided South Africa and rejected apartheid. The first democratic government was elected in 1994.

During the apartheid era, residential segregation was enforced with areas of African housing typically located outside of the main town area. This housing for Africans, which lacked many basic services, is just outside the small town of Grahamstown in eastern Cape Province. (William Norton)

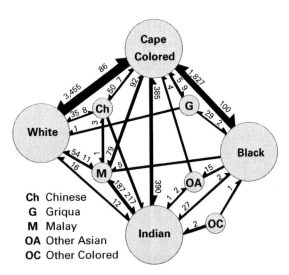

Ch Chinese
G   Griqua
M   Malay
OA Other Asian
OC Other Colored

Figure 7.2 Changes in Race Classification, South Africa, 1983–1990. Classification was based on physical appearance and social acceptability, with the general intent of ensuring that those who were definitely not White not be classified as White. Appeals were permitted, and this figure indicates the about 7,000 successful appeals between 1983 and 1990.

Source: A.J. Christopher, *The Atlas of Apartheid* (New York: Routledge, 1994):104. (Reprinted by permission of Taylor & Francis Books Ltd)

process, which was still incomplete when apartheid collapsed in the early 1990s. When it was clear that the apartheid system was crumbling, right-wing White political groups made several proposals to partition South Africa; Figure 7.4 illustrates one of these proposals.

The second spatial expression of apartheid affected urban areas. The aim was to restructure these to ensure complete segregation and to restrict Blacks' access to urban areas. Certainly, the colonial period and the Union period before 1948 witnessed the evolution of a partially segregated city, but the aims of apartheid were much more comprehensive. One proposal for the structure of a segregated urban area is shown in Figure 7.5.

Third, personal/social apartheid included a myriad of regulations intended to minimize contact between members of the White group and others. Two examples of many suffice to make the aim of this form of apartheid clear. Figure 7.6, the floor plan of a typical White home, shows separate quarters (usually one small room and a washroom) for a Black servant, while Figure 7.7, showing the standard zoning for a beach, illustrates just

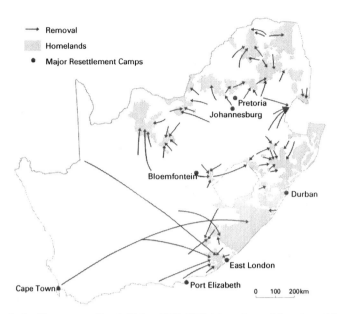

Figure 7.3 Forced Population Movements, South Africa, 1960–1983. It is estimated that about 1.7 million people were resettled in designated homeland areas between 1960 and 1983. Most of those moved were Blacks who were regarded as surplus to the labor requirements of White farming areas.

Source: A.J. Christopher, *The Atlas of Apartheid* (New York: Routledge, 1994):104. (Reprinted by permission of Taylor & Francis Books Ltd)

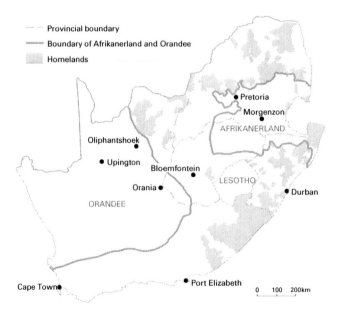

Figure 7.4 A Proposed Political Partition of South Africa. Proposals by right-wing White political groups for the creation of specifically Afrikaner states appeared towards the end of the apartheid era. This figure shows two proposed states, Orandee, an isolated desert state, and Afrikanerland, a state that included much of the agricultural and industrial heartland. Neither of these came to pass.

Source: A.J. Christopher, *The Atlas of Apartheid* (New York: Routledge, 1994):99. (Reprinted by permission of Taylor & Francis Books Ltd)

how far apartheid legislation was taken to ensure that Whites were kept separate from others, and just how dehumanizing this legislation was.

Overall, Whites manipulated space in order to manipulate South African society. It is clear that apartheid policies were introduced by and for Whites, who derived many benefits from measures such as the Group Areas Act. While Whites benefited from such measures, the several other designated groups experienced real disadvantages. The apartheid era was one of ever-increasing hostilities between Whites and others. By the early 1990s it was clear to the Nationalist government that the experience was not sustainable, and the end came both peacefully and quickly. The first genuinely democratic South African government was elected in 1994.

As a concluding comment, it is a mistake to think that the extreme conditions of apartheid could only have occurred in South Africa. Box 7.3 details some provocative ideas concerning the possibility that the United States might have adopted segregationist and apartheid policies similar to those employed in South Africa had the demographic circumstances been comparable to those in South Africa.

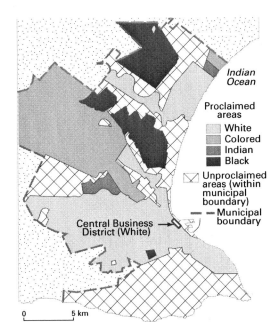

Figure 7.5 Proclaimed Group Areas, Port Elizabeth, 1991. The final proclamation concerning group areas in the city of Port Elizabeth. Together, Figures 7.3 and 7.5 give an indication of the enormity of the apartheid project.

Source: A.J. Christopher, *The Atlas of Apartheid* (New York: Routledge, 1994):111. (Reprinted by permission of Taylor & Francis Books Ltd)

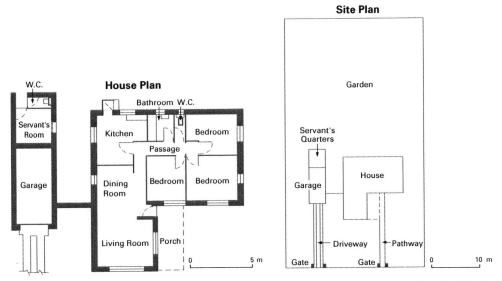

Figure 7.6 Typical South African House Plan in the Apartheid Era. Residential segregation was enforced within the confines of each individual White residential lot.

Source: A.J. Christopher, *The Atlas of Apartheid* (New York: Routledge, 1994):142. (Reprinted by permission of Taylor & Francis Books Ltd)

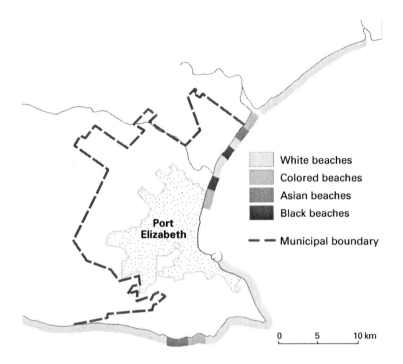

Figure 7.7 Beach Zoning Under Apartheid, Port Elizabeth. Most of the beach areas, including the better beaches, were reserved for Whites. Together, Figures 7.6 and 7.7 show both the absurdity of the apartheid project and the humiliations it involved.

Source: A.J. Christopher, *The Atlas of Apartheid* (New York: Routledge, 1994):146. (Reprinted by permission of Taylor & Francis Books Ltd)

Pause for Thought

A distinguished South African historian published a magisterial volume, The Afrikaners: Biography of a People *(Giliomee 2003), that attempted to answer the seemingly unanswerable question:* Why apartheid? *Afrikaners were a deeply religious people, proud of their values and cherishing their independence, but why did they implement such a perverse ideology that had such devastating effects on others? Through a detailed historical analysis, Giliomee argues that the Afrikaners fell victim to a delusion—that they had to retain their cultural identity, specifically their presumed racial purity, at all costs. Through time, he argues, they developed a persecution complex, accumulating grievances against others and becoming increasingly embittered. Their motive for implementing apartheid was survival of their group, or what they understood as their race. If this argument is correct,* *then similar concerns might be articulated for other groups in other parts of the world today. There seems to be every reason to worry if a group talks about the need for survival and is then placed in a position to seemingly ensure their survival through diminishing the lives of others.*

Genocide

Overall, remarkably little work has been accomplished by geographers, cultural or otherwise, on the geographic dimensions of **genocide** or **ethnic cleansing**. This is despite the very substantial transformations of national, regional, and local geographies that result from the selective population deportations and exterminations involved in these horrendous activities. Most of the relevant literature is historical, political, or sociological.

DEFINITION

The term *genocide* was coined during the Second World War and formally defined in the 1948 United Nations Convention on Genocide as follows:

In the present Convention, genocide means any of the following acts committed with intent to destroy, in whole or in part, a national, ethnical, racial, or religious group such as:

(a) killing members of the group;

(b) causing serious bodily or mental harm to members;

(c) deliberately inflicting on the group conditions of life calculated to bring about its physical destruction in whole or in part;

(d) imposing measures intended to prevent births within the group;

(e) forcibly transferring children of the group to another group.

Box 7.3 What Might Have Been? Part II

Separation of population groups was a feature of many areas of European overseas expansion during the colonial era and also subsequent to that era if a European group achieved political dominance. In all cases a key rationale for separation was the maintenance of European privilege and the furthering of European interests at the expense of other groups. A brief consideration of the southern United States as it compares to South Africa is interesting in this context.

In both South Africa and the American South the term *segregation* came into popular usage in the early twentieth century, but the details of the way the idea was implemented were quite different. In the case of the southern US, there were two principal elements to segregation before the 1960s: legalized social separation, comparable to the unfortunately named petty apartheid of South Africa, and exclusion of one group from the electoral system. There was no American equivalent to the acts identified in Box 7.1. Indeed, after the Civil War the southern United States, despite much discrimination and both *de facto* and *de jure* separation, was characterized by considerable integration. The main reason for the different experience in the southern United States was that the subordinate group was a minority and not an overwhelming majority, was culturally similar to the dominant group, and was a part of a democratic state and not a conquered people.

There were, however, some proposals in the southern US to adopt a segregation system similar to the one developed in South Africa. One proposal following the Civil War called for an explicit spatial separation of groups, while a second proposal favored implementing policies along the lines of the South African 1913 Native Lands Act. The circumstances of the two cases point to the value of coun-terfactual thinking as this line of reasoning was outlined in Box 5.6. Frederickson's comments with reference to American Indians are interesting here:

Assume for a moment that the American Indian population had not been decimated and that the number of European colonists and immigrants had been much less than was actually the case—creating a situation where the Indians, although conquered remained a substantial majority of the total population of the United States. After the whites had seized the regions with the most fertile and exploitable resources, the indigenes were consigned to a fraction of their original domain. All one has to envision here are greatly enlarged versions of the current Indian reservations. Then suppose further that Indians were denied citizenship rights in the rest of the country but nevertheless constituted the main labor force for industry and commercial agriculture. It is hardly necessary to continue; for one immediately thinks of the kind of devices a white minority might then adopt to insure its hegemony under conditions where a majority of the Indians in fact work off the reservation and even outnumber the whites outside these designated areas. The twin objectives of white supremacy would then be, as in South Africa, to maintain direct minority rule over most of the country and some kind of indirect rule over the reservations, while at the same time providing for a controlled flow of Indian workers for industry and agriculture in the white regions. (Fredrickson 1981:246–7)

In brief, what happened in South Africa might easily have happened elsewhere had the demographic circumstances been comparable.

For many scholars, this definition is inadequate both because it is too broad, including acts other than mass murder, and because it is too narrow, excluding such possible victim groups as the mentally ill, homosexuals, a particular class, or a particular political group. Chalk and Jonassohn (1990:23) proposed the following definition, which is more useful for our purposes: 'Genocide is a form of one-sided mass killing in which a state or other authority intends to destroy a group, as that group and membership are defined by the perpetrator.' The mass killing is 'one-sided' in that there is no reciprocity: the victim group has no wish to destroy the perpetrator group. The reference to 'group' is intentionally non-specific, with no indication of how group membership is determined.

UNDERSTANDING THE INCOMPREHENSIBLE?

Why are genocides committed? One important reason was discussed in the earlier section on the illusion of race: one of the preconditions for genocide is a distancing or separation of one group—the perpetrator 'in-group'—and another, the victim 'out-group'. The victim group is often given a derogatory label, further emphasizing that they do not belong. Victim groups may be perceived by perpetrator groups as something less than human; they are dehumanized, vilified, stereotyped, scapegoated, and stigmatized. All victim groups are seen in negative terms, certainly as outsiders, but possibly also as worthless, immoral, a threat to the perpetrator group, and perhaps even as sub-human or animal-like. In the eyes of the perpetrator group, the victim group's identity is predetermined and fixed, with individuals slotted into distinct categories. Necessarily, all such attempts to fix identities are flawed. Although specific genocides have specific causes, it is clear that the illusion of race combined with the rise of national identity foster the insider–outsider perceptions that have contributed to many of the genocides of the twentieth and twenty-first centuries. Identities are often linked to place, and in most cases genocide is justified by the perpetrators in terms of their group right to occupy a particular place without others present.

For genocide to occur, the perpetrator group must be able to exercise power over the victim group. Totalitarian states have been especially prone to commit genocide because they function without real constraints on their ability to exercise power. The worst mass killings of the twentieth century were committed by dictatorial states in which most aspects of cultural, political, and economic life were controlled by the government. It is much more difficult for a democracy to commit genocide because this would require the perpetrators to act in secret. Genocide requires the participation of many people in the perpetrator group, and this level of participation is more likely to occur when the actions taken are formally authorized by the state, such that participating in mass killing is legitimized. Further, much evidence suggests that most of those who participate in genocide are 'relatively ordinary people engaged in extraordinary behaviors that they are somehow able to define as acceptable, necessary, and even praiseworthy' (Alvarez 2001:20).

Victims of genocide are killed because of who they are, not in an individual sense, but in terms of the group to which they are seen to belong. Furthermore, their group identity is typically determined by the perpetrators. In many cases, how victims see themselves is irrelevant; it is how they are viewed by the dominant group that matters. They are not killed because of what they have done, although they may often be blamed for all kinds of ills.

Most cases of genocide occur at times of severe national tension. In the cases discussed below, genocides are associated with wars, post-revolutionary circumstances, tensions stemming from the defeat or withdrawal of a colonial power, or collapse of a larger political unit.

A Century of Genocide

Although the fact is rarely highlighted in conventional accounts, much of human history has been characterized by violence and brutality, and the world has witnessed many horrendous events that could be characterized as genocides (Chalk and Jonassohn 1990:5–8, 32–9). This account focuses on those genocides that involved assumptions about race. Together, the myth of race and colonial activity enabled nineteenth-century Europeans to see themselves as naturally dominant over others, Africans especially. The assumption of natural superiority is evident in Winston Churchill's 1880 report of a British defeat of the Sudanese as one of those 'spectacular conflicts whose vivid and majestic splendour has done so much to invest war with glamour', and his description of such battles generally as 'only a sporting element in a splendid game' (quoted in Weitz 2003:47). Brutal colonial regimes were established especially by Belgium in the Congo and by Germany in South West Africa, with the latter killing about 65,000 of the 80,000 Herero people between 1907 and 1911 (a first formal apology was offered by Germany in 2004). But the most tragic genocides resulted from the combination of the myth of race, the concept of national identity and related desire to create uniform societies, and the rise of powerful and corrupt totalitarian states capable of imposing horrific policies.

ARMENIANS IN TURKEY

With the decline of the Ottoman Empire prior to the First World War, the new state of Turkey, motivated by nationalism and related ideas about creating an exclusive and homogeneous Turkish population, initiated genocide against the Armenians, the largest non-Turkish group in the country. Armenians and Turks had shared the area for several thousand years, but Armenians were different to Turks both linguistically and religiously, having been one of the first groups to adopt Christianity. In 1916 an American diplomat in the region described forced movements and mass murders, recognizing that the issue was 'nothing less than the extermination of the Armenian race' (quoted in Weitz 2003:1). Estimates suggest that about 1 million Armenians out of a total population of 2.1 million were killed.

THE SOVIET UNION

Following the 1917 revolution, the Soviet Union attempted to modernize rapidly in order to establish a utopian nation with great-power status. One aspect of this objective was an obsession with cat-egorizing people as a means to distinguish between friends and presumed enemies of the state; the category of 'enemy of the people' was basic to this process. There was never a well-developed racial ideology; instead, people were categorized according to other identifying characteristics that could not be objectively determined, such as, class, national background, and political view. As a result of the complex set of officially recognized social classes that was created, people came to be seen not as individuals but as members of a class—the favored class was the proletariat. Nationality, the second basis for classifying people, reinforced then widely accepted assumptions about the reality of nations, and Russian national identity was valued above others. The third principal basis for classifying people was presumed political ideology. Needless to say, only those who accepted the principles and policies of the governing Soviet regime were tolerated. In addition to these three identifying characteristics, particular behaviors such as alcoholism and vagrancy were seen negatively and typically resulted in exile or execution.

With these bases for classifying people in place, the Soviet Union engaged in population purges, forced starvations, forced labor, and deportations from the 1920s to the 1950s, substantially changing the cultural geography of the state in terms of population distribution and local cultural identities. Cossacks were the first victims in 1919–20, with perhaps 500,000 of a population of 3 million either killed or deported. But most acts of genocide occurred between 1929 and 1953, when Stalin was leader. Between the late 1920s and the mid-1930s, a campaign against small peasant proprietors resulted in about 6 million deaths from famine. The Great Terror of 1936–8 involved concerted assaults on a wide range of undesirables, including people of aristocratic descent, minority national groups, and those with contacts outside the Soviet Union. Also in the 1930s most national identities—the list is far too long to itemize—other than Russian were targeted for purges, with millions of people deported or killed either intentionally or by neglect. The number killed through execution or neglect and in prison camps is not known; traditional estimates range between 20 and 40 million, although some recent historical research suggests a lower total. These activities were part of what today is called *ethnic cleansing* and resulted in a dramatic transformation of the cultural geography of the state. What is important to remember is that in all cases, individuals were targeted because they were seen as members of groups, not because of any individual characteristics or because of what they had done.

THE HOLOCAUST

While Stalin's Soviet Union was based primarily on the slippery concept of class, Hitler's Germany was premised on the even more slippery myth of race and especially on the belief that Jews were a racial enemy that had to be eliminated in order to permit the German state to progress. Although anti–Semitism prevailed throughout much of Europe, both the intent and scale of the Holocaust are beyond rational understanding. As an ideologically motivated genocide, it is without precedent. Overall, between 6 and 8 million Jews, Gypsies, and other undesirables were murdered between the late 1930s and the end of the Second World War.

Race thinking in general and anti–Semitism in particular were typical of the times, but Hitler was especially vitriolic, using the basest of terms to describe Jews in his political treatise *Mein Kampf*. The great tragedy is that someone holding these beliefs achieved a position of power enabling him to surround himself with like-minded people to carry out such ruthless policies.

The principal goal of the Holocaust, articulated in numerous pseudoscientific writings, was

Box 7.4 **The Holocaust**

Concentration and death camps were located in Germany and in areas of German occupation during the Second World War, including six major camps in Poland. The comments here give details about some of the camps shown on the map in Figure 7.8.

Dachau was the first Nazi concentration camp, established in 1933. It was used mainly to imprison German political prisoners until late 1938, after which members of many other groups that were seen as enemies of the state were imprisoned here, including Jews, Gypsies, Jehovah's Witnesses, and homosexuals. Dachau was liberated in April 1945.

Buchenwald was a concentration camp opened in 1937 and liberated in April 1945.

Auschwitz-Birkenau was a complex consisting of concentration, extermination, and labor camps. It was established in 1940 as a concentration camp and included a killing center in 1942. Between 1 million and 2.5 million people were killed here, mostly Jews but also Poles, Gypsies, Soviet POWs, and others.

Chelmno was set up in 1941 as an extermination camp. Between 170,000 and 360,000 Jews were murdered here between late 1941 and 1944. Chelmno was technologically primitive, employing carbon monoxide gas vans as the main method of killing. The Nazis dismantled the camp in early 1945.

Majdanek was a camp and killing center opened in late 1941. Originally a labor camp for Poles and a POW camp for Russians, it was reclassified as a concentration camp in April 1943. At least 200,000 people were killed here. The Soviet Army liberated it in July 1944, and a memorial was opened there in November of that year.

Belzec was initially a forced labor camp but became an extermination camp in 1942. About 600,000 Jews were murdered here in 1942 and 1943. The Nazis dismantled the camp in the fall of 1943.

Sobibor was an extermination camp opened in May 1942 and closed the day after a rebellion by its Jewish prisoners on 14 October 1943. At least 250,000 Jews were killed here.

Treblinka was an extermination camp opened in July 1942. Between 700,000 and 900,000 persons were killed here. A revolt by the inmates on 2 August 1943, destroyed most of the camp, and it was closed in November 1943

Bergen-Belsen was set up as a concentration camp in 1943. Thousands of Jews, political prisoners, and POWs were killed here. It was liberated by British troops in April 1945, although many of the remaining prisoners died of typhus after liberation.

to save Germany and the purported Aryan race from racial pollution. Aryans, or Germans—the terms tended to be used interchangeably at the time—were represented as the great creators of culture, while Jews were the great destroyers. Jews were also portrayed as carriers of various diseases, and German society was to be purified through the elimination of all Jews. Some other groups were also judged expendable, including Gypsies, Blacks, the mentally and physically handicapped, homosexuals, communists, and others variously seen as outsiders. Indeed, a program of euthanasia, in place in the 1930s, was responsible for the deaths of several thousand mentally and physically handicapped Germans. Further, during the war, German military successes in eastern Europe involved distinguishing different types of people and eliminating those

who were judged physically or culturally inferior. Strategies used included enforced birth control, sterilization, and abortions, together with forced movements, neglect, and murder.

Following his appointment as chancellor in 1933, Hitler introduced a series of policies designed to limit the daily lives of Jews. Implementing these policies required clarification as to who was and was not Jewish, a vexed question given that Jews had been a part of larger German society for many years. Definitions of degrees of 'Jewishness' were first formulated in the Nuremberg laws of 1935; like any attempt to place individuals into identity categories, this exercise was flawed from the outset. But neither practical nor scientific absurdity prevented the Nazi regime from implementing the presumed racial classification.

Figure 7.8 The Holocaust. This map shows some of the concentration and death camps located in areas of German occupation during the Second World War.

Source: Various.

By early 1941, 'Jewish life in Germany had been reduced to a confined, impoverished, frightened experience' (Weitz 2003:124). But the prevailing discrimination and random acts of terror were soon to become genocide. By mid-1941, Jews were explicitly targeted for extermination, with mass killings becoming commonplace in areas of German occupation, including Poland, Lithuania, Ukraine, and Serbia, where these killings often met with the approval of local non-Jewish populations. But with deportations proving difficult in wartime and with the rate of killing seen as far too low, the Third Reich looked for other mechanisms for exterminating Jews. At the Wannsee Conference, held in January 1942, procedures for the organized mass murder of Jews in concentration camps were developed, and the policies were implemented over the next three years (see Figure 7.8).

It is not easy to understand either the chaotic population displacements or the mass murders that were routine in Germany and German occupied areas during the late 1930s and the Second World War. Armed with the illusion of race, certain as to the need to preserve the purity of an Aryan racial identity, embittered by perceived grievances against non-Aryans and especially Jews, empowered by widespread acceptance and even active involvement by many ordinary Germans, the National Socialist party took advantage of their position of power to perpetrate the unthinkable.

CAMBODIA

The communist Khmer Rouge, led by Pol Pot, ruled in Cambodia (renamed Kampuchea in 1976) from 1975 to 1979 and engaged in mass killings designed to purify Cambodian society and create a pure race, a unified nation, and a classless society. At least 1 million people were killed, with some estimates as high as 2 million. Groups were identified as undesirable based on racial criteria and communist logic about classes. The first group targeted was the intelligentsia, including artists and ballet dancers. Subsequent groups targeted included people of Vietnamese and Chinese origin, Muslims, and selected social classes. As with the other genocides described, people were defined not by

their individual characteristics or their behavior, but by their presumed group membership. Strategies for eliminating undesirable groups included deportation, intentional starvation, forced labor, and murder.

The first major purge designed to help homogenize the population was the forced evacuation of the capital city, Phnom Penh. This was followed by a series of ethnic purges against the close to twenty ethnic minorities in Cambodia; these were intended to produce a single nation with a single language. The attempt to purify the population, to eradicate all difference, was doomed to failure and ended with members of the Communist party turning on each other.

RWANDA

Rwanda comprises two principal tribal groups, Hutus and Tutsis. Belgian colonial rule, which lasted from 1916 until 1962, favored and ruled through the minority Tutsi group. This caused considerable resentment among the majority Hutus. Crucially, all Rwandans were issued with an identity card stating their tribal affiliation.

Since independence, Rwanda and several adjacent countries have been beset by ethnic tensions. Following the 1994 murder of the Hutu president, Juvénal Habyarimana, extremist Hutu militia initiated a systematic massacre of Tutsis. In less than four months, about 800,000 Tutsis and moderate Hutus were killed, sparking massive refugee movements. The violence was so severe and the country so divided that news reports at the time described situations in which members of a Catholic Church congregation murdered others who attended the same church. The situation was further complicated by instabilities in neighboring Democratic Republic of Congo (known until 1997 as Zaire). Although there are many differences between this genocide and the others discussed, the basic circumstance is the same, with people being killed because of their presumed group membership, in this case Tutsi or Hutu. Tragically, this pattern of tribal conflict leading to genocide has occurred in recent years in a number of African countries, including Burundi, the DR of Congo, Ethiopia, Liberia, Nigeria (Biafra), Sierra Leone, Somalia, and Sudan.

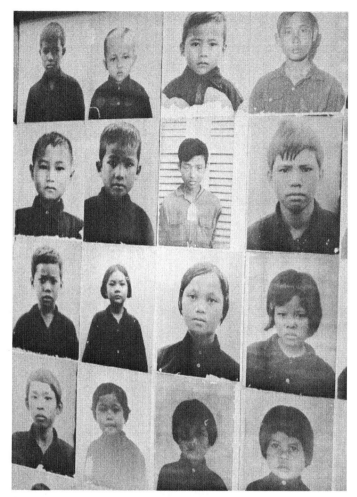

This is one of many disturbing images inside the Tuol Sleng Museum, also known as the Museum of Genocidal Crimes, in Phnom Penh, Cambodia, which stands as a most painful reminder of the holocaust directed by Pol Pot's Khmer Rouge. The building that houses the museum was a school used by the Khmer Rouge in the late 1970s as a secret detention center, where enemies of the state were held, interrogated, tortured, and exterminated. Exhibits include paintings, photographs, cells, and torture devices used to extract confessions. It is claimed that only seven of the 17,000 people imprisoned here left alive. The illustration shows photographs of some of the many child victims. Among the museum's other displays was the 'skull map', composed of 300 skulls and other bones. Although the map is now dismantled, the skulls of some victims are still on display on the museum's shelves. This museum is a most painful reminder of the Cambodian Holocaust directed by Pol Pot's Khmer Rouge. (Claire Huguet/ABACA/CP)

SERBIA

Yugoslavia was established in 1945 as a socialist federation of six culturally diverse republics with Tito as leader. Following Tito's death in 1980, various nationalist sentiments began to surface, anticipating the political collapse that began in 1991 with the secession of Slovenia, which was quickly followed by that of Croatia. During this tumultuous period there were numerous conflicts between members of different cultural groups, and atrocities were common. Serbs, the most powerful group, were embroiled especially in two bitter conflicts: first in Bosnia, beginning in 1991, where Bosnian Serbs clashed with Croatian Muslims, and later in Kosovo, in 1999, where Serbs fought with Albanian Muslims. The extremely complicated pattern of ethnic distribution in the former Yugoslavia is suggested in figures 7.9 and 7.10 and in Table 7.1.

As Yugoslavia disintegrated, each republic sought to clarify the nationality of its members, but it was Serbia, the largest of the six republics, that developed a particularly virulent form of nationalism. Serbian leader Slobodan Milosevic resolved to preserve socialism and turned to a nationalist and racist ideology to justify creating a homogeneous and racially pure Serbian state. In retrospect, many commentators suggest that the pivotal event in the rise of a virulent Serbian nationalism was a 1987 speech in the southern autonomous Serbian province of Kosovo, whose population was about 80 per cent Albanian. In this speech, given at a time when tensions between local Serbs and Albanians were rising—and near the site of the powerfully symbolic 1389 battle of Kosovo Polje, where Ottoman Turks defeated Serbs—Milosevic stressed the presumed purity of Serbian identity and the right of Serbs to occupy what he claimed was their homeland. Subsequently, Serbian leaders began the complex, indeed impossible, process of categorizing people into fixed identities in order to facilitate creation of a pure state.

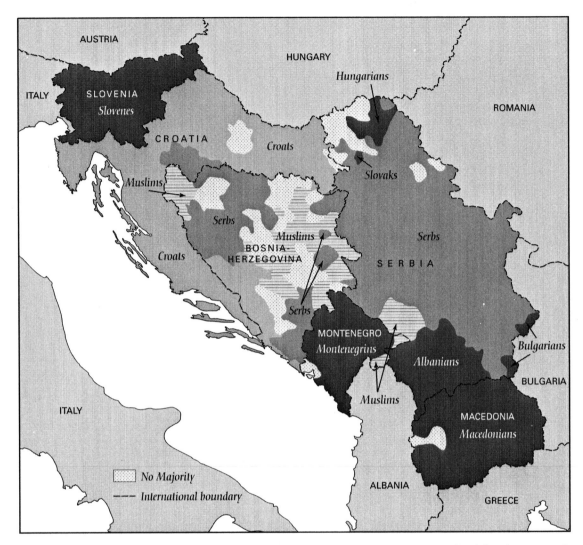

Figure 7.9 Ethnic Groups in the Former Yugoslavia. Although this map may appear very complex, it is only an approximation of a much more complex distribution. Indeed, it is not possible to accurately map cultural distributions at this scale because the patterns are so remarkably complex.

Source: A.J. Christopher, *The Atlas of States: Global Change, 1900–2000* (New York: Wiley, 1999):191.

The several conflicts going on as republics strived to assert their distinctiveness and gain independence, the racially motivated belief in a pure Serbian identity, and the fact that Serbia was the most powerful republic together provided the necessary preconditions for several genocidal incidents that were motivated by the desire to make specific locations racially pure. The Croatian town of Vukovar, located on the border with Croatia, was the scene of a massacre of about 300 Croats in 1991. In Srebenica, a sup-

posed United Nations safe haven, Bosnian Serbs massacred about 7,000 Bosnian Muslims in 1995. Further atrocities were committed by Serbs against Muslims in several areas of Kosovo in 1998 despite United Nations intervention, including air strikes. Ethnic violence resurfaced in 2004 as Kosovo moved painfully towards becoming a viable multi-ethnic region.

The examples of genocide presented above all follow a common pattern, which can be summarized as follows. Genocide is an extreme

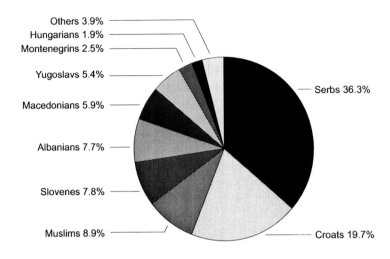

Others 3.9%
Hungarians 1.9%
Montenegrins 2.5%
Yugoslavs 5.4%
Macedonians 5.9%
Albanians 7.7%
Slovenes 7.8%
Muslims 8.9%
Serbs 36.3%
Croats 19.7%

Figure 7.10 Ethnic Composition in the Former Yugoslavia. This diagram showing the relative sizes of cultural groups is a simplification because the specific cultural identities are only some of those that might be included.

Source: Various.

behavior, usually taken as one part of larger efforts to dramatically reshape the geography of peoples and places. Racist and nationalist ideas form the basis for categorizing populations into groups, with perpetrators of genocide labeling victim groups in the most negative of ways. Victim groups are typically blamed for any and all social and economic problems and are generally seen as impediments to progress and prosperity. The perpetrators, using powerful metaphors of

Table 7.1 Ethnic Composition in Regions of the Former Yugoslavia, 1981 (Percentage)

Ethnic Group	Slovenia	Croatia	Bosnia and Herzegovina	Montenegro	Serbia (including Kosovo and Vojvodina)	Macedonia
Slovenes	91					
Croats	3	78	17			
Serbs	2	12	33		66	
Muslims	1		44	13		
Montenegrins				69		
Albanians				7	14	20
Hungarians					4	
Macedonians						67
Turks						4
Others	3	10	6	11	16	7
Source: Various.						

'purifying' and 'cleansing', justify genocide on the grounds that victim groups are something less than human. A powerful state is usually a prerequisite to the introduction of the policies designed to categorize people and to the subsequent practices of deportation and mass murder.

Pause for Thought

Although it is not difficult to summarize the character of genocides, any summary is obviously going to fall short of answering all of our questions. One question that still seems necessary to ask is why? It isn't enough to point to a few sadistic leaders, for after all, there is much evidence to suggest that many ordinary people become directly or indirectly involved in genocides. And although individual perpetrators are almost always men, this may not be significant since it is mostly men who are asked or ordered to commit these acts of cruelty and murder. Wilmer (1997:4) asks: 'How is it that individual people are persuaded to abandon civility in favor of brutality? Why were elites able to so readily mobilize them to undertake acts of inhumanity on the basis of appeals to identity?' Hobsbawm (1993:62–3) responds by noting that 'history is the raw material for nationalist or ethnic or fundamentalist ideologies, as poppies are the raw material for heroin-addiction.' There is a vast social science literature that tackles the difficult questions of why some people want to engage in acts of genocide and why so many others participate, but might it be sensible to conclude that such actions are indeed inexplicable?

Ethnicity and Nationality

A central consideration in contemporary social science is the issue of group membership as it relates to aspects of identity. This issue has been raised earlier in this chapter as well as on several prior occasions in this textbook. The tendency in earlier chapters was to treat group membership in essentialist terms—that is, as a given (an exception to this is the discussion in Chapter 6 of the Hispanic group initiated by Blaut [1984]). The current interest in identity raises a very challenging question that reintroduces the debate about essentialism and constructionism. Recall that essentialism involves the claim that a specific identity is basically unchanging, whereas

constructionism involves the claim that identity is fluid, contested, and negotiated. Much work in contemporary cultural geography accepts that identities are socially constructed, an idea that was at the forefront of the preceding accounts of the myth of race and the related discussions of apartheid and genocide. Although the perpetrators of apartheid and genocide employ essentialist ideas of identity, it is clear in the discussions of these topics that such essentialist assumptions are necessarily flawed.

In the process of inventing and reinventing group identities it has been quite usual for political leaders and others to use the words *race*, *ethnic*, and *nation* interchangeably. Perpetrators of apartheid and genocide, prone to sloppy thinking designed to serve their own purposes, have abused these terms to suit their objectives. Given our explicit intent not to adopt the inappropriate term *race*, it makes sense to consider *ethnicity* and *nation* as useful alternatives. Both, however, need to be carefully defined.

Understanding Ethnicity

Several of the studies that identified the unscientific status of race use *ethnicity* as an alternative. But this is also a term that many scholars and others use interchangeably with *race*. So what does *ethnicity* mean?

Unfortunately, but not unsurprisingly, the concept of ethnicity has a rather difficult scholarly origin, and there remains considerable debate about the meaning of the term. According to Rupesinghe (1996:13): 'ask anyone to define ethnicity and the problem begins. We are left with a host of interpretations. The difficulty in defining ethnicity is that it is a dynamic concept encompassing both subjective and objective elements. It is the mixture of perception and external contextual reality which provides it with meaning.' Given these difficulties, and the longstanding links to the myth of race, it is hardly surprising that an analysis of 65 studies of ethnicity noted that 52 of them offered no explicit definition (Isajiw 1974). Broadly speaking, minority groups are often defined as 'ethnic', especially in the North American immigrant context, and Raitz (1979:79) notes that in this capacity, 'ethnics are custodians of distinct cultural traditions.' This is the interpretation traditionally used by some governments; in Canada, for example, the Census includes a question on ethnic ancestry, not ethnic identity.

SIX COMPONENTS OF ETHNICITY

For cultural geographers, the six components of ethnicity identified by Smith (1986:21–31) are most helpful. First, an ethnic group has an identifying name. This is a means by which people are able to distinguish themselves from others and to confirm their existence as a group. The symbolic importance of naming is clear, suggesting the reality of an identity and its difference from others. (The related question concerning the importance of naming places as one means of indicating ownership and authority is discussed in Chapter 9.)

Second, an ethnic group has a common myth of descent. This may take the form of myths of temporal origins and spatial origins. The various perpetrator groups in the preceding accounts of genocides, especially Germans in the 1930s and Serbs in the 1990s, each invented and reinvented a myth of descent. Germans under the Nazi regime promulgated the idea of an original Aryan race, while Serbs portrayed the 1389 battle against Muslims as a decisive historical moment that justified their territorial claims in the 1990s.

Smith's third component of ethnicity, closely related to the second, is a shared history. An ethnic group is a historical community that typically relies on shared memories to unite successive generations. Each new experience is added to the previous history and interpreted and understood in terms of that history. Necessarily, such a history is but one reading of the past, a reading designed to reinforce the identity of the group.

Fourth, an ethnic group has a distinctive and shared culture that is reflected in lifestyles and values of the group. This may take the form of a common language and religion along with, for example, particular customs, institutions, folklore, architecture, dress, food, and music.

Fifth, an ethnic group is linked to one or more places. This link is most powerful when it is to a particular territory, a place that is seen as a homeland. The group may occupy this territory, may aspire to occupy this territory, or may

only have recollections of this territory. Especially meaningful places may be contained within, or may even lie outside, this homeland.

Sixth, an ethnic group is a community that possesses a real and meaningful sense of self-identity and self-worth, such that other ways of dividing people—for example, according to social class—are not able to create meaningful divisions within the ethnic group. Group identity also involves a belief that all members share a common fate, and that individual identity is inseparable from larger group identity—circumstances that facilitate the process of mobilizing group members and strengthening group solidarity as needed. Such ethnic solidarity is most evident during times of stress, when the identity, perhaps even the very existence, of the group is challenged.

Smith (1986:32) synthesized these six criteria to define ethnic groups as 'named human populations with shared ancestry myths, histories and cultures, having an association with a specific territory and a sense of solidarity'. Incorporating the conventional cultural variables of way of life, language, and religion, along with a spatial component and a sense of identity, ethnicity emerges as a useful, almost catchall, concept. Needless to say, however, the term does have some unfortunate overtones. The most obvious of these, already stressed, is the fact that it tends to be used as a surrogate for the myth of race. Related to this is the often explicit use of the term to delimit groups in opposition to others. Most ethnic groups are ethnocentric, seeing the in-group beliefs and behaviors as natural and desirable and the out-group beliefs and behaviors as unnatural and perhaps as abhorrent. Box 7.5 discusses the complex example of ethnic identities in Macedonia.

The genocidal circumstances discussed earlier are products of states constructing national identities for themselves and to exclude undesirables. But there are also examples of regional groups constructing identities of nationhood for themselves. In many cases, these constructed identities are manipulated by people in positions of power in order to seek national status and thus achieve territorial status, as Knight (1982:523) explains:

Ethnic or cultural integrity and uniqueness, minority status, the desire for national freedom, the natural right to independence through self-determination—these are the terms used by regional ethnic group elites as they formulate separatist political ideologies, which are, in turn, often rooted in a variety of emotional, historical (generally from a revisionist perspective), and politicoeconomic realities, as perceived and understood by the regional (ethnic) group itself.

Language and Religion

Language and religion are two of the bases that groups employ to assert their ethnic distinctiveness and, in some instances, their national identity. Certainly in cultural geography it has been usual to regard language and religion as the most important variables that unite groups (for a brief overview of language and religion in a global context, see Box 6.6). The presumed importance of language as an indicator of cultural, ethnic, or national identity is noted in Box 7.6 in the larger context of the rise of English as a global language. Language is one of the principal means by which certain cultures identify themselves through time, while religion is often the key repository for important beliefs, attitudes, and traditions that serve to unite a group.

There is often a close relationship between a religious creation myth and the origin myth

Box 7.5 Ethnic Identities in Macedonia

Today, Macedonia is a small landlocked country in southeastern Europe, with a population slightly over 2 million. It borders on Albania, Bulgaria, Greece, and Serbia and Montenegro. Part of the border shared with Serbia and Montenegro separates Macedonia from the Serbian province of Kosovo.

Historically, the region has always been of strategic importance, both because it is located on major thoroughfares connecting different parts of Europe and connecting Europe and Asia, and because it has often been seen by major powers as a key to the larger Near Eastern region. These factors have con-

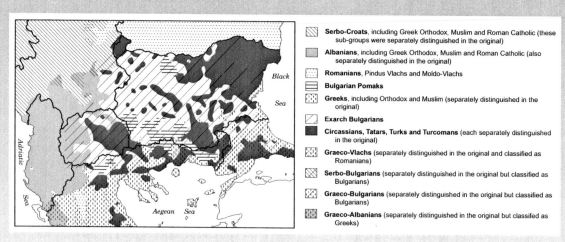

Figure 7.11 One Interpretation of Ethnic Identities in Macedonia. This is one of the about 200 maps reproduced by Wilkinson. It was drawn by an Austrian official, Sax, in 1877, at a time when Austria wished to consolidate political power in the region. In compiling this map, Sax emphasized religious rather than linguistic facts and incorporated what he described as an element of group consciousness, by which he did not mean the myth of race or folklore but rather a deep-seated sense of community. This map served Austria's interests, stressing heterogeneity at the expense of simplification and thus combating claims by other powers, especially Bulgaria at that time, to have some rights to the region.

Source: Adapted from H.R. Wilkinson, *Maps and Politics: A Review of the Ethnographic Cartography of Macedonia* (Liverpool: Liverpool University Press, 1951):80–1.

of an ethnic group, and as noted in Box 6.6, there are some religions that are classed as 'ethnic', meaning that they are associated with a specific group of people; principal examples include Hinduism, Judaism, Taoism, Confucianism, and Shinto. Further, the structure and organization of a religion may provide mechanisms to communicate ethnic myths and symbols. There has been little explicit discussion of these matters in the cultural geographic literature, although Stump (1986:2) has noted that religion was 'important in identifying and understanding the diffusion, distribution and character of groups defined by their social, ethnic or regional identity'.

LINKS TO ETHNICITY?

It is interesting that although language and religion have long been seen as distinguishing markers of ethnic groups, it can be argued that they are becoming less relevant. This suggestion is in accord with the emphasis that contemporary cultural geographers place on the idea of multiple and flexible identities, but it is not in accord with the comments about Welsh and Irish languages and identities included in Box 7.6. It is also a suggestion that implies that policies favoring a minority language at the expense of some other language, as in the case of Quebec, may be misguided if the ultimate goal is to retain group identity. In brief, although traditionally they have

tributed both to a very complex pattern of ethnic distribution and, critically, to numerous different and contested interpretations of the specific pattern and identities of people.

With these facts in mind, Wilkinson (1951) conducted a remarkable survey of about 200 ethnic maps of Macedonia, most of which were prepared after 1840. In total about 150 different groups were identified on these maps, a fact that makes any attempt at ethnic mapping perilous to say the least. In describing ethnic maps prepared by representatives of a number of interested parties, including Greeks, Bulgarians, Serbs, Albanians, Romanians, Germans, Austrians, British, French, and Americans, Wilkinson (1951) showed that the maps reflected the interests of their creators, with each map 'becoming the vehicle of the ideas of its particular compiler—ideas which were largely the product of their age and environment and which differed accordingly. It is patent that even if the population of Macedonia had remained unchanged and static, the variety of ethnographic maps would still have remained comparatively rich' (Wilkinson 1951:321). Figure 7.11 is an example of one such map. Wilkinson's work is very detailed and full of insights that anticipate many of our current interests in the social construction of identity and, especially, the ways in which people in positions of power manipulate identities.

Table 7.2 shows the size of different national groups in Macedonia in 1948, using Yugoslav census data; these data offers an interpretation of the ethnic composition of Macedonia very different to the one depicted in Table 7.1.

Table 7.2 Ethnic Composition in Macedonia, 1948

Ethnic Group	Number
Orthodox Serbs	29,335
Orthodox Croats	2,680
Slovenes	777
Macedonians	788,889
Montenegrins	2,329
Moslem Serbs	417
Moslem Croats	24
Moslem, undetermined	1,565
Bulgarians	890
Other Slavs	1,331
Albanians	197,433
Vlachs	9,508
Turks	95,987
Greeks	1,013
Gypsies	19,500
Other nationalities	1,308
TOTAL	1,152,986

Source: H.R. Wilkinson, Maps and Politics: A Review of the Ethnographic Cartography of Macedonia (Liverpool: Liverpool University Press, 1951):313. (Reprinted by permission of the publisher)

played an important role in establishing the identity of ethnic groups, both language and religion may not be critical components of ethnic identity in the contemporary world.

Sharing a common language is not always essential to a sense of ethnic community. In the case of Wales, for example, English-speaking people in southern Wales may sense a Welsh identity similar to that of the Welsh-speaking people in parts of northern Wales. Similarly, Scotland has a distinctive English dialect in the Lowlands, known as Lallans and sometimes considered to be a separate language, and a Gaelic-speaking Highlands, such that a continuing sense of Scottish ethnic identity may be explained in terms of the Presbyterian religion and the particular Scottish legal and educational systems rather than in terms of language.

Further, recent census evidence for Scotland suggests that there is currently a decline of the Gaelic language. But this decline in the number of Gaelic speakers—the apparently objective indicator of ethnic identity—is accompanied by a Gaelic cultural revival that can be seen in television programs, educational changes, and attendance at Gaelic festivals. This seeming contradiction might be explained by reference to measures of cultural identity other than language: in other words, it is unclear whether language really is an appropriate indicator both of the health of Gaelic culture and of identity with the cultural group.

Similarly, religion is not necessarily a good indicator of the strength of a group's ethnic identity. Although it is important as a symbolic code of communication and as a focus for social organization, particularly prior to the onset of an

Box 7.6 English as a Global Language?

Why a language becomes a global language has little to do with the number of people who speak it. It is much more to do with who those speakers are. (Crystal 1997:5)

For a language to achieve global status, it needs to be spoken by large numbers of people as a mother tongue in some countries, to be an official language in other countries, and to be a priority in foreign-language teaching in other countries. English qualifies on all three counts: it is spoken by many people as a mother tongue in Australia, Britain, Canada, Ireland, New Zealand, South Africa, the United States, and some Caribbean countries; it is an official language in the sense that it has some special status in over seventy countries (more countries than for any other language), including India, Ghana, and Nigeria; and it is taught as a foreign language in over 100 countries, again more countries than for any other language. Together, these three circumstances are producing increasing numbers of English speakers, about 1.5 billion today.

Language, of course, does not have some existence independent of those who speak it, such that any decline or increase in numbers of speakers has much to do with the cultural, economic, and political success of the speakers of the language. In the case

of English, Britain was a leading world power during the critical period of global migration, when capitalism and industrialization were spreading. English, it can be suggested, was simply in the right place at the right time.

Is it possible that the rise of English as a global language is contributing to the death of some other languages? The evidence here is contradictory. Through the course of human history, as many as several thousand languages have appeared and disappeared, but most of these disappearances were related to the demise of a particular group of people rather than to the spread of a global language. Certainly, it has often been argued that retaining a language is central to the survival of a distinctive cultural or ethnic group, and there has been a great deal of interest in the circumstances of a **minority language** in this context. It is in this context that Aitchison and Carter (1990) have stressed the threat to the Welsh language posed by English. Similarly, with specific reference to Irish, Kearns (1974:86) has claimed: 'That a separate language contributes to the imprint of a distinctive nationality is widely accepted.' However, a rather different interpretation of the role played by language in the preservation of a distinct identity is possible, as noted in the text discussion of Scottish Gaelic.

industrial way of life, religion often transcends conventional ethnic boundaries. This is especially so in the case of Christianity, a universalizing religion that has spread spatially and grown numerically, notably in association with the larger political and economic contexts of imperialism and colonialism. With over 2 billion Christians worldwide, the religious identity is characterized by ethnic diversity. Indeed, few Christian denominations, large or small, are limited to a particular ethnic group identity.

UNDERSTANDING OTHERS

In some instances, a particularly sensitive issue concerning religion and ethnicity is the perception that one religious group has of another. Recall the suggestion in Chapter 6 that the contemporary world is characterized by a clash of civilizations, especially as these are associated with religious beliefs and attitudes. In particular, many commentators contend there is currently a conflict between Islam and Christianity, a conflict that assumes cultural, political, and economic dimensions.

According to Mazrui (1997:118), the Islamic world is perceived by the Christian world, especially by the United States, 'as backward-looking, oppressed by religion, and inhumanely governed', although it is in fact 'animated by a common spirit far more humane than most Westerners realize'. In this context, recall the reference in Box 3.5 to the ideas of Said concerning the way in which dominant groups perceive other groups. Of course, the myth of race is also worth considering here.

It can be argued that some differences between the Christian and Islamic worlds con-

This extraordinarily long name of a Welsh town means 'St Mary's Church in the hollow of the white hazel near a rapid whirlpool and the Church of St Tysilio of the red cave'; it is often shortened to Llanfair P.G. and is known to locals as Llanfairpwll or Llanfair. Welsh is known for highly descriptive place names, and they contribute to what is widely perceived by English speakers as part of the charm of the language. But in the case here, all is not what it appears to be. The original name of the village was Llanfair Pwllgwyngyll, which means 'The Mary church by the pool near the white hazels'. The village council coined the much longer name in the mid-nineteenth century, in an attempt to encourage tourism. Today, the place name is the longest in the United Kingdom and the village is a popular tourist destination. The fact that the place name is inauthentic means that that this small Welsh village has as much in common with Disneyworld as it does with rural Wales. (Wales Tourist Board Photo Library)

cerning, for example, population fertility policies reflect the specific timing of cultural change rather than fundamental disagreements about quality of life. Mazrui identified numerous ways in which the Islamic world might be considered more humane than the democratic Western world: for example, Islam has always been more protective of minority religions; it has not been hospitable to either fascist or communist governments; it has not witnessed policies of genocide at the same scale as has the Christian world; and, it has been resistant to racist policies such as apartheid. Certainly, it can be argued that there is a difference between democratic principles and humane principles.

Pause for Thought

From a Western perspective these comments about Islam and Christianity are important because they challenge what is essentially a taken-for-granted ethnocentric world. As highlighted by James (1964), as noted repeatedly in this chapter, and as discussed as well in other contexts in the following chapter, one of the seemingly inevitable but unfortunate—indeed, often tragic—consequences of stressing bases for group identity such as ethnicity is that it quite often results in a ranking of groups. In his account, Mazrui (1997:118) concluded that: 'In the end, the question is what path leads to the highest quality of life for the average citizen, while avoiding the worst abuses.'

National Identities

Since becoming, along with race, a dominant concept in the nineteenth-century European world, nationalism and its corollary, the **nation state**, have become influential globally. The idea of the nation state is rooted in the belief that the world can be conveniently divided into national groupings, each associated with a particular territory. The concept of nationalism is the belief that each of these groups has some inalienable right to occupy and govern its territory. One possible consequence of these two related ideas is that the world cannot be without conflict until all national groups have their own nation states. However, the contemporary world is a long way from fulfilling this nationalist goal. A mid-1990s account identified 210 nations without states but noted that

these were only a fraction of the total, which could be as high as 9,000 This account asserted that: 'In varying degrees these people resent their subservience, and they struggle, sometimes covertly, often overtly, to achieve political independence and hence, in their view, to control their own lives and destinies' (Doob 1996:xiii).

But of course the nationalist endeavor is fundamentally flawed. Put simply, the world is not conveniently organized into groups of people, nations, waiting to be packaged into nation states. The very idea is nonsensical given the long history of human migrations, European overseas expansion, the diffusion of languages, religions, and other cultural criteria, and, of course, the fact that there are numerous other ways that people use to identify themselves and to associate with others. Reflecting a constructionist rather than an essentialist view, nationalism is now usually seen as a concern with an imagined community (Anderson 1983). Hudson and Bolton (1997:1) described the Australian national identity as a 'fabulous beast', noting that any one supposed identity was necessarily exclusionary because it ignored the reality of multiple identities, including those of groups defined ethnically and sexually. To give credence to the concept of national identity alone is wrong. But, of course, this does not prevent nationalism continuing to be a potent force in the contemporary political world, as evidenced by the plethora of separatist and independence movements. Figure 7.12 maps some of the major ethnic conflicts for the period 1917–98.

The idea of nations as imagined communities, while attractive, is usually obvious and often empirically unsatisfactory—obvious in the sense that all feelings of belonging are socially constructed, and empirically unsatisfactory as it appears to ignore the fact that many nationalisms, such as those of the United States and Japan, are products of shared history and shared projects. But there is another problem with thinking about nations, whether viewed as real or imagined, which is that the nation state may not be in a privileged position as a unit of analysis, as Wilmer (1997:3) suggests:

The state appears to be more movable, malleable, and contestable than ever. The state consists of a

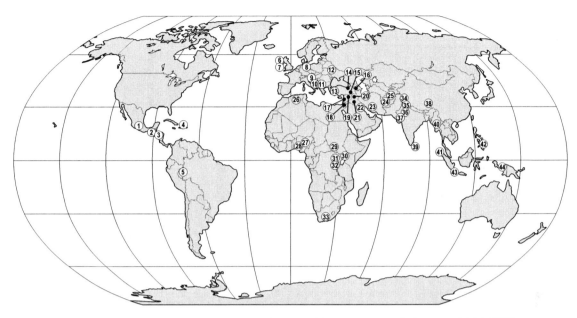

Figure 7.12 Ethnic Conflicts, 1917–1998. Numerous groups throughout the world continue to be persecuted because of their perceived ethnic identity. This map locates conflicts related especially to ethnic differences, including language and religion, although in most cases ethnicity cannot be conveniently separated from other issues such as quality of life, income, and availability of land.

1 Zapatistas v Govt 1994
2 Govt v Maya 1980s
3 Sandinista Govt v Miskito 1980s
4 Govt of Dominican Republic v Haitians 1937
5 Govt-supported colonization of Indian territory mid 1960s
6 Catholics v Protestants 1969-98
7 Catholics v British Govt 1919-21
8 Nazis v Jews 1939-45
9 Serbian Orthodox Christians v Croat Roman Catholics 1991-92
10 Christians v Muslims 1992-95
11 Serbs v ethnic Albanians 1998
12 White Russian Army v Jews 1917
13 Greeks v Turks 1922
14 Turks v Armenians 1916

15 Turkish Govt v Kurds 1984-
16 Christian Armenians v Muslim Azerbaijanis 1990-95
17 Shiite Muslims v Sunni Muslims v Maronites 1982-90
18 Jews v Muslims 1948
19 Syrian Govt v Muslims 1982
20 Iraqi Govt v Kurds 1988, 1991-92
21 Kurds v Christians 1933
22 Shiite Muslims v Iraqi Govt 1992
23 Muslims v Shah of Iran 1978-89
24 Muslim factional fighting 1994-
25 Communists v Muslims 1992-95
26 Muslims v Govt 1992-
27 Muslims v Govt 1980-81, 1984
28 Hausa & Fulani v Igbo 1967-70
29 Muslim Govt v Christians and others 1983-

30 Govt v Lango & Acholi 1971-78
31 Hutu v Tutsi 1949, 1963, 1994-95
32 Tutsi v Hutu 1988-95
33 Zulu Inkhata v ANC 1983-94
34 Muslims v Indian Govt 1990
35 Sikhs v Indian Govt 1982
36 Muslims v Hindus 1947-49, 1992
37 Hindus v Christians 1998
38 Chinese Govt v Tibetan Buddists 1950-
39 Tamils v Govt 1983-
40 Karens v Govt 1948-51
41 Muslims v Govt 1953, 1990
42 Muslims v Govt 1972-95
43 Muslims v Govt 1953, 1990
44 Indonesian Govt v Papuans 1962-

Source: Adapted from P.K. O'Brien, ed. *Oxford Atlas of World History* (New York: Oxford University Press, 1999):269.

set of institutions that are authoritative only because other sets of institutions recognize them; it has been socially constructed within historical and interpretive social processes and practices.

Given this debate about the identity of the nation, two further ideas might be helpfully added to the six components of ethnicity identified earlier in this discussion. First, the idea that a group has an identifying name implies some-

thing more than that the group has a sense of self, of identity; it also implies that the group is in some sense secure enough that any threats to the sense of identity may elicit a defensive response in order to avoid psychological or physical annihilation. Some psychoanalytic ideas suggest that the self is bounded and identified with reference to positive images of who we are along with negative images of who we are not. Because this is an idea that can be applied at both individual and group scales of analysis, it is important in the context of ethnicity. Thus, national and other more local identities may be framed in terms of opposition to others, an idea that is discussed further in Chapter 8. In addition to this group perception of self, there is also tension between national ethnic identity and other more local ethnic identities, a circumstance that is reflected in some accounts of English and Scottish identity.

Second, nationalism can also be seen as a form of cultural capital—that is, the knowledge, attitudes, behaviors, and styles that characterize a group. This cultural capital occupies a space, a homeland. It might be argued that homelands can be seen as *motherlands*, in patriarchal terms, because they offer group members protection and emotional security—interestingly, the concept of a *fatherland* is more likely to be used with reference to matters of law and order, not homeliness.

A Cultural Geography of Our Unequal World

This section builds on the discussion in Chapter 6 of the shaping of the modern world and ways of devising a set of global regions. Here the concern is with the evolution of the current world system, especially as it includes a fundamental division between more and less developed countries. This section explores some of the reasons for this division and examines current changes related to processes of globalization.

Creating Global Inequalities

EUROPEAN MIRACLE?

One conventional answer to the question of why some nations are rich while others are poor builds on the idea that Europe is in some ways different from other parts of the world, and that the particular characteristics of the European region sparked a 'European miracle' that began in the fifteenth century (Jones 1987). Forceful arguments along these lines were made by Landes (1998)—who stressed that Europe is distinctive in two respects: its physical geography and its culture—and by Dia-

mond (1997), who outlined major global historical and economic changes over the past 13,000 years. The basic claim made by both authors is that physical geography and location do matter and are more important in determining the fortunes of a population than any purported genetic differences.

At the beginning of the Holocene period, all humans were pre-literate hunter–gatherers capable of fashioning and using tools of stone. Advancing beyond this way of life required humans to begin domesticating plants and animals, and in terms of the physical geography favoring this kind of development, different parts of the world were not equal in the opportunities that they offered. The basic logic runs as follows: in Europe, mild summers permitted physical activity, cold winters reduced the dangers of disease, and adequate rainfall permitted agricultural development. In many other parts of the world, the climate was not as conducive to physical activity. Further, not all plants and animals are suited to domestication. Indeed, only a few parts of the world had a concentration of suitable plant and animal species, and it was in these areas that agriculture developed, to be followed by increasing population densities, sedentary living, urbanization, and technological and social changes. Agricultural innovations diffused from hearth areas into environmentally suitable areas, including Europe and throughout China.

It was the technological and social changes following domestication, including the evolution of the concept of private property and the emergence of several separate political units, that allowed European guns, germs, and steel to eventually conquer the world. In semiarid areas, on the other hand, water had to be controlled with vast irrigation projects that could only be carried out by a powerful centralized state, the existence of which lessened the likelihood of private property and individual initiative.

Three other factors—culture, money, and knowledge—help to explain why Europe was the first area to industrialize, to experience **development**. The Industrial Revolution in Britain was initiated in a society with a sense of national cohesion, a competitive ideology combined with a respect for merit, and a desire to enhance knowledge. As outlined by Weber, some of these characteristics were inherent values of Protestantism, the dominant religion in northern and western Europe; other religious beliefs, notably Catholicism and Islam, were seen as elevating authority and limiting individual initiative. In short, culture does indeed make a difference. For Landes (1998) this cultural argument as it applies to the development of Europe can be extended to address the question of contemporary global differences in degree of industrialization. Poor areas are poor because they possess particular cultural characters that proved, and continue to prove, inimical to industrialization. In some cases the cultural character is combined with a physical environment that is not conducive to physical activity.

Notwithstanding criticisms that some of these arguments about the distinctiveness of Europe reflect a narrow-minded environmental determinism and are Eurocentric, these studies are important contributions to the debate about the way in which the contemporary world came into being. Indeed, Martis (2003:119) argued that Diamond (1997) is 'one of the most fascinating, controversial, all-encompassing, and thought provoking books I have ever read. . . . a must read for all geographers.'

EUROPEAN MYTH?

These ideas about the singularity of Europe and European technological achievement have been challenged from several directions. Most generally, critics of the 'European miracle' argue that viewing history and economic change from a multicultural perspective does not involve seeing Europe as being more advanced than other parts of the world, at least not in the period prior to the Industrial Revolution, a phenomenon that can be explained by reference to European exploitation of other areas. According to this argument, it is the process of European overseas expansion, colonization, and exploitation of resources, the spread of European disease, the institution of slavery, and the specific character of European scholarship that can be

considered causes of the poverty in the less developed world. This line of argument is a fundamental part of world systems theory as this is discussed in Box 7.7. From this perspective, ideas about the singularity of Europe are seen as reflecting Eurocentrism.

Of course, such fundamental arguments cannot be resolved in the present context; the purpose here is merely to identify what is a very important area of disagreement, both conceptually and empirically. Especially relevant for cultural geographers is the fundamental question of whether or not it is legitimate for one culture to identify weaknesses in another; the consensus view in much contemporary cultural geography, informed by postmodernist and related ideas, is that it is not. It is worth noting, however, that in one important respect the disagreement on this matter is one of belief systems—a clash of conservatives and radicals. Thus, questioning why some nations are rich while others are poor, and speculating about the existence of a European miracle or myth, produces answers that are colored by the preferred **ideology**, the intellectual baggage, of those who tackle such questions. A particular argument about the contemporary importance of culture is presented in Box 7.8.

These comments raise the much larger and more contentious issue of the legitimacy of European scholarship. It can be argued that any claim that Europe is in some way different is itself a part of the ideology of colonialism and imperialism—one of the stories that Europeans like to tell about themselves. For Blaut (1993b:14–17) this fundamentally flawed European view of the world identifies an 'Inside', namely Europe, and an 'Outside', namely the rest of the world. In this picture of the world, the Inside is naturally progressive because it is seen to possess specific mental factors favoring rationality and innovation, while the Outside is viewed as naturally backward because it lacks these European factors; further, Blaut explains, this flawed European view sees the Outside as progressing only as a consequence of contact with the Inside. Blaut (1993b) criticizes this 'Colonizer's Model of the World' for being a tale about Europe, told by

Europeans, that needs to be rethought and retold from a broader cultural perspective.

Pause for Thought

With regard to the relative merits of, on the one hand, a postmodern concern with multiculturalism and, on the other hand, the more deterministic and possibly Eurocentric work of authors such as Landes (1998), Eichengreen (1998:133) claimed: 'that the great questions of economic history that occupy Landes tend to be regarded as fair game for pundits rather than scholars is one of the intellectual tragedies of our times. . . . The postmodernism and multiculturalism that run rampant in history departments are fundamentally incompatible with the approach taken by Landes.' Sowell (1994:225) offered a similar view: 'The a priori dogma that all cultures are equal ignores the plain fact that cultures do not present a static tableau of differences, but rather a dynamic process of competition.' This is an important debate that remains unresolved.

Globalization Processes

'Something new, or the same old thing? A process impossible to control, or something that needs to be managed? A corporate conspiracy (a con trick pulled by the rich and powerful), or a brave new world, pregnant with emancipatory opportunities?' With these words, Short (2001) highlighted some of the uncertainties and debates about globalization. Certainly, it is common to assert that the contemporary world is undergoing a set of unprecedented changes associated with what typically are labeled as globalization processes. As Dicken (1992:1) observes, the 'notion that something fundamental is happening, or indeed has happened, to the world economy is now increasingly accepted.'

There are two principal components to current economic change. First, economic activity is continuing to be *internationalized*, meaning there is a spread of trade and flows of capital. Second, economic activity is becoming *globalized*, mean-

---

Box 7.7 **World Systems Theory**

As noted in Chapter 6, a world system is a large social system that possesses three principal distinguishing characteristics:

- It is autonomous, meaning that it can survive independently of other systems.
- It has a complex division of labor, both economically and spatially.
- It includes multiple cultures and societies.

Although historically world systems covered only certain parts of the world, the current capitalist world system encompasses the entire world. Two types of world system can be identified: a *world empire* is an area unified in military and political terms, while a *world economy* is a more loosely integrated economic system.

The capitalist world economy was initiated around 1450 and since then has both broadened (that is, expanded spatially) and deepened (that is, evolved into a more complex system), such that it now dominates the world. The modern world system has broadened to cover most of the globe, but this process has involved an increasing separation of the world into three parts. Politically and economically,

the core of the system is the dominant part, while the periphery is the weakest; the semiperiphery lies between these two extremes. The three parts are dependent on each other. Especially significant is the fact that, in order to thrive, the core needs to maintain the underdevelopment of the periphery. **Dependency theory** revolves around the idea that the more developed world, in order to become even more developed, must create a less developed world. This process illustrates the consequences of the Marxist idea of ceaseless capital accumulation.

There are five mechanisms involved in deepening of the system, or making it more complex. These are

- *commodification*, a shift from use values to exchange values in both social and economic life;
- *proletarianization*, the transformation of subsistence labor into paid labor;
- *mechanization*, the application of ever-increasing technologies to productive activities;
- *contractualization*, a formalization of human relationships; and
- *polarization*, the increasing disparity between different parts of the world system.

ing there is some functional integration of dispersed activities. The increasing importance of **transnational corporations** that operate across political borders, with the result that states are no longer containers of the production process, and ongoing industrial **restructuring** are both components of these ongoing changes. However, without denying the reality of current economic changes, it is useful to recognize that, in many respects, they can be interpreted as evolutionary rather than revolutionary. Bayly (2003) uses a wealth of world historical data from different societies to demonstrate just how interconnected the world was already prior to the current globalization processes.

It is also commonly argued that the contemporary political world is one of considerable uncertainty, a circumstance that is related to the nature and pace of change since about 1945. The period from 1945 to the end of the twentieth century witnessed a substantial expansion of the global economic system, an expansion that placed much stress on traditional political units; this occurred along with a process of decolonization that resulted in the creation of a number of independent nation states. There have also been environmental impacts linked to increasing levels of technology and population growth. Further, the fundamental structure of global political power changed dramatically in the early 1990s with the demise of the former Soviet Union and the more general transition from communist rule to capitalist democracy throughout eastern Europe. Another dramatic transition was sparked by the terrorist attacks on the United States in 2001, on Bali in 2002, and on Madrid in 2004. All of these changes have had far-reaching implications for the contemporary world, and there is a broad consensus that the world is entering a new epoch. Box 7.9 identifies some of the key terms and ideas related to this apparent transition.

In addition to these five essentially linear evolutionary changes there are two types of cyclical change. These are the Kondratieff cycle of economic prosperity and decline, which typically lasts about fifty years, and the hegemonic cycle, in which a single political unit dominates a system politically, economically, and culturally for a period before giving way to another political unit. A penetrating political geographic analysis argued that there have been three dominant political units (or *hegemons*) so far, namely the seventeenth-century Dutch, the nineteenth-century British, and the twentieth-century Americans (Taylor 1996). The future is, of course, uncertain, but it is possible that the twenty-first century will be the 'Asian century', with three possible candidates for the position of dominant political unit: China, India, and Japan.

The deepening of the modern world system, like the broadening into core, periphery, and semiperiphery, can be understood in terms of the Marxist concept of ceaseless accumulation of capital. Certainly, as Hopkins and Wallerstein (1996:4) have argued, the 'process of accumulating capital on a world scale required the continual development of the world's forces and means of production. This process was a very uneven one, and thereby continually reproduced and deepened what we call the core-periphery zonal organization of world production, the basis of the axial division and integration of labour processes'. This is why the modern world system is the capitalist world system and also why it cannot remain unchanging. World systems theory explicitly refutes the concept of **developmentalism**, the idea that countries are autonomous in terms of their potential for development and that all countries proceed through a set of similar stages as development proceeds.

A conventional criticism of world systems theory is that it overestimates the importance of the world economy, thus underestimating the role of cultural and ethnic identity. Accordingly, world systems theory has been developed in a number of directions that have taken it away from this dependence on economics. Especially interesting are attempts to provide 'the beginnings of a unifying framework defining a world of more than political states and economic zones', with a consideration of 'people, landscapes, cultures, social institutions, economic capacities and political tenets' (Straussfogel 1997:128).

Despite the fact that both economic and political structures are changing dramatically, it is the issue of cultural change that seems to be most threatening. Many critics of globalization see the greatest threat as being one of a loss of local and national cultural identities leading to a less picturesque, less colorful, and increasingly homogeneous world.

With the idea that economic, political, and cultural transformations are necessarily interwoven, three topics are now considered. First, is the gap between rich and poor increasing or decreasing? Second, what is the future of the political state, our current dominant basis for organizing political life? Third, will our cultural future be characterized by a single global culture? Three important questions, each of which prompts several and varied responses.

The Gap between Rich and Poor

Debates about globalization often focus on whether inequalities in quality of life are increasing or decreasing. Specifically, there is a substantial debate concerning whether or not the gap between rich and poor parts of the world is widening. Again, this is a debate with powerful ideological overtones, with catastrophists viewing the contemporary world in essentially negative terms and cornucopians arguing that global problems are exaggerated and that human technology will solve any existing problems. Certainly, there is evidence to suggest that the gap between rich and poor—with respect to both countries and individuals—is increasing; at the same time, however, there is evidence that the gap is decreasing. Much depends on what data are used and how they are interpreted. Cultural geographers have not focused much attention on this important topic, but there seems every likelihood that this situation will change given the growing recognition of the need to consider economic, political, and cultural issues together rather than separately.

Box 7.8 Culture as Cause of Economic Growth?

A review of economic data for Ghana and South Korea shows that in 1960 the two countries had similar levels of per capita GNP and other conventional economic measures. And yet, by 1990, the Ghanaian economy was largely unchanged, while South Korea had become an economic giant. Huntington (2000:xiii) uses this example to introduce the argument that culture matters when it comes to explaining economic growth. For the purposes of the argument, culture is understood as comprising the values, attitudes, and beliefs of a group of people.

Rejecting world systems theory and related claims about links between colonialism and dependency, Harrison (2000) focuses on five themes that are key to understanding the links between cultural and economic growth. First, there is a need to identify and emphasize links between values and progress. Second, a small number of universal values are effectively stated in the United Nations Declaration on Human Rights. Third, physical geography is an important factor—though not a decisive one—in economic growth. Fourth, culture encourages the creation of particular institutions that may promote or impede economic progress; in this context, recall the ideas of Mogey outlined in Box 4.3. Fifth, cultures do change, but slowly. With these five themes in mind, Harrison argues that governments and development institutions must take culture into account when planning strategies for economic growth.

The experience of the Arab world might be seen as supporting these arguments (United Nations Development Programme 2002). The economic progress of this region is often regarded as being impeded by the shortage of three essential circumstances: freedom, knowledge, and gender equity. The shortage of freedom is evident in the continuing presence of absolute autocracies and a social environment that is generally intolerant by Western standards. The shortage of knowledge is evident in the generally low quality of education. The absence of gender equity is seen as a monumental waste of human resources.

The geographer Spencer (1960) proposed a comparable argument more than forty years ago, but this is not an approach favored by most cultural geographers today because it tends to see culture as cause, accepts the reality of culture in a national context, and also implies that there is a need for ideas to be imposed on countries from outside.

It is not enough, of course, to talk of such matters as income and wealth strictly in terms of economic criteria. The gap between more and less developed worlds can also be measured in terms of such fundamental matters as population fertility and mortality. Here again, much depends on how data are collected and interpreted. On the positive side, it is clear that differences in fertility are decreasing as the less developed world proceeds through a fertility transition. On the negative side, deaths from AIDS and related reductions in overall quality of life are especially concentrated in parts of Africa.

Taylor (1992:20) stated: 'We need to construct a new global geography that focuses on the world map of global inequalities. We should very consciously proclaim this to be the most important map in the world and focus our Geography accordingly.' Klare (1996:354) argued for a new cartography that reflects the 'increased discord within states, societies, and civilizations along ethnic, racial, religious, linguistic, caste, or class lines',

while noting that these stresses, along with ever increasing environmental damage, represent the greatest threat to global security. Kaplan (1996) offers what is perhaps the most compelling argument along these lines, although it has been criticized for making unwarranted assumptions about the maintenance of historical patterns of conflict.

The Demise of The State?

Is the institution of the state and the related conception of nations at risk? Certainly, there are three general reasons why it can be argued that the nation state is increasingly becoming a dysfunctional unit for organizing human activity and economic endeavor.

The first is the rise of multinational organizations that, in this era of increasing globalization, serve to replace the state in some specific matters. Political groupings such as the United Nations, defense organizations such as the North Atlantic Treaty Organization, and economic groupings such as the World Trade Organization and Euro-

---

Box 7.9 Naming the Transition

There is general agreement that the world is a changing place. Webber and Rigby (1996:1–5) have identified the following new circumstances as representing relatively sharp breaks from earlier circumstances:

- Industrial economies are in disarray, with many having huge debts.
- A transition from manufacturing to service employment has forced increasing numbers of people into unemployment and many others into unstable, part-time, or contractual work.
- Major corporations are expanding globally, resulting in increasing amounts of money being transferred across the globe.
- There are several new industrial countries, especially in South and East Asia, with southern China emerging as a dominant manufacturing region.
- Finally, the very value of growth is being questioned in light of damage to the environment.

Recall that in Box 3.6 Jameson (1991) described postmodernity, using Marxist terms, as a mode of production, a period of late (taken-for-granted) capitalism characterized by multinational capitalism and the commodification of culture. Numerous terms have been

proposed to identify the current global shift in economics, culture, and politics. The most general terms introduced so far in this chapter are *modernity* and *postmodernity*. But it might be useful to describe the transition from modernity to postmodernity as one

- from **Fordism** to **post-Fordism**, or
- from **organized capitalism** to **disorganized capitalism**, or
- from mass production to **flexible accumulation**, or
- from national economies to a single global economy, or
- from growth to stagnation, or
- from nation states to region states or ethnic territories, or
- from local cultures to a global village, or
- from the golden age to the less-than-golden age.

Of course, labeling in this way increases the danger of creating a reality, such as the golden age, which never really existed. It is with these linguistic concerns in mind that, as noted in Box 3.5, post-structuralist philosophers, such as Derrida, Deleuze, and Foucault, emphasize the heterogeneous and plural character of reality.

This hut is in Kerala, a small coastal state of western India. The rather jarring juxtaposition of tradition and modernity might be seen as evidence that the gap between rich and poor is decreasing because of communication technologies. Certainly, many people in parts of the less developed world have unprecedented access to information from an increasingly diverse set of sources. Whether or not such access is a meaningful indicator of a more equal world can of course be debated, but it certainly highlights one of the changes that is currently taking place and that will undoubtedly enable other changes. (Philip Dearden)

pean Union all function to reduce the role played by individual states. Second, while many nations are joining together in political and economic associations that diminish the power of individual states, many groups in many parts of the world, basing their nationalist arguments on claims of ethnic distinctiveness, are laying claim to subnational territory. Third, the state is challenged by current attitudes to environment, both because environmental impacts do not honor national boundaries and because resources are limited. Taylor makes this argument:

The threat to the state comes not from the cause of globalization, an economic one world, but the consequence, the destruction of the environmental one world. It is not only the fact that pollution is no respecter of boundaries: the whole structure of the world–system is predicated on economic expansion which is ultimately unsustainable. And the states are directly implicated as 'growth machines'—it is unimaginable that a politician could win control of a state on a no-growth policy. (Taylor 1994:184)

Similarly, Wallerstein notes that ecology is now a principal political issue:

Over 500 years, the accumulation of capital has been predicated on the vast externalization of costs by enterprises. This necessarily meant socially undesirable waste and pollution. As long as there were large reserves of raw materials to be wasted, and areas to be polluted, the problem could be ignored, or more exactly considered not to be an urgent one. (Wallerstein 1996:225)

There are no obvious answers to questions about the future of the nation state or about cultural identities as these relate to the presence of nation states. It may be that nations will become less than, but also more than, nation states, meaning that nations are of increasing importance in the world, but that they do not need to be states in order to be meaningful units. From this perspective, nation and state can be delinked without one being subordinate to the other. With reference to Canada, Penrose (1997) detailed three stages in the process of state evolution culminating in a delinking of nation and state. First, the state is constructed through the combined processes of reinforcing state structures, building a nation, and the actions of hegemonic groups. Second, the state is deconstructed through the impacts of both globalization and fragmentation. Third, it is possible that the state might be reconstructed in a form that is more appropriate to contemporary circumstances. Croucher (2004:112) reaches a similar conclusion, noting that 'it is the very malleability of nationhood, both in content and form, that explains its persistence.'

Interestingly, the value of counterfactual thinking—that is, of proposing alternative scenarios—is widely acknowledged as a way of understanding the emergence and transformation of global systems. 'The increasing complexity emerging with the transformation of world-systems through time means that there was a great deal of chance involved in what types of structures eventually emerged. That is to say, just because a particular sequence of historical events did happen, does not mean it had to happen' (Straussfogel 1997:128). This argument is similar to the ideas presented in Boxes 5.6 and 7.3.

Cultural Globalization

Globalization is broadly understood as a set of processes that increase links between people and places. Thus, 'information, ideas, and people are moving and interfacing worldwide more freely, rapidly, frequently, and at greater distances than ever before' (Croucher 2004:15). Certainly, for cultural geographers, the single most interesting aspect of current global changes concerns the oft-stated claim that the world is becoming culturally homogenized, along with the counterclaim that it is not. Needless to say, then, there are varying assessments concerning what is happening to cultures globally. If we think of culture in terms of who we think we are—our identity—then it is helpful to note that such perceptions of identity have changed through time in accord with larger cultural changes.

SCALES OF IDENTITY

It has commonly been suggested that prior to the rise of the nation state, cultures and identities were essentially local, community-based, but that the rise of the nation state gave individuals the option of viewing themselves as part of either a local or national identity. It can be argued that the nation state acted as a type of cultural integrator, with a common political system and, in some cases, a common civil religion creating a cultural harmony. In some other cases, however, the creation of a nation state resulted in much cultural uncertainty, and the fact that this circumstance has not played itself out satisfactorily in much of the world is evident from the earlier discussion of ethnicity and also, more generally, from any consideration of contemporary political events. Finally, there is the possibility that a global culture is now emerging, which necessarily adds another layer of potential identity confusion or, expressed in a more postmodern constructionist tone, contributes to the creation of multiple identities. Any discussion of this possible third stage is complicated by the presence of several quite different claims and interpretations. Three possibilities are noted.

TOWARDS A GLOBAL CULTURE?

A first position on the existence of a global culture focuses on the claim that an emerging global culture is related to the erosion of local cultures. This argument is premised on the belief that the diffusion of industrialization outwards from the West involves a diffusion also of Western culture and Western values, and is typically expressed as an argument against Westernization or Americanization.

To date, the principal areas affected by industrialization are in Asia, and so the question as to whether parts of Asia are becoming more Western in cultural terms is often raised. Zelinsky (1992:156) discussed this general issue using the heading, 'Transnationalization of Culture'. Of course, such questions are not entirely innocent—as Blaut (1993b) stressed, there is little doubt that people in the European world are used to thinking of the West and the rest, and that some of the traditional distinctions that those in the West identified are becoming less and less clear. Economically and politically, Western hegemony is being challenged, especially by Asian economic growth. But what about culture? And what about values? Are Western character-

istics being accepted elsewhere such that it might be proper to speak of an emerging global culture? It is not difficult to recognize that any suggestion of Western values being accepted elsewhere is a controversial suggestion.

Interestingly, the spread of American values in particular is challenged even in some Western countries, such as France and Canada; these challenges are premised on the idea that there is and should be something that can be defined as a national culture. What is possibly distressing about policies that try to prevent supposedly national cultures from being diluted is that they can be seen as just one step away from the policies of totalitarian states that, 'fearful that any opening will destroy them, close themselves off

This photograph was taken during a 2003 tour of Asia by members of the Real Madrid soccer team, generally regarded as the most famous and glamorous of all soccer teams, both because of a long history of sporting success and because of a policy of attracting many of the world's best players. Three of the greatest players are in this photograph: on the left is Zinedine Zidane, also captain of the French national team; in the centre is Raul, a preeminent Spanish player in the glamour striker position; on the right is the then newly signed David Beckham, captain of the English national team and an iconic figure not only inside but also outside the world of soccer. The players are holding up Real Madrid shirts with their names in Chinese characters, a powerful indicator of the all-important role of marketing the Real Madrid brand as a global, rather than merely a Spanish or European, commodity. (Guang Niu/Reuters)

and issue all types of prohibitions and censures against modernity' (Llosa 2001:68).

'JIHAD VERSUS MCWORLD'?

A rather different interpretation of the cultural consequences of the diffusion of technology suggests that as technology brings humans closer together, the cultural barriers between groups become more evident (James 1964). This second position sees a confrontation between the two processes of parochial ethnicity and global commerce. It can be argued that some cultures choose to be excluded from the globalization process because of the association with science and technology, and that such exclusion results in an assertion of cultural identity that is expressed through various localisms. According to Barber (1995:6), the tendencies of local ethnic affiliation (which he labels 'Jihad') and the tendencies of global integration (which he labels 'McWorld') are both at work, both visible sometimes in the same country at the same time and working with 'equal strength in opposite directions, the one driven by parochial hatreds, the other by universalizing markets, the one re-creating ancient subnational and ethnic borders from within, the other making national borders porous from without'. Both of these forces negatively affect the nation state.

For Croucher (2004) and many other commentators, the 2001 terrorist attacks confirm this cultural division, with the conflict being not between states but between different sets of values.

OVERLAPPING IDENTITIES?

A third position is based on the belief that global processes are more likely to produce ambivalence or multiple identities than any coherent identity, a situation that will result in overlapping scales of culture and identity. According to this argument, 'spatialized communities are the real "containers" of culture, of meaning and identity, not the "virtual" communities created by forms of electronic communication, and the networks built around flows of goods and services' (Axford 1995:164). This argu-

ment attempts to distinguish between, on the one hand, the possible creation of global or world spaces in technological and economic terms, and, on the other hand, the possible globalization of meaning structures, of identity.

Clearly, this third interpretation can be regarded as one aspect of the modern-to-postmodern transition. It is proving particularly influential in cultural geography and is considered further in several parts of Chapter 8, with discussions of sexuality and emerging cultures. However, it may be useful to note that most of these discussions are themselves products of a particular time and place, namely the English-speaking academic setting at the end of the twentieth century.

Pause for Thought

Although most commentators agree that the twenty-first-century world will be less culturally varied than previous centuries, there is little agreement as to whether or not this is a good thing. One way to think about this question is to ask whether or not culture ought to be static. If the answer is no, which seems eminently reasonable as cultures were never static in earlier times, then it is worth asking precisely what is wrong with the current changes. The answer, for many people, is that current changes are not best described as changes but rather represent losses of cultural identity. A counterclaim might suggest that some components of culture might be lost, but surely there is little likelihood that entire cultures will be erased if they are truly a reflection of a real and rich tradition. Complex issues indeed.

Concluding Comments

This chapter includes some critical comments about the myth of race that merit repeating. All humans are members of a single species, *Homo sapiens sapiens*, and there are no biological subspecies or races. The idea that there are races, which has been highly influential in Western thought especially beginning in the nineteenth century, is factually incorrect. Logically, then, the idea that presumed races can in some way be ranked in terms of cultural and other abilities is also factually incorrect. Notwithstanding that race is an illusion, the assumption that races exist and can be placed in a hierarchy has contributed to the social construction of identities, leading in turn to such horrors as apartheid and genocide. One way to interpret the illusion of race is to note that it was the belief that human difference is rooted in the body.

The concept of ethnicity, although like so many other social science concepts difficult to pin down, is a useful term. As discussed by Smith (1986), it includes characteristics—such as language, religion, and nationality—and circumstances—such as a shared history and vision for the future—that together often prove to be meaningful bases for identity formation. Nevertheless, it is an unfortunate term in some respects, as it is closely tied to the myth of race.

Discussions of the myth of race and of the concept of ethnicity highlight the important role that group identities play in our understanding of the world. It is worth asking, however, whether or not the recognition of group identities is problematic in the sense that such identities are necessarily reductionist and dehumanizing. This tendency towards reductionism and dehumanization was painfully evident in the accounts of apartheid, genocides, and racism generally. But, of course, to point to such abuses of identity is not to deny the very real importance that most people place on a sense of belonging to a group or, in many cases, to several different groups. Ideally, we might conceive of a world where people associate with others for positive reasons, without such associations also involving negative perceptions of those who favor other group identities.

The discussion of our unequal world highlights some critical differences, but there are, of course, many other directions that such a discussion might take. Population geographers highlight global patterns of, for example, fertility, mortality, and life expectancy, and it is clear that these matters are closely tied to cultural circumstances; declining fertility, for example, is tied to the weakening of male authority in patriarchal societies. Economic geographers address global inequalities in analyses of, for example,

employment opportunities and income levels, while political geographers identify differences in, for example, global patterns of human rights and democratic institutions; all such matters are also linked to larger cultural contexts. Certainly, any comprehensive review of our unequal world needs to take all of these, and other, circumstances into account, and it might be argued that cultural geographers, keen as they are to incorporate more than just cultural content in their studies, might usefully extend their analyses in such directions.

The basic logic of this chapter, namely that identities are socially constructed, typically reflects prevailing power relations, and evidence of these in landscape is further discussed in the next chapter. The spatial scale changes, with increased focus on local and community scales, while the basis for identity formation is broadened to include gender, sexuality, and various forms of disadvantage. There continues to be emphasis on landscape, not only as it reflects identity but also as it contributes to identity.

Further Reading

The following are useful sources for further reading on specific issues.

Montagu (1997) is the seminal statement on the fallacy of race, while Hannaford (1996) and Weitz (2003) provide excellent overviews.

Jackson and Penrose (1994) provide discussions of race and nation that challenge the social construction of these identities while at the same time working uncritically within the socially constructed discipline of geography. In their social geography text, Pain et al. (2001) include a chapter on race (used without quotation marks) and ethnicity (see also Panelli 2004).

Nash (2003) considers the need for anti-racist geographies. There is a substantial literature on the related topic of anti-racist education (for example, Dei 1996).

Bonnett and Nayak (2003) comment on the cultural geographies of racialization.

Bonnett (2000) stresses that the colors white and black are labeled with positive and negative connotations respectively. See also Sibley (1995).

Arendt (1951) and Steiman (1998) describe the nineteenth-century rise of anti-Semitism.

Lester (1996:1–14) outlines the debate between liberal and Marxist scholars concerning the rise of apartheid. Lester (2000) provides a succinct historical geography of South Africa.

Western (1981) analyzes the impact of apartheid on Colored populations in the Cape Town area.

Dodson (2000) provides a clear account of the dismantling of the apartheid landscape (which is described as a dystopia, a bad or evil place) and outlines the need for the cultural geography of South Africa to be understood in terms of power relations and social constructions, not maps and material artifacts.

Geographers have not conducted major regional studies of genocides, using the spatial perspective to record and explain population movements, location of prison camps, and regional and national consequences of mass murders. One insightful article, by Wood (2001), compares geographies of the Bosnian and Rwandan genocides. Most regional textbooks, for example Jordan-Bychkov and Bychkova Jordan (2002) on Europe, refer to relevant genocides, but necessarily do not include substantial primary research. A similar observation applies to political geography textbooks, for example Taylor and Flint (2000). There are, however, various articles and book chapters that consider specific aspects of genocides, most of which reflect interests in such issues as contested and memorial landscapes. Charlesworth (for example 1992, 1994, 2003), Cole and Smith (1995), Ó Tuathail (1996, chapter 6), and Doel and Clarke (1998) are examples. Smith (2000) discusses the Holocaust from a moral geographic perspective.

Goldhagen (1996) wrote about *Hitler's Willing Executioners*, stressing the complicity of ordinary Germans in the Holocaust.

Wilmer (1997) provides a thoughtful discussion of identity and culture in the Balkan region.

Bell-Fialkoff (1993) outlines the history of ethnic cleansing.

Sowell (1996) looks at the historical migrations of selected culture groups.

White (2000) outlines nationalist tendencies in southeastern Europe.

Gil-White (2001) tackles the difficult question of why it is that, despite the myth of race and social construction of ethnicity, ordinary people routinely accept that these categories are real.

Marger (2003) provides a discussion of ethnicity based on the idea that ethnic groups are groups within larger societies that are characterized by some unique set of traits, a sense of community, ethnocentrism, membership that is basically fixed, and a link to territory. This book includes detailed discussions of ethnic circumstances in several regional contexts, namely the United States, South Africa, Brazil, Canada, and Northern Ireland. All of this is accomplished in the context of an account of 'race and ethnic relations'.

Hutchinson and Smith (1996) include discussions of ethnicity in conceptual, theoretical, historical, and contemporary terms, along with accounts of the relationships between ethnicity and religion, language, the myth of race, and nationalism. The book concludes with several chapters on the possibility of moving beyond ethnicity as a marker of identity.

Heffernan (1998) provides a good account of European identity through a discussion of the historical experience.

Eller (1999) reviews ethnic conflict from an anthropological perspective, with discussions of various regional examples including Sri Lanka, the Kurdish region, Rwanda and Burundi, and Quebec.

Zelinsky (1992:149–53) discusses the rise of a common civil religion.

With reference to national identity, Dijkink (1996) refers poignantly to 'maps of pride and pain'.

Taylor (1991), Withers (1995), B.J. Graham (1997), Brace (1999), Pittock (1999), Darby (2000), Morley and Robins (2001), and Hardill, Graham, and Kofman (2001) discuss aspects of geography and identity in a British context. Rogerson and Gloyer (1995) discuss Gaelic culture.

Juergensmeyer (1995) describes some dangers of religious nationalism.

Anderson (1983) is the seminal statement on nations as imagined communities; McLeay (1997a) describes the construction of the Australian national identity; Castells (2004) offers a rather different interpretation.

Hage (1996) discusses homelands as motherlands. One of the ways in which one group establishes authority over another group during times of conflict is by assaulting women in order to demonstrate the inability of the fatherland to protect its daughters (also see Ó Tuathail 1995:260). Agnew (1997a) provides a clear account of the geographies of nationalism and ethnic conflict.

Herb (2004) analyses the construction of national identities in post–Second World War divided Germany.

In a discussion of the breaking of nations, Cooper (2003) argues that the twenty-first century might be the worst in European history.

Subnational movements in South Asia are discussed by Mitra and Lewis (1996).

Hopkins and Wallerstein (1996:1–10) describe six 'institutional domains "vectors" of the world-system' that are 'distinguishable but not separable'. These are the interstate system, structure of world production, structure of the world labor force, patterns of world human welfare, social cohesion of states, and structures of knowledge.

Bayly (2003) presents a world history that explicitly avoids a Eurocentric perspective that sees Western values as diffusing around the world. He achieves this through a variety of strategies, including showing how many of the supposedly original Western ideas had antecedents or equivalents elsewhere. Nevertheless, he does point out that the process of European overseas expansion served to make much of the world a vast agricultural hinterland for western Europe.

McLynn (2004) identifies 1759 as a climactic year with the defeat of the French in North America facilitating the rise of the British empire.

Nisbett (2003) is an intriguing and controversial book that reports a substantial body of research highlighting the different ways in which Westerners and Asians think. Rankin (2003) provides a detailed overview of globalization processes through an account of anthropological and geographic perspectives. Kofman and Young (1996) consider aspects of globalization including the development of theory, the role of the state, and gender implications.

Some accounts of globalization tend to be highly critical. Thus, Renner (1996) argues for an increasing gap between rich and poor countries. Other accounts interpret globalization in a positive light. Thus, Norberg (2001) and Legrain (2002) argue that poverty, the environment, and democracy all benefit from globalization, while Dollar and Kraay (2002) focus on details of the poverty reduction resulting from globalization.

Turner (1994), Axford (1995), Albrow (1997), and Lee and Wills (1997) discuss the need to consider economic, political, and cultural issues as one.

Smil (1993:5) identifies the all too common ill-informed reporting of environmental issues.

Dalby (1996) outlines some criticisms of Kaplan (1996).

An original survey of the geopolitics of the world system is provided by Cohen (2003). Ohmae (1993) identifies economic challenges to the nation state.

Various scholars argue that the key characteristic of globalization is not economic but rather the political process of deterritorialization (Ruggie 1998; Scholte 2000).

For arguments that interpret cultural globalization as a form of creative destruction, see Cowen (2000), and for a critical review see Geertz (2003). The edited volume by Berger and Huntington (2002) focuses on issues of cultural diversity in the contemporary world stressing, for example, that global culture may be diffusing but that it always does so with localized modifications.

8

Other Voices, Other Landscapes

Studies of cultural identity formation and reformation highlight three different circumstances concerning how cultures are viewed by, and in turn view, others. First, as described in the previous chapter, there are many deplorable circumstances involving dominant societies identifying and labeling minorities, typically thinking in terms of 'us' and 'them', of self and others. Apartheid, acts of genocide, and racism generally are extreme versions of this tendency, but there are many equally distressing examples, including prejudicial attitudes and discriminatory behaviors based on identity formations such as gender. Second, there are circumstances in which groups that are seen as different by the dominant society actively choose to embrace and celebrate that difference, often notwithstanding hostility from the larger society. This tendency is especially evident in cultural geographies of sexuality. Third, there is evidence of an emerging trend towards a voluntary loss of identity, or a fusing of identities. This circumstance seems most prevalent among ethnic minorities. All three circumstances are evident in contemporary cultural geographic research into identities and into the landscapes groups occupy or aspire to occupy.

Today, the categories that are acknowledged as foundations for identity construction are both more numerous and more varied than in the past, with, for example, age, gender, sexuality, body type, style, image, and subculture added to the traditional sociological categories of class and community and the traditional cultural geographic categories of language, religion, and ethnicity. Rather than thinking in terms of some monolithic culture, there is interest both in diverse cultures and also in fluid and negotiable cultures that may overlap and that are continually being socially constructed and reconstructed.

This chapter discusses ways in which people use landscapes to structure identity and, in turn, construct landscapes that reflect identity. Some of this work builds on the Sauerian tradition by emphasizing the visible and material aspects of landscapes as a part of the physical and built environment; most of this work, however, builds on newer theoretical formulations and emphasizes the ongoing cultural construction, representation, and interpretation of landscape. Particularly important to the current discussions are five topics developed within the feminist and postmodern literature, namely

- discourse and power relations,
- identity creation,
- relationships between identity and behavior,
- the practice of research, and
- the mode of representation.

The first of these topics has to do with the idea that power is exercised through the production of knowledge and truth in the context of a particular discourse. This circumstance plays out in two important ways: first, as the dominant discourse changes through time, so do understandings of knowledge and truth; second, at any one time, competing discourses offer different understandings of knowledge and truth. The idea that knowledge and truth are changeable and uncertain is challenging, but it is fundamental to accounts of identity creation. Consider, for example, the way in which ideas about men, women, children, or disabled people are constructed by those in positions of power or influence, such as governments or advertising agencies.

The second topic concerns the notion that identities are fluid and negotiable, and are formed and reformed through the processes of social interaction. This means that any one individual may have multiple identities or, more appropriately expressed perhaps, one multifaceted identity. Different aspects of this identity surface in different circumstances.

The third topic addresses the belief that behavior is related to difference. Cultural groups, sharing identities defined by socially constructed categories, have sets of behaviors that have been deemed appropriate for them by those in positions of power. Simply put, cultures are centers of meaning and behavior, with behavior linked especially to matters of power that affect the way the world is interpreted. Further, the constructed categories are spatially configured and are subject to change and negotiation. Why are these categories subject to change and negotiation? One answer draws on the idea that identity is performative, meaning that it is not what a person *is* but rather what a person *does*. This challenging claim is applied especially in the context of gendered and sexual identities.

The fourth topic looks at how research activities are situated. It is not possible to conduct research in some neutral and value-free manner, meaning that researcher positionality—in other words, who the researcher is—must be considered in any evaluation of the research product. An especially influential body of ideas related to this topic examines the practice of research from a postcolonial perspective. For example, it is evident that geographies of racism produced by a white, male, middle-class, British cultural geographer need to be interpreted within the context of that personal background.

The fifth topic surrounds the belief that the mode of representation that cultural geographers favor today is different to that used in the Sauerian tradition. According to this view, there is no one correct way of writing cultural geography, no one correct way of representing the world.

In this chapter you will find the following:

- a discussion of cultural identities, with references to the politics of identity, to ideas about others and other worlds, and to instances of emerging cultural ambiguity;
- discussions of gender and sexuality, with emphasis on questions of insider and outsider status and related circumstances of landscape inclusion and exclusion. In both cases, the politics of identity is a fundamental consideration, with heterosexual assumptions about the norming of people and place challenged. Debates about links between sex and gender are noted, and the conceptual sophistication of much feminist-inspired research emphasized;
- an elaboration of these accounts that includes other marginalized voices and related landscapes, especially those of groups defined in terms of age (both children and the elderly) and ablebodiedness;
- a discussion of geographies of resistance associated with identities that challenge the dominant identity. Again, there is concern with geographies of exclusion rather than with the more usual geographies of belonging;
- a consideration of the sensitive philosophical and political question of individual and group rights;
- a brief concluding section.

Socially Constructed Identities

Recall that in the context of identities, the social constructionist argument rejects the essentialist claim that a specific identity or group membership is fixed and unchangeable, arguing instead that identities are fluid, contested, and negotiated. This position was well stated by Lewis (1991:605): 'Cultures, societies, communities, ethnic groups, tribes, and nations are coming to be viewed as contingent or even arbitrary creations rather than essential givens of human existence.' There is reference to the social construction of Aboriginal groups in Chapter 5 and more detailed accounts of the myth of race and ethnicity in Chapter 7; in these examples the emphasis is on groups having identities imposed on them by dominant societies. In this chapter there are additional examples along these lines, but more important, there is an explicit focus on groups that are constructing their own identities, often as part of a larger resistance process that, in turn, results in landscapes being contested between groups.

Although it is common in the cultural geographic literature to locate social constructionism in feminist and postmodernist approaches that were becoming influential in the 1970s and later, the concept was first fully articulated in the 1960s by Berger and Luckman (1966). Berger and Luckman's argument was itself based on earlier philosophical ideas that were part of the humanist approach of phenomenology (see Box 3.2) and the philosophy and practice of symbolic interactionism (see Box 3.7). Social constructionist logic reflects a belief—shared by phenomenology, symbolic interactionism, feminism, and postmodernism—in the relevance of the socially constructed aspects of human experience. Together, the various approaches informing social constructionism question everyday assumptions about the nature of reality. From this perspective, self, identity, community, and social reality are all creations of the human mind and should not be regarded as objective entities in some way separate from ourselves.

Conceptual accounts of the social construction of human identity are usually based on either Freudian psychoanalysis or symbolic interactionism. According to Freudian identification theory, a child gradually assimilates external persons and objects, effectively taking over the features of another person, as a means of reducing tension. The child first identifies with parents and subsequently with others who seem to be successful in gratifying the child's needs. In symbolic interaction theory, the self emerges in such a way that the individual is able to reflect on his or her position in the social world; identification is the process whereby individuals place themselves in socially constructed categories.

Notwithstanding widespread acceptance of social constructionist logic, cultural geographers frequently argue that landscapes are indeed given meaning by people and that, in turn, peo-ple are constituted through landscape. Following this logic, people interpret themselves and also are interpreted by others according to the landscape they live in, belong to, or originate from. Arguing in this way necessarily implies that there is some stability both to identities and to landscapes, or at least to some identities and some landscapes. Thus, accepting the logic of social constructionism does not preclude recognizing and analyzing some relatively stable identities and landscapes. Thus, this chapter includes accounts of both relatively fixed and highly changeable identities and landscapes.

Sameness and Difference

Debates about identity are common today. In many Western countries, for example, some conservative thinkers worry about immigrants—about who they are, how they think, and what values they hold. One English politician proposed a 'cricket test', asserting that only those immigrants who supported the English team, rather than the team from their home country (such as the West Indies, Pakistan, or India), had become sufficiently assimilated. Some English commentators, meanwhile, worry that schoolchildren are no longer taught to be proud of their English heritage. In Canada, there are seemingly never-ending debates about Canadian identity, fuelled especially by concerns that it is being diluted because of increasing American influence. In the United States, some observers worry about the multiculturalism that they see as running rampant through the educational system. In France, a ban on wearing religious symbols in schools was prompted by concerns about the wearing of Muslim headscarves; the implicit suggestion was that one cannot be both Muslim and French.

What all of these concerns share in common is a belief in the existence, the reality, and the need for a national culture, indeed of the need for an essentially unchanging national culture. But is there, or ought there to be, any such thing? Are there any cultural identities, national or otherwise, that have some inalienable right to exist and to remain relatively unchanged?

Agreeing with the legitimacy of static, or relatively static, national identities raises a myriad of questions concerning who has the right to define that identity.

Most cultural geographic analyses of identities are premised on the twin claims that there is no such thing as a static, unchanging national or any other identity and, furthermore, that such a scenario is undesirable. With these claims in mind, there is a surge of interest in cultural difference rather than cultural sameness.

Excluding Others

Unfortunately, our contemporary culturally diverse world is characterized by many unwelcoming and uncelebratory trends that typically have long and frequently distasteful heritages. Consider, for example, that a 2003 poll in France reported that 10 per cent of respondents admitted disliking Jews while 23 expressed prejudice against North Africans. Similarly, in Switzerland in 2003, as in some other European countries, far rightwing political parties experienced electoral success following campaigns against immigration, while a 2004 report described Britain as institutionally Islamophobic. Recognizing that, for many people, the world is comprised of self and others, cultural geographers pay attention to the way that identities are constructed and maintained, always stressing the importance of power relations in these processes.

CREATING OTHERS

The postcolonial theoretical work of Said (1978, 1993) focuses on the way in which, in a colonial situation, identities of both colonizers and colonized are a product of the process of colonization. Known as Orientalism (see Glossary), this idea implies a relational concept of culture involving colonizers seeing those who were colonized as others, different and somehow less than those who were doing one the colonizing. For Said, Europeans constructed the concept of the Orient as a counterpoint to the West. As Shurmer-Smith and Hannam (1994:19) explain, 'It is the function of the Orient to play "Other" to the West's "Same"'. Westerners developed an image of themselves as having the right to inter-

vene elsewhere and to use others and their places as they chose. Through this imperial image, places outside of Europe were colonized both conceptually—for example, through a renaming process—and practically, with the image manifesting itself in such ways as slavery, resource extraction, and plantation agriculture. And the concept of otherness may continue to be an important part of many Western states' foreign policies, especially when a conservative-minded leader is in power, such that a geopolitical strategy of national security may revolve around the intentional exclusion of others.

Interestingly, there is evidence today suggesting that a competing trend, Occidentalism, is emerging. According to Buruma and Margalit (2004), Occidentalism is the view of the Western world—especially the United States—that has come to the fore in much of the Middle East in recent years. Those viewing the West from an Occidentalist perspective see it as corrupt, arrogant, militaristic, materialistic, and secular; most critically, they see the West as an enemy. Such a view shares the fundamental flaws of Orientalism, as it comprises a set of generalizations that, although not without some factual basis, combine to paint an oversimplified and distorted picture of a complex and varied Western world. It is possible to interpret Occidentalism either as the latest development in a longstanding history of conflict between Islamic and Christian worlds or as a more recent product of American foreign policy that involves support for Israel as well as political and military involvement in Afghanistan and Iraq.

UNSAVORY OTHERS

The logic of Orientalism applies, more generally, in many other contexts in which dominant groups impose identities on minorities. Recall, for instance, the common practice of relegating others including different cultural groups, the insane, poor people, children, or women to a lesser animal status and, in some cases, to inferior places. Sibley (1995), in an influential conceptual and empirical study, attempted to explain why and show how such circumstances of discrimination arise, to show why some people are viewed as others and treated negatively. This concern with geographies of exclusion builds on selected work in psychoanalytic theory that sees human behavior and experience as deriving at least partly from the unconscious mind. The argument here is that humans have an unconscious desire to establish order in their world and purify their environment in order to minimize personal unease and discomfort. This argument is seen as helping to explain the importance of a series of dualisms between, for example, clean and unclean, tidy and untidy, private and public, inside and outside, and, most crucial for our purposes, self and other.

Following this logic, it is natural for people to devalue, reject, marginalize, and even exclude those who are different in any one of a number of ways, including skin color, supposed race, ethnicity, language, religion, social class, gender, sexuality, mental ability, and ablebodiedness. Of course, it is often the case that only those people in positions of power are able to build on their unconscious desires to create worlds of their choosing. Also, what is viewed as tidy or untidy, for example, is not fixed; rather, such ideas vary across cultures and through time.

Much of the work inspired by these ideas considers the way boundaries are established and challenged. As Cresswell (1996a:149) states: 'The geographical ordering of society is founded on a multitude of acts of boundary making—of territorialization—whose ambiguity is to simultaneously open up the possibilities for transgression'; he adds that the 'geographical classification of society and culture is constantly structured in relation to the unacceptable, the other, the dirty'.

Although these ideas have been well received within cultural geography, they are, of course, open to many of the same criticisms leveled against the larger body of psychoanalytic theory. Most of these criticisms focus on the psychoanalytic conceptualization of what it means to be human, because it asserts that it is not possible for individuals to fully know themselves, given there is a part of the human mind that is unconscious and therefore necessarily unknown and unknowable.

Pause for Thought

There is little doubt that our appreciation of others and the worlds they occupy has been greatly enhanced by recent developments in social theory. But there is also little doubt that some of the key ideas have long been acknowledged, albeit without sophisticated conceptual support. For example, in light of the considerable attention given to the ideas discussed above, it is interesting to recall a discussion of geography and development by Spencer (1960:36), who, more than forty years ago, remarked on the 'conventional language employed by Occidentals who judge Malaya and Malayans, not from the Malayan point of view but from their own particular biases' and that 'contributed to an impression of Malaya that can be made to fit the Occidental generalization "underdeveloped"'. Although is it understandably much less sophisticated in conceptual terms, Spencer's basic argument anticipates the later work.

CONSTRUCTING EXCLUSIONARY LANDSCAPES

Dominant societies do not just construct identities of others: they also construct the landscapes in which others might be obliged to live. In doing so, they construct landscapes that take certain characteristics for granted. For example, urban areas in the Western world have typically been built with the assumption that families are heterosexual and nuclear, that women are dependent on men, and that people are ablebodied. These assumptions, and the effects they have had on landscape formation, have reinforced the notion that those who do not conform to societal expectations are different. Difference translates into exclusion and disadvantage.

Intrusions of others into landscapes not constructed for them have proven to be very controversial, both culturally and spatially. Certainly, a distinctive feature of the contemporary world is the unsettling effect that expressions of other identities and their crossing of spatial boundaries have on dominant groups. Indeed, landscapes built with a particular group in mind may become sites of contestation. Further, these contested landscapes can play an important role in identity creation. For example, many collective identities can be interpreted as products of transgressions of normative spatial behavior. This arises when landscapes that are constructed with certain assumptions concerning who uses those landscapes and how they are used are being used differently in that the behavior that occurs in landscape is not considered usual. Such behaviors in landscape may be seen as challenges to authority and as expressions of resistance that result in the creation of contested landscapes.

IDENTITY POLITICS

Identifying and categorizing difference is an inherently political process. For groups struggling to have their identities recognized and legitimized within larger societies, it is also a crucially important and often arduous process. Consider, for example, the struggles experienced by labor movements to have unions established

as legitimate entities. Or consider women's efforts to have basic democratic rights acknowledged. This process is caught up in identity politics, which Agnew (1997b:249) describes as being about 'struggling to establish the recognition of collective differences in identity within a society in which those differences are either not acknowledged or involve negative evaluations and sanctions'.

Studies of identity politics stress the often symbiotic links with place. Just as members of cultural groups living in close proximity to one another contributes to the rise of regional landscapes, as cultural geographers have long recognized, so are other identities associated with specific places. Indeed, much of the interest in place focuses on political meanings, with places understood as sites of power struggles, displacements, and contestation.

In May 2004 this memorial to French Jewish soldiers who died during the 1916 Battle of Verdun was defaced with swastikas and Nazi slogans. The 80-foot high monument is located in Fleury-devant-Douaumont in eastern France and is adjacent to France's principal site of homage to those killed during the First World War. Such acts of vandalism targeted at Jews and other groups are all too common today. Monuments and cemeteries are often favored by those responsible for these actions as they are important symbolic locations linking the past with the present—desecration of a memorial site is also an assault on the memory and identity of the group being attacked. (AP Photo)

Chouinard (1997:379) observes that 'If one were to try to identify a single theme that resonates throughout intellectual and political debates in the late 20th century, it might well be "the difference that difference makes". Although it is still often the case that those who share an identity also share a place, there are many examples of identities that do not share a space. Examples of both place-based and non-place-based others and resistance identities are discussed later in this chapter, where it is acknowledged that both the identities and 'the meaning ascribed to the role of place in identity politics are highly contestable' (Agnew 1997b:254).

CLASSIFYING IDENTITIES

Williams (1977) distinguished among dominant, residual, and emergent forms of culture, and these distinctions are helpful in understanding identity politics. Residual forms of culture represent some earlier institution or tradition, often an ethnic identity, while emergent forms are newer cultural expressions, such as a sexual

identity; both are in a process of continuous tension with the dominant culture and may explicitly oppose that culture (see Box 8.1).

Also relevant to our understanding of identity politics are the circumstances surrounding identity construction. Regarding this Castells (2004:7) stated: 'I propose, as a hypothesis, that, in general terms, who constructs collective identity, and for what, largely determines the symbolic content of this identity, and its meaning for those identifying with it or placing themselves outside of it.' Given this hypothesis, Castells (2004:8–10) distinguishes three forms of identity construction, each having a separate source or origin. First, there is the *legitimizing identity*: this form of identity construction is used by the dominant institutions in society to rationalize their domination of others, and leads to the creation of civil society. Second, there is the *resistance identity*: this form, often discussed in terms of identity politics, is initiated by those who are in some way excluded and/or disadvantaged, and leads to the formation of communities.

Box 8.1 Emergent Alternative Identities

The tendency for groups of people who may or may not be associated with a specific place to engage in a struggle to establish a distinctive identity that is in opposition to some dominant identity is a relatively recent phenomenon. These emergent alternative cultures develop from processes of explicit resistance against the larger societies of which they are a part. It is a trend that may be seen as reflecting globalization in general and, more specifically, a weakening of such established bases of identity as kinship or religion. Thus, groups identified in the context of, for example, environmental concern, ethnicity, gender, or sexuality can all be interpreted as instances of new cultural movements, what might be called emergent alternative cultures. These may be described as 'instances of cultural and political praxis through which new identities are formed, new ways of life are tested, and new ways of community are prefigured' (Carroll 1992:7).

A substantial amount of sympathetic conceptual support for these ideas is included in the Gramscian concept of hegemony as well as in various accounts

of new social movements. But there is, of course, more than one way to view this new identity politics, with some theorists viewing the emergence of new identities based on such variables as religion, ethnicity, and sex as a form of tribal mania that is tolerable only if these alternative cultures are not treated seriously. Others have noted the clear link between these emergent cultures and the Marxist model of class-consciousness, in which some subordinate group develops an identity and pursues some political path accordingly. However, most emerging identities are based on considerations other than class. What is happening is that there is a dialectic of culture, identity, and politics that acts to initiate social change.

Perhaps the most substantive concern about analyses that tend to focus on emergent cultures is that, rather like the dominant discourse that equates culture, ethnicity, community, and place to produce stereotypes, they tend to reify culture and to assume a single monolithic cultural identity for a group, and thus may well omit some critical nuances of individual identity.

Third, there is the *project identity*: this form of identity creation is initiated by a group seeking to redefine its position in the larger society by means of a new identity, with the ultimate goal of transforming that society. The gay community, for example, is traditionally viewed as typifying a resistance identity; however, the Gay Pride movement, in which gay men and lesbians assert their distinctiveness proudly, represents the emergence of a project identity.

Embracing Diversity and Ambiguity

Some current work in cultural geography, rather than studying presumed national or other cultural identities or focusing on geographies of exclusion and identity politics, revels in studying what we might describe as the bewildering abundance and interdependence of cultures. From this perspective, cultural worlds are disordered, not easily labeled and categorized.

There is evidence to suggest that identities once considered separate are no longer quite so different, that we are, perhaps, inching ever closer to sameness. Take, for example, the national dish of England. In terms of popularity, it is no longer roast beef, nor even fish and chips, but rather chicken tikka massala, a hybrid meal that combines English and Indian culinary traditions. Consider also the rise of what has been described as a new ethnic category—*ethnically ambiguous*. Many young people, immersed in aspects of popular culture, reject such labels as Black, White, or Asian, seeing them as crude and outmoded. In the United States especially, the value of being seen as ethnically ambiguous is acknowledged by entertainers and advertising agencies, and many popular entertainers are experimenting with ethnic ambiguity. David Beckham, a white English soccer player, was voted the most famous black man in Britain in 2003 because of his musical preferences and taste for designer clothes and jewelry.

Although these examples may not appear substantive, they do confirm what a growing number of social scientists are acknowledging, namely that identities can be what we wish them to be, and not something fixed that we are and always will be. It is possible that these fluid and hybrid identities are most likely to emerge in circumstances of cultural privilege and are not a reflection of larger trends. On the other hand, a detailed analysis of identities in multi-ethnic London confirmed the fluidity, if not necessarily the hybridity, of cultures, with many individuals seeing themselves as members of several different communities at one and the same time (see Box 8.2).

Pause for Thought

As these comments suggest, cultural geographers studying sameness and difference, while needing to take account of prevailing patterns of power relations, also need to be mindful of and sensitive to the implications of their work. This point was stressed in the previous chapter in connection with academic discussions that, while denying the reality of race, nevertheless continued to use the term, often simply eliding it with ethnicity.

In the current chapter, the emphasis on difference and 'the difference that difference makes' might be interpreted as implying that difference is static and fixed. Certainly, it is not. But there is perhaps an inbuilt contradiction in, on the one hand, highlighting the postmodern notion of fluid and negotiable cultural identities while, on the other, describing and discussing identities as though they were fixed. This is especially problematic in those situations where defining an identity in opposition to a dominant society is strategically and politically important, as is often the case when questions of human rights, such as gay marriage, are on the agenda. There is no easy resolution to this dilemma, which reflects a complex combination of ideas about identity as these are affected by larger social and political circumstances.

Gender

Inspired by feminist concepts, cultural geographers have recognized gender as an important marker of identity, at the same time identifying numerous and varied landscape consequences of patriarchy. The key contribution in the study of landscape has been to uncover what was always there but was typically not seen because of the tendency to ignore (more appropriately expressed perhaps as take for granted) the patriarchic power relationships that underpin landscape formation.

Gendered Identities

Feminist geographers have usually accepted that gender is a social construction deriving largely from the natural category of biological sex. Gender is formed initially through the differential treatment of girls and boys and continues to be reinforced through the life cycle. This differential treatment is accompanied by different societal expectations of the values, attitudes, and behavior of boys and girls. Feminist theorists emphasize that the differential treatment of boys and girls is a nearly universal, cross-cultural circumstance that applies regardless of place and time. Thus, gender is formed based on sex.

But it is not sufficient to note that girls and boys are raised differently and expected to be different throughout their lives. It is also clear that they are raised unequally, with boys socialized to be aggressive and to assume leadership roles, while girls are socialized to be passive and to be compliant followers. It is possible that such characteristics have some initial biological cause, but even if this is so, it is clear that the socialization process emphasizes and increases any natural differences. Overall, this process works to the advantage of men. In the context of work, for example, women are typically relegated to the private domain or to limited opportunities in the public domain.

Working with gender, as with any other basis for identity formation, is problematic because it can be taken to imply that one identity characteristic is always more important than any other characteristic. Specifically, focusing on gender alone may wrongly imply that all members of a gender are more alike than they are different. Recognizing this danger, feminist geographers have increasingly concerned themselves with bases for identity formation in addition to gender, especially ethnicity, class, and sex. In this context, Radcliffe (1994) stressed the need to

Box 8.2 Contesting Culture

There is much evidence to suggest that many people do not regard themselves as members of one group, but rather as members of many groups, with the specific affiliation at any one time varying according to circumstances. Although it may be that such a situation is characteristic of any plural society, it is also well established in the sociological literature that each person has many selves and plays many roles—an idea that is also central to much postmodern thought. Yet this simple observation has implications for any cultural geographic analysis that stresses the links between culture, ethnicity, community, and place.

In a discussion, based on extensive fieldwork, of the multiethnic town of Southall, London, Baumann (1996) observed that

> the vast majority of all adult Southallians saw themselves as members of several communities, each with its own culture. The same person could speak and act as a member of the Muslim community in one context, in another take sides against other Muslims as a member of the Pakistani community, and in a third count himself part of the Punjabi community that excluded other Muslims but included Hindus, Sikhs, and even Christians. (Baumann 1996:5)

This observation is important because it is usual for the dominant discourse to rely on equating culture, ethnicity, and community, and sometimes place. But if this is not possible—as it clearly is not in the case of Southall—then many of these simple associations literally fall apart. To describe the tendency for people to create new communities, or to subdivide existing communities, or to fuse existing communities, Baumann (1996) introduced the term 'demotic discourse', a concept that serves as an alternative to—yet that often overlaps with—the notion of dominant discourse.

In short, new ethnic identities are continually being created, while old identities are being subdivided or fused with other identities. In much of this work there is a celebration of diversity or ambiguity. This tendency is highlighted by some cultural geographies of sexuality, which are discussed later in the chapter.

consider how different femininities, and masculinities, are created in different times and places. Feminist theorists continue to debate these issues, as there is clearly a contradiction between arguing the importance of gender while at the same time acknowledging that gender is interwoven with other bases for identity.

Researchers who choose to stress the similarities of women's experiences regardless of ethnic and other differences are said to be practicing **strategic essentialism**. Although it is the constructionist view and not essentialism that frames much contemporary social thought, it is clear that such constructionism weakens the political commitment and agenda of some groups. Strategic essentialism offers a conceptual resolution to this difficulty by setting aside— perhaps only temporarily—the constructionist argument in order to benefit politically from the essentialist argument. Whether or not it is appropriate, intellectually speaking, to reject the constructionist position for pragmatic political reasons is clearly open to debate.

Feminist ideas and practice have contributed to widespread social change, especially in the Western world, where these ideas have brought about increased involvement of women throughout the workforce and greater understanding of the need for women- and family-friendly workplaces, to cite just two examples. Nevertheless, at the global scale most observers agree that the basic structures of patriarchy continue to function. As genders are defined in terms of each other, changing ideas about women necessarily imply a rethinking of men, and so there is an ongoing reevaluation of masculinity as well (see Box 8.3).

Gendered Landscapes

A fundamental argument initiated by feminist geographers is that the cultural landscape is made by men and for men, both as a reflection of patriarchy and as a means of maintaining the privileged position occupied by men. There is little debate about the claim that landscapes reflect patriarchy, but the assertion that these landscapes are created with the deliberate aim of maintaining male societal privilege is clearly more controversial. Certainly, however, although

relationships are typically complex, all landscapes reflect gendered identities.

HOMES

Consider the design of homes. The home is an essentially private domestic space, constructed at least partly as a reflection of the assumption that heterosexuality prevails, that men interact with the public domain through their work while women focus their attentions on maintaining the home and raising children. As such, the home can be understood as a warm and welcoming place, filled with love. Traditional humanist understandings of home emphasize positive interpretations that focus on family, security, shelter, and companionship.

But some feminist geographers have stressed that a home can also be understood as a site of unpaid domestic work, oppression, and sometimes violence. Indeed, for many women the simple distinction between home and work is fundamentally flawed because their work is in the home. It can further be argued that space inside the home is gendered. Certainly, it has been quite usual in the Western world to locate the kitchen, designed as a separate area for the unpaid work of women, at the rear of the home, and this might be interpreted as devaluing the work of women. In many other cultural traditions, homes are explicitly structured with different spaces for different genders.

On the other hand, feminist-inspired research on such gendered social behaviors as childcare and eldercare identifies the home in more positive terms, as does some work on domestic open spaces. Of course, homes are neither all good nor all bad; rather, the meanings attached to home tend to result from social interactions and are generally characterized by fluidity.

SUBURBS

Consider the design of suburbs. The suburban expansion so characteristic of urban areas since the Industrial Revolution, and especially in the second half of the twentieth century, involved a separation of workplace and home that can be understood as reflecting gender. Suburban expansion involved several factors that limited the ability of women to work away from home, including an absence of childcare facilities and limited public transportation. Rose, Kinnaird, Morris, and Nash (1997:149) observe that 'As well as the design of individual living units, then, the design of housing estates also articulates particular assumptions about who will be living in this environment and what they will be doing there.'

One component of suburban expansion is the shopping mall. Today, in the Western world, the limited participation of women in the paid workforce, increasing personal mobility, additional discretionary income, and conformity with gender expectations results in women being the principal consumers at shopping malls. As a result, the typical shopping mall landscape reflects the gendered identity of consumers in the way the mall is laid out, in the way many individual stores display merchandise, and more generally in advertisements directed at women.

MONUMENTS

Landscapes in urban areas are also gendered, explicitly reflecting and reinforcing the domi-nant role of men through monuments and other structures that record the successes of men. Most such structures are of political and military figures, communicating a message of male power and achievement that both reflects and reinforces patriarchal gender relations. Monk (1999:154) notes the example of a statue in Central Park, New York, that captures the way gender is often represented. The statue is of General William Sherman Tecumseh, a hero of the Union Army, sitting astride a horse in a commanding position of power and authority. In front of, but not leading, the horse is a winged female holding a palm frond. Representing victory, the female is delicately posed and subservient to the much larger horse and man.

Although surveys of urban areas confirm that most monuments commemorate men, notable exceptions in former British colonial areas are the numerous monuments to Queen Victoria. In those cases where women have occupied traditionally male positions of power and have been successful as judged by the standards of the time, their roles may be reflected in

Box 8.3 **Changing Understandings of Masculinity**

In a discussion of the cultural politics of masculinity, Jackson (1990) identifies the rise of new versions of masculinity as a logical consequence of recent political and social changes associated with feminism and recent rethinking of femininity. New versions of masculinity arise, Jackson explains, because masculinity, like femininity, is an identity that is constructed through gender relations. But according to Jackson, these identity changes have not disturbed the fundamental imbalance of power between men and women. Some other authors disagree, however.

According to Nathanson and Young (2001), hatred of men—'misandry'—prevails in contemporary Western society, with men increasingly patronized, even demonized. Prototypical female characteristics, such as sensitivity and caring are now seen favorably, while characteristics traditionally ascribed to males, such as aggression and discipline, are seen negatively. This has led to men being judged as dysfunctional in the important arena of emotional expression. Furthermore, Nathanson and Young argue, the increased involvement of women in traditionally male pursuits has caused men to be emasculated through loss of their traditional functions of protecting and providing for women. These arguments also imply that men lack choices: they can do only what they have always done, while women are able to choose between traditionally female work and the new arena of male work; men are restricted to the briefcase, while women are free to choose between briefcase and baby. Perhaps most fundamentally, as Nathanson and Young observe, men are blamed for social ills and conflicts generally.

Of course, all social change poses challenges, and some people find their values—perhaps even their lifestyles—threatened at the same time as others feel emancipated. Current interpretations of the changing world of men are very much guided by larger ideological perspectives, with more conservative thinkers likely to be uncomfortable with change and more liberal thinkers welcoming change.

Most monuments and memorials record the accomplishments of men, with women either absent or peripheral. A principal exception to this generalization is the very considerable presence of Queen Victoria in the landscapes of Britain and its former colonies. Located in Liverpool, England, this impressive monument was built in 1901 on the site of a church and an old castle. The structure represents the patriotism of Liverpool's citizens as well as the national self-confidence that was a feature of the long reign of Queen Victoria. Built in neo-Baroque style, the monument includes a large standing figure of the Queen who is surrounded by allegorical groups recalling some of the achievements of her reign. (Dave Wood/Liverpool Pictorial, www.liverpoolpictorial.co.uk)

monument construction, as well as, of course, in place names.

WOMEN AS OTHER IN LANDSCAPE

One way to interpret the gendering of human-made landscapes is to appreciate that women are often used by men as the other, as a backdrop against which men are able to articulate and reinforce their masculine identity. Raglon (1996) argues along these lines in a discussion of the Canadian wilderness, in which she stresses differences between women and men—that is, between domesticity and the home on the one hand and the outdoors and wilderness on the other. In this instance, men are seen as framed within nature, which is traditionally female.

Agricultural landscapes are sometimes seen as masculinist landscapes, being depicted as examples of men's success in mastering or taming what was previously wild. In some cases, especially in areas such as New Zealand that have emerged from relatively recent settlement experiences, it may be possible to read the agricultural landscape as masculine, marginalizing women and constructing a subservient femininity. Rather confusingly, this coding of rural landscapes as masculine and urban landscapes as feminine contradicts the characteristic landscape gender codings found in the United States, where the rural landscape has typically been associated with nature rather than with culture and has, accordingly, been coded as feminine with urban landscapes coded as masculine. It is, of course, possible that both arguments are correct and that the differences reflect particular differences in settlement and the larger culture.

Landscapes of Fear

One cause of differences in the attitudes, feelings, and behaviors of men and of women is the disparity in the likelihood of personal assault. Specifically, the threat of sexual assault by a man is one factor that constrains the behavior of women, both inside and outside the home. This threat can be interpreted as a means by which patriarchy is reinforced, since women may consider it necessary to become dependent upon a man to reduce the risk of being assaulted.

It is clear that the design of urban environments has not taken into account women's personal safety, with public spaces in particular being effectively masculinized spaces. Areas of risk for women are those areas where the behavior of men sharing that space is difficult to predict and control, including less-frequented locations such as parks or even rural areas in general, and also small relatively enclosed locations such as subway stations and underground or multilevel parking garages.

In a study conducted in areas of Edinburgh, Scotland, Pain (1997:235) found that 70 per cent of women worried about being sexually assaulted by a stranger outside of their home, and 59 per cent worried about being physically assaulted in similar circumstances. Fears of assault are much higher during hours of darkness, when it is not unusual for young males to dominate urban public spaces. One result of such concerns is that women develop mental maps of the areas they need to negotiate; these maps reflect their assessments of the degree of danger of different places, and they behave accordingly by avoiding the places where they can be least certain of their safety. With reference to Reading, England, Valentine (1989:386) notes that the 'predominant strategy adopted by women I interviewed is the avoidance of perceived "dangerous places" at "dangerous times"'. Dangerous places can include women's own residences, as 'a significant proportion of younger women display certain levels of fear in their own homes at night' (Pawson and Banks 1993:61).

It is not difficult to appreciate that the meanings and experiences of many places vary considerably for men and for women. The possible different understandings of home noted earlier are but one example of the way in which places are bound up with gender.

Pause for Thought

The practice of feminist-inspired research is rarely neatly organized and clearly structured, and it is not value-neutral. Like Marxist research, feminist research usually aims to identify a social problem, to uncover the (usually patriarchal) causes of the problem, and to propose solutions.

In this way, feminist work demonstrates a complex inter-weaving of academic, practical, and political concerns. In addition, discussions of gendered identities and land-scapes are supported by a substantial, changing, and often contested conceptual literature. Most notably, feminist geographers—and it is feminist geographers, usually women, who conduct most of the relevant cultural geo-graphic research—recognize the problematic understand-ing of gender.

As is the case with any discussion of identity char-acteristics, there is a need to continually acknowledge that people cannot be neatly labeled in terms of a single char-acteristic. At the same time, much of the logic of feminist approaches is premised on the importance of gender as a marker of identity. Reconciling these competing claims is an important aspect of contemporary feminist social the-ory. Acknowledging these circumstances, feminist geogra-phers increasingly center their attentions on two or more of gender, class, ethnicity, and sex.

Sexuality

Since about 1990, cultural geographers have shown much interest in studying **sexuality**—particularly sexualities other than heterosexual-ity—as an expression of identity, as a basis for recognizing urban landscape regions, as one way in which the dominant heterosexual land-scape can be challenged, and as a basis for help-ing to understand violence against those seen to be different.

Work on the diversity of sexual identities and related landscapes has been conducted mostly within the larger framework of feminist geography, but researchers have increasingly turned to a set of ideas labeled **queer theory**. This is an uncertain and rather controversial term that refers to a concern with all people who are seen as and/or who have been made to feel different, and this has produced a common identity on the fringes. Queer theory also emphasizes the fluidity and even hybridity of identities, and is concerned with empowering those who lack power. From this perspective, a queer epistemology is needed in order to chal-lenge the way in which sexual dissidents have

been treated within geography. Several studies consider the ways in which particular sexual-identity groups attempt to challenge the main-stream perception of space as heterosexual. All of this work, whether inspired by feminist or by queer theory, is both scholarly and politicized.

Sexual Identities

Traditionally, there are two recognized sexual identities: **heterosexuality** and **homosexual-ity**. Heterosexuality is acknowledged to be the dominant sexuality, and the minority homosex-uality has been viewed differently by the major-ity in different places at different times. In some circumstances, homosexuality has been regarded as acceptable behavior, while in other circum-stances it has been seen as unacceptable, even as abhorrent and deviant. Attitudes toward homosexuality and homosexual relationships are thus determined within the context of dom-inant discourses.

TERMINOLOGY

Terminology is important in discussions of sex-uality, as certain terms tend to be loaded with emotive meanings. In particular, the longstand-ing negative interpretation of homosexuality has encouraged those labeled as homosexual to dis-tance themselves from that term. The favored term today is *gay*, a term that first functioned in an in-group context but that, by the 1960s, became established in a wider social context. The term is acceptable when applied to both women and men, though by convention it is not typical to use *gay* to refer to women, and to avoid ambi-guity, it is better to use the term *gay men* than simply *gays*. Thus, the terms preferred—most importantly from the perspective of those so labeled—are *gay men* and *lesbians*.

DEFINING SEXUAL OTHERS

It is now generally accepted that sexual behav-ior is not simply the product of some instinc-tual drive designed to ensure the continuity of the species; rather, sex is a multifunctional behavior that can be fully understood only in a larger cultural context. This is why a particular sexual behavior may be unacceptable in one cultural context and yet may be acceptable in another, and why cultural attitudes toward a sexuality other than heterosexuality change through time. Distinguishing between what is in some sense 'right' or 'wrong' clearly depends upon what criteria are being used to make such judgments; same-sex relationships, for example, may be dysfunctional in terms of reproduction, but they may serve to enhance a sense of well-being. Most critically, it is clear that consensus judgments about particular sexual behaviors usually reflect power relations, with most feminists arguing that the critical power relation in this context is patriarchy, which is in turn based on heterosexuality.

By the early twentieth century, the dominant view of homosexuality in the Western world was negative. A desire to be involved in same-sex relationships was seen as symptomatic of a medical and mental disorder and, furthermore, indicative of moral inadequacies. This negative view was evident in society generally and reinforced by both the legal and the medical establishment—indeed, the American Psychiatric Association identified homosexuality as a sexual deviation. That formal designation was eliminated in 1973 on the grounds that same-sex relationships are just one of the ways in which sexual preference is expressed.

Today, societal attitudes to homosexuality are changing, as are relevant legal issues. Increasingly, relationships involving same-sex partners are being legally recognized as legitimate and socially acceptable lifestyle choices. There is, of course, considerable opposition to these changes, especially from right-wing conservatives, from certain Protestant denominations, from Catholicism, and from Islam. Two important conferences sponsored by the United Nations in the mid-1990s—the third International Conference on Population and Development, held in Cairo in 1994, and the fourth World Conference on Women, held in Beijing in 1995—exposed substantial differences of opinion on the matter of new forms of family structure. But in the Western world

especially, it is clear that the status of sexual minorities is changing, although it has not by any means changed so much that sexual identities other than heterosexuality are considered equally acceptable.

In many parts of the world, lesbians and gay men continue to experience homophobia, homosexual sex remains a criminal activity, prejudice and discrimination are routine in many workplaces, and people in same-sex relationships have reduced rights to pensions and inheritance. One cultural geographer, Valentine (1998), chronicled a personal history of harassment and persecution that may be all too typical of the experiences of gay men and lesbians in many walks of life. More generally, with reference to geography, Binnie and Valentine (1999:175) contend that 'sexuality as an object of study has become assimilated into the discipline while homophobia remains deep-seated.'

CHALLENGING HETEROSEXUALITY

There are two rather different processes at work when groups that are excluded from some larger dominant and legitimized identity develop their own identity. The first process—and the one most usually discussed—involves the exclusion of minority populations by the dominant group. But many of these groups that experience exclusion have intentionally reinforced that exclusion through a reinterpretation of their identity as positive, not negative. This reinterpretation effectively inverts the perception of their identity by placing a high value on an identity that was previously, and perhaps continues to be, disparaged by the dominant group—this is the project identity referred to earlier. Castells (2004:9) describes this second process as *the exclusion of the excluders by the excluded*. The principal example of this phenomenon is lesbians, gay men, and people with other non-heterosexual identities, such as bisexuality.

Increasingly, sexual liberation is understood as a new form of self-expression. Gay men and lesbian populations especially—groups that typically have been excluded and disadvantaged—have begun the process of community formation

Landscapes of sexuality are commonplace features of many contemporary urban areas. Displays of sexual preference through annual gay pride parades have become enormously popular, both for those who participate and for those who come to observe what is often promoted as a tourist spectacle. The photograph *left* captures some of the spirit and sense of occasion of the parade held in Toronto as it makes its way along Church Street in the heart of the city's gay and lesbian neighborhood. It is abundantly clear that while many problems remain, gays and lesbians are now able to be more open about who they are (Rick Madonik/CP Photo, Toronto Star). The photograph *below* shows some of the businesses that used to be located on 42nd Street in New York City. Serving a predominantly male heterosexual clientele, landscapes such as the one shown are usually viewed in an unfavorable light and are prone to displacement as changing city regulations prompt frequent relocation. Forty-second Street is an especially intriguing example: as the home of many famous theaters and businesses associated especially with the world of music, it was the city's entertainment center from about 1899 to the 1930s, but it transformed into a major retail pornographic center, as shown in the photograph, until it was mostly demolished and then rebuilt in the largest urban renewal project in New York City (Bebeto Matthews/AP Photo).

in explicit opposition to the social norm. Probably the most important consequence of these social movements is the increased challenge to patriarchy as a central structure of contemporary societies, because patriarchy is, of course, founded in the traditional family structure. It is inevitable that increased social acceptance of other family structures based, for example, on two adult females or two adult males leads to a weakening of patriarchy.

RETHINKING GENDER AND SEX

As noted earlier, the conventional understanding of gender is that it is rooted in the biological distinction between female and male, with femininity and masculinity being the social constructions imposed respectively on female and male. But another interpretation of gender and sex is possible, one that asserts that the two sexes, female and male, are the primary social constructions. The argument follows primarily from consideration of three unusual circumstances:

- intersexuality, when a child is born of uncertain sex;
- transsexuality, when body and gender do not appear to match; and
- instances where a person has XXY chromosomes (men have XY and women have XX chromosomes).

The typical social and, perhaps, medical response to these circumstances is to follow appropriate procedures to make that person either biologically male or female. With these circumstances in mind, it can be argued that

The notion that we all fit into either a male or female body is just that, notional. It cannot be sustained, either over time or space. Simply put, there is nothing 'natural' about 'male' and 'female' bodies. There is nothing natural about everyone being forced into one sex or the other. Rather, our belief in the existence of two, and only two, sexes is structured by our ideas about gender. What this means is that our understanding of gender (man and woman) is not determined by sex (male and female) but that our understandings of sex itself are dictated by an understanding that man and woman should inhabit distinct and separate bodies. So, sex does not make gender; gender makes sex. (Gregson, Rose, Cream, and Laurie 1997:195)

This constructionist view of sex and the body involves what may be described as a destabilizing of sexual identities. Much work in this vein proceeds to develop these ideas further by building on Butler's (1990) work regarding the idea of sexual identity as performance. This is a means of explaining why some people do not conform to the presumed heterosexual norm, choosing instead to engage in different behaviors at different times and in different places. As discussed in Box 8.4, such an approach goes beyond that of conventional constructionist arguments with the emphasis on the fluidity of identity.

Sexual Landscapes

Some of the earliest analyses of sexual landscapes were studies, often in the urban ecological tradition, of prostitution and of indirect attempts to delimit the landscapes occupied and modified by gay men and lesbian groups. A principal concern today is with the landscapes and lifestyles of gay men and lesbians within larger, usually urban, areas designed for heterosexuals. This emphasis is different from earlier work as it stresses sexualities as identities and is typically based on ethnographic research. Bell and Valentine (1995:5) note that the earlier form of research can be criticized for being 'patronising, moralistic and "straight"' in comparison to the latter form, which was seen as 'sex-positive'.

Central to much of this work on geographies of sexuality is the recognition that heterosexuality dominates, both in terms of preferred sexual behavior and in terms of power relations in everyday environments. Those who are different tend to feel out of place in most environments that simply assume that all who use the environment are heterosexual. Costello and Hodge (1999:152) summarize the vision that underpins much of the research into sexual landscapes as being one that does not privilege heterosexuality: 'We envisage spaces where same-sex desire becomes an unquestioned possibility.'

CLAIMING SPACES

Historically, because of the typically prevailing circumstances of social prejudice, legal constraints, and threat of violence, lesbians and gay men have been relatively invisible, a fact that made leading an open lifestyle extremely difficult. Necessarily, then, claiming public and private spaces, both in the short term and the long term, has long been a key goal.

The best short-term examples of claims to space are parades that serve to proclaim the reality and legitimacy of non-heterosexual identities. These parades are commonly staged in major cities in the Western world, but although they are increasingly accepted by larger communities, they have given rise to numerous local conflicts. For example, L. Johnston (1997) describes the contested site of a gay pride parade in Auckland, New Zealand, noting that the first two gay pride parades, in 1994 and 1995, took place outside of an area that is generally accepted to be gay, and hence were seen as challenges to the larger heterosexual community, as what might be described as a queering of the

streets. However, the 1996 parade took place within an area widely perceived to be gay, and proved to be less controversial. This is a good example of the unsettling effect, for those who are not gay, of boundary crossing by those who are gay. Similarly, with reference to the participation of a gay, lesbian, and bisexual group in the annual St Patrick's Day Parade in Boston, Davis (1995:301) notes that: 'Symbolically, the very existence of alternative sexualities was a threat to the locally prevailing notion of what it meant to be Irish.' Although the group was eventually allowed to participate, they experienced much hostility and abuse.

In the longer term, as part of the larger but spatially uneven cultural acceptance of gays and lesbians during the 1960s and later, both commercial and residential landscapes associated specifically with these groups developed. In accord with the claim that sexual identities and sexual landscapes are mutually constituted, gay neighborhoods are increasingly common in large cities. By the early twentieth century, parts of New York, notably in Greenwich Village,

Box 8.4 The Body

Cultural geographic studies of identities and of human relationships with places traditionally incorporated accounts of bodily characteristics, such as skin color, but these studies were not framed in terms of bodies themselves. The current cultural geographic interest in the **body** was stimulated especially by Bourdieu, Foucault, and Butler.

For Bourdieu (1994) social class and the body are inextricably related. Thus, compared to poor people, wealthy people are able to display their bodies in more varied and exotic ways, especially through choice of clothing and ornamentation, and they are able to engage in a greater variety of bodily activities. Class distinctions are also evident in bodily mannerisms and communication skills.

Foucault (1978) focuses on ways in which the body is produced through prevailing discourses, noting, for example, that attitudes to different sexual preferences change through time and are different in different places. Foucault (1977) also stresses that the body is subjected to discipline, a set of rules of con-

duct that carry different expectations for the body in different places and social contexts; consider, for example, what are seen as acceptable body displays in different workplaces, or in a typical workplace as compared to a vacation setting.

Butler (for example, 1993) discusses the body in relation to the concept of **performativity**. This is a particularly influential idea in much feminist geography focusing on the production of gendered and sexual identities. From this perspective, these identities are not only contested and unstable, but they are also performative, meaning that rather than determining our behaviors, these identities are formed by our behaviors, by our bodily performances. Put another way, gendered and sexual identities are what we do, and not what we are. In this sense, the body is an act, a style.

How does this idea play out in practice? There are at least two possibilities. First, performances are affected by the dominant discourse, specifically by the understandings of how different gendered and

Times Square, and Harlem, were identifiably gay in terms of land uses, social facilities, and events staged. However, the rise of such neighborhoods has been most evident since the 1960s as part of larger social changes, with well-known examples including the Castro district in San Francisco, West Hollywood in Los Angeles, and Soho in London. Most such neighborhoods involve gay men and not lesbians. Figure 8.1 shows the spread of the gay residential area in San Francisco; Figure 8.2 shows the distribution of lesbian and gay commercial establishments in Philadelphia; Figure 8.3 maps gay and lesbian landmarks, sex-trade areas, and the locations of AIDS organizations in a part of Vancouver. In all cases, there is significant clustering.

Two observations about these landscapes emphasize the complexity of sexuality and landscape relations. First, and perhaps consistent with traditional ideas about the social construction of gender, there are fewer identifiable lesbian areas and, where they do exist, they are often primarily residential rather than commercial. Second, analyses of gay landscapes often highlight the fact that such areas do not necessarily welcome all gay people, but perhaps only those who are deemed appropriate in terms of other identity markers, such as ethnicity and class.

Needless to say, despite growing acceptance of visible gay and lesbian landscapes, these claims to public and private space have not come without considerable effort to overcome prejudice and have provoked severe backlash by those seeking to preserve heterosexual spaces. The most extreme version of the heterosexual norming of space involves acts of violence against those who are not heterosexual, including lesbians, gay men, bisexuals, and transgendered people. Namaste (1996:221) states that 'a perceived transgression of normative sex–gender relations motivates much of the violence against sexual minorities, and that an assault on these "transgressive" bodies is fundamentally concerned with policing gender presentation through public and private space'. It is apparent that a specific identifiable, spatial presence does make it easier for those prone to violence against gay men and lesbians to target victims.

sexual identities ought to behave. For example, McDowell (1995) describes the workplaces of female merchant bankers in London as being defined by a dominant form of male masculinity that affects the performances of those who do not conform to that dominant identity; specifically, female merchant bankers use their bodies in gendered performances, playing out a role that is expected of them. The second possibility is that performances are freely chosen by the individual. Thus, a person might wake up each morning, make a decision on what their gendered or sexual identity will be for that day, and then behave—perform—accordingly. Regardless of the specific interpretation, gender and sexuality are not only constructed identities, they are fluid and negotiable. In principle this idea of performativity applies also to identities other than gender, and to the production of place, though it has yet to be applied in these other contexts.

As noted earlier in this chapter, although much cultural geography has centered on relationships between humans and environment, and between humans and other humans, it is only recently that some of the work has been situated in the context of the body. In accord with the three inspirations noted, much contemporary cultural geography understands bodies in at least two fundamental ways. First, the body is each individual's personal space and contributes to identity construction in that it incorporates such characteristics as sex, physical appearance (including skin color), physical ability, and mental ability. Second, the body functions as a boundary between self and other, contributing to circumstances of social inclusion or exclusion and to related struggles and contestations (Valentine 2001:15).

Viewed in these two ways, the body is important in discussions of identity characteristics generally—recall the accounts of apartheid and genocidal actions that involved dominant societies constructing the identities of others with explicit focus on bodily characteristics. But while this focus on body has been widely applied in cultural geography, it has proven especially popular in studies of gender and sexuality.

Figure 8.1 Spread and Growth of Gay Residential Area in San Francisco, 1950s–1980. This map was produced from five different sources of information, including evidence from informants in the gay male community and fieldwork; the five sources reinforce each other, providing confirmation of the details shown. The map shows clear evidence of sequential and cumulative development similar to the neighborhood diffusion process discussed in Chapter 5. By 1980 the gay residential area also included gathering places for social, business, and political activities.

Source: Adapted from M. Castells, *The City and the Grassroots: A Cross-cultural Theory of Urban Social Movements* (Berkeley: University of California Press, 1983):146.

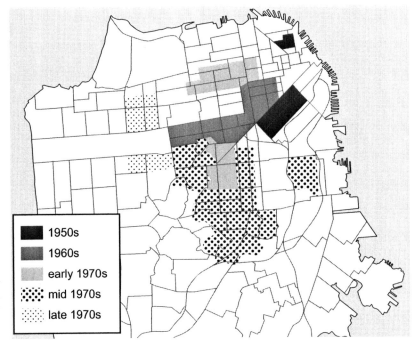

■	1950s
■	1960s
▨	early 1970s
░	mid 1970s
░	late 1970s

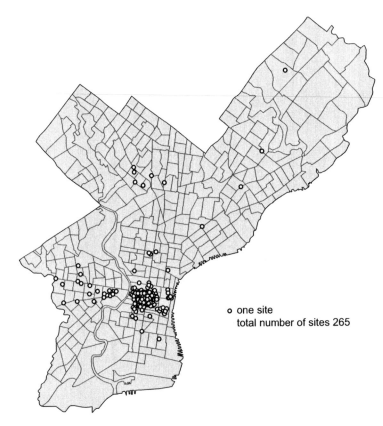

Figure 8.2 Lesbian and Gay Commercial Establishments in Philadelphia, 1945–1974. The map shows that most of the known lesbian and gay commercial sites are concentrated in one area of Philadelphia (Center City). The establishments include bars, bathhouses, bookstores, clubs, hotels, movie houses, restaurants, theaters, and other businesses.

Source: Adapted from M. Stein, *City of Sisterly and Brotherly Loves: Lesbian and Gay Philadelphia, 1945–1972* (Chicago: University of Chicago Press, 2000):56.

o one site
total number of sites 265

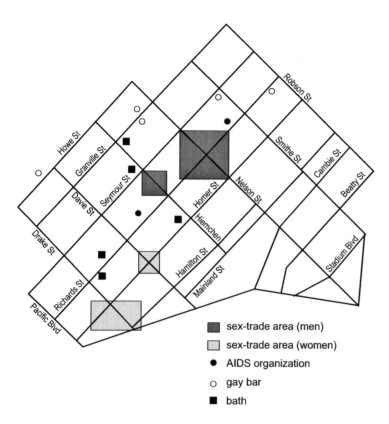

Figure 8.3 Sexual Geography of Yaletown, Vancouver, 1990s. During the 1990s, Yaletown was a rapidly changing, gentrifying urban area located close to both the gay West End and the downtown core of Vancouver. It was also a leading area for sex-trade workers and the center of local responses to AIDS.

Source: M. Brown, 'Sex, Scale, and the "New Urban Politics": HIV Prevention Strategies From Yaletown, Vancouver'. In *Mapping Desire: Geographies of Sexuality*, edited by D. Bell and G. Valentine (New York: Routledge, 1995): 250. (Reprinted by permission of Taylor & Francis Books Ltd)

- ▨ sex-trade area (men)
- ▦ sex-trade area (women)
- ● AIDS organization
- ○ gay bar
- ■ bath

BLURRING SEXUAL AND SPATIAL BOUNDARIES

Of course, ongoing debates about destabilizing sexual identities, especially their uncertainty and fluidity, mean that it is not necessarily appropriate to formally identify specific sexual landscapes. This argument is pursued by Costello and Hodge (1999) in analyses of a neighborhood in Melbourne, Australia, that is widely understood as a lesbian space and of the gay and lesbian Mardi Gras celebrations in Sydney, Australia. It is argued that, because sexual identities are neither stable nor knowable, specific sexual landscapes are also neither stable nor knowable. This is a line of reasoning that might, in principle, be applied to any other identity characteristic—ethnicity, for example—and the fact that the argument has been developed within the geography of sexuality literature is an interesting indicator of the conceptual sophistication of that work.

Pause for Thought

Much of the content of this section refers, implicitly or explicitly, to a fundamental question about human identity, namely, can each of us be conveniently labeled as basically just one thing—male or female, for example? For most cultural geographers working on topics of sexuality, the answer is an unequivocal no. More so, perhaps, than in other areas of cultural geography, studies of sexuality stress that identities are not monolithic, emphasizing instead the instability, fluidity, and even uncertainty of any particular sexual identity. These claims are challenging, not just in the present context but more generally for the larger realm of identity and landscape studies. It may be helpful to draw a parallel with ethnicity: recall the earlier reference to Muslim identity in France, where it was noted that there are competing claims as to whether or not it is possible to be both Muslim and French.

Other Peoples and Landscapes

The cultural geographic concern with others and their identities and places, linked as it is to a diverse body of postmodern and other theory, has expressed itself primarily in the study of ethnicity, gender, and sexuality. But there are, of course, numerous additional groups that have been regarded as others, such as the disabled, children, youths, elderly people, the homeless, and the unemployed. Some of these groups have a visible presence in the landscape, while others may be less visible. Some of these groups may be controversial to the majority and viewed negatively, while others may be viewed with indifference. Some groups may be seen as threatening, while others may feel threatened. These comments reflect four circumstances that prevail in contemporary Western society.

First, wherever people are collected together—for example in an urban area—there is a diversity of individuals. Individuals tend to be labeled by others, and also to label themselves, and there is often a proliferation of quite bewildering, often uncertain, and usually fluid identities.

Second, when people label other people, they tend to do so according to a particular identifying marker, and then they make judgments about individuals according to assumptions that they make about the larger group. Of course, such assumptions represent an unsatisfactory, reductionist, and dehumanizing way of looking at people.

Third, labeling and then making assumptions about people is an inherently political process that frequently results in discrimination, increased levels of fear, and even in violence. Racism and sexism have been discussed earlier, but there are many other –isms that might be identified, including ageism and ableism, both of which are discussed in this section.

Fourth, because people are labeled not just according to their appearance but also on the basis of where they are, different places are assigned different meanings. In particular, some parts of a larger area may become landscapes of fear. Many public spaces are unsafe for some members of the larger population if they are seen as not conforming to a particular identity that some members of the dominant group judge to be appropriate. It is important to acknowledge, however, that those individuals who are responsible for making public places unsafe for others are not typically representative of the dominant society by virtue of the very fact that they are prepared to resort to harassment, intimidation, or violence. Notwithstanding this fact, crime or the threat of crime is one of the means by which the behavior of some people is severely constrained.

Ethnicity in the City

As noted earlier, one of the principal means by which people are identified by dominant societies and also by which they choose to identify themselves is on the basis of the myth of race and/or ethnicity. In Chapter 7 the discussion focused on global and national issues, but ethnic differences are also a principal means used to label and group people at more local scales. As a result, there is a great deal of work in urban and social geography focusing on ethnic patterns in urban centers and on the experiences of minority ethnic groups. In general, North American cities tend to include ethnic enclaves, whereas in British and European cities ethnic groups tend to be more dispersed. In all cases, there is much evidence to suggest that stereotyping and discrimination combine to diminish the quality of life of ethnic minorities while creating unease because of increased fear of crime and violence. Conceptually, some recent studies have sought to uncover aspects of ethnic stereotyping by highlighting the claim that whiteness tends to be understood as a 'natural' category, with other skin colors seen as problematic.

Class

It has been common in social science to link resistance and conflict with social class, although, not surprisingly, cultural geographers working in the Sauerian tradition have shown little interest in class as a basis for delimiting groups and describing landscapes. Rather more surprisingly perhaps, class has not emerged as a key concern in the current postmodern fascination with identity and place. It might be argued that the rise of feminism effectively displaced the traditional Marxist concern with class in many sociological accounts of conflict, and similarly functioned to limit any significant geographic interest to a few conceptual discussions and empirical analyses.

Arguing from a Marxist perspective, Blaut (1980) notes that classes are an appropriate variable for delimiting cultures, and that certain individuals or classes typically exert power over others, a situation drastically affecting the behavior of groups. Most of the work concerned with class focuses on conflict between classes. Harvey (1993) details a powerful example of a geography of resistance, interpreted in terms of class power relations, especially the way in which industry was prepared to exploit rural poverty and to minimize expenses at any cost. This analysis was of an industrial accident, a fire that occurred in a chicken-processing plant, the exit doors of which were locked, killing twenty-five workers and seriously injuring fifty-six.

Age

Cultural geographers concerned with the identities and behavior of people in particular age categories do not have any meaningful body of conceptual work to turn to, nor have they developed relevant concepts. The contrast with cultural geographies of ethnicity, gender, and sexuality is clear. Most of the relevant research has uncritically accepted and applied the traditional psychological and sociological distinctions that identify such categories as children and elderly people. This focus on groups positioned at the beginning and end of the life cycle reflects the interest in studying those who are understood to be on the margins and perhaps excluded from larger society as a consequence. Although there are obvious and fundamental physical differences between age categories, most research on groups identified by age focuses on the social construction of identities, arguing that prevailing stereotypes and assumptions about children and elderly people reflect **ageism**. In terms of the relationship between age and place, there are links that are worth examining, not just because any person's experience of a place is affected by their age but because some places are associated with—may even be built for—particular age groups.

CHILDREN

How children's identities develop depends on a complex interplay of personal characteristics including their degree of introversion/extraversion; their home setting, including location; their class and ethnic background; and their social interactions with peers. Institutions put in place by society will also play an important role in children's identity formation. For example, the institution of compulsory schooling is usually explicitly designed to teach the norms and values of the dominant society, a fact that has led many groups wanting to preserve some sense of their distinctiveness to control their own schools. Canada, for example, has numerous language immersion schools, most of which are designed to preserve cultural traditions and promote a sense of loyalty to various minority ethnic groups. Schools also contribute to a child's sense of individual identity through presumed norms of behavior that children are expected to conform to; traditionally, these expectations have included the assumption of heterosexuality, though in many places this assumption is no longer explicit.

Although schools play an important role in identity formation, social interaction between peers, especially during adolescence, contributes to the formation of a range of dynamic and often short-lived identities, many of which are in opposition to the dominant society. Though a person's association with any one such identity may be ephemeral, it can nevertheless be a formative experience. Of particular significance during adolescence are the emerging sense of self-worth and the clarification of such important personal characteristics as sexuality. Unfortunately, it is sometimes the case that children learn inappropriate values and behaviors through their interactions with peers, with much evidence suggesting that racist and sexist attitudes are learned in this way. Overall, psychologists and sociologists have been concerned with the details of such situations, but it is clear that cultural geographers also need to be aware of these circumstances, especially as they relate to an emerging sense of identity and of belonging to groups that in turn impact on spatial behavior and landscape formation.

The spatial behavior of children is similarly affected by peer interaction, as well as by the complex interplay of personal, domestic, and institutional circumstances. In particular, the institution of schooling plays a significant role in limiting the spatial behavior of children by restricting mobility for extended periods during much of the year. Some spatial behavior outside of school might be seen as either in accord with or in opposition to the constraints imposed while in school. In this context, there is interest in the formation of groups that may pose threats to larger society. In some areas, such groups, usually called gangs and sometimes associated with a particular ethnic identity, create a sense of unease or fear among others. Indeed, as is the case with some groups defined in terms of sexual identity, there are examples of young people who feel excluded from larger society actively creating identities and places, such as clubs and other entertainment places, in opposition to that larger society.

The Elderly

The idea of an elderly identity is gaining attention from cultural geographers, especially as a consequence of the demographic changes of falling birth rates and increased life expectancy that are resulting in rapidly increasing numbers of elderly people throughout most parts of the world. Increasing numbers mean that elderly people are collectively playing more active roles in society and in politics—in many countries, elderly people constitute a significant special interest group.

The perception that the elderly represent a problem group is usual in Western societies, where they are often viewed in terms of their financial and physical dependence on the larger population. Instances of ageism are prevalent in Western media, which tend to portray the elderly as immobile, incapable of coping, helpless, and vulnerable. Certainly all humans experience some loss of physical well-being as they age, and

some elderly people are supported by the state, but the typically negative perception of the elderly—for example, the presumed association of aging with a dramatic loss of physical ability—is often inappropriate and merely serves to reinforce misunderstandings. The fact is that, as with other identity categories, there is no one elderly identity, but rather many and fluid identities. Indeed, age is a relative term, one that can be understood in chronological, physiological, and social terms. Aging means different things for different individuals and is interpreted differently in different social contexts.

Geographers have focused attention on the spatial behavior of elderly people as it is enabled and constrained by personal and social circumstances, including available funds, degree of mobility, and contacts with family members and peers. These accounts have paid particular attention to the geography of fear that elderly people may experience both inside and outside the home. Most of these studies also take into account gender, ethnicity, and other relevant characteristics. Concerning crime, Table 8.1 shows that elderly people have higher levels of fear, worry, and unsafe feelings than do those of other age groups; note also that the levels in these categories are higher for women than for men. Table 8.2 reports women's concerns about the possibility of physical or sexual assault and shows that younger women are the most concerned.

Traditionally, elderly people lived with other family members in extended family units, but increasingly in the Western world the elderly live as couples or alone. One geographic stereotype about the elderly is that they 'gradually become prisoners of space as physiological deterioration and environmental constraints necessitate physical, social, psychological and, by implication, spatial withdrawal' (Rowles 1978:xv). A presumed consequence of imprisonment is the tendency to remain in an old and possibly inappropriate house and neighborhood. This stereotype is rooted in fact, as many elderly people are constrained in their residential mobility for essentially negative reasons, such as financial limitations. But, the reason that many elderly people have for not moving may be related to a

Table 8.1 Concerns About Crime as Related to Age and Gender

Gender	Fear		Unsafe		Worry	
	16–25 years	61+ years	16–25 years	61+ years	16–25 years	61+ years
Male	0.0	19.5	3.4	31.7	31.0	63.4
Female	8.6	39.1	27.8	70.8	51.4	79.7

This table presents the results of questionnaire surveys that focused on concerns about crime in parts of Stoke-on-Trent in the English Midlands. The data in the table are the percentage of respondents within the two age-groups who are the most fearful, feel most unsafe, and are most worried. There are clear links between both age and gender on the one hand and fear of crime, feelings of safety, and worry about crime. Use of the three categories of fear, safe, and worry derives from earlier work suggesting that these are three separate but related aspects of concerns about crime.

Source: D.J. Evans and M. Fletcher, 'Fear of Crime: Testing Alternative Hypotheses', *Applied Geography* 20 (2000):406. (© 2000, reprinted by permission of Elsevier)

more positive consideration, namely love of place, or topophilia, as discussed by Tuan (1974).

Disability

Earlier accounts of gendered and sexual land-scapes cite as one of the basic premises behind much cultural geographic research the claim that landscapes in general are made by and for heterosexual men; indeed, much feminist work

asserts that this is intentional in order to per-petuate patriarchy. But it is clear that landscapes also 'presume able-bodiedness, and by so doing, construct persons with disabilities as marginal-ized, oppressed, and largely invisible "others"' (Chouinard 1997:380). In this account, the term *disabled* is used in a conventional way to refer to people who are not able-bodied, and as such it includes, for example, those whose mobility,

Table 8.2 Women's Fear in Public and Private Spaces as Related to Age

% very worried or fairly worried about each incident	Age in Years			
	18–30	31–45	46–60	60+
Sexual assault outside by a stranger	83.2	71.9	53.5	39.2
Physical assault outside by a stranger	65.1	63.8	48.0	44.4
Sexual assault in your home by someone you know	28.0	28.0	18.4	21.7
Physical assault in your home by someone you know	23.7	29.5	18.2	20.1

This table, presenting results of questionnaire surveys that focused on fear of violent crime in parts of Edin-burgh, Scotland, shows relationships with age, gender, and type of space.

Source: Adapted from R.H. Pain, 'Social Geographies of Women's Fear of Crime', *Transactions of the Institute of British Geographers NS* 22 (1997):240.

Table 8.3 Constructing Disabled Identities

Able-bodied	Disabled
Normal	Abnormal
Good	Bad
Clean	Unclean
Fit	Unfit
Able	Unable
Independent	Dependent

Source: R. Imrie, *Disability and the City: International Perspectives* (London: Paul Chapman, 1996):37. (Reprinted by permission of Sage Publications Ltd)

ently is known as **ableism**. Table 8.3 identifies some stereotypical perceptions about able-bodied and disabled people.

There are two ways of thinking about disability. The traditional approach focuses on medical issues, by arguing that disabilities are specific and individual departures from normal bodily and/or mental conditions. It is this perspective that is used, implicitly or explicitly, when disabled people are defined as different and inadequate others. A more recent approach focuses on social issues and sees disability not in terms of individual medical circumstances but rather in terms of society's response to those who are 'differently abled'. From this perspective disability is understood as those disadvantages and restrictions on activity that are imposed by dominant society on those who have physical and/or medical impairments. One problem with

vision, or hearing is impaired. The assumption that disabled people are in some sense inferior to others and might need to be treated differ-

This photograph highlights what is likely one of many daily challenges faced by people confined to wheelchairs. Movement that most people take for granted, such as using stairs, may be impossible for others. Even a cursory appraisal of the landscapes that we have constructed make it clear that the builders have taken certain characteristics for granted. Although much is being done to remove such impediments by constructing wheelchair access ramps, many obstacles remain. The photograph is suggestive with the person in the wheelchair partly included and partly excluded; certainly, until recently, those restricted to wheelchairs were effectively invisible to those who built the urban environment. (Gary Salter/CORBIS Canada)

this social perspective is that there is a danger of trivializing the very real problems that some disabled people face in their everyday lives.

As the earlier quote from Chouinard indicates, built environments have not typically taken into consideration the problems faced by disabled people and might be appropriately described as hostile, unwelcoming, and exclusionary landscapes. At least implicitly, the assumption made by those responsible for the built environment is that it is the responsibility of those who are disabled to find ways to overcome their disabilities rather than the responsibility of dominant societies to accommodate disabled others. This attitude no longer prevails in many parts of the world and, increasingly, both public and private spaces are being modified. The most obvious example is the construction of ramps to facilitate wheelchair access to buildings. But, of course, disabled people continue both to be treated differently and to have different experiences in many landscapes. Table 8.4 highlights the fears of a sexual attack experienced by those women who have a physical disability that limits mobility. As with many of

the other cases of difference discussed in this and the preceding chapter, stereotyping, discriminatory attitudes, and hostile behavior continue to confront disabled people.

Exploring Different Identities

Some other groups, such as Gypsies, have been identified by cultural geographers as marginalized, stigmatized, and both socially and spatially excluded in much the same way as have the physically impaired and mentally ill (see Box 8.5). And yet, there are so many other bases for group formation that have not received significant attention from cultural geographers, including those personality traits that might diminish a person's sense of self-worth and ability to use space fully. An obvious example of this kind of trait is **agoraphobia**, the fear of public spaces, which can severely limit a person's social interaction and spatial mobility. It is clear that cultural geographers choose to study some identities but not others.

The identities selected for study by cultural geographers and discussed in this book have three general features in common: first, they are

Table 8.4 Effects of Fear of a Sexual Attack on Women With and Without a Disability

Always or sometimes do the following because of fear of sexual attack	% of respondents with a physical disability	% of respondents with no physical disability
do not answer the door	45.5	35.0
put off routine calls	23.8	10.6
feel unsafe with strangers	59.1	61.8
feel unsafe with people I know	18.2	4.8
do not go out	27.3	8.3
do not go out alone	59.1	33.7

This table, presenting results of questionnaire surveys that focused on fear of violent crime in parts of Edinburgh, Scotland, indicates some of the effects of fear of a sexual attack as these relate to whether or not women have a walking or other physical disability that limits their mobility.

Source: Adapted from R.H. Pain, 'Social Geographies of Women's Fear of Crime', *Transactions of the Institute of British Geographers NS* 22 (1997):241.

identities that are often discriminated against by the majority; second, they are usually associated with a political agenda; third, they are often identified with by their researchers (it is not unusual for researchers, especially those studying sexualities other than heterosexuality, to explicitly state their insider status). As a result, it is possible that cultural geographers, in their desire and willingness to focus on particular disadvantaged others, neglect those others who tend to discriminate rather than be discriminated against and who either do not have a political agenda or whose political agenda is less palatable to researchers.

For example, it is possible that conservative groups, including some religious cults and certain North American militia groups, are less likely to be studied because their different identities are not in sympathy with the identities of cultural geographers attracted to this type of work, which typically involves insider research. Similarly, there is very little evidence of a concern in cultural geography with certain identities, behaviors, and personalities, such as introvert and extravert, or authoritarian and submissive. Consider those who suffer from phobias or any of a wide range of anxiety states, stresses, mood disorders, schizophrenic states, and personality disorders recognized by psychologists. All of these might be interpreted as differences of interest to cultural geographers, yet they have received scant attention in the literature of the discipline. Certainly, it can be argued that, largely for ideological reasons, cultural geographers have chosen to focus on those different identities that can be classed as others, usually disadvantaged groups. Overall, the interest shown in difference is selective rather than all encompassing.

Pause for Thought

It is worth stressing that accounts of the myth of race, ethnicity, nations, gender, sexuality, and disability all build on the idea that identities of non-dominant groups have traditionally been determined by those in positions of power. This process of social construction has typically involved dominant societies identifying those who are different as

inferior others. The idea of labeling people according to some perceived primary characteristic is of dubious value, notwithstanding the fact that minority groups sometimes actively choose to label themselves for pragmatic reasons (this is the strategic essentialism noted earlier). The key equation in all cases is that difference merits unequal treatment. This is, all too clearly, factually flawed logic, but more importantly, it is unethical. The assumption that difference equals inequality has been used to justify all kinds of unequal treatment, including unequal provision of services and unequal access to places. These inequalities contribute to the rise of what might be called resistance identities and to related competition for landscapes.

Identity, Resistance, Landscape

It is clear that if some groups wield power, then other groups will emerge that resist that power. By definition, both the identities and the landscapes of emergent and resistance identities can be interpreted by the dominant society as challenges to their authority precisely because resistance is designed to oppose power. Recall that the concept of an emergent identity refers to some new cultural expression by those who, feeling in some way excluded by and/or disadvantaged within larger society, introduce a resistance identity.

Expressing Resistance

In any society, there are always ongoing and emerging tensions associated with emergent and resistance identities and related landscapes; several examples are noted in the accounts of gender and sexuality. In some instances, tension over landscape is an incidental component of larger social and political contestations, while in other instances, occupation of and/or control over landscape is the main source of conflict. In this respect it is useful to recognize that the most potent sources of power in much of the contemporary world are rooted in the capitalist mode of production, in racism, and in patriarchy, and it is these sources that resistance identities typically oppose (Katz 2003). In recent years, the most outspoken resistance identity of this kind is that of the anti–globalization movement.

EMERGENT AND RESISTANCE IDENTITIES

Identifying emergent and resistance identities is one part of the larger cultural geographic concern with recognizing that different places mean different things to different people; indeed, as Cresswell (1996a:59) explains, 'places are the results of tensions between different meanings' and 'are also active players in these tensions'. In some cases, different meanings might comfortably coexist, while in other cases they result in confrontation. Inevitably, some places are contested sites because they are valued in different and incompatible ways by more than one group.

Many disputes over landscape arise from the emergence of identities in the form of new social movements. Unlike many earlier social movements, these identities are not typically based on class but on a variety of other foundations. As one part of a larger discussion of new social movements, Castells (2004:171) identifies five principal types of environmental groups 'as they have mani-

fested themselves in observed practices in the past two decades'. Each of the five, noted in Table 8.5, is defined by some combination of three characteristics that are seen as defining social movements generally, namely identity, adversary, and goal.

The traditional type of movement is concerned with the conservation of nature; the Sierra Club is a leading example of this. The fastest-growing type is local community involvement, sometimes seen as the 'not in my back yard' syndrome, although the concerns of many such movements often extend beyond the purely parochial. Concern about environment has also been associated with the emergence of groups that expressly oppose some traditional social beliefs and practices; the ecofeminism discussed in Chapter 4 is an example. Greenpeace, founded in Vancouver in 1971 and now an international organization, is the leading example of the save-the-planet type of environmental movement. Finally, the rise of political green

Box 8.5 A Landscape of Exclusion

Throughout much of Europe, Gypsies are recognized as a distinctive group living on the margins of a dominant culture; they are often labeled an ethnic or cultural group. Although there are many different versions of Gypsy culture today, they share a distaste for waged labor and are associated with a lifestyle that this aversion to waged labor implies.

But beyond being a minority cultural group in several European countries, Gypsies are also a group that does not fit easily into the standard classifications imposed on people by a dominant culture. This uncertain status is related to their characteristic mobility and to the fact that they are usually viewed as different, as outsiders. But this status is also related to the discrepancy between their romantic rural image and their typical lifestyle. For example, in Britain the stereotypical romantic and rural perception of Gypsies remains, in spite of the fact that the spaces they occupy are as likely to be urban wasteland as they are to be rural. Certainly, the places that Gypsies occupy are, like the people themselves, judged to be different and outside. They are landscapes of exclusion that are the consequence of and contribute to the outsider status of the group.

Sibley (1995:68) explains the popular association of Gypsies and dirt: 'Here, the problem is Gypsies' dependence on the residues of the dominant society, scrap metal in particular, and their need to occupy marginal spaces, like derelict land in cities, in order to avoid the control agencies and retain some degree of autonomy.' Places occupied by Gypsies are avoided by members of the dominant population because they are seen as threatening. As Sibley (1999a) notes, a fear of the other becomes a fear of place. Further, the apparently disorganized character of Gypsy space compared, for example, to a homogeneous suburb results in a devaluing by the dominant culture of that space and of the people occupying that space. In the city of Kingston upon Hull, England, where Gypsies have lived for about 100 years, numerous conflicts eventually prompted the building of two permanent locations for what was widely regarded as a deviant group. One was in a heavily polluted industrial area and the other was in an old quarry that was used for dumping garbage. Outsiders are relegated to the outside; deviants to residual places.

Table 8.5 Typology of Environmental Movements

Type	Identity	Adversary	Goal
conservation of nature	nature lovers	uncontrolled development	wilderness
defense of own space	local community	polluters	quality of life/health
counter-culture, deep ecology	the green self	industrialism, technocracy, patriarchialism	ecotopia
save the planet	internationalist eco-warriors	unfettered global development	sustainability
green politics	concerned citizens	political establishment	counterpower

Source: Adapted from M. Castells, *The Power of Identity* (Malden. MA: Blackwell, 2004):172.

parties, most notably in some European countries, represents a fifth type of environmental social movement.

Most types of resistance identities are appropriately interpreted in terms of their opposition to powerful identities. Many ethnic separatist and nationalist movements and anti-war movements are obvious examples. Many of these movements have widespread applicability, but some are more local. The Baliapal social movement in India, for example, arose in opposition to a decision to build a military establishment, which would involve evicting or relocating some of the local population. The emergence of the movement resulted in the area becoming a 'terrain of resistance' (Routledge 1992:588). A similar contested space developed as a result of protest by women outside the Greenham Common military base in England during the early 1980s. What is interesting about this case is that the media raised concerns about some of the behaviors of protesters. For example, women protesters were portrayed as being out of place both because they had deserted their homes and because they had occupied an especially unfeminine place. For the media, the fact that many of the women present were reported to be indulging in lesbian behavior further added to their oth-erness. Naturally, these media reports contributed to a popular belief that the women were participating in inappropriate behavior. Cresswell (1996a) uses the term 'heretical geographies' to refer to such behaviors that are seen to be in some way different from those considered normative for the place.

Other contested landscapes are associated with subcultures rather than more formally defined new social movements, especially those subcultures created by youthful populations. Again, these identities are characterized by opposition to mainstream lifestyles and mainstream places, with groups choosing to display their difference, thereby expressing their resistance to and rejection of dominant cultural values through the adoption of alternative lifestyles, including musical and clothing preferences. Certainly, one of the ways in which it is possible to interpret many social trends is in terms of their apparent or sometimes explicit opposition to the conventional.

Contesting Landscapes

Accepting the argument that culture and power are inseparable leads to the conclusion that all identities have a political component, and that all landscapes are, at least to some degree, contested. Societies are made up of collections of

For several years during the 1980s, an old World War II airbase on Greenham Common in southern England was developed into a high-security area housing 95 cruise missiles, their transporters, and other support vehicles, but it became better known as the site of a massive and prolonged anti-nuclear protest, mostly by women. This photograph shows women protesters being removed by police while the media watch and record what is happening. Tens of thousands of women either lived at or visited the Common during the 1980s. Among the highlights of the extended protest were 30,000 women joining hands to embrace the base in December 1982, 70,000 people protesting in April 1983, and 50,000 women encircling the base and holding up mirrors in December 1983. Numerous arrests were made during this quite remarkable expression of civil disobedience. One way to interpret what was happening at Greenham Common is that it was a contest for a landscape: those in positions of authority determined that it should be put to particular use, but many others saw the world and this particular place very differently and expressed their disapproval in no uncertain terms. Although the cruise missiles were removed in 1990 as part of the larger reduction of Cold War tension, the original Women's Peace Camp closed only in September 2000, after 19 years continuous presence outside the airbase. (Barry Batchelor/CP Photo)

diverse cultural groups such that there is always some degree of conflict over who is responsible for exercising control over both places and the production of culture.

For example, many cities contain at least one area—usually an inner-city area—that is seen as different and treated accordingly because it lacks some of the services that the larger urban area enjoys; in such cases difference can be disadvantage. Indeed, the 'inner-city' label carries negative connotations. It is often the case that disadvantage is perceived negatively by others, and this contributes fur-

ther to the disadvantage. Boundaries may be drawn and redrawn both spatially and culturally as a means to distinguish between those who belong and those who do not belong.

Morehouse (1996:6), examining this issue in a much different context, looked at issues of changing power, influence, and control surrounding the boundaries of the Grand Canyon National Park. Her study reveals 'a story of the contests that have taken place regarding how the area should be shared, inhabited, protected, and used'. Among the many groups that have contested this matter with the Park Service are various Aboriginal populations, government agencies, transportation companies, specific land use interest groups, and recreational interest groups, all vying for control of this contested landscape.

Cultural geographers have also been interested in how cultural groups and the state differ in their evaluations of, for example, particular buildings and lifestyles. Kong (1993), for instance, in an analysis of religious buildings in Singapore, focused on 'the oppositional meanings and values invested in religious buildings by individuals on the one hand and the state on the other'. She found that the attachment of a cultural group to a particular religious building is strengthened when the group is threatened with relocation or with having its building demolished by the state. She found also that the related tensions between group and state did not usually result in conflict, although some resistance was involved because individuals adapt the meanings that they invest in religious buildings to avoid conflict (Kong 1993a:342).

CONTESTING RELIGIOUS SPACE

Most cultural geographic work on competing identities vying for control of contested landscapes focuses on situations in which emergent and resistance identities arise in explicit opposition to some characteristic of dominant society. But there are, of course, many other bases for arguing about rights to, ownership of, and use of landscapes. Jerusalem, for example, contains sites that are sacred in Judaism, Christianity, and Islam, and access to these sites at particular times, and even ownership of these sites, is a contributing factor to the ongoing difficulties of the larger region.

A less well-known example is the competition between different Latter-day Saint (Mormon) churches. There are today about 75 different Mormon organizations—and there have been many others—each claiming that their church is the legitimate successor to that founded by Joseph Smith Jr in 1830. This has resulted in competing claims about the history of places of Mormon settlement. Following the 1844 death of Smith, there was much uncertainty

about his rightful successor. After some intensive debate, the vast majority of church members accepted Brigham Young as leader. The church headed by Young moved from Nauvoo, Illinois, west to the Salt Lake area and has thrived both numerically and politically, such that it is by far the largest Mormon church today. But several other Mormon churches founded in and shortly after 1844 also laid claim to being the one true descendant of the original church.

James Jesse Strang was one of several claimants to the leadership of Smith's church. Strang had joined the church in early 1844 and was 225 miles northeast of Nauvoo in Voree, Wisconsin, at the time of Smith's murder, having been sent there by Smith to look into the possibility of establishing a major settlement in an area that was already home to a few Mormons. Strang claimed leadership of the church based on revelations and on an appointment letter purportedly from Smith. The number of Mormons who accepted Strang as leader and moved to Voree to be with him is uncertain: some newspapers at the time reported a settlement of 10,000, while others suggested the area was virtually uninhabited; a number between 500 and 2,000 seems probable. The settlement lasted only from 1844 until 1847, when Strang and his followers moved to Beaver Island in Lake Michigan. Whatever the numbers involved, it is clear that Voree was a Mormon settlement and that it continues to be of symbolic importance to at least two Mormon churches today, both of which acknowledge Strang as the rightful successor to Smith. One of these two churches has a church building nearby that is in regular use.

Both churches descended from Strang are numerically small and politically weak, and the Voree landscape today contains only a few reminders of past Mormon settlement. These include a few stone buildings—one of which is on the National Register of Historic Places—and a number of historical markers erected by the local historical society. Nevertheless, Voree is a contested landscape today, with two very different interpretations of the history of the place evident. For the two small churches descended from Strang, Voree is the site where Strang received his revelations, and it is thus symbolically important. For the largest of the Mormon churches, now headquartered in Salt Lake City and for which Strang is an insignificant figure, Voree has no special meaning.

The contrast between these two views of landscapes came to the fore in 1992 when the Wisconsin State Historical Society, in association with the largest Mormon church (the one that Young headed after Smith's death), dedicated a historical marker in a downtown park in nearby Burlington. From the perspective of churches descended from Strang, the marker is incomplete and misleading as it excludes any specific reference to Strang, to Voree, and to the between 500 and 2,000 settlers who came there from Nauvoo. The marker refers only to Wisconsin Mormons forming 'separatist churches here' after 1844. Further, the reference on the marker to Moses Smith, who organized the first Wisconsin Mormon community in the area in 1837, neglects to mention that he became a supporter of Strang, not Young. Clearly, the marker is one selective interpretation of Mormon history, the interpretation favored by the principal Mormon church that by virtue of size and influence is able to impose its identity on landscape at the expense of smaller Mormon churches.

This photograph shows the contested marker that is located in Echo Veterans Memorial Park in the town of Burlington, close to the Voree site. Erected in 1992 despite some opposition, the marker reflects and reinforces one particular interpretation of the history of the area. Most notably, dispute over the marker reflects disagreement as to who was the rightful successor to Smith. Most historical markers can be interpreted in several different ways, and one reading of this marker is that it denies the past reality of a brief but real challenge to the identity of the principal Mormon church, thus minimizing the identity and sense of place that the small churches descended from Strang wish to retain. Certainly, this marker is about more than history and religion: it is about the politics of landscape. (William Norton)

Individual and Group Rights

Difference is problematic. At one and the same time it demonstrates the need for, but threatens the very existence of, universal principles of justice. In some ideal world, would we wish to emphasize difference or strive to focus on what we all have in common? Consider the following argument.

The Dangerous Notion of Cultural Identity?

According to Llosa (2001:68) the notion of cultural identity is dangerous. His argument can be summarized as follows:

- From a social point of view, cultural identity is a doubtful, artificial concept; but from a political perspective, it threatens humanity's most precious achievement, namely freedom.
- Of course, people who speak the same language, were born and live in the same territory, face the same problems, and practice the same religion and customs have common characteristics.
- Nevertheless—and this is the critical point—any collective denominator can never fully define each individual member of a collective identity, and it only abolishes or relegates to a secondary plane the sum of unique attributes and traits that differentiates one member of the group from all other members.
- Accordingly, the concept of identity, when not applied on an exclusively individual scale, is inherently reductionist and dehumanizing: it is a collectivist and ideological abstraction of all that is original and creative in each human being, of all that has not been imposed by inheritance, place, or social pressure.
- Indeed, true human identity springs from the capacity of human beings to resist the influences of inheritance, place, and social pressure, and to counter them with free acts of their own invention.

Building on these claims, Llosa argues that the possible current dilution of cultural identity related to processes of globalization has a very positive outcome:

> Now citizens are not always obligated, as in the past and in many places in the present, to respect an identity that traps them in a concentration camp from which there is no escape—the identity that is imposed on them through the language, nation, church, and customs of the place where they were born. (Llosa 2001: 69)

In light of our earlier accounts of apartheid and genocides, this is a powerful argument that merits consideration. But there are, of course, other perspectives on this issue.

Debating Multiculturalism

Multiculturalism is a term that refers both to a fact—the reality that some countries are pluralist societies—and to a social ideal—a value that views pluralism positively and as worthy of preservation. The philosophy behind multiculturalism is contested, with some seeing multiculturalism as offering a vision of national identity based on pluralism, and others seeing it as promoting divisiveness. This is a fundamental debate that has been behind much of our discussions on the geographies of identity. Box 8.6 presents an overview of the current situation in Canada.

As noted consistently in this and the previous chapter, it is clear that, when groups regard

Pause for Thought

It might be that a principal reason many minority groups in larger societies have willingly embraced the reification of their culture by the dominant group has to do with their attempts to gain some distinctive rights for their group. Such group rights have often been justified on the grounds that the group was treated unfairly in the past, usually in the context of colonialism, and on the grounds that the principle of individual equality of opportunity— the liberal individualism claim—is not adequate to over- come the earlier period of discrimination against a group. Accepting this argument—and it is indeed the dominant discourse in many countries—has resulted in the target- ing of specific groups for specific rights that are additional to those enjoyed by the majority population. In cultural geographic terms, what is happening here is that cul- ture—group identity—is being reified and used to justify some form of political and intellectual separatism. It is possible that the cultural geographic interest in identify- ing different and distinct cultural landscapes has con- tributed to this discourse. Critics of multiculturalism, such as Turner (1993:414), have argued that this domi- nant discourse has come to favor 'cultural nationalists and fetishists of difference'. It can also be argued that the most disadvantaged individuals in a society tend to miss out on many of the opportunities available to their group.

Concluding Comments

Much of the content of this chapter reflects the cultural turn. The traditional Sauerian view has been modified and complemented—some might even say replaced—by new conceptual inspira- tions. Key components of the cultural turn include a rejection of all theories that purport to offer some single correct answer and an appre- ciation of human differences.

The emphasis on identity that is at the heart of this chapter is a key component of contem- porary cultural geography, but it is not without criticism. In addition to those cultural geogra- phers who prefer to conduct research along more established Sauerian lines, perhaps accept- ing particular identity formations and focusing

on questions of landscape change and regional- ization, contemporary social geographers have raised concerns about the focus on identity. Thus, Gregson (1995:139), raising concerns about the cultural turn, worries specifically about the preoccupation with matters of meaning and identity and the related lessening of interest in the facts of social inequality. Certainly, there is a possibility that some accounts of identity that are intent on identifying differences lose sight of pragmatic issues.

Notwithstanding these disagreements, it is evident that humans are both individual and social beings. It is also evident that people are social with regard to a plurality of groups, so that no one group is the sole determinant of our iden- tity. Certainly, although many people have a pri- mary allegiance to one group, they are likely to have allegiances also to a number of other groups. Further, the identity of all these groups is not fixed but is rather subject to ongoing change.

Discussions of human identity necessarily raise questions about how people associate with each other and about how they construct and reconstruct their identities in opposition to oth- ers. As we have seen, answers are based on the idea that identities are not given to people at birth. Rather, they are created by people for themselves, or they are created by people and imposed on others. Regardless of the specific process, identities can and do change and also can and do have meanings that are open to dis- pute. It is increasingly popular for cultural geog- raphers to study human identities as the people involved understand them rather than as they are understood and assigned by the researcher. According to Hall (1996:5), the process of iden- tity creation is a strategic and positional activity such that identities are 'the products of the marking of differences and exclusion'.

Related to the issue of identity, and also cen- tral to much of the content of this chapter, are the links between identity and landscape. The humanist interpretation of place stresses that place is not so much a location, a thing, but rather a setting for human behavior, a relation- ship. From a humanist perspective, the meaning

of a place cannot be understood without awareness of the identity or identities of those who
occupy the place, but there are also more recent
interpretations of place that are conceived in
terms of the way places are controlled by those
in authority and challenged by others. Declaring
a place to be a 'no go' area or to be 'out of
bounds' results in that place having an increased
symbolic significance. Both of these interpretations of place recognize that, because we segment the world, we believe that there are proper
and therefore also improper places for some
things and some behaviors—we and our behaviors can be *in* place or *out of* place.

Pause for Thought

---

*Much of the material in this chapter suggests that land-
scapes reflect a complex interweaving of dominant and
subordinate cultural identities, with processes of inclusion
and exclusion at play. Overall the specifics of inclusion
and exclusion processes have long been determined by
those in positions of power, with that power based on any
one or more considerations—political, social, linguistic,
religious, gender, sexuality, or class-based. However, our
cultural world is changing. Other non-dominant identities, however defined, are becoming increasingly prominent as a consequence of two circumstances. First, the
voices of other groups are becoming more assertive, challenging dominant and legitimizing identities in many
different arenas. Beginning in the 1960s, environmental
and women's movements were among the first effective
challenges, and since then most Western countries have
seen an explosion of special interest groups. Second, dominant identities are more receptive to hearing the voices of
others than was the case in the past. This new willingness to listen to others is related to increased acceptance
both of democratic principles and of basic human rights.*

Further Reading

The following are useful sources for further reading on specific issues.

There are good accounts of postcolonialism by Blunt
and McEwan (2002), Shurmer Smith (2002, chapter 7), Clayton (2003), and Nash (2004).

Barnett (2005) uses the work of Levinas and Derrida
to consider relations between proximity, identity, otherness, and caring.

Dalby (1991) discusses the concept of otherness as it
applies in a political context, and McEwan (1996)
focuses on the gendered colonial process.

Harris (2004) discusses reasons for the rise of anti-
Americanism in the early twenty-first century.

In addition to the seminal work on geographies of
exclusion by Sibley (1995), see also Hodge (1996),
Wilton (1998), and Sibley (1999a).

Keith and Pile (1993), Pile and Thrift (1995:13–51), Pile
(1996), Longhurst (1997), Sibley (1999b), and
Kingsbury (2004) discuss selected recent
advances in psychoanalytic theory that offer
insights into the relations between the social
and the individual; there is particular focus on
object relations theory as it facilitates accounts
of geographies of difference.

Jackson (2000b) provides an account of contemporary cultures of difference in the British context, including reference to emerging hybrid
identities.

Ford (1998) reports on a British survey that determined that 3 per cent of the white-skinned population, 4 per cent of the Indian population, and
10 per cent of the black-skinned population
avoided soccer games because of the fear of violence, a fear that was also reflected in activities
such as movie theater attendance.

Hodge (1996) identifies the western part of Sydney,
Australia, as an area of neglect, disadvantage,
unemployment, lack of services, and high levels
of criminal activity, while Dear and Wolch
(1987:8–27) discuss the 'social construction of the
service-dependent ghetto'.

Dunn and McDonald (2001) describe geographies of
racism in New South Wales, Australia.

Bonnett (2000) discusses the implications of taking
whiteness for granted.

There are several social geography textbooks that
include substantive content on gender and sexuality, including Blunt and Wills (2000), Holloway and Hubbard (2001), Pain et al. (2001),
Valentine (2001), and Panelli (2004).

Bondi and Davidson (2003) discuss gender in cultural
geography.

Hanson and Pratt (1995) focus on the way local labor

markets discriminate against women, and Ehrenreich and Hochschild (2003) focus on nannies, maids, and sex workers in the emerging global economy.

Examples of feminist geographic studies of home include those by Gregson and Lowe (1995), Domosh (1998), and Ahrentzen (1997). Dowling and Pratt (1993) and Wagner (1996:80) specifically reinterpret home as a site of oppression for women.

Hallman and Joseph (1999) analyze the gendered geographies of eldercare.

Johnson (1999) discusses the gendered geographies of Australian private domestic spaces.

Winchester (1992) and Monk (1999) discuss gendered landscapes in urban areas.

Berg and Kearns (1996) and Liepens (1996) discuss gendering of the New Zealand agricultural landscape.

Valentine (1989) and Pain (1999) provide analyses linking gender and landscapes of fear, while Day (1999) maps fear in public spaces in Orange County, California, as it relates to gendered, ethnic, and classed identities. Wekerle and Whitzman (1995) include many examples of women working to make neighborhoods more women-friendly.

Bell and Valentine (1995:23) and Gregson et al. (1997:74) discuss the merits of strategic essentialism.

Bell and Valentine (1999) and Phillips (2004) review geographies of sexuality.

Boone (1996) and Binnie (1997) discuss geography and queer theory; see also Pain et al. (2001, chapter 6). Brown and Knopp (2003) discuss queer cultural geographies.

For the argument that sex is socially constructed, see Cream (1995).

Accounts of the body and performativity by cultural geographers include Valentine (1993a), Crang (1994), McDowell (1995, 1999, chapter 2), Longhurst (1997), Callard (1998), Teather (1999), Gregson and Rose (2000), Nash (2000), Hubbard, Kitchin, Bartley, Fuller (2002), Moss and Dyck (2003), and Landzelius (2004).

Symanski (1974, 1981) discusses the geography of prostitution.

Lyod and Rowntree (1978) is an early study of gay landscapes; more recent studies of gay and/or lesbian landscapes include those by Knopp (1995), Myslik (1996), Valentine (1993b, 1996), Costello and Hodge (1999), and Visser (2003).

Hemmings (2002) has studied bisexual landscapes.

Hubbard (2000) describes the way that heterosexuality has created landscapes of oppression in Western cities. Sumartojo (2004) analyzes contested landscapes with reference to gay and lesbian hate crimes.

Pain et al. (2001) provide an overview of children and the elderly in the context of social geography.

Matthews and Limb (1999) observe that cultural geographers have typically ignored children as one of the identities to be recognized and analyzed. Holloway and Valentine (2002) and Gagen (2004) provide reviews.

Two comprehensive reviews of the geographies of youth cultures are the book edited by Skelton and Valentine (1998) and that by Aitken (2001). Leary (1999) provides an Australian example of spaces created by youths.

Evans and Fletcher (2000) analyze the spatial and social distribution of the fear of crime with reference to one English city. See also Pain (1997, 1999) and Evans (2001).

Mcpherson (1998) provides a detailed sociological perspective on the social construction of the elderly.

Rowles (1978) provides a classic humanist understanding of elderly people with the key idea that they are prisoners of space.

Kitchin (1998) and Gleeson (1999) discuss geographies of disability, and Golledge (1993) describes the experiences of disabled people in landscape. The edited work by Butler and Parr (1999) includes studies of disability in varied geographic contexts.

Bastian (1975) looks at class and traditional cultural geography.

Touraine (1981) and Geyer (1996:xiv) describe various new social movements.

Pile and Keith (1997) provide examples of resistance movements, especially as they are tied to place, and Routledge (1997) discusses the subculture of resistance that emerged surrounding the construction of a freeway in Glasgow, Scotland. Cresswell (1993) discusses mobility and resistance through a reading of the book *On the Road* (see also Cresswell 1996b; McDowell 1996).

Norton (2003b) discusses competing Mormon identities in Voree, Wisconsin.

Baumann (1996) discusses the idea that societies comprise distinctly different groups of people.

Howe (1998) debates Afrocentrism.

Niezen (2004) discusses attempts to define a global identity with reference to the contrasting ideas of cultural universalism and cultural particularism.

Mitchell (2004) reviews evidence suggesting that state-sponsored multiculturalism is losing ground in several countries. Kobayashi (1993) and Hutcheon (1994) describe the official policy of multiculturalism in Canada.

Mitchell (2003b) considers the importance of studying landscapes in a political context in order to focus on issues of justice.

Kymlicka (2001) is a collection of fifteen essays concerned with minority group rights; emphasis is on the need for ethnocultural justice in a liberal democracy based on the claim that protection of individual human rights is insufficient to ensure justice between ethnic groups and that minority rights must supplement human rights.

Ignatieff (2000) discusses individual and group rights by referring to the pool table on one hand and the patchwork quilt on the other. See also Crick (1996), Henderson (1996), and Pulvirenti (1997) for consideration of questions relating to individual and group rights.

Jackson (2002) discusses two models of multiculturalism. Young (1993) and Wilson (1987) discuss some alternatives to multiculturalism.

Lee (1997) outlines multicultural education in geography in the United States.

Zelinsky (1973, 1988, 1997) describes the impacts of a national culture on landscape; much recent work, sometimes informed by postmodernism, identifies the impacts of local or regional identities (for example, Hayden 1995).

9

Living in Place

This chapter focuses on everyday lives and places, some ordinary and some extraordinary. Ideas about image and perception resurface in discussions of imagined places and vernacular regions, but the emphasis here is on the links between images and everyday lives and on current rather than past circumstances. Ideas about identity and power resurface as well, in discussions of landscapes as texts and as places of consumption and spectacle. However, unlike the earlier focus on markers of identity, the concern in this chapter is with the typically close relationships between people—both individuals and groups—and the places they occupy. Once again it is apparent that many place-to-place differences result from the way in which power is exercised through the production of knowledge and truth in the context either of a particular discourse or of competing discourses.

Some of the broad issues tackled in this chapter relate to how people understand the places they create and how they reflect themselves in their places. Throughout the reader is reminded that relationships between people and place are complex and reciprocal. Further, it is clear that people, because they vary enormously in their attitudes and behaviors, build landscapes that reflect and may reinforce these differences. Within North America, for example, consider the differences between, on the one hand, a folk-culture landscape of Amish settlement, characterized by relative simplicity, lack of change, and conservative values, and on the other hand, a popular-culture landscape such as that of Las Vegas, characterized by fluidity both

spatially and temporally. Consider also the many different ways in which people express themselves in places, for example through music, art, and literature. Certainly, the contemporary world is full of very different places associated with the very different aesthetic, social, economic, and political interests of groups of people. In general, place-to-place differences result from the abundance of cultures that are expressed in the everyday lives of people.

In this chapter you will find

- a discussion of people and place that draws on humanist approaches and on the concept of place;
- accounts of two approaches to understanding human behavior: habitat and prospect-refuge theories and the Geltung hypothesis. These accounts, which are objectivist in focus, are somewhat idiosyncratic in the context of the subdiscipline, but they are stimulating contributions that merit consideration;
- examples of imagined places. This section focuses on perception, images, mental maps, and cognitive maps, and is conceptually similar to earlier discussions of historical imagined landscapes;
- a brief account, building on the discussion of imagined places, of vernacular regions;
- a discussion of how places are reflected and created through such sources as literature and music;
- a wide-ranging discussion of places that emphasizes the idea of landscapes as texts than can be both written and read. Included in this

discussion are an account of ways in which humans express their identity through naming places, a discussion of sacred places, and a series of brief overviews of rural, urban, and national places;

- a discussion of issues relating to consumption and spectacle, with particular reference to tourism, retailing, food and drink, and music. Much of this discussion observes a distinction between folk-culture and popular-culture groups and identities;
- a brief concluding section.

Revisiting Place and People

Place

Cultural geography has long been concerned with the related concepts of landscape and *pays*, or local regions. This concern stems most notably from the Sauerian interest in landscapes in a regional context and from the Vidalian interest in *pays*. But for many cultural and also social geographers today, the preferred concept is that of place. Since being introduced in the context of humanist approaches in the early 1970s, the concept of place has generated a tremendous amount of interest, especially among those approaching the concept from feminist and postmodern perspectives.

The current interest in place persists in spite of the general weakness of the humanist tradition in geography, a weakness related to the failure to build on the logic of the social construction argument during the 1970s. In this regard, the fortunes of other social sciences are instructive. For instance, sociology, as early as the 1960s, developed an influential humanist tradition that was linked to social constructionism. Psychologists similarly introduced and applied humanist approaches. Cultural geographers, on the other hand, largely failed to pursue social constructionist logic explicitly until feminism and postmodernism provided them with the concepts and practice to take social constructionism further. In this context, recall from Box 3.2 that humanist geography explored various philosophical avenues without developing a commitment to one in particular. Thus,

humanist geography neither built on the broad interdisciplinary idea of social constructionism nor demonstrated some singular philosophical focus. Perhaps the principal achievement of the humanist approach as it developed after 1970 was a new phenomenological understanding of the geographic concept of place, an understanding informed by the idea that place is where space and time come together.

Certainly the phenomenological concept of place added another dimension to established concerns of humanist geography. Rather than using the term to refer to where something is located—as a container of things—or emphasizing visible features of a landscape, humanist geographers used place to refer to a territory of meaning and to where we live. In this sense, *place* is qualitatively different from terms such as *landscape*, *space*, and *region* in that it involves the idea of being known by and knowing others. In this tradition, places are interpreted as experiential and social phenomena: place is *intersubjective*, that is, shared in the sense that the meaning of place can be communicated to others. Relph (1985:26) explains the meaning of place in this way: landscape and space 'are part of any immediate encounter with the world, and so long as I can see I cannot help but see them no matter what my purpose. This is not so with places, for they are constructed in our memories and affections through repeated encounters and complex associations.' Understood in this way, places are created through the human occupation of space and the use of symbols to transform that space into place. And, as Box 9.1 explains, they serve to help people address both local and global problems.

HOME

An especially useful interpretation of the concept of place concerns the idea of home—seen as both a symbolic and a physical space—as a familiar and usually welcoming setting within some larger, more uncertain world. In this sense, homes may be as small as a single room in a house and as large as a regional or even national homeland.

Tuan (1991a:102) built on the popular definition of geography—the study of the earth as the home of humans—when he defined *home* as 'a unit of space organized mentally and materially to satisfy a people's real and perceived basic biosocial needs and, beyond that, their higher aesthetic–political aspirations'. This understanding of place invites consideration of the related ideas of sense of place, topophilia, and placelessness.

Although it is usual to interpret home in a positive sense, there are two issues that complicate this understanding. First, it is possible to interpret the meaning of home at the scale of the dwelling place rather differently for women and for men. Specifically, feminist geographers have raised the idea that, for women, the home is in some instances most appropriately interpreted as a center of oppression and confinement such that it may be quite unwelcoming. Women can be seen as constrained in their experiences and their behavior because men try to restrict them

to just the one world—home—while reserving for themselves the right to experience two worlds—home and outside home. Second, as evident in the earlier discussions of ethnicity and nation, the concept of place informs studies of the sense of belonging to and perhaps being excluded from a home or place.

People

The discussion of human behavior in Chapter 3 introduced two quite different conceptions of humans, with the central distinction concerning whether humans are better viewed as passive objects or as active subjects. Broadly speaking, scientific perspectives view humans as passive objects, while humanist perspectives view humans as active subjects. This simple distinction reflected the contested character of geogra-

Box 9.1 Focusing on Place

Of course, 'no one lives in the world in general' (Geertz 1996:262). Although most people are able to live in local landscapes in which they are comfortable, many minority identities are obliged to lay claims to place that compete with other more dominant claims. This simple observation has had a powerful impact on social science generally in recent years, causing several disciplines to 'discover' geography and undergo what is sometimes called a 'spatial turn'. Reflecting on this development, Agnew (1989:9) could not help but wonder 'why we should have taken so long to arrive at a *theoretically coherent* concept of place as a defining element of geography and key idea for social science as a whole'. The answer, he concluded, had to do with place being overlooked in favor of social science's traditional focus on class and community, which was in turn related to the longstanding preference given to evolutionary and naturalist emphases.

Places, as Johnston (1991:67–8) explains, structure 'how people tackle problems, both the small and usually trivial problems of everyday life and the large, infrequently met, problems which call for major decisions'. Given this importance of place, Johnston proposed a reorientation towards a concept of place that comprises the following six points:

- The creation of places is a social act; places differ because people have made them different.
- Places are self-reproducing entities because they are the contexts in which people learn, and they provide role models for socialization and for nurturing particular sets of beliefs and attitudes.
- No regional culture exists separately from the people who remake it as they live it.
- Within a capitalist world economy, places are not autonomous units whose residents have independent control over their destinies.
- Places are seldom the unintended outcomes of economic, social, and political processes but are often the deliberate products of actions by those with power in society, who use space to create places in the pursuit of their own interests.
- Places are potential sources of conflict.

Johnston's interest, as reflected in these points, is primarily with places as sites of conflict rather than with the humanist understanding of place. The conceptual basis for these ideas as they have been developed in geography and other social sciences, especially anthropology, is the postmodern concern with others. This general idea is evident in the earlier accounts of racism and of landscapes of exclusion.

phy in the early 1970s, when the positivist-inspired spatial analysis that had dominated the discipline began losing ground to other approaches, including those based on humanist philosophies. The dispute was often framed in the context of the debate about structure (society) and agency (individual), and occasionally in the comparable humanist terms of context (society) and intentionality (individual).

The conceptual underpinnings for a discussion of the model of humans have, however, shifted considerably with the emergence of postmodern concerns. Indeed, for many cultural geographers there is no longer a meaningful debate about the relative merits of more scientific and more humanist emphases, as the former are now widely viewed as lacking credibility. Nevertheless, it would be misleading to simply conclude that the humanist model of humans is now seen as correct, when in fact a revised and as yet unresolved debate has developed around the question of who or what is the subject. Thus, Philo (1991) comments that cultural geographers might profitably consider 'the temporal and spatial variations in what selves are and in how selves conceive of themselves', and this view is echoed by Pile (1993:122), who argues that a 'search needs to be instigated into alternative models of the self, as a means of understanding the position of the person within the social'.

In addressing this question, cultural geographers have turned to a variety of inspirations, but especially to some recent advances in psychoanalytic theory that offer insights into the relations between the social and the individual. Pile (1996), for example, argues that psychoanalysis provides a basis for understanding the spatiality of everyday life. Essentially, the psychoanalytic argument is that identity in place is explained principally by reference to the subconscious. In particular, object relations theory is used to unravel connections between individuals, social worlds, and material environments. A related source of inspiration from psychoanalytic theory is the work of Jacques Lacan (1901–81), which recognizes three different types of space—real, imaginary, and symbolic—and which sees the self formed in relation to otherness.

UNDERSTANDING HUMAN BEHAVIOR

At one and the same time, authors sought a law-like science of society and upheld belief in individual freedom and dignity. To put it bluntly, they denied and asserted individual freedom of action. This contradiction correlated with the subject–object distinction—the split between the knowing mind, the sphere of reason, and the object known, the sphere of physical events—embedded in Cartesian dualism. (Smith 1997:261)

Given the larger uncertainty in social science, it is not surprising that cultural geographers have struggled in their attempts to understand human behavior. Although it can be argued that geographers have had a sustained interest in seeking causes of human behavior—witness the logic of environmental determinism, possibilism, probabilism, and the landscape school, each of which assumes a particular cause of human activities and landscape change—it is appropriate to note that prior to the 1960s, geographers only occasionally acknowledged the importance of behavior in geographic studies, and that there were few theoretical or empirical advances specifically concerned with behavior. Indeed, a review of surveys of twentieth-century human geography will show that the study of behavior has not been treated as core subject matter in the discipline. For example, the encyclopedic *Dictionary of Human Geography* (Johnston, Gregory, and Smith 2000) includes only incidental references to behavior. Further, most of the accounts of the subdiscipline of behavioral geography interpret it, in retrospect, as an episode in the history of the discipline rather than as an approach that has strengthened through time and become established as a key subdiscipline of human geography.

The bulk of material discussed in this chapter builds on subjective understandings of people, as these understandings are the most popular and influential. But before moving on to these subjective approaches, it is worth noting that some cultural geographers favor more scientific approaches that focus on the behavior of people. Three such avenues are currently being explored; all three are best seen as tentative and quite controversial contributions. The first of these more scientific contributions involves applying the concepts and principles of behavior analysis to studies of landscape change. As discussed in Chapter 6, this trend is rooted in some of the behavioral geography of the late 1960s, as it highlights the importance of imitative behavior and of learning in landscape (recall as well the discussion in Box 5.5). The two other contributions, discussed in the following sections, do not focus on learning; both are original and thought-provoking contributions.

Habitat and Prospect-Refuge Theories

Habitat and prospect-refuge theories build on the claim that some aspects of human behavior result from spontaneous responses to stimuli, a claim that incorporates the suggestion that humans are innately programmed to behave in a certain way. The key argument is that some aspects of human behavior in landscape can be profitably studied with reference to animal behavior, emphasizing the importance of biological drives while downplaying the significance of human imagination and creativity. From this perspective, understanding behavior in landscape is comparable to understanding why humans mate, protect their young, and eat. This essentially Darwinian approach borrows concepts used in the study of animal behavior not to seek direct parallels in humans, but to locate clues to understanding human behavior. The approach involves two principal ideas: habitat theory and prospect-refuge theory.

Habitat theory—relying directly on studies of animal behavior showing that species seek optimal environmental conditions—'asserts that aesthetic satisfaction, experienced in the contemplation of landscape, stems from the immediate perception of landscape features which, in their shapes, colours, spatial arrangements and other visual attributes, act as sign–stimuli indicative of environmental conditions favourable to survival' (Appleton 1975b:2). This hypothesis stresses the idea of spontaneous human response to, rather than rational appraisal of, landscapes encountered. Learned patterns of behavior are considered to be secondary to inner needs. In those circumstances where there is a correspondence between human habitat and human inner needs, an aesthetic sensibility of landscape results. If such a correspondence does not prevail, then anxiety is evident. Simply put, humans seek a natural habitat.

Prospect–refuge theory moves beyond the first hypothesis in an effort to identify those innate sign–stimuli that continue to be important for humans. The basic argument is that the ideal environment is one that humans are able to retreat to in safety (a refuge in which they cannot be seen) and that provides the opportunity to observe surroundings (in other words, a prospect). As Appleton (1975b:3) explains, 'It is a part of our nature to wish to be able to see and to hide when the occasion demands, and the capacity of an environment to furnish the opportunity to do these things therefore becomes a source of pleasure.' Prospect has to do with perceiving, with gaining information, while refuge has to do with shelter and seeking safety. Simply put, humans have an innate desire to see without being seen.

Using these two ideas to inform analyses of landscapes involves classifying both physical and built landscape features that serve as habitats in terms of their prospect and refuge characteristics. Thus, in a comparison of British and Australian symbolic landscapes, Appleton (1975b) stressed the importance of such physical characteristics as light, landforms, and vegetation, along with such elements of the built landscape as houses and settlements.

The Geltung Hypothesis

The Geltung hypothesis, introduced by the distinguished cultural geographer Wagner, also incorporates the idea that humans are innately programmed to behave in a certain way. Consider two important questions. First, why do we behave as we do? Second—and more specifically—how might we begin to behave more appropriately, both toward each other and toward the environment? Various proposed answers to these questions were noted in the Chapter 4 discussion of environmental ethics, but the Geltung hypothesis produces a rather different type of answer, outlined here.

The Geltung hypothesis is based on the original claim that 'human beings are innately programmed to persistently and skillfully cultivate attention, acceptance, respect, esteem, and trust from their fellows' (Wagner 1996:1). In introducing the hypothesis, Wagner (1996:12) explained: 'I shall contend that human behavior responds universally to the interpersonal effectiveness in communication that I have termed Geltung and that the latter reflects a special feature acquired through evolutionary selection'. In short, we are born to show off, to strive for what is labeled *Geltung*.

Wagner developed the hypothesis from ideas about diffusion and communication, recognizing that the 'opportunity for transmitting an (always innovative) message with success depends upon the Geltung of its sender compared with that of the recipient' (Wagner 1988:190). In this respect, recall the reference in the Chapter 5 discussion of cultural diffusion to variations in innovativeness, especially as this involves the presence of opinion leaders. It is personal Geltung that is the principal explanation for our social relationships, integrating individuals as members of larger groups. It is personal Geltung that explains our activities in environment. It is personal Geltung that allows humans to both co-operate and compete with others in order to attain positions of influence and power. Conceptually most interesting, it is personal Geltung that clarifies the meaning of culture as the dynamic consequence of a series of imitative diffusions.

Given the basic argument that humans are genetically programmed to behave in a particular way, this hypothesis is quite different from most current social science approaches to the understanding of human behavior. The idea that humans put on displays and that other people are important to us is a well-established one in some social psychological theory, but the suggestion that such behavior is instinctual, a part of our animal heritage, is relatively novel.

Pause for Thought

Wagner's elaboration of the Geltung hypothesis leads to a series of prescriptive recommendations having to do with the need to respect the importance of place, respect the importance of people, expose and deconstruct vanities,

challenge spatial monopolies of power, oppose war, enhance personal development, respect environment, and modulate population processes. Given these implications, it is clear that the hypothesis introduces an argument that merits the attention of cultural geographers, not simply because it has been proposed by a distinguished practitioner but because it includes some stimulating suggestions about, first, how humans behave toward each other and how humans treat the earth, and second, how humans ought to behave toward each other and how they ought to treat the earth. As this book consistently notes, these are central issues in cultural geography.

Imagining Places

Rather than pursuing scientific approaches, such as those discussed above, to understanding people and place, most cultural geographers favor humanist, feminist, and postmodern methods. The remainder of this chapter reflects work along these lines.

Images and Mental Maps

How do we see and comprehend the world? To help answer this question, recall the following statement by Humboldt: 'In order to comprehend nature in all its vast sublimity, it would be necessary to present it under a twofold aspect: first objectively, as an actual phenomenon, and next subjectively, as it is reflected in the feelings of mankind' (von Humboldt, in Saarinen 1974:255–6). Following postmodern logic, many contemporary cultural geographers reject the claim that it is possible to achieve knowledge of an objective world, but there is widespread agreement on the need to understand nature and places as people interpret them.

The single most important concept in behavioral geography as it developed beginning in the late 1960s is likely that of environmental **perception**, applied in conjunction with the related concepts of image, **mental map**, and **cognitive map**—terms that were not clearly distinguished at the time they first came into use. Cultural geographers took an immediate interest in these ideas, with Mikesell (1978:6) noting that, for cultural geographers, the 'recent

LIVING IN PLACE   339

development of greatest potential interest has been the proliferation of work on environmental perception'. With a clear focus on subjectivity and the role played by both individual and cultural appraisals, the essential logic of perceived environments, images, and mental maps is certainly most attractive to the cultural geographer. The concept of a subjective, imaged environment different from the presumed objective, real environment was fundamental to the development of a humanist-inspired behavioral cultural geography, and cultural geographers were encouraged to turn to such books as *The Image* by Boulding (1956) for conceptual content and *The Image of the City* by Lynch (1960) for both conceptual and empirical content.

The concept of cognitive map was derived from earlier work in psychology. Downs and Stea (1973:7) defined cognitive mapping as a 'process composed of a series of psychological transformations by which an individual acquires, stores, recalls, and decodes information about the relative locations and attributes of the phenomena in his everyday spatial environment'. This concept raises some important questions, especially concerning whether or not people represent their environments as maps. Evidence from psychology is contradictory on this question, with descriptions of human spatial memory as both propositional and analogical in form (*propositional* refers to the idea that knowledge of place is stored in lists, and *analogical* refers to the idea that the knowledge stored corresponds directly to the depicted objects).

Despite its popularity among cultural geographers, the approaches based on environmental perception proved somewhat controversial. In particular, Bunting and Guelke (1979:453) advanced important criticisms of two of the fundamental claims of this type of behavioral geography. They questioned the claim that 'identifiable environmental images exist that can be measured accurately' and the claim that 'there are strong relationships between revealed images and preferences and actual (real-world) behavior'.

It is fair to say that the current status of humanist-inspired behavioral geography is uncertain. In cultural geography there has been much interest in this more humanist version of behavioral geography, although most of the principal analyses date from the 1970s and 1980s rather than from more recent years. Another approach to the study of subjective environments focuses on vernacular (often called perceptual) regions; this is the subject of the following section.

Vernacular Regions

Recall the definition of vernacular cultural regions in Box 6.1. Vernacular regions are areas that are presumed to have a regional identity and that are the product of the spatial perception of average people. They may exhibit some distinct visible material landscape, but their distinguishing characteristics have more to do with a perceived sense of identity than with a visible sense of identity. Not surprisingly, there is much overlap between formal and vernacular regions, with many cultural regions being both formal and vernacular. Many small, relatively local tourist areas qualify as vernacular regions.

Some vernacular regions, such as the French Riviera, comprise what might be called elitist space. The emergence of a distinct French Riviera came as a response to the requirements of a particular class in French society and their desire to identify with a given environment. In much the same fashion, some urban neighborhoods and communities might be regarded as vernacular regions created by particular groups to help express their identity. Most generally, vernacular regions are named places, and it is the fact of being named that implies that the identity of the place is important, that the area has meaning to the occupants. But some vernacular regions demonstrate an especially clear link between people and place. For instance, Miller (1968), in a study of the Ozark region of the United States, used folk materials to delimit a distinct region characterized by the predominant way of life, the self-sufficient family farm. The inhabitants of a vernacular region may thus share distinct values and an especially clear link between people and place is implied.

If we think of vernacular landscapes as regions that are perceived by those both outside and inside the region, then it seems likely that these might be marked in many different ways. In North America, billboards are regularly used to identify characteristics of a place and the people who live there, for example affiliation with a local sports team or a religious identity. Those who wish to identify with a particular perspective on a controversial moral question can also claim landscapes. The billboard shown here (*above*), located in rural northern Wisconsin, represents the expression of moral opinion and helps identify the population of the area as a group of people sharing some fundamental beliefs. Such statements are not uncommon in the American rural landscape, and cultural geographers might map these landscape features as one means of facilitating the recognition of vernacular regions (William Norton). Some of the best-known vernacular regions are those created by, or at least reinforced by, tourist boards promoting a place for largely commercial reasons. Such identifications are evident in the visible landscape, often through billboards again or through informative directional signs that highlight the attractions of, for example, 'Sunset Country' or 'Land of 1000 Lakes'. They are evident also in pamphlets and other tourist literature. Increasingly, a major vehicle for promoting a region as a consumption site for tourists is the World Wide Web. The example shown here (*below*) appears at the top of web pages promoting 'Custer Country' in southeastern Montana. The pages are authored by Custer Country Tourism Region, located in Forsyth, Montana; the address is <http://custer.visitmt.com/contact.htm>. A principal attraction in the region is the Little Bighorn Battlefield National Monument, site of 'Custer's Last Stand', where Lt. Col. George Custer led the 7th Cavalry against a force of Plains Indians headed by Sioux chiefs Crazy Horse and Sitting Bull. Other attractions in the region include a sandstone butte known as Pompeys Pillar National Monument, where Captain William Clark carved his name along with the date, July 25th 1806. The promotional image on the web pages highlights these various attractions through a collage accompanied by an invitation: 'Custer Country, Montana. It's where legends such as Custer and Lewis & Clark roamed. And it's where today's adventurers continue to roam.'

Because vernacular regions are largely self-defined—both by those inside and by those outside the region—many attempts at delimiting vernacular regions are based on data collected from individuals, with the idea that if the goal is to understand regions as they exist in the minds of the people, then one must ask those people. Thus, Jordan (1978) gathered data from 3,860 students in an analysis of regions in Texas, and Lamme and Oldakowski (1982) collected responses from 356 people attending the Florida state fair in a study of regions in Florida. The most extensive data collection of this kind was conducted by Hale (1971, 1984) involving 6,800 responses from weekly newspaper editors, county agents, and postmasters.

Although the results of these studies are suggestive, the validity of such data-gathering exercises is debatable, as the samples, though large, may not be representative of entire populations. They are often restricted to a particular subset of people, such as students. A different procedure for collecting data involves calculating the ratio between a specific regional term, such as 'Dixie', and a term such as 'National' or 'American' as these appear in the names of businesses listed in telephone directories, with the assumption that the greater the relative incidence of the regional term, the more likely that the location belongs to the region (see Figure 9.1).

Pause for Thought

In addition to region delimitation, the exercise of identifying vernacular regions may produce some useful generalizations about regional identity. There is a tendency to reflect values in regional commercial names: for example, there appears to be a Texan preoccupation with adjectives such as 'Big' and 'Golden', and there is a specific inclination among businesses in the Appalachian Mountains to avoid the term 'Appalachia' because of its association with regional poverty and peculiar folkways. Comparisons of

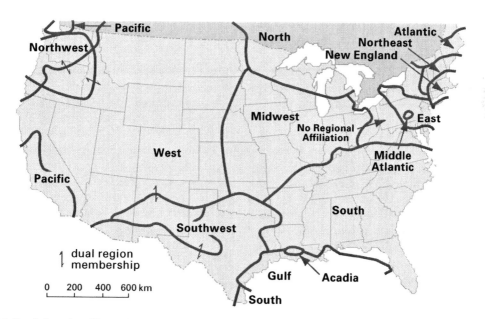

Figure 9.1 North American Vernacular Regions. Zelinsky presented this map of fourteen subnational vernacular regions as a summary statement. Note that two areas are given dual region membership. The correspondence between this map and the map of cultural regions as distinguished using the concept of first effective settlement (also presented by Zelinsky and included as Figure 6.4) is unsurprising, although six of the vernacular regions are locational and not cultural in character (Atlantic, Gulf, North, Northeast, Pacific, and East). A study of regions in Kansas combined the same procedure—that is, comparing the incidence of regional and national terms—with questionnaires (Shortridge 1980).

Source: W. Zelinsky, 'North America's Vernacular Regions', *Annals of the Association of American Geographers* 70 (1980):14. (Reprinted by permission of Blackwell Publishing)

vernacular regionalizations with other attempts at delimiting formal cultural regions often show a close similarity between the two. More generally, vernacular regions may have differing levels of intensity: in the United States, for example, there appear to be much stronger regional affinities in the Southeast than in the Northeast. Certainly, recognizing that a given area has a particular identity by designating it with a name implies that the identity is culturally important and that the area has meaning to the occupants.

APPALACHIAN COGNITIVE MAPS

An attempt to delimit an Appalachian vernacular region used the strategy of gathering individual cognitive maps and then producing generalized maps for each of a number of groups (Raitz and Ulack 1981b). The term 'cognitive map' as it is used in this research is analogous to the term 'mental map' as used in much other research in that it refers essentially to a place-preference surface. The study was not designed to carry out any detailed investigation of the psychological concept of cognition, but the results reveal a great deal about the importance of location as a factor affecting spatial cognition.

Cognitive maps were acquired from 2,397 students at 63 colleges in and close to the American Appalachian region. Respondents were asked to outline Appalachia on a base map that included only state boundaries and names. Composite maps were then produced for each of two groups in each state: one group included those who lived in Appalachia and the other those who lived outside Appalachia (in the case of West Virginia, there was only one group, as the entire state is within the region). The regional delimitation on which these distinctions were made was a conventional physiographic demarcation.

For each group, the composite map produced showed the core, defined as that area identified by 80 per cent or more respondents, and the larger Appalachian region, with the regional boundary being the 20 per cent isoline. Most of the composite maps indicated that groups shifted both the core and the regional boundary toward their home area, and that

there was a consistent decline in recognition of the region with increasing distance—what might be called cognitive distance decay. Respondents from the central area of Appalachia tended to increase the size of the core and, relatively, to reduce the total area. Figure 9.2 shows the core of Appalachia (defined as the most frequently occurring location identified as being in Appalachia) for a selection of the groups identified in the analysis.

Overall, the results of this research suggest that proximity affects cognition, and that groups residing in a region are likely to have a significantly different view of a place than groups residing outside the region. To properly understand these kinds of cognitive maps, it is clearly helpful to distinguish between the perceived environment that is accessible and known, and the presumed or inferred environment, which is some larger area about which detailed knowledge is lacking. In this context, it is useful to distinguish between those areas that are familiar, where individuals can rely on cognitive representation, and those that are less familiar, where external information is needed.

Two of the most important issues that arise from this analysis are the question of the origin of regional stereotypes and the implication that differing cognitive maps may have for behavior. For Appalachia, the former issue was addressed by Hsiung (1996) using a theoretical framework to examine the internal connectedness of the region and external links to American society more generally. The latter was considered, in a historical context, in the discussions of the Swan River Colony, the Great Plains, and the Canadian Prairies in Chapter 5.

Imagining Through Literature

Since first being introduced as a source of geographic understanding by regional geographers in the early twentieth century, creative writing has generated an uneven amount of interest. By the 1970s, humanist geographers were attracted to the idea that literature 'calls up within the reader essential images of the world' (Salter and Lloyd 1977:1), while more recently, cultural geographers inspired by the cultural turn have taken

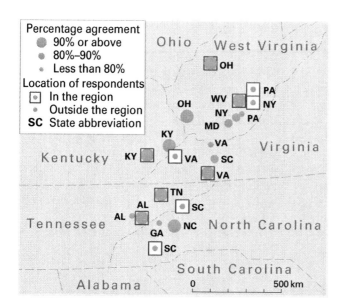

Figure 9.2 The Core(s) of Appalachia. This figure shows the location of the Appalachian core for selected state groups. Note the differences in core location as identified by those groups comprising respondents resident in the region and those groups comprising respondents from outside the region, and also the tendency for the core to shift towards the state in question. These results suggest that most individuals relate to an environment that is too complex to permit any full understanding, that near areas are better appreciated than distant areas, and that individual members of relatively cohesive groups share similar cognitive maps. Overall, cognitive map distortion reflects the preferences, attitudes, and values of group members.

Source: Adapted from K.B. Raitz and R. Ulack, 'Cognitive Maps of Appalachia', *Geographical Review* 71 (1981):209. (© The American Geographical Society; reprinted with permission)

an interest in literature as a 'material artefact which fulfils a role designated by its position in various social and economic processes' (Sharp 2000:327). Academic uses of literature provide insights into the worlds, both people and place, that are imagined or created by writers.

From both a regional and a humanist perspective it can be asserted that literature contains universal truths that speak to the human condition, such that no other sources of knowledge are necessary. As Salter contends,

There is no need to write additional textbooks in cultural geography. All the messages of the profession are already committed to ink. The motivations, processes, patterns and the consequences

of human interaction with the landscape have all been discovered and chronicled with grace and clarity. Authors dedicated to the comprehension and elucidation of order within the overtly haphazard flow of human events have given academics the materials needed to profess the patterns which illustrate this order. We fail, however, as scholars to make adequate use of these data for the simple reason that this material is labelled 'fiction'. (Salter 1981:142)

Salter's claim may be over-ambitious, but it does capture the motivation and spirit of much of the work in geography that uses literature to help capture the essence of place. In Britain, places studied include Thomas Hardy's Wessex, the Brontes' Yorkshire, D.H. Lawrence's Nottinghamshire, James Joyce's Dublin, Arnold Bennett's Potteries, Charles Dickens's London, and Jane Austen's rural England. For all of these authors, the setting is integral to the story being told, not merely an incidental backdrop. Readers learn not only about the pleasures and perils of London from Dickens and about the green and pleasant landscape of Wessex from Hardy but also about social relations and working conditions in these areas during times of significant change. More recently, Catherine Cookson and James Herriott respectively have made Northumberland and the Yorkshire moors and dales popular tourist areas through their vivid evocations of these places. In Canada, Stephen Leacock nostalgically sees small-town Ontario as a haven from the hustle and bustle of the big city, while for W.O. Mitchell, the small Prairie town is a refuge from the harsh and unforgiving environment beyond. In a rather different manner, John Steinbeck's *Grapes of Wrath* provides a wealth of geographic insight into the experience of human mobility from place to place.

LITERATURE IN CONTEXT

A possible difficulty with geographic studies based on literature is that they tend to treat literary source material as unproblematic, as being in some way able to reflect the essence of place truthfully. Such an assumption is dangerous, as our earlier discussions of situated knowledge (see Glossary) and writer positionality made clear. Literary texts are no different from academic texts in that both are part of larger discourses that are imbued with power relations. Literature needs to be interpreted with an awareness of both the author's biases and the circumstances of the time. It is not possible to assess novels, or other creative writing, for use in geographic studies without taking context into account. Box 9.2 expands on this idea with a discussion of two English poets, William Wordsworth (1770–1850) and John Clare (1793–1864), whose works offer insights into both the poetic imagination of landscape and the way in which poetry might reflect larger social, political, and economic contexts.

Inevitably, as part of a larger discourse, literature both reflects and creates people and place. For example, novels written for boys during the late Victorian and Edwardian periods 'leave us in no doubt as to what manly boys were supposed to think and how they should act' (Hugill 1999:335). Imperialism, racism, social advancement, new technologies, individualism combined with team spirit, Christian values, and chaste sexual behavior were all character traits valued in young men and promoted in the literature of the time. Contemporary literature for young people reflects a different set of values, tending to emphasize social equality, respect for others, and environmental awareness.

The way that literature is necessarily bound up with prevailing attitudes and behaviors is exemplified by the writings of authors such as the late nineteenth–century British writer H. Rider Haggard. Haggard's fiction features soldier heroes that offered readers the 'assurance of a clearly recognizable gender identity and, through this, the security of belonging to a gendered national collectivity that imagines itself to be superior in strength and virtues to others' (Dawson 1994:282). Through the construction of an imagined Africa Haggard's novels also rein-

Box 9.2 Poetic Imaginations—Wordsworth and Clare

A leading figure of the English Romantic movement, Wordsworth wrote many poems concerned with nature and landscape, typically portraying humans and nature as unified. For example, in 'The Thorn', he compares a thornbush, in its size and appearance, to a two-year old child: 'No leaves it has, no thorny points; / It is a mass of knotted joints'. Similarly, in 'Michael, a Pastoral Poem', which includes the lines 'Upon the Forest-side in Grasmere vale / There dwelt a shepherd, Michael was his name', the title character makes his home the natural place of the forest side. In 'I Wandered Lonely as a Cloud' the title—also the sonnet's first line—identifies the poet with nature in the form of a cloud. The first verse describes a landscape made up of 'A host of dancing Daffodils; Along the lake, beneath the trees, / Ten thousand dancing in the breeze'. These lines are imaginative rather than faithfully descriptive, as evidenced by his sister Dorothy Wordsworth's journal entry for 15 April 1802:

When we were in the woods beyond Gowbarrow park we saw a few daffodils close to the water-side. We fancied that the lake had floated the seeds ashore, and that the little colony had so sprung up. But as we went along there were more and yet more; and at last, under the boughs of the trees, we saw that there was a long belt of them along the shore, about the breadth of a country turnpike road. (in Keith 1980:14)

Wordsworth thus transforms the reality of at first a few and only later many daffodils, to 'When all at once I saw a crowd, / A host, of dancing daffodils'. The landscape is at least partly imagined, an interpretation of reality, rather than an attempt to faithfully reflect reality.

In addition to his poetry, which established his reputation as one of the greatest English poets, Wordsworth wrote several prose pieces about his

forced the key Victorian British perception of Africa as the other, reflecting the prevailing view of the 'dark continent' as the 'White Man's Burden' in an age of imperial expansion. MacDonald (1994:113), outlining the invention of Rhodesia by the British in the late nineteenth century, notes that 'the historical facts of conquest were rearranged, dramatised, and mythologised almost as soon as they happened, epitomising, for imperial apologists, the triumph of civilisation over barbarism'. It was precisely this image of Africa as a savage land that was perpetuated by Haggard. But the Victorian British also maintained a romantic image of Africa, as seen in both fiction and exploratory literature of the time, which contains descriptions of Africa's natural beauty, limitless natural resources, and peaceful population.

Recently some African authors have produced influential writings critical of African people and places. One of these, Ngugi wa Thiong'o (b. 1938), is a black African writer from Kenya. His father was a dispossessed peasant farmer, his mother was arrested and tortured in 1955 by the British, and one brother was a guerilla fighter during the 1950s. In 1977, Ngugi was arrested and detained without trial and without any formal charges for almost a year. During this time he wrote *Matigari* (1980), a highly acclaimed novel about searching for the past and for truth. The central character of this tragic novel, a former freedom fighter, searches for his children, and his people, only to find that his people have changed beyond recognition, suffering more after independence than during the colonial period.

J.M. Coetzee (b. 1940), a white Afrikaner South African writer, is a liberal highly critical of the racist policies implemented during the apartheid era, who has written intellectual challenges to the oppression and violence inherent in the apartheid system. His novel *Foe* is a rewriting of *Robinson Crusoe* that speaks about those voices that are silenced under apartheid. Although Ngugi and Coetzee use very different writing strategies to express their criticism, both make uncovering corruption, injustice, violence, and searching for truth central themes of their work. Ideas about place are important to both authors as they are especially concerned with

home area, the Lake District in northwestern England. His *Guide to the Lakes* (first edition in 1810 and a definitive edition in 1835) is in many respects similar to the wrings of von Humboldt. It includes a portrait of a dynamic physical landscape changing through fluvial activity and description of a local region—*pays*—resulting from human occupance.

Clare wrote about his home area, the village of Helpstone in the rural English Midlands, during a time of substantial landscape change from an open field system of landholding to an enclosed system, a process that meant removing trees and planting hedgerows to delimit individually owned areas of agricultural land. In 'Helpston Green', Clare writes: 'Ye injur'd fields ye once were gay / When natures hand displayed / Long waving rows of willow grey / And clumps of awthorn shade / But now alas / awthorn bowers / All desolate we see / The woodmans axe their shade devours / And cuts down every tree'. Clare expresses concern over this development but is unable to prevent change that is part of a larger

economic and social transformation. An especially compelling example of his concern is a journal entry in which he describes the destruction of a favorite whitthorn tree in terms suggestive of the loss of a truly meaningful place, indeed even of a home (Bate 1991:106–107).

Although both poets offer intimate descriptions of and insights into landscape, they do so from rather different vantage points. Wordsworth was well educated and wrote for a privileged class, whereas Clare was born into a poor family and worked at a series of rural laboring jobs while writing his poetry. Wordsworth viewed the Lake District as being only for those who could appreciate the picturesque, but for Clare the Helpstone area was home and workplace. Both poets were representative of particular classes, and their poems need to be read with these backgrounds in mind. Of course, neither poet is able to provide objective accounts, as both are emotionally bound up with the places they are imagining and describing.

the way in which the places they occupy are being abused. Furthermore, both convey the passion shown for people, place, justice, and truth in powerful and compelling fashion.

Box 9.3 takes ideas discussed in Chapter 8 about identities based on resistance and applies them to larger society along more clearly humanist lines, recognizing that one of the basic means by which disadvantaged groups might articulate resistance is through literature and other artistic endeavors.

This account of literature reflects but one aspect of the interest being shown in non-traditional source materials. Cultural geographers may also refer to advertisements, magazines, photographs, film, art, cartoons, websites, and other evidence. Analyses of these sources often refer not only to what is represented but also to what is omitted, since this can provide information about the use of power as it plays out in inequality and oppression. Of particular interest as a source material and a subject for study is music.

Music, Identity, Place

Studies of musical styles associated with particular cultural groups and their places are abundant in the cultural geographic literature, dating back to the late 1960s. It has been suggested that there is a continuing need for cultural geographers 'to open their ears to the auditory components of culture—the sounds of people and places' (Carney 1990:45). An edited volume concerned with American folk and popular music incorporated three basic research themes—regional and ethnic studies, questions of cultural hearths and cultural diffusion, and the role of place—and included examples of such musical styles as country, gospel, jazz, rock, folk, Latin, blues, and zydeco (Carney 1994a).

The earliest analyses of the geography of music were informed by the Sauerian tradition. The most researched musical style is country, and among the least researched are the various forms of military, classical, and religious music. The focus on country music is understandable given the especially close links between this kind of music and local cultural groups and the places they occupy. The diffusion emphasis has considered the spatial spread and temporal growth of such indicators as music festivals, music contests, and radio station programming. Increasingly, however, there are studies of the

Box 9.3 The Soul of Geography

Humanist cultural geographers study places and people as they are represented in various fields of human endeavor, notably literature, art, and music. Most of these studies are based on the premise that impersonal, objective research cannot meaningfully reveal the experiences of identity and place; only an astute observer or, especially, a participant in the making of that identity and place is able to achieve such understanding. But cultural geographers are unable to experience everything that they might choose to research, especially the lives and places of those who live outside the mainstream, who lack power, or whose identities and places are not valued by the majority. As a result, some researchers investigate the way that people express their feelings about identity and place through their artistic activities. Indeed, for Macphail (1997:36), the 'entire spectrum of artistic human endeavour has the potential to be embraced by humanistic geography', not only

popular sources such as novels and landscape painting, but also poetry, music, sculpture, and dance.

Responding particularly to the suggestion that some cultural geography was failing to represent many ordinary people, cultural geographers have increasingly turned their attention to literary and other sources. Their aim is to allow those whose experiences are not typically heard to become participants in research by looking at how these people are represented in literature and other fields of human endeavor, and by studying their literature and other artistic works. For example, in an overview of Canadian prairie literature, Avery (1988:272) stresses that a careful reading of certain texts can help in understanding the geography of the region and especially the contribution made by women to that geography: 'The female voice may not have had an impact on the actual settlement of the prairies, because of the relatively powerless position of its

geography of music that are informed by the cultural turn.

Many musical styles have local or regional origins and reflect a particular place identity, hence the concern with place, diffusion, and the images conveyed by music. Some well-known American examples of place-based musical styles are those of New Orleans jazz, Memphis blues, the Detroit Motown sound, and Nashville country, but there are also cultural geographic analyses of, for example, Pacific Northwest rock from 1958 to 1966, 'indie' music in Manchester, England, a single Singaporean artist, and western North Carolina bluegrass music, all of which stress how regional identity is able to play a key role in the evolution of musical style. In some cases music serves as an important marker of difference, one means by which people are able to assert an identity, while in other cases music serves to break down barriers as exemplified by the idea of ethnic ambiguity. Both of these circumstances are especially evident in the case of popular music styles favored by young people that often arise outside of official institutions.

Of course, recognizing the role that ethnic identity and place can play in the evolution of a particular musical style is not to deny the often overwhelming importance of commercialism and global marketing in the music industry. Much popular culture is part of what might be considered global culture. Indeed, performers that are generally associated by consumers worldwide with a specific place may not have been readily embraced by that place; according to Cohen (1997), for example, this is the situation with the Beatles and Liverpool. In most cases of artists who have achieved international success, there is no real association with place, except perhaps for a national identity, and this situation, as with other cultural phenomena, may result in part from globalization processes, which do not serve merely to eradicate difference and introduce some dominant style, but to aid in the dispersal of previously local styles.

Box 9.4 discusses two related themes in the cultural geographic study of country music, namely the origins and diffusion of bluegrass and the rise of Branson, Missouri, as a country music center. The analyses of these themes share a concern with the relevance of place and local identity, but also need to be understood with reference to larger commercial strategies.

speakers at the time of settlement, but it can be heard in prairie literature.' Of course, such sources are recognized as being but one representation and as potentially misleading.

Another discussion of the need to acknowledge literature as an important source of information emphasizes the value of poetry as a way of listening to those who are oppressed and not normally heard. The need to listen and respond to voices that are not a part of the mainstream prompted Watson (1983:393) to assert that 'the soul of geography is the geography of the soul' and to describe the poetic expressions of the polluted landscape of the mining settlement of Kirkland Lake, Ontario, and of the oppressed lives of mine workers as 'primal knowledge' (398). These thoughtful claims must be understood in the context of humanist interests rather than in terms of any involvement with the cultural turn.

Similarly, Macphail's humanist study of Soweto poetry during the apartheid era—Soweto is the major Black residential area outside of Johannesburg—emphasizes the inability of Whites to understand the Black experience, such as the ongoing fear, the lack of a sense of belonging as a result of some of the apartheid legislation, and most critically the dehumanizing effects of that legislation. This poetry is a form of resistance, and White academic authors 'cannot capture the same experiential passion which black urban residents have embodied in Soweto poetry' (Macphail 1997:40). Similarly, with reference to ethnic groups in large urban areas Watson (1983:397) asserts:

Their eye may have been full of prejudice, and therefore on a purely impersonal, objective basis, not to be trusted, but that prejudice dominated the places they valued, sought, fought for, were displaced from, or held. This should be an axiom in geography: People generate prejudice; prejudice governs place.

During the late 1970s and 1980s, punk was an influential musical style that allowed a predominantly White subculture to express its difference and otherness within the larger White population. Originating in London, New York, and Brisbane, the music was characterized by harsh, offensive lyrics that attacked conventional society and popular culture through expressions of alienation and anger. Often intending to shock, groups adopted such names as the Sex Pistols and the Dead Kennedys, wrote and sang profane lyrics, and entertained fans with confrontational stage acts. Fans adopted the fashions favored by performers, such as intentionally mutilated clothing and exotic body presentation that included use of cosmetics and Mohawk hairstyles. The music and related fashions diffused widely and rapidly, helped by a thriving underground press; the photo here, taken in the 1980s, shows youths in Tallinn, Estonia (then part of the USSR). (Peter Turnley/CORBIS Canada)

MUSIC AND NATIONALISM

Recall the idea of nations as imagined communities. Music is one of the ways in which national identity is imagined and cultivated, with the most obvious examples being anthems and songs of national praise. Both the British national anthem and the French 'Marseillaise' date from the eighteenth century, and both captured ideas about identity and patriotism during crucial periods of national identity formation. More generally, nationalist aspirations have been, and continue to be, played out in larger musical worlds. Handel's oratorios are widely regarded as the first real expression of British musical nationalism, having been used to stir up patriotic feelings during the Jacobite Revolution (1745–6) and the Seven Years War (1756–3)—this despite the fact that Handel was born in Germany. The oratorios, which were about the victory of Israel over various enemies, were interpreted as symbolizing Protestant triumph over Catholicism.

Oppressed groups that aspire to national status favor music that recognizably reflects and reinforces cultural identity, with an early example being the rise of eighteenth-century Scottish songs and fiddle music quite different to the Italian style that dominated much of Europe, including Britain, at the time. Probably the most successful example of aspiring, or resistance, nationalism, so successful that it transformed into hegemony nationalism, was that of Germany. Folk tales were used in an explicitly nationalist fashion by Weber in the 1820s and later, and most effectively, in Wag- ner's operas. By the end of the nineteenth century, German musical styles dominated much of Europe, in turn sparking many other aspiring nationalisms such as those in Bohemia, Norway, and Spain.

Box 9.4 Country Music Places— Bluegrass Country and Branson, Missouri

There is a substantial body of cultural geographic writing concerned with country music. Much of this literature is concerned especially with origins in particular folk lifestyles, with mechanics of diffusion, and with the cultural infrastructure that facilitates diffusion. There are also analyses of links between the music industry and tourism as these relate to urban growth and larger commercial issues. The American cultural geographer George Carney is the most prolific writer on these issues, and this box summarizes two of his studies that reflect these interests.

Studies of origins and diffusion of particular musical styles tend to be framed explicitly within the Sauerian diffusion tradition. In the case of the distinctive bluegrass sound, the geographic origin or culture hearth has long been debated, with traditionally suggested source areas including the Pennyroyal Basin of western Kentucky in Appalachia and the Bluegrass region of western Kentucky. However, it is likely that the true source area is the mountain and piedmont region of western North Carolina during the period from about 1925 until 1965. Groups and individual musicians in this region—Carney (1996:66) referred to 'human innovators and place incubators'—pioneered the use of the distinctive combination of instruments, playing techniques, and vocal styles that together make up the bluegrass sound. The period of innovation was followed by growth during which bluegrass spread locally through such mechanisms as homes, schools, and churches, and regionally through fiddle contests, music festivals, radio broadcasts, and recordings.

The story of Branson, Missouri, 360 miles (579 km) west of Nashville and the new 'Mecca' of country music, is one of a fascinating mix of local country music activities and the larger commercial world of tourism and consumption. As a small resort town located in the Ozark country music region, Branson in 1959 began to host local bands to perform evening shows for tourists, and the town gradually emerged as a center for showcasing country music talent. There was a close association between this growing musical business and the established tourist attractions in and adjacent to the town, with performances held at the Silver Dollar theme park and a local commercial complex of caverns. There was also a steady growth of theaters along Highway 76, the main street known as '76 Country Music Boulevard' (Carney 1994b:20). By 1981, when the Ozark Country Jubilee moved to Branson from Springfield, Branson had become established as a major regional center with related resort and musical tourist attractions.

The transformation of Branson into something more than a regional center began in 1983, when a major national recording star, Roy Clark, opened a theater on the main street, which subsequently served as a venue for other nationally known performers. Branson proved attractive for major stars for several reasons: it offered them the opportunity to maintain a semipermanent location (in addition to the major center of Nashville) that audiences came to rather having to visit audiences through a series of one-night stands at various locations; the Branson theaters were seen to be healthier and more secure venues than many of those encountered during tours; the Ozark region was an attractive place both physically and culturally; and, for many performers interested in settling, Branson offered excellent opportunities to invest in land, facilities, and media. During the 1990s, the style of Branson country entertainment broadened, as many younger artists were drawn to the new theaters with larger capacities.

Branson today is more successful in the group tour market than is Nashville as a result of the small-town atmosphere that appeals to many fans, the large number of theaters and shows that are in close proximity, and the low cost of food, lodging, and entertainment. Effective marketing has, of course, also played a large role in Branson's development.

Writing and Reading Places

There is today a persistent concern with landscapes as places, as containers of meaning. As we have seen, the general acceptance within twentieth-century North American cultural geography of Sauer's definition of landscape as the cultural transformation of the natural world weakened after about 1970, owing in large part to the increasingly diverse scholarly interest in landscape as place. Use of the term 'explosion' is doubtless something of an exaggeration, but there is certainly an ongoing and seemingly ever-growing concern with the meanings and symbolism of cultural landscapes, both within and beyond cultural geography. As Thompson (1995a:xii) asserts, 'It is important to understand that landscape—as revealed in place—is not the province of one, two, or three academic disciplines, but is the concern of at least a score of art forms and academic fields'. Or, as expressed by Lowenthal (1997:180): 'Suddenly, landscape seems to be everywhere—an organizing force, an open sesame, an avant-garde emblem, alike in fiction and music, food and folklore, even for professors and politicians'.

Social scientists have discovered place often as a consequence of particular readings of post-

modern theorists, and Yaeger (1996:18) was prompted to ask: 'Why are scholars from a range of disciplines suddenly reinvested in the energetic pursuit of geography?' Wilson (1998:xvi) identified a postmodern inspiration for what was called the new regionalism in history when he claimed that cultural identity 'hardly exists apart from social relations in specific places and contexts'. In sociology, Cuba and Hummon (1993:112) observed that 'place identities are thought to arise because places, as bounded locales, imbued with personal, social, and cultural meanings, provide a significant framework in which identity is constructed, maintained, and transformed'.

This popularity of landscape and place studies is related to humanism and the cultural turn, as these prompted sociologists and other scholars to turn from studies stressing culture as cause to studies emphasizing identity and symbolism. As we have seen, the superorganic claim that cultures authored landscapes was challenged by Duncan (1980) and replaced by an abundance of other interests. Landscape studies since about 1970 have covered a range of ideas, three of which come up in the following discussions. First, Cosgrove (1985, 2003) proposed an influential humanist perspective of landscape as a particular 'way of seeing'; this implies that there are also other ways of seeing, leading to the suggestion that we might talk about the 'duplicity of landscape' (Daniels 1989). Second, Duncan and Duncan (1988) identified the important role of power relations in shaping landscapes, which could be understood as texts and as outcomes of discourses. Third, Cosgrove (1997a) argued that landscapes could also be viewed as theater, spectacle, or carnival. Commenting on these three developments, Demeritt (1994:167) jokes that cultural geographers today are setting 'aside the hiking boots preferred by Sauer for the patent leather shoes more appropriate to fieldwork in the cafés and art museums now of empirical interest'.

Six Tenets of Cultural Landscape Studies

Observing that—despite the centrality of Sauerian and related *Landscape* magazine traditions—

there is no one monolithic approach to studies of the cultural landscape, Groth (1997) identified six widely held tenets of such studies.

- First, the **ordinary landscape** is important and worthy of study because our everyday experiences are essential in the formation of that cultural meaning. Logically, then, the landscapes in which we are interested are not restricted to those that are in some way different, but include those in which we live and work. While it is common to trace this idea back to the introduction of *Landscape* magazine in 1951, the traditional Sauerian approach by no means excluded the ordinary, although it certainly emphasized the rural and did not encourage studies of the ordinary in urban areas. As Groth (1997:6) comments, 'For every forty studies of barns and fields, there has been only one about urban factories, workshops, offices, or corner stores as workplaces.'

- Second, contemporary research into cultural landscape is as likely to be urban as it is to be rural, and as likely to be concerned with consumption as with production. These are substantive changes to traditional interests, and they represent more than just a changing scene: the interest in urban landscapes particularly is linked to questions of power and inequality in landscape.

- Third, cultural landscape studies sometimes choose to stress the uniformity of a given landscape as a reflection of some overarching national considerations, and sometimes stress local identities and differences from place to place as a reflection of ethnic variations. These two emphases are best seen as complementary. Relevant questions here concern whether or not a particular national culture, perhaps interpreted in superorganic terms, is able to override the many more local variations in culture and identity as these are reflected in landscape. Is the landscape best viewed as 'one book' or as 'multiple, coexisting texts' (Groth 1997:7)?

- Fourth, writing about landscape emanates from both scholarly and more popular sources, reflecting both academic and literary styles.

Indeed, many of the articles that appeared in *Landscape* magazine exemplified the literary style, even when written by an academic, and the collection of landscape readings edited by Thompson (1995b) continues this tradition of incorporating writing by both academics and non-academics. It is not difficult to understand that landscape attracts writers with both scholarly and popular interest in landscape, just as landscape also attracts artists, photographers, filmmakers, poets, and novelists.

- Fifth, the fact that there are numerous and quite different approaches to the study of cultural landscape is a consequence of the interdisciplinary character of the enterprise. Some writers elect to use a theoretical framework, while others favor a more explicitly empirical focus. Some studies of landscape stress description while others stress interpretation.

- Sixth, most landscape analyses are based on spatial and visual data, meaning that landscape is often studied directly. Landscapes are, of course, accessible to us and can be experienced differently from place to place. One longstanding complaint about the landscape studies carried out by cultural geographers relates to their apparent overemphasis on what can be observed as opposed to what is recorded in written texts. Notwithstanding such concerns, the emphasis on seeing, and recording and interpreting what is seen, remains fundamental.

Landscape as Text

Regardless of conceptual inspiration, there is general agreement that cultures express themselves in the landscapes they create. This idea invites an interpretation of landscape as something more than material and visible, namely, as something symbolic. All landscapes have a meaning and value for their creators and for those who subsequently occupy and recreate the landscape.

Cultural groups assert their identity on landscape, intentionally and otherwise, and the following discussion of **iconography**—the description and interpretation of landscape based on uncovering its symbolic meanings—is premised on the idea that the identity of a landscape is expressed through symbols. The lengthy and detailed examples of the Hispano and Mormon homelands illustrated how cultural groups adapt to physical environments, change those environments through the impress of their culture, and create a sense of place—their landscape. In this sense, homelands are landscapes that are laden with meaning, reflecting the attitudes, beliefs, and values of the occupying cultures. Clearly, such landscapes can also be read as texts.

Reading the landscape is an important tradition in cultural geography, and the metaphor of landscape as text—a manuscript that has been written upon by people over time—is not a new one. Before about 1970, it was quite common for historical and cultural geographers to conceive of landscape as a document, written on by successive groups of people with each writing obliterating some of what was written previously while adding something new. The approach of sequent occupance implied such an interpretation of landscape—recall in particular the idea of landscape as palimpsest. In this Sauerian tradition, the writing was accomplished by cultural groups and was typically seen as resulting in the formation of cultural regions. But this traditional view of landscape as text has been further informed by humanist ideas, with landscapes being studied as 'our unwitting autobiography' (Lewis 1979:12), and by postmodern ideas, with landscapes regarded as systems of communication that may be appropriated by a dominant group.

Given the renewed emphasis on the metaphor of landscape as text, studies by cultural geographers are increasingly being informed by work in literary theory that considers how texts are read. In particular, there is interest in **semiotics**, in the study of the way landscape meanings are built, and in uncovering the various writings through deconstruction. Recall that deconstruction theory is used to uncover what is not said in a text, such as the role of power relations, as well as what cannot be said because of the constraints imposed by language, especially given the limitations imposed by a particular discourse. For some cultural geographers, reading a landscape is often

an exercise in reading social relations and in understanding individual and collective action.

The metaphor of landscape as text, though well established, continues to be a source of critical comment and debate. A particular source of contention is the relative emphases placed on the metaphoric and symbolic content of landscape on the one hand and on the 'real' content of landscape on the other hand. On this, Mitchell (1996:95) comments that 'Because landscape is partially a text or a representation we have often tried to understand it entirely as a text or representation.' Certainly, as Peet (1996b) observes, much contemporary cultural geography emphasizes the reading of landscapes 'as though they were texts' (Peet 1996b). But for many contemporary cultural geographers, reading a landscape involves both reading the text and also reading interpretations of the text—not necessarily a straightforward task. Further, understanding landscapes in these terms requires us to appreciate that they are related to a number of usually competing discourses, defined by Duncan (1990:16) as the 'social framework of intelligibility within which all practices are communicated, negotiated, or challenged'.

Naming Places

It is often claimed that language is central to individual and group cultural identity, as it is both a means of communicating and also a symbol or emblem of groupness. Although there is debate about this matter, the symbolic function of language can make it a powerful tool for expressing ethnic and nationalist sentiments, and in some instances language may operate primarily as a symbol and be of only minimal use as a means of communication. In Ireland, for example, survival of the Gaeltacht, the area where Irish is spoken, was argued to be 'synonymous with retention of the distinctive Irish national character' (Kearns 1974:85). In Wales, the Welsh language and culture are often seen as inextricably entwined, to the extent that 'the language may be considered the matrix which holds together the various cultural elements which compose the Welsh way of life' (Bowen and Carter 1975:2).

Some cultural geographers inspired by the Sauerian school have analyzed place names as examples of the ability of a group, usually in their capacity as first effective settlers, to impress their identity on landscape; place names have also been used as a surrogate for culture spread and as a basis for region delimitation. There are numerous studies of place names in the Sauerian tradition. For instance, in a series of analyses of American place names, Zelinsky (for example, 1967, 1983) considered various aspects of place name geography, noting especially that the presence of ethnic place names cannot disguise the essentially Anglo–American national identity of most of the landscape. With reference to Finnish place names in Minnesota, it is clear that the number of names a group could bestow was related especially to variations in physical landscape and to the arrival date of the group—the first effective settlement concept. This example, detailed in Box 9.5, stresses the important role of place names as one means of both helping to build environments in which groups feel comfortable and confirming ownership of a region.

Other cultural geographers inspired more by identity politics have analyzed the naming of places in terms of unequal power relations and related identities. These cultural geographers show a concern with cultural identity, with place names as symbols—one of the ways by which people attach meanings to places—while recognizing that the process of naming places is not entirely an innocent one. Place name studies informed by identity politics build upon earlier cultural geographic work to emphasize that the naming of places is one means by which a group strives to dominate place and, by extension, dominate other groups. Most of this work addresses the naming of places by Europeans in overseas areas, centering on the contestation of place and the way in which European place names assisted both in creating a landscape that is symbolically and materially European and in asserting state ownership and control of place.

RENAMING PLACES

The subject of renaming places in former colonial areas is a source of considerable debate.

Accepting the argument that 'to name is to claim' leads to the conclusion that an attempt to *rename* can be viewed as an attempt to *reclaim*. Berg and Kearns (1996) used discourse analysis to study attempts to reinstate Maori place names in favor of European names in one area of New Zealand in the late nineteenth century. An analysis of proposals for and subsequent opposition to renaming highlighted the perceived importance of names to both groups involved. This study of New Zealand also stressed the masculinity of the naming process, with many places named for male and often military figures.

But the situation may be more complex than this suggests, as Nash explains:

Returning to a precolonial name can be read as a simple postcolonial strategy of legitimating forms of knowledge and experience demeaned or suppressed under colonialism. Alternatively, this act of reinstating the 'original' or 'authentic' name can be read as a return to problematic notions of cultural purity and authenticity. (Nash 1998a:1; see also Nash 1999)

Regardless of the interpretation, place-renaming is an example of place contestation, such that in some cases, place names can become sites of resistance. Nash (1998b:74) interprets the Irish landscape as being 'in deliberate opposition to English landscape aesthetics', noting that 'the cultural significance of the Irish landscape often resides in the stories associated with particular places and in the place names that connect language and locality.'

In addition naming places as a means of asserting authority over, and potentially ownership of, a landscape, dominant groups achieve these goals through the mapping of landscapes, a circumstance that has been central to the cartographic enterprise for many years, as evidenced, for example, by the fact that in early China map making was the responsibility of government officials. Naming and mapping and the related production of knowledge can be thought of as exercises of power. Names and maps can direct our attention to things that we might not have seen before and provide an indication of the circumstances of the various groups occupying a landscape.

Regardless of conceptual inspiration, it is clear that place names contribute to the social construction of place and that they are able to generate powerful emotions. To name places is to write upon the world.

Sacred Spaces

Ideology, often in the context of religion, is one of the principal ways in which identity is constructed. In some areas, particularly those experiencing an upsurge in fundamentalism, religion is of increasing importance. In other areas experiencing an increasing secularization of society, religion in the conventional sense of the term may be of diminishing importance. Where the latter is the case, as in much of Canada, it may be that another bonding factor, notably a civil religion or nationalism, is replacing religion.

The concept of **sacred space** is one aspect of the important idea that all places have cultural meaning. Specifically, sacred space is imbued with some particular meaning for an individual or a group. It can be contrasted with profane or ordinary space, which, although it conveys

---

Box 9.5 Naming Places—Finns in Minnesota

According to Kaups (1966:381), the principal factors responsible for the number, distribution, and types of Finnish place names in Minnesota are the 'Old World background of the immigrant group, the occupational structure of the immigrants, the time and place of their arrival, the nature of the physical environment in which they settled, and decisions made by American officials'. Figure 9.3 locates 92 such names, primarily in northeastern Minnesota and secondarily in the west-central part of the state; all of these names are in rural areas.

The first Finnish settlers in the state arrived in the mid-1860s and settled in areas where others were already present; as a result, there was no opportunity for the incoming Finns to name places. During the 1870s, other Finns settled in the west-central area, but the majority arrived during the 1880s and 1890s, settling in ethnic enclaves in the northeastern area as pioneer agriculturists. It was here that Finns were able to dominate the rural scene as first effective settlers. But numerical dominance was not enough to ensure that the Finns were always able to bestow place names of their choosing; for example, a 1912 petition by Finnish settlers proposed the Finnish name 'Salmi' for a new township, but local county officials decided, without any explanation, to name the township 'Vermilion'.

Kaups devised a classification of both physical and cultural place names that distinguished three types, namely, *possessive* (further subdivided into *personal* and *ethnic*), *descriptive*, and *commemorative* (see Table 9.1). The majority of names belong in the possessive category, and most of these are personal.

Figure 9.3 Finnish Place Names in Minnesota. Finnish place names are located in the northeastern and west-central parts of the state and in a few other scattered locations. Of the 92 place names identified, 59 are names of physical features, mostly lakes and creeks, and 33 are names of cultural features, mostly hamlets. The relative profusion of physical place names in the northeastern area reflects not only the number of Finns in that region and the timing of their arrival but also the varied physical environment that affords numerous place-naming opportunities.

Source: M. Kaups, 'Finnish Places Names in Minnesota: A Study in Cultural Transfer', *Geographical Review* 56 (1966):379. (©The American Geographical Society; reprinted with permission)

meaning, lacks particular symbolic quality. Indeed, spaces that are sacred for some may be ordinary for others. Sacred space implies some special attachment to place, perhaps spiritual devotion, reverence, affection, pride, nostalgia, or a more general sense of belonging, and it is often made special through the performance of rituals. As these possible implications suggest, the degree of sacredness of sacred spaces is highly variable, and a sacred space may be long-lasting or quite ephemeral, depending on the type of space, the number of people involved, and the degree of sacredness. Sacred spaces also vary in size and

impact (if any) on the visible landscape. Given such a wide variety of circumstances, Jackson and Henrie (1983) proposed a classification that recognized three types of sacred space.

Mystico-Religious Space

Mystico–religious space, Jackson and Henrie's first kind of sacred space, includes those spaces associated directly with religious beliefs and experiences. Needless to say, there are many such places. Some are sacred because a specific religious event occurred there or is believed to have occurred there; in some cases, they are sites of

Possessive names are used to indicate an association between a Finnish individual or group and a particular place. The descriptive names identify the character of a place through incorporation of a Finnish word in the place name; for example, the place name Kumpula describes the location through use of the Finnish word *kumpu*, meaning 'hill' or 'mound'. Most of the commemorative names are those of important Finnish literary figures.

The majority of possessive names refer to physical features, the majority of commemorative names refer to cultural features, and the descriptive names are evenly divided between the two. In addition, 14 abandoned names were identified, 13 of which referred to cultural features.

Not included in Figure 9.3 or in Table 9.1 is another group of Finnish names, labeled as *in-group names*. These names are used by Finns but have not received official sanction or appeared on official maps. Of these, 53 were noted, all located in the same areas of the state as the official names. These in-group names arise in part because ethnic groups are not usually free to name places as they choose, and such names are one means of creating a local environment that is supportive of group identity in new and possibly difficult surroundings.

Table 9.1 Classification of Finnish Place Names in Minnesota

Number of Place Names	Possessive (personal)	Possessive (ethnic)	Descriptive	Commemorative	Other
Physical	37	8	10	3	1
Cultural	5	2	9	13	4
Total	42	10	19	16	5

Source: M. Kaups, 'Finnish Places Names in Minnesota: A Study in Cultural Transfer', *Geographical Review* 56 (1966):377–9. (©The American Geographical Society; reprinted with permission)

Photo Essay 9.1 Mystico-Religious Sacred Space. There are intimate relationships between people and place, with different groups of people valuing different parts of the earth's surface. There are, for example, places of literary pilgrimage. The grave of Anne Bronte (*top left*) is in the cemetery of St Mary's Church in the English coastal town of Scarborough. A popular tourist attraction, the gravesite is planted with primroses, and admirers regularly leave bouquets of carnations or single red roses. Today the headstone is almost wind-blasted to illegibility (William Norton). Many other mystico-religious sacred spaces are directly associated with religious beliefs about the physical landscape. Rising 335m above the Australian desert floor with a circumference of 9 km, Uluru/Ayers Rock (*bottom*) is located in Kata Tjuta National Park, Northern Territory, and is owned and run by the local Aboriginal Anangu people. Depending on the time of day and atmospheric conditions, the rock changes color, from blue to glowing red. Uluru is a sacred place for the Anangu, for whom the natural world is both a physical and spiritual reality, and is one of three major locations in central Australia where the tracks of several ancestral groups cross (William Norton). Other mystico-religious sacred spaces are parts of the built environment. In the entrance court of the Qairiwiyin Mosque in Fès, Morocco (*top right*), a wor-

shipper washes his hands, face, and feet at the fountain before entering for prayer (William Oxtoby). The Hindu Temple of Sri Swaminarayan Mandir (*top*) in London, England, opened in 1995 as the first traditional Hindu Mandir in Europe. Bulgarian limestone and Italian Carrara marble were shipped to India, carved by over 1,500 craftsmen, and reshipped to London. The construction process involved contributions from more than 1,000 volunteers. Widely seen as a masterpiece of exotic design and workmanship, the temple is an important tourist site, a place for visitors to learn about Hinduism, and a site for religious festivals (Philippa Lewis/Edifice, CORBIS Canada). Completed in 2001 and surrounded by peach and apple orchards, hay pastures, and snowcapped peaks in the heart of Mormon Utah, this Hare Krishna temple (*bottom*), which is modeled on an ancient temple in India, is part of a growing worldwide movement that has placed temples in such cities as Stockholm, São Paulo, and Miami. Although the two faiths are very different, local Mormons helped construct this temple with donations and labor. The Mormon co-operation and support stems from a moral dictate requiring them to be in service, respect the religious beliefs of others, and build bonds with other communities (William Norton).

events that are believed to have occurred and that are inexplicable without reference to religion. Most religions create sacred spaces through the expression of their identity in landscape; buildings, shrines, and symbolic markers are examples of this process. 'Indeed,' Park (1994:199) observes, 'symbols of religious worship are woven into the very fabric of many areas and give them a special and sometimes unique identity,' an identity that is sacred to members of the religious group and that is often readily recognizable by others.

Waitt (2003) details the intriguing example of the building of a Buddhist temple—the largest in the southern hemisphere—in the industrial city of Wollongong in eastern Australia. Although there are few Buddhists in this area, the location was selected because it is an auspicious location in terms of **fengshui**. Formally opened in 1995, the temple today is a sacred space for Buddhists, but it is also socially constructed as a tourist attraction by local authorities, who interpret it and promote it as exotic and symbolic of the 'mysterious' East. Many similar attempts to construct religious buildings in non-traditional locations have met with much opposition. A Hindu temple was opened in northwestern London, England, in 1995, but only after many expressions of concern and failed attempts at building it in preferred locations in the city.

Parts of the physical environment are often considered sacred by various religious groups, especially rivers such as the Ganges, which is considered sacred in Hinduism, and mountains such as Mount Fuji, which is considered revered in Shintoism. Members of the religious group visit almost all such places, but some are so frequently visited that they may be considered pilgrimage sites. For members of the Islamic faith, Mecca and Medina are so sacred that pilgrimage as least once is required of all whose circumstances permit—some of these sites even have their sacredness enhanced by the exclusion of others from those places. Figure 9.4 locates principal pilgrimage places for Hindus in India.

HOMELANDS AS SACRED SPACE

The second kind of sacred space identified by Jackson and Henrie is the homeland. Some homelands represent sacred space to some groups because they represent the roots of both the individual members of the group and the group as a whole. In some cases such homelands are imagined as a result of being created by those in positions of power; many supposed national homelands fall into this category. More locally, some individuals may perceive their place of birth and/or their area of upbringing as sacred while others may not.

Vernacular regions can be interpreted as an example of a particular type of sacred space, recognized as different both by those inside and by those outside the region. As with many homelands, it is common for the meaning of a vernacular region to be contested. Typically, for those living in the vernacular region, that region is home and is valued positively, while for those living outside the region, it is someone else's home. The vernacular designation 'Bible Belt', which is used to refer to parts of the southeastern and south-central United States, is a regional name in which many residents take great pride—thus, Oklahoma City proclaims itself the 'Buckle on the Bible Belt'. However, for some who live outside the region, the name is used in a derisive manner—thus, there are disparaging references to a 'Bible and Hookworm Belt' or a 'Bible and Lynching Belt'.

For many Aboriginal groups, the homeland concept of sacred space takes on a more spiritual meaning. Throughout much of the Canadian North, for example, there is a conflict of understanding between the European-derived idea prevalent in southern Canadian of the North as wilderness frontier and the Aboriginal idea of the North as homeland. This conflict first arose explicitly following the publication of the Mackenzie Valley Pipeline Inquiry, titled *Northern Frontier, Northern Homeland* (Berger 1977). In this context, the idea of sacredness applies not simply to a few specific locations but to the whole landscape, which is seen by Aboriginal communities as sacred because it is a place containing crucial, well-known resources. Aboriginal movement through the homeland and related resource use has traditionally been understood in spiritual terms: Aboriginal people see them-

Figure 9.4 Hindu Pilgrimage Sites in India. Visiting pilgrimage sites, often on a regular basis, is one means by which individual members of a religion reinforce their identity as group members. Certainly, a dominant Hindu identity is generally considered to be one of the factors that help promote unity among an otherwise diverse Indian population. This map shows principal pilgrimage sites, some of which, such as Varanasi on the Ganges, are visited by more than a million people annually. At many of these locations, Hindus seek help from deities in their personal and social lives. Brahma, symbolizing creation, Vishnu, symbolizing preservation, Shiva, symbolizing dissolution, and the Mother Goddess, symbolizing energy, are four of the more important deities.

Source: J.O.M. Broek and J.W. Webb, *A Geography of Mankind*, 2nd edn (New York: McGraw-Hill, 1978):143.

selves as integrated with the physical world. This same logic applies generally throughout Aboriginal North America and Aboriginal Australia. Such understanding of sacredness is rather different to the European-derived idea of a sacred homeland that is seen as a place to occupy rather than a place to live in and be with.

Aboriginal understandings derive from traditional indigenous ways of knowing and are connected with what Europeans see only as natural features of a landscape, such as rivers and mountains. In a traditional Aboriginal understanding, such features are sacred rather than natural and are associated with a body of beliefs and rules concerning how humans are to behave in that place. It is this difference in interpretation, this different way of knowing, that contributes to issues of contested ownership when sacred lands are used or occupied by others.

HISTORICAL SACRED SPACE

Jackson and Henrie's third kind of sacred space is historical sacred space. Typically historical sacred spaces are places assigned sanctity as a result of some important event either occurring there or being remembered there. Most national and ethnic groups identify several such places. Indeed, landscapes may include places that play an important role in the construction of identity and that have accordingly 'been denoted as essential codes in the national signifying system, while other sites and landscapes, or elements attached to them, have been consciously or unconsciously excluded from this, being either overlooked or denied' (Raivo 1997:327–8).

For many Canadians and Americans, historical sacred spaces may be associated more with national circumstances than with culture-specific circumstances. In both countries church membership is common, a commitment to religion is less common, and religious places and religious occasions are evidently becoming less and less sacred. Of course, it is possible to interpret religion very broadly, so that nationalism or civil religion can be thought of in similar terms to the more conventional understanding of religion and can therefore be associated with the creation of sacred space. Examples of such sacred spaces include military cemeteries, monuments, and preserved historical sites, for example the Plains of Abraham in Canada, and Gettysburg, the Alamo, and Arlington cemetery in the United States. In many instances sacredness is emphasized through use of national symbols, most usually a flag. Although this circumstance is best documented by cultural geographers in the context of the United States, Zelinsky (1988:175) suggests that 'the workings of the state have set their mark upon the land' in all countries.

Jackson and Henrie (1983) applied this tripartite classification of sacred space in a study of spaces sacred to Mormons; Figure 9.5 locates principal Mormon sacred sites in the United States, and Table 9.2 ranks these spaces.

Reading Rural and Urban Places

Places are read routinely through larger discourses, and the readings that dominate are, of course, the interpretations favored by dominant groups. But all places are multivocal, with different people hearing only the voice that they understand because it is the voice that accords with their feelings.

Table 9.2 Ranking of Selected Mormon Spaces

Place	Mean
Salt Lake Temple	1.92
Future City of Zion	2.12
Sacred Grove	2.39
Utah	2.76
Temples other than Salt Lake City	2.95
Bethlehem	3.02
'Holy Land' (Israel)	3.48
Joseph Smith birthplace	3.70
Nauvoo and Kirtland	3.70
Chapels	4.63
Carthage Jail	4.65
Present home	4.74
Present state of residence	4.79
Utah's mountains	4.95
Regions surrounding Utah	4.95
Childhood home	5.41
Present day Jackson County	5.41
Lincoln Memorial	5.86

Notes: 1. In this research, a first open-ended questionnaire was used to identify all-important places, and a second questionnaire required respondents to rank places.

2. Values represent how respondents valued each place on a scale of 1 to 7, with 1 as most sacred and 7 indicating no sanctity.

Source: R.H. Jackson and R. Henrie, 'Perception of Sacred Space', *Journal of Cultural Geography* 3, no. 2 (1983):102.

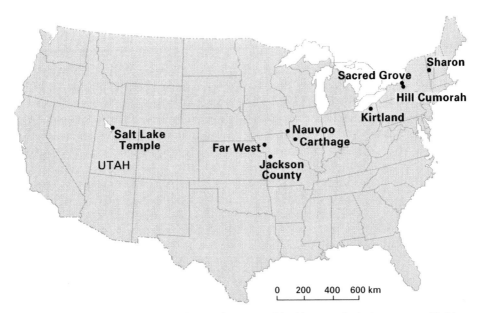

Figure 9.5 Mormon Sacred Sites in the United States. Sites sacred for Mormons include some specific Mormon places and some places associated with Christianity generally. As noted in Table 9.2, the three highest ranking sites are all mystico-religious places specific to Mormonism, namely the Salt Lake Temple, the future city of Zion, and the Sacred Grove. The homeland space seen as most sacred was the state of Utah, with such spaces as present and childhood home considered significantly less important by Mormons. Sacred historical sites were those associated with the history of Mormonism, such as Sharon, the birthplace of Joseph Smith Jr, and Carthage Jail, the site of his death. The importance of historical sites reflects the Mormon concern with historical and religious roots. National sites, such as Lincoln Memorial, are less sacred.

Worldwide, Salt Lake Temple is the principal religious symbol of the Church of Jesus Christ of Latter-day Saints. There are six major towers and finial spires that signify the restoration of priesthood authority, as well as numerous other symbols. It is a major tourist attraction for Mormons and others. Jackson County in Missouri is considered a possible location for the Future City of Zion (Zion refers to the idea of a gathering place for those who are pure in heart). The Sacred Grove is a treed area on the Smith farmstead where Joseph Smith Jr experienced his first vision, and the Hill Cumorah is the site that Smith was directed to in order to retrieve the gold plates from which the Book of Mormon was translated. Far West, Missouri, was settled by Mormons in 1836 and is important as the place where a temple site was dedicated and seven revelations received. Kirtland, Ohio, and Nauvoo, Illinois, were church headquarters from 1831 to 1838 and from 1839 to 1846 respectively.

In this research, a first open-ended questionnaire was used to identify all of the important places, and a second questionnaire required respondents to rank places. This innovative study concluded that, for Mormons, the most sacred of spaces are those associated with their religion.

Source: R.H. Jackson and R. Henrie, 'Perception of Sacred Space', *Journal of Cultural Geography* 3, no. 2 (1983):101.

RURAL PLACES

An example of the way in which places are multivocal is the English countryside, which is characteristically seen nostalgically as a rural idyll, a place that provides opportunities to escape from the pressures of urban life. This is an image that was evident in the works of Romantic writers and was pivotal to the classic work of Hoskins; recall that, for Hoskins (1955:231), 'especially since the year 1914, every single change in the English landscape has either uglified it, or destroyed its meaning, or both.' This way of seeing the rural is a social construction that today is increasingly commodified and purchased.

But such a representation is necessarily partial and misleading for at least three reasons. First, the countryside is very much associated with only one particular segment of English society, namely White people. It is also especially associated with men rather than women and with people of upper- or middle-class backgrounds. Second, the countryside is contested space, with ongoing debates about fox hunting and the right that people have to roam unimpeded. Third, for many rural dwellers the countryside is far from idyllic as it is a place of poverty and deprivation. Thus, the image of the English countryside as rural idyll is a constructed place identity that is but one of several possible readings.

BUILDINGS

Building styles are visual texts. The most popular approach to the reading of these texts and related understanding of place is iconography, an approach that originated in art history as a means of interpretation and has accordingly often been employed in studies that stress visual landscape. One example is a pioneering analysis by Harvey (1979), in which he presented a Marxist historical geography of the powerful political symbolism associated with the Basilica of Sacré Coeur in Paris.

In some instances, landscape symbolism may seem to be self-evident and therefore relatively easy to read. This is the case with many religious structures and with tall buildings in urban areas that are home to transnational companies. In other cases a much more subtle and detailed reading may be required if there are small variations in housing styles, coloring, and ornamentation. Regardless, buildings speak to us. As Domosh (1992:475) observes, 'The city, like a painting, is a representation, an image formed out of the hopes and ideas of the cultural worlds in which we live. As a cultural construct, the meaning of the city can be deciphered by closely examining its complex relationship with the culture of which it is a part.'

Although it is clear that tall buildings in an urban landscape symbolize economic achievement, it may be that landscape meaning, the sense of place, is a consequence of historical processes that are not quite so visible. For example, each of three buildings in Hong Kong represents a different claim about political and economic power. The Hong Kong and Shanghai Banking Corporation headquarters has remained at the same address since 1864—the fengshui of the location is excellent, and this bank has functioned as an agent of British political and economic interests. The Bank of China—Hong Kong branch, a Chinese national project completed in 1988, symbolizes the uncertainties relating to the 1997 Chinese takeover of Hong Kong. Central Plaza, the tallest building in Hong Kong, has reoriented the skyline, having been built in the previously socially marginal area of Wanchai; the building has multiple tenants and, most significantly, was built by a local group and not by the British or Chinese. Three buildings, three different interpretations of power and authority.

URBAN PLACES

Many landscapes and their meanings need to be understood as parts of larger discourses. Schein (1997) conceptualizes the landscape of one neighborhood in Lexington, Kentucky, as the intersection of several discourses, namely those of landscape architecture, insurance mapping, zoning, historic preservation, neighborhood associations, and consumption. Each of these discourses includes ideals, guiding principles, and often even explicit and institutionally mandated rules that function to discipline the residents of the neighborhood.

In an account of urban redevelopment in central Christchurch, New Zealand, McBride (1999) outlines a series of different readings of the city. A traditionalist reading sees the city as English in character, with Cathedral Square as the symbolic heart, and argues for landscape preservation in the face of changes prompted by capitalism. An alienated reading sees central Christchurch as having been taken over by tourists, with others excluded. In explicit opposition to the traditionalist reading, a market reading interprets urban change in solely economic terms, without any consideration given to questions of preservation and heritage. Those who are responsible for selling the city to tourists provide yet another reading, as they typically strive to find a balance between preservation and development.

In many instances, powerful individuals and institutions, including the state, act as agents to shape the landscape, and landscapes are being read accordingly. In a discussion of a planning dispute surrounding proposals for urban expansion in Lexington, Kentucky, McCann (1997) identified two interested groups. Opposing expansion were environmentalists, preservationists, and some horse-farm owners; favoring expansion were developers, members of the construction trade, and the chamber of commerce. These two groups—textual communities—read the symbolic landscape surrounding Lexington differently, the former seeing 'an interconnected, living entity, the existence of which benefited the population of the entire region', and the latter seeing 'a series of individual, privately owned land parcels, the use of which was primarily a concern of individual landowners' (McCann 1997:647). The debate between these two groups—involving two very different readings of the landscape—played out in the context of the planning process, and McCann (1997:660) concluded: 'The morphology of landscape in capitalist societies is, therefore, mediated by the contexts and discourses of the planning process which make use of the legitimacy and power associated with institutional sites in order to maintain a liberal ideology of equality and fairness.'

National Places

With reference to the national scale, Meinig (1979c:164) notes that a 'mature nation has its symbolic landscapes. They are part of the iconography of nationhood, part of the shared set of ideas and memories and feelings which binds a people together.' These places may be monuments, government or religious buildings, sites where key events occurred, or larger regions that in some way capture the presumed essence of the national being. They may be intentionally constructed to reflect and reinforce national identity, or they may be places that become symbolic through time.

New nations also strive to promote places that reflect some desired national unity. Israel achieves a sense of identity at least partly through the Zionist rediscovery of places that were important in ancient Jewish history and that are interpreted as foundational to the state (Azaryahu and Kellerman 1999). In the notorious case of Nazi Germany, the town of Rothenburg ob der Tauber exemplified new ideas about landscape and national identity through efforts to reorient tourism and also to cleanse the town of alien influences (Hagen 2004).

Other nations that experience a transformation in their political circumstances, for example through decolonization or the removal of an oppressive regime, frequently make explicit attempts to highlight the change. They may do this by renaming places and destroying monuments. The celebratory toppling of a statue of Saddam Hussein in Baghdad captured the sense of relief and new-found freedom experienced by many Iraqis following the American-lead intervention in 2004. More generally, after about 1990 most eastern European countries that were previously under Soviet influence made conscious efforts both to eradicate the evidence of an oppressive past and to build signposts for the future. Similarly, aspiring nations identify places that are interpreted in national terms. Raento and Watson (2000) discuss the contested meanings underlying the town of Gernika/Guernica as a center of meaning for Basques in northern Spain; the town was significant in the Spanish Civil War, was represented in a famous Picasso painting, and is commemorated in the Americas by Basque immigrants.

Many monuments are constructed as a means of creating or reinforcing some ethnic or national identity. War memorials are often both a readily visible and an easily understood symbolic landscape feature and, in the case of Australia, the building of numerous memorials to those who died during the First World War has made a significant contribution to national identity and unity. However, while war memorials are dominantly interpreted as celebrations of bravery and national identity, they can also be seen in terms of a national failure to protect citizens, so, as with so many other landscape features, their meaning is contested through competing discourses.

Box 9.6 discusses the representations of landscape promoted by the Australian Tourist

Commission in order to help maintain a particular national identity.

Monuments and Memorials

Landscapes reflect a complex blend of dominant and subordinate cultural identities, with processes of inclusion and exclusion constantly at play and largely determined by those in positions of power. As noted earlier, a popular strategy to assert identity and authority in landscape is the erection of memorials in the form of monuments, statues, or informational plaques—recall the Voree example from the previous chapter. A part of the iconography of landscape, such structures represent a particular version of the past: they tell the story that those erecting the structure wish to tell. But there are always other versions of a story, and indeed, there may well be other quite different stories. Monuments are key points of hegemonic meaning in landscape, and for this reason the interpretation of monuments is often contested. Many monuments commemorate leading political and military (usually male) figures, thus memorializing a sense of the past that is presumed to be meaningful to the present. This is why changing political circumstances, such as the defeat of an oppressive regime, can prompt monument destruction.

In some cases monuments not only reinforce some story about the past but also contribute to the making of the future. For example, the Voortrekker Monument in Pretoria, South Africa, tells a story of heroic Afrikaner migrations north from the Cape into an often difficult interior. At

Box 9.6 Representing Australia

It is often said that image is everything. Certainly, the importance of creating a favorable image of a product through advertising and other means is well accepted. Landscapes and identities are no exception to this principle, as evidenced by the abundance of literature produced in many areas designed to attract tourists. Understanding such literature requires a knowledge not only of the landscape and identity being represented but also of the authors of the texts and of their specific motivations—the discourses of which they are a part. In this context, recall the discussion in Chapter 5 of images, image makers, and image distortion during the period of European overseas expansion.

In the case of Australia, official representations of identity prior to the 1970s 'embraced mateship, the bush and egalitarianism' with the principal symbolic landscape being the Australian semiarid interior (Waitt 1997:49; see also McLeay 1997a). This image was reinforced by leading Australian painters, and the principal human component was the White male, with women and other groups excluded or serving merely as background. More recently, the image of Australian identity conveyed by the Australian Tourist Commission is more in accord with a multicultural focus, although such an identity continues to be defined by the dominant group. The image of landscape commonly conveyed today is one aimed at fulfilling the tourists' desire for adventure and paradise.

Adventure is most clearly suggested in images of the bush, which stress the isolation and frontier conditions. In these images, the Aboriginal population serves as an important component, as Waitt (1997:50) points out: 'Aboriginality is socially constructed as the other to serve a particular political context and a deliberate economic strategy, that of selling Australia as an escape from civilisation to a primordial, timeless world, and/or a return to Nature where Aboriginals as the "original conservationists" live in perfect harmony with their environment.' White males are depicted in positions of power and females in more submissive and accepting roles.

Paradise is signified through the depiction of parts of the Australian landscape, especially coastal and subtropical areas, as luxuriant and fertile, with emphasis on rainforest, waterfalls, beaches, and coral reefs; it may be that such depictions are designed to stress the femininity of landscape.

Clearly, because these and all other images are texts to be read, they can be interpreted in more than one way, and it is interesting to think about any one interpretation in terms of the positionality of the interpreter. Certainly, as the overall content of this chapter indicates, it is quite usual in much current cultural geography to interpret from a cultural studies perspective, through the eyes of the cultural turn, and thus to stress aspects that relate to issues of power, authority, and otherness.

Completed in 1949, the Voortrekker Monument was built to honor the courage and determination of the Afrikaner pioneers who left the British-controlled Cape Colony in the mid-1830s on the 'Great Trek' north. Intended as a national icon by the Afrikaner South Africans, this monument sent a very different message, one of displacement and oppression, to the majority South African population. Today, it remains as a relict, a reminder of the Afrikaner dominated past, and is far removed from being the national icon that was intended. The main structure includes a granite cenotaph placed such that a ray of sunlight will fall on the inscription 'Ons vir jou, Zuid-Afrika' ('We for you, South Africa') at noon each 16 December, to mark the anniversary of the 1838 Battle of Blood River, in which a small party of Afrikaners defeated a much larger Zulu force. There is also a Hall of Heroes, with walls lined with a frieze consisting of 27 marble panels depicting the Great Trek's main events. A 'laager' of 64 granite ox-wagons surrounds the main block of the Monument. This illustration shows a detail of one of the panels that aims to represent the group solidarity necessary to survive the many challenges faced during the trek.

the same time, it helped create a particular version of Afrikaner identity, which in turn facilitated the rise of the apartheid state. Neglected in this selective Afrikaner interpretation of the past are other groups, especially Africans, and their stories.

Controversy may also erupt when changes to a memorial landscape are proposed. As home to a series of Confederate statues, as well as many expensive houses, Monument Avenue in Richmond, Virginia, is a powerful symbolic location. In 1994, a proposal to erect a statue on the Avenue of the Black tennis player Arthur Ashe provoked uproar among some members of the White community on three grounds. First, there

were aesthetic objections, based on the claim that the symbolism of the Avenue would be destroyed, especially as a casually dressed tennis player would simply not fit in with Confederate soldiers in full military dress. Second, there were objections based on doubts that Ashe's status was sufficient for him to merit a statue alongside Confederate war heroes. Third, there were explicitly racist objections. But there were also objections from members of the Black community: some argued that Ashe's accomplishments as a Black man were far superior to those of Confederate soldiers who had fought to preserve the institution of slavery in the South; others believed that the impact of the statue as an inspiration for young Black people would be diminished if it were placed on Monument Avenue. After a long public meeting with about a hundred people speaking, City Council voted in favor of the Monument Avenue site, and despite subsequent attempts to have that decision reversed, the statue was unveiled in July 1996. This example highlights the complex intersection of the power of place, the social construction of identities, and landscape iconography. Figure 9.6 shows the 1996 location of statues along Monument Avenue.

DEATHSCAPES

There are also complex debates surrounding the meaning of deathscapes, especially as these reflect tensions between individuals, groups, and the state, as well as tensions between secular and religious interests. In the case of roadside memorials to accident victims in a part of eastern Australia, Hartig and Dunn (1998:5) observe that, despite the fact that such private intrusions into public landscapes are generally tolerated, even respected, 'contradictory discourses condemning

and condoning youth machismo' circulate around these memorials.

Some densely populated places face a large and growing problem with how to dispose of the dead. In the case of Hong Kong the problem has been addressed through two compromises. First, the region's citizens have had to all but abandon the traditional Chinese practice of coffin burial rather than cremation. Second, prior to Chinese takeover in 1997, the British authorities accepted the Chinese practice of 'second burial', which involves exhuming a body after about six or seven years and then either cremating it or storing the bones in a pottery urn (*jinta*). Figure 9.7 identifies the choices available and spaces required for disposing of dead people.

Consuming Places

Some places are created as cultural resources and as sites of consumption, while other places—and sometimes people—are recreated or manipulated in order to increase their appeal to consumers. Cultural geographers are taking an

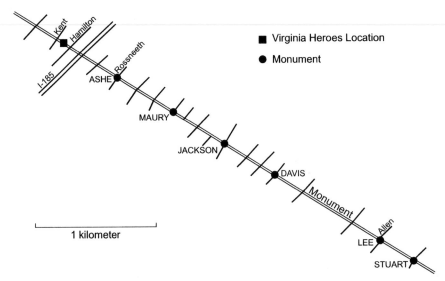

Figure 9.6 Location of Monuments Along Monument Avenue. Monument Avenue is one of Richmond's most prominent streets. The Ashe statue is located along with statues to five Confederate heroes (Maury, Jackson, Davis, Lee, and Stuart) that were erected between 1890 and 1929.

Source: Adapted from, J.I. Leib, 'Separate Times, Shared Spaces: Arthur Ashe, Monument Avenue and the Politics of Richmond, Virginia's Symbolic Landscape', *Cultural Geographies* 9 (2002):297.

This roadside shrine in rural northern New Mexico commemorates a road accident victim. In this Catholic area, such visible signs of mourning are not uncommon, and rather than being seen as intrusive by others, they appear to be widely accepted as symbolically important albeit ephemeral landscape features. Indeed, shrines such as the one shown blend naturally into the surrounding physical and human landscape, being made with local materials and bearing a resemblance to memorials in cemeteries. (William Norton)

increasing interest in the consumption of places and other cultures as a part of tourism, the consumption of the activities of others by attending sporting events, festivals, and carnivals, and the consumption of food and drink. Many sites of consumption—places such as hotels, malls, and theme parks—falsify place and time to create both other places—what might be called *elsewhereness*—and other times. Consider, for example, the many 'Polynesian' hotel lounges outside of Polynesia, or 'Beefeater' hotel lounges far removed from London. Consider a place like Disney World, where children are transported into a world they have visited only in books and videos. Consider the Abraham Lincoln Presidential Museum in Springfield, which has been criticized for being more like 'Abeworld' or 'Six Flags over Lincoln' than museum and is a place 'where the unreal blurs with the real and ultimately upstages it' (Kalin 2005). Certainly, much consumption is increasingly exotic in character, a feature that is evident generally in tourism, retailing, and the food and drink industry.

All consumption is cultural (Slater 2003). The seemingly mundane act of eating, for example, involves numerous cultural decisions such as what constitutes food and how that food needs to be prepared, while shopping similarly involves decisions about identity formation and needs. As a cultural practice, consuming has two theoretical implications. First, consumer culture needs to be understood by reference to both the institutions of consumption, such as the shopping malls and tourist sites where consumption occurs, and the advertising that mediates between our needs and wants. These worlds of consumption are continually being socially created, and the pleasures of consumption are social pleasures. Second, consumption is a symbolic practice that may be interpreted. Much contemporary cultural geography is building on these ideas to generate analyses of consumption and the sites of consumption.

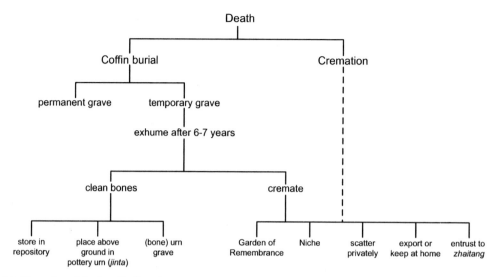

Figure 9.7 Disposing of the Dead in Hong Kong. This figure identifies the choices available for, and spaces associated with, the disposal of human remains in Hong Kong.

Source: E.K. Teather, 'Themes from Complex Landscapes: Chinese Cemeteries and Columbaria in Urban Hong Kong', *Australian Geographical Studies* 36 (1998):24.

Folk and Popular Culture

The discussions in this section are informed by a convenient, but by no means absolute, division of people and place into the two categories of *folk* and *popular*. Although these two terms are not clearly defined in cultural geography, this division is conventional in the discipline and continues to be useful in a world characterized both by global processes of homogenization and by more local processes that stress the distinctiveness of people and place.

Broadly speaking, popular landscapes are the everyday, taken-for-granted landscapes that reflect ever-changing cultural circumstances, while the concept of folk refers to groups that choose to retain at least some aspects of the traditional. A **folk–culture landscape** is, for most of us, different because of this emphasis on tradition, which might include the oral transmission of songs, a distinctive local history, an integration of nature and culture. This sense of tradition may be expressed through ritual. Folk cultures include cultures that prevailed in the preindustrial and precapitalist period but that remain evident today, albeit in a limited manner. In some cases where ethnic and national identities are

traced to their folk roots, folk is taken to mean some original unadulterated cultural condition that is evident usually in song, dance, crafts, and dress; this is comparable to the fourth of the six components of ethnicity referred to in Chapter 7. Folk-culture landscapes are occupied by groups that intentionally choose to reject at least some of the cultural and technological changes that are typically embraced by most people, and in this sense there is a distinction between folk cultures as local, premodern cultures and a single, global, modern or postmodern popular culture.

Several examples of folk cultures and related landscapes have been referred to earlier in this textbook, especially in the discussions of the traditional approach to diffusion and of ethnicity and cultural islands. Cultural geographic analysis of folk culture, especially as reflected in housing styles and building materials, is closely associated with the works of Evans and Kniffen. Indeed, there is a voluminous and continuing cultural geographic literature on folk–culture landscapes as these are evidenced by preferences in food and drink, music, sports, and material landscape features. Many of these studies belong in the Sauerian tradition.

A **popular–culture landscape** usually implies culture for the people, and it is therefore associated with change and with everyday life for the masses. Popular culture is the culture of the majority. In particular, most landscapes in urban areas, including sites of consumption such as malls and sports stadiums and sites of spectacle reflect popular culture, and thus may be similar from place to place with only minimal evidence of any local characteristics.

One interesting debate about popular–culture landscapes concerns the extent to which they are so similar from place to place that they are effectively placeless, lacking meaning for those who frequent them. This situation is seen as resulting primarily from the fact that such landscapes reflect non–local processes of investment and consumption. From a critical humanist viewpoint, Relph makes the claim that

The postmodern logic of places is that they can look like anywhere developers and designers want them to, and in practice this is usually a function of market research about what will attract consumers and what will sell. In postmodernity it is as though the best aspects of distinctive places have been genetically enhanced, then uprooted, and topologically rearranged. (Relph 1997:220)

It is helpful to think about this claim in a larger context. For example, as Sutter (1973) points out, traveling through North America often means that the 'next place you come to' is essentially the same as the place you came from. Recognizing the sameness of small–town North America also highlights the powerful bond between those towns and the people who live there or who visit on a regular basis.

Ameriquest Field in Arlington, home of Major League Baseball's Texas Rangers, is a 49,115-seat open-air ballpark designed with tradition and intimacy in mind. It includes a granite and brick facade, exposed structural steel, an asymmetrical playing field, and a home run porch in right field. Although much of the popular-culture landscape shows little relation to local or regional identity, in some cases efforts are made to reflect tradition. Here, Texas architecture is featured throughout, from the outer facade to the Lone Stars in the concourses and on the seat aisles. A venue designed primarily for sporting events, the stadium is supplemented with a shopping mall and commercial area, with office buildings, stores, and a restaurant. (William Norton)

Apparent sameness does not necessarily translate into placelessness.

Although this contrast between folk and popular culture is helpful, popular culture is also often contrasted with high culture, or with mass culture, and the term has a very different meaning depending upon the contrast implied. When contrasted with mass culture, popular culture refers to a particular consumer niche, for example the market for a novel or film; when contrasted with high culture, as in the early Marxist-oriented British cultural studies tradition, popular culture implies the interests and activities of the working class.

Heritage Tourism

Tourism is increasingly important as a 'means by which modern people assess their world, defining their own sense of identity in the process' (Jakle 1985:xi). Tourists are primarily spectators, consumers rather than producers—but what is it they are consuming? The answer is other cultures and other places. Through this process, tourism may help to reinforce the notion of the cultural other, setting up a distinction between observers and those being observed that is essentially a power difference. This is especially the case where folk cultures are the object of tourist interest and scrutiny.

Heritage tourism is a popular device of both folk-culture and popular-culture contexts. The typical strategy is that of commodifying some selected past to present characteristics of people and place, creating heritage products that are then purchased and consumed by tourists. In many cities, areas no longer used for their intended purposes, perhaps wharf areas or industrial sites, are packaged as tourist attractions that often incorporate opportunities for education and participation. Most such heritage sites are necessarily contested, both in the sense that there may be debates about the merits of preservation and/or reconstruction and, more importantly, in the sense that most heritage sites are partial and involve an intentional directing of the **tourist gaze** (see Box 9.7).

In North America, a popular strategy used by many towns to encourage tourism is to highlight some distinctive ethnic identity: Dauphin in Manitoba, Fergus in Ontario, Solvang in California, New Glarus in Wisconsin, Pella in Iowa, and Lindsborg in Kansas are just a few examples, respectively emphasizing their Ukranian, Scottish, Danish, Swiss, Dutch, and Swedish heritage. Two intriguing apparent examples of celebrating heritage are those of Helen in Georgia and Leavenworth in Washington, both of which market a Bavarian identity—they are *apparent* examples because in both cases the identity has no basis in reality but is simply an invention designed to enhance tourism. There is a parallel here with more obviously false landscapes, including those built in theme parks such as Disney World and in entertainment centers such as Las Vegas.

In some places that are marketed as folk or ethnic tourist destinations there is clearly tension within the community, as some members display a folk culture—putting on a performance for tourists—while other members reflect a more contemporary lifestyle. This point may apply especially to those expressions of cultural identity expressed through festivals.

Ironically it is often the case that folk-culture tourism is capable of destroying the very thing being sought by visitors, namely folk-cultural authenticity. Principal exceptions to this generalization are those tourist sites and activities that do have meaning in that they genuinely reflect values, traditions, and a sense of place. For example, some Amish groups in North America stage events for tourists that reflect their lives but do not intrude on their privacy.

Postmodern Consumption

SPECTACLES

The most prominent sites for contemporary tourism and recreation are those landscapes of consumption and **spectacle** associated with postmodernity and related rise of a global culture. Among the many such places there are culturally enhanced physical attractions such as Niagara Falls, culturally enhanced historical sites such as Gettysburg, and cultural constructions including world fairs and theme parks. Many of these places involve an intentional removal of the everyday along with an emphasis on the far-

Today, the small town of Helen is the most popular tourist destination in northern Georgia, and the third most popular in the state behind Atlanta and Savannah. Set in a high mountain watershed, Helen was earlier a gold town and then a lumber town, but by the 1960s, it was economically depressed and rundown. In 1969, in what turned out to be an inspired decision, a group of local businessman proposed to reinvent Helen as an Alpine theme town. As shown in the illustration, buildings were remodelled and painted to resemble Alpine or Bavarian structures, while events such as Oktoberfest and Alpine Lights became annual celebrations. Georgia's new Alpine village was an immediate commercial success. Tourists to theme parks such as Epcot Center are aware that they are visiting simulations, but towns such as Helen implicitly suggest that the ethnic landscape has foundations in the settlement of the area. (Raymond Gehman/CORBIS Canada)

away, remote, and exotic. In all cases these leisure places are constructed with the interests of consumers in mind.

There is intense competition to stage major world events, such as world fairs and the Olympics, as these contribute to local economies, provide opportunities for urban renewal, and enhance place images. Sporting events especially attract both visitors and television viewing audiences. Major sporting occasions, such as the Summer Olympics, the World Cup of Soccer, and the Super Bowl, are spectacles or theater that, at least temporarily, unite diverse cultures in a shared activity. The 2002 World Cup of Soccer final match between Brazil and Germany is (as of 2004) the most viewed event in sporting television history, having been watched in virtually all countries of the world by an audience of 1.1 billion.

More generally, television permits large numbers of people to share time and place without having to move from their homes. Perhaps the first event that might be classified as truly global was Neil Armstrong's 1969 walk on the moon, which was watched by 723 million people. Sixteen years later, 1.6 billion people shared one experience, watching the Live Aid concert in support of Ethiopian famine victims.

RETAILING

Retailing sites in the contemporary developed world offer much more than the opportunity to purchase goods and services—increasingly they offer opportunities for leisure, recreation, and other activities. The development of the shopping mall is one consequence of the rise of consumer culture and the success of advertising that promotes the shopping experience as something more than merely acquiring needed goods and services. Many retail outlets emphasize lifestyle themes, such as the outdoors, that appeal to consumers and increase the attractiveness of their products. Although their existence is bound up with global movements of capital and their inspiration is commercial, malls are not all similar to one another and placeless. They may lack some traditional and local identity, but it is evident that consumers often quickly identify with these new environments; as cultural institutions they offer a comfortable and secure environment for consumers.

Malls highlight one of the dominant features of the contemporary world, namely the tendency to commodify everything that can be commodified. This trend complicates an already complex politics of identity and raises questions about what is authentic and what is inauthentic. But, as the discussion of heritage tourism indicated, such questions are inherently complex and far from easy to answer.

The Mall of America is the largest in the world. First proposed in 1985, it is located in the Minneapolis–St Paul urban region on a 78-acre site close to the airport, two freeways, and numerous motels. The mall opened in 1992 with a 70 per cent retail occupancy rate, but by 1997 the rate was 99 per cent and it was attracting more visitors annually than Disneyland, Disney World, and Sea World combined. The mall houses extensive and varied retailing and recreational facilities, like the much-discussed West Edmonton Mall in Edmonton, Alberta, which was the first mall to explicitly incorpo-

Box 9.7 Directing the Tourist Gaze

In some places, folk-culture characteristics are artificially preserved to create a tourist attraction. Folk cultures especially may be intriguing to dominant cultures precisely because such groups are different, or at least are seen to be different, from the majority, and the tourist industry and cultural producers may sometimes manipulate a local cultural identity. In the case of Nova Scotia, for example, folk identity has been emphasized through an erasure of those components of modern identity related to urbanization and industrialization, as well as those identity differences based on ethnicity, class, or gender. Indeed, tourists only rarely have access to authentic cultural experiences.

Stratford-upon-Avon in the English Midlands, the birthplace of William Shakespeare, is a prime example of a town that is conserved and commodified for tourists such that it is not easy to disentangle what is authentic from what is inauthentic. Packaging the Shakespearian past for visitors involves 'a selective commodification of place identity in the interests of capital accumulation and local economic development'—in other words, a careful directing of the tourist gaze (Hubbard and Lilley 2000:231). As is often the case with historical reconstructions, those responsible for reconstructing and marketing a place are likely to concentrate on characteristics of people and place that are marketable, even if they are not necessarily typical.

Nauvoo in Illinois, the 'Beautiful Place', is one of the best-known Mormon heritage sites and similarly makes a concerted effort to direct the tourist gaze. As the principal Mormon settlement between 1839 and 1844, it had more than 2,000 homes and businesses. Today it is a major tourist attraction, receiving about 220,000 visitors each year. Restoration began in the 1960s, and there are 27 reconstructed buildings, primarily homes and businesses. But, as with Stratford, the tourist gaze is carefully directed, and the image presented is not representative of the past. For example, all the reconstructed buildings are brick, suggesting prosperity and permanence, whereas most of the original buildings were log.

The meaning of many tourist sites is contested such that there are debates about how and where the tourist gaze ought to be directed. De Oliver (1996), addressing the case of the Alamo historic site in San Antonio, argues that it has been culturally refashioned as a site emphasizing the heroic behavior of Anglos rather than as a site of a Spanish mission, thus reaffirming a dominant Anglo identity and a subordinate Spanish identity.

Malls in large cities are about much more than shopping, with the largest also providing a wealth of other consumption opportunities. Inside the Mall of America in Bloomington, southwestern Minneapolis, Camp Snoopy is the largest indoor family theme park in North America. A Kite Eating Tree swing ride that hurls people around its trunk is just one of the many attractions visitors can enjoy at the 7-acre amusement park. (Owen Franken/CORBIS Canada)

rate substantial recreational activities; this example, the prototypical megamall, is detailed in Box 9.8.

Pause for Thought

Are some landscapes of postmodernity—such as shopping malls—really placeless and lacking in identity? Do you think that such a circumstance might simply seem to be the case to those who do not live in and experience those landscapes? For those who frequent malls for shopping or other activities, these places may be full of meaning and significance. The fact that there may be other very similar places elsewhere is not relevant to people frequenting one particular mall. Certainly, one of the ideas evident in the Landscape *magazine tradition, namely the claim that there is no such thing as a dull landscape, does seem compelling in this context.*

FOOD AND DRINK

In the contemporary more developed world, the consumption of food and drink, like retailing, has become much more than some necessary activity. It reflects who we are, where we wish to be, with whom we wish to be, and how we wish to express ourselves to others. It is a social activity and a leisure activity. In short, while it is often said that we are what we eat, it is clear that we are also *where* we eat, and *with whom* we eat.

The role of franchising is of increasing importance in the context of the emerging global culture: 'Today, McDonald's golden arches are one of the most recognizable symbols of American popular culture throughout the world, from Tokyo to Moscow' (Carney 1995b:95). The American landscape of fast food restaurants is discussed in detail by Jakle and Sculle (1999), who stress the virtue of sameness both in terms of the places that sell food and the food that is sold.

Many cultural geographic studies of food, place, and ethnicity are in a Sauerian tradition, but there are also studies that center on food and drink consumption as a process imbued with symbolic meaning, especially when it is seen as an increasingly exotic activity. Building on the work of Said (1978) concerning Western fascination with the Orient as other, May (1996) discussed the tendency of some Westerners to consume exotic foods and drinks, and indeed other exotic goods, as one means of declaring who we are, both to ourselves and to those around us. Viewed in this way, the consumption of exotic food and drink is yet another means of creating self–identity, not dissimilar to such considerations as ethnicity and sexuality; it is a part of our cultural capital. Of course, such ideas are well established in the social psychological literature, where there has been much discussion of how people present themselves, especially through their choice of clothing. But the consumption of exotic foods might also be interpreted as an act of power, with the other being the object of consumption.

Given the increasing interest in consumption and related presentation of self, it is not surprising that there is also an emerging concern with advertising and promotion. Jackson and Taylor (1996:356) note that advertising 'is an inherently spatial practice, playing a crucial role in an increasingly mediated world as part of the national and international expansion of markets; creating uneven patterns of demand across space; and striving for universality but constantly subject to local variations in meaning and interpretation.' In many areas, such as the English West Country, a regional image and related ideas about production are being con-

Box 9.8 The Mall as a Postmodern Consumption Site

The West Edmonton Mall Map and Directory (West Edmonton Mall n.d.) leave little doubt that this place is intended to be more than a collection of retail establishments:

How to shop West Edmonton Mall. It's a lot easier than you might expect . . . and even more fun. Whether you're shopping in a hurry or in the mood to explore, it's all here under one roof. With over 800 stores, there's always one nearby that has what you're looking for. And there's always something exciting to see or do.

Certainly, the inclusion of a hotel, an ice rink, a deep-sea adventure setting that includes submarine rides and dolphin presentations, a waterpark that has artificial waves, a bungee jump, a large amusement park, theaters, clubs, a casino, a bingo hall, numerous eating places, and financial and other services—all in addition to the over 800 stores— leaves little doubt that this is indeed much more than a shopping mall. It is also a tourist attraction, an entertainment center, and a business center, and it is therefore concerned with more than one aspect of consumption.

How are cultural geographers whose research is inspired by the cultural turn studying the mall? One answer is that they are reading it as a communicative text. According to Goss (1993:19), developers are 'designing into the retail built environment the means for a fantasized dissociation from the act of shopping' as one means of manufacturing 'the illusion that something else other than mere shopping is going on'. The mall is being interpreted as a postmodern consumption site, 'a landscape of myths and elsewhereness' (Hopkins 1990:2). As with theme parks such as Disneyland, the mundane is seen as unacceptable in the megamall.

Gregson (1995:137) has expressed concerns about the direction of such research, questioning specifically 'the omission of the fundamental and the mundane (notably shopping) from these readings'. Noting that most shopping is accomplished by women and that women comprise the majority of retail workers, Gregson (1995:137) contends: 'When we look at geographers' readings of the megamall what we find then are masculine and masculinist representations masquerading as universal and homogeneous tendencies in the world of consumption.'

structed with particular reference to the quality of food and drink items.

Concluding Comments

Building on questions of human and place identity, this chapter stresses the symbolic character of landscape—landscapes as representations of cultural values and ideas and as outcomes of competing discourses. Both people and place can be understood in material and symbolic terms, and they can also be meaningful in the sense that one either belongs or does not belong to a group of people and to a place. Places are complex things, and the folk/popular distinction provides a useful measure of, on the one hand, the powerful forces of global capitalism and, on the other hand, the reassertion of the local.

Much of what we understand about people and place is mythical: stories are told and images created that serve to set different groups of people and different places apart from one another. Cities, for example, are often seen as beacons of progress and opportunity attracting migrants at ever-increasing rates, although increasing congestion and crime might summon up quite different images. Rural areas in many parts of the Western world are nostalgically seen in idyllic terms, but at the same time rural depopulation processes are often ongoing.

Although this chapter focuses on some of the more recent approaches to the study of cultural landscapes as places, it must not be forgotten that there remains a substantial interest in the evolution of landscape, in regionalization, and in landscapes as expressions of human and land relations (recall chapters 4, 5, and 6, respectively). Perhaps most significantly, the concern best articulated within the *Landscape* magazine tradition continues to be evident in the humanities and social sciences generally.

It is significant that only a small portion of content of this chapter reflects a more scientific focus. This circumstance is evident also in previous chapters and is an interesting commentary on how cultural geographers view their subdiscipline. Most strikingly, this absence of scientific content contrasts markedly with current interests in the closely related area of social geography, which both acknowledges and incorporates scientific—that is, positivist—ideas and work (see for example Panelli 2004).

Pause for Thought

The places discussed in this chapter are meaningful contexts for particular behaviors. Although they are imagined somewhat differently by different people, it is acknowledged that there are required behaviors for most places. Consider, for example, the behavioral expectations of classrooms, churches, fast food restaurants, and sports arenas. Some of the places we frequent, such as our homes, are always available to us, but many, such as parks and restaurants, open and close on some regular basis and our behavior is adjusted accordingly. Overall, places are so much more than containers of things: they are parts of who we are. All behaviors take place.

Further Reading

The following are useful sources for further reading on specific issues.

Ley and Samuels (1978) identify the humanist tradition in geography, and Berdoulay (1989) provides an account of Vidalian cultural geography.

Tuan (1977, 1989) provides seminal statements on the humanist concept of place. Smith, Light, and Roberts (1998) discuss philosophies and geographies of place. Terkenli (1995) outlines processes through which places become homes. Dominy (2001) discusses home, place, and identity in the context of New Zealand.

Entrikin (1991) stressed the need to conceive of place in both scientific and humanist terms.

Tuan (1982) discusses relations between behavior and place.

In a discussion of the Cameron Highlands, Malaysia, Freeman (1999) provides an interesting example of how a small number of individuals influence the process of landscape creation.

Accounts of habitat and prospect-refuge theories are in Appleton (1975a, 1975b, 1990, 1996).

With reference to mental maps, Downs and Stea (1973) reported that there was a tendency to straighten curves and to generally simplify real-ity; Gould (1963) proposed that each individual has an image of environments that serves as a preference surface, with different locations being varyingly attractive; for various examples, see Gould and White (1986); the most comprehen-sive account of these and related ideas is in Golledge and Stimson (1997).

Downs (1979), Rushton (1979), and Saarinen (1979) include responses to some of the criticisms of research on images and mental maps.

Gade (1982) on the French Riviera as élitist space.

Miller (1968) studied the Ozark region with reference to folk materials, identifying the self-sufficient family farm as the principal way of life. Lamme and Oldakowski (1982) reported a close link between vernacular regions and an established cultural divide in Florida. Zdorkowski and Car-ney (1985) identified past vernacular regions through the inclusion of a question on name changes through time. See also Raitz and Ulack (1981a) on Appalachia.

Stein and Thompson (1993) discussed the distinctive identity of Oklahoma, as viewed from both inside and outside, in terms of what they called psychogeography, with the emphasis on atti-tudes rather than material landscapes.

Donaldson (2001) shows how poetry can be used in the context of elucidating Pattison's four tradi-tions of geography (spatial, area studies, human-environment, and earth science).

Statements about or examples of geographic studies of literature are provided by Salter and Lloyd (1977) and Noble and Dhussa (1990).

Calder (1998) uncovers environmental determinist content in creative writing set in the Canadian Prairies.

Studies of geography and literature that emphasize a humanist perspective include those by Birch (1981), Barrell (1982), Herbert (1991), and Hudson (1992). See also the edited books by Pocock (1981) and Simpson-Housley and Mallory (1987).

Not all authors associated with places have prompted a reevaluation of that place. According to Spooner (2000), Philip Larkin, a poet closely asso-ciated with Kingston upon Hull in eastern Eng-land, has not prompted the creation of a Larkinland comparable to, say, Bronte country.

Studies of geography and literature influenced by the cultural turn include those by Sharp (1994, 1996, 2000), Kong and Tay (1998), Hugill (1999), and Kneale (2003).

Discussions of Wordsworth and the English Lake Dis-trict include those by Squire (1988), Whyte (2000), and Kay (2000). Barrell (1972) provides an insightful interpretation of the poetry of Clare.

Rodgers (1997) on the writings of H. Rider Haggard; Cook and Okenimpke (1997) on Ngugi; Gal-lagher (1991) on Cootzee.

For a discussion of the cultural geographic use of cin-ematic environments see da Costa (2003), Hor-ton (2003), and Peckham (2004). For art see Rees (1984), Daniels (1994, 2004), and Hall (2003). For exhibitions see Crang (2003). For photographs see Schwartz (2003).

Although there is rarely any meaningful integration of the two traditions, many of the music studies informed by the cultural turn do acknowledge the Sauerian tradition. One exception is Leyshon, Matless, and Revill (1995:423) that explicitly sets out to introduce the study of 'the place of music' and to 'discuss previous work on music by geographers' and yet makes no refer-ence to studies conducted in the Sauerian tradi-tion or to the research of leading American writers in the field such as Carney, Ford, Fran-caviglia, Gritzner, Lehr, and Nash, to cite but a few (see Carney 1990).

Examples of studies of the geography of music include those by Gill (1993), Halfacree and Kitchin (1996), Kong (1996), Carney (1996), and Connell and Gibson (2004). A Special Edition of the *Journal of Cultural Geography* (1998, vol. 18, no. 1) includes an overview and additional studies. See also Back (2003).

The music of the group U2 is used by McLeay (1995:1) to 'show how geographic imagery is used for political purposes.' See also McLeay (1997a) on regional identity and the development of a par-ticular musical style. Kong (1995) discusses the production of popular music and the way in which it can be a form of cultural resistance that can be used by the state to maintain and rein-force particular ideologies.

Connell and Gibson (2003) is the first book on geog-raphy and music that pays detailed attention to the way in which music is interconnected with cultural and economic capital.

The Oxford Companion to Music (Latham 2002) includes an informative entry on nationalism.

Discussions of landscape as informed by humanism and the cultural turn include those by Olwig (1996b) and Kobayashi (1989). Walton (1995) on the idea of landscapes as texts. Muir (1999) discusses the current diverse yet incomplete geographic approaches to landscape. Granö (1997) outlines a traditional European interpretation of landscape.

Hirsch and O'Hanlon (1995:22), in an apparent desire to establish an anthropological interest in landscape seen to be different from the interests of cultural geographers, proposed the idea of landscape as process—'one which relates a "foreground" everyday social life (us the way we are) to a "background" social existence (us the way we might be)' (see also Bender 1993).

Both *Landscape* and the *Journal of Cultural Geography* are important outlets for discussions of ordinary landscapes. See also Zelinsky (especially 1994). An atlas of North American societies and cultures edited by Rooney, Zelinsky, and Loudon (1982) is a comprehensive description of one continental region. The wealth of interest in ordinary and sometimes not quite so ordinary landscapes is reflected in the introductions to and some of the contributions included in the edited volumes, *Landscape in America* (Thompson 1995b), *Understanding Ordinary Landscapes* (Groth and Bressi 1997), and *Everyday America* (Wilson and Groth 2003). See also Jakle (1987) and Hart (1998).

Tuan (1991b) on the naming of places.

Howitt and Suchet-Pearson (2003) discuss the need to relate claims to land to different knowledge systems.

Baker (1992) on ideology and landscape.

Eliade (1959) distinguishes sacred and profane space.

Naylor and Ryan (2003) discuss new religious sites in England, including the example of a Hindu temple in northwest London. Biswas (1984) on sacred buildings.

Examples of aboriginal sacred spaces are included in Carmichael, Hubert, Reeves, and Schanche (1994), and in Oakes, Riewe, Kinew, and Maloney (1998).

Francaviglia (2003) provides an insightful personal discussion and interpretation of the spiritual geography of the Great Basin with reference to Native peoples, Mormon settlers, and scientific observers.

Halfacree (2003) on the idea of the British rural landscape as idyllic.

Duncan (1990) on the early nineteenth-century landscape in Sri Lanka; this is an influential landscape study in the newer tradition of landscape arguing that accounts of the world and the world itself are both intertextual.

Cosgrove and Daniels (1988) discuss iconographic analyses. Also Cosgrove (for example, 1984, 1997b) on the landscapes of Renaissance Venice and Vicenza within their larger cultural framework. Domosh (1989) discusses the example of a single newspaper building, the New York World Building constructed in 1890 in New York City, in terms of both personal egos and industrial capitalism (see also Domosh 1999). Eyles and Peace (1990) provide an iconographic analysis to help understand two prevailing but contradictory images of Hamilton, Ontario, namely those of smokestack city and cultured city. Kong (1993b) on religious buildings in Hong Kong. Cartier (1997) on the symbolism of bank buildings in Hong Kong.

Carter (1987) on the exploration and mapping of Australia as this related to the production of knowledge, and Ryan (1996) on the exploration and description of social space by European explorers.

Osborne (1988) considers the iconography of national identity that was forged by painters, stressing the role played by the early twentieth-century 'Group of Seven' and their concern with the Canadian North (see also Osborne 1992).

Taylor (2000) discusses Perth, Australia, in terms of the dominant reading of the city as a place where the sun always shines.

Zelinsky (1988), Jeans (1988), and Johnson (1995) discuss monuments as a reflection of ethnic or national identity. Bell (1999) outlines the redefinition of national identity in Uzbekistan. Foote, Toth, and Arvay (2000) discuss the example of Hungary after 1989. Crampton (2001) considers one iconic South African monument and

Crampton (2003) reflects on political imagery at the South African National Gallery. Marshall (2004) explores the power of war memorials in everyday English landscapes.

Featherstone (1995) distinguishes local folk cultures and a global popular culture. Carney (1998) and Francaviglia (1996) discuss folk-culture landscapes. Carney (1995a) discusses popular culture as the culture of the majority.

Frenkel and Walton (2001) and Schnell (2003) discuss examples of ethnic heritage tourism.

McKay (1994) identifies ways in which the tourist industry and cultural producers have manipulated the cultural identity of Nova Scotia.

Jackson (1992), Getz (1995), Abram (1997), and Waterman (1998) consider the tensions that may be evident in a community if some but not all members perform for tourists.

Discussions of heritage sites include those by Young (1999), Graham, Ashworth, and Tunbridge (2000), and Hubbard and Lilley (2000). Markwick (2001) outlines the cultural commodification of Ireland. Boyd, Cotter, and Gardiner (2001) describe the interesting example of a small Australian town that is 'more Scottish than Scotland.' Kneafsey (1998) discusses tourism and place identity in Ireland. Ringer (1998) outlines the cultural landscapes of tourism.

Analyses of memorials in landscape include those by Bender (2001), Crampton (2001), and Olsen and Timothy (2002). Leib (2002) discusses Monument Avenue in Richmond, Virginia; see also Monument Avenue (2004).

Teather (1998) considers the symbolic significance of Chinese cemeteries in Hong Kong. Auster (1997) focuses on several different interpretations of a memorial column on a hillside outside Armidale,

Australia. Kong (1999) discusses deathscapes and the new cultural geography.

Schnell (2003a, 2003b) describes Lindsborg, Kansas, as a Swedish-themed tourist town. Lovato (2004) discusses Santa Fe as a town that is both genuinely different but that also strives to be different in order to encourage tourism.

Slater (2003) outlines cultures of consumption. Goss (2004) reviews studies of geographies of consumption.

Ley and Olds (1988) and Squire (1994) provide examples of landscapes of consumption and spectacle. Yeoh and Teo (1996) discuss the example of Dragon World in Singapore.

Bale (2003) provides an overview of sports geography. Pillsbury (1974, 1989) discusses stock car racing in the American South; Rooney (1974, 1993) and Rooney and Pillsbury (1992) discuss golf and football. Hague and Mercer (1998) examine ties between the small Scottish town of Kirkcaldy and the local soccer team, Raith Rovers.

Miller et al. (1998) discuss shopping as it relates to place and identity in a British context.

Hopkins (1991), Goss (1993, 1999a, 1999b), Gerlach and Janke (2001) describe malls as sites of consumption. See also the Special Issue of *Canadian Geographer* (1991, vol. 35, no.3) on the subject of West Edmonton Mall and megamalls generally.

Rooney and Butt (1978), and Shortridge and Shortridge (1995, 1998) on food as it relates to place and ethnicity. Bell and Valentine (1997) provide a pioneering account of food and drink consumption as part of our cultural capital. Valentine (1999) on eating as a bodily practice. May (1996), Kneafsey and Ilbery (2001), and Cook and Harrison (2003) on specialty foods. Gade (2004) looks at the role of place and tradition in French viniculture.

10

Cultural Geography— Continuing and Unfolding

This book is a journey through a body of changing ideas and changing practices, and the aims of this final chapter are to assess where cultural geography stands today and where it may be heading. Both aims are difficult to achieve.

Recall that cultural geography has been described as a 'scholarly discourse that has shifted from the comfort of placid marginality toward the overheated vortex of ferment and creativity in today's human geography' (Zelinsky 1995:750). There is much evidence to support this assertion. For example, there are questions about integrating physical and human geography in general, questions that have real implications for cultural geographers, especially those concerned with ecological issues. There are ongoing debates about the relative merits of Sauerian approaches and more socially and theoretically informed approaches, especially those associated with the cultural turn. These debates center on the understanding of culture, the meaning of landscape, the merits of searching for causes or seeking understanding, the appropriateness of particular writing styles, and the interweaving of culture, politics, and economy. In short, cultural geography is contested terrain, with different scholars putting forward different interpretations. This book is an attempt to reflect, incorporate, and, on a few occasions, reconcile these differences.

This chapter begins by commenting on ongoing work centered on cultural landscapes, then suggests that it is necessary to focus more on global issues and to broaden analyses of difference and others. The chapter concludes with comments on the contested status of contemporary cultural geography.

Cultural Landscapes

The study of cultural landscape has been a persistent concern of cultural geographers from the late nineteenth century to the present. Since it was first explicitly formulated by Ratzel, further developed by Vidal and Schlüter, and then decisively introduced into English-language geography by Sauer, the idea of cultural landscape has been an almost endless source of conceptual inspiration and a rich vein for local and regional understanding. The impact of Sauer's methodological statements and the example set by the empirical studies conducted both by Sauer and by his students is difficult to exaggerate. For Sauer, at least in principle, there was a universal model that could be used to inform analyses of cultural landscapes, whereas the contemporary interest in cultural landscape incorporates a wide variety of approaches. Most of the earlier studies focused on landscape evolution, regional landscapes, and landscapes as outcomes of human-and-land relationships.

A principal concern in this tradition is with understanding the impact of humans, typically seen as members of cultural groups, and this concern broadened, beginning in the 1970s, through studies that stressed the need to read that human impact and to interpret it in appropriate larger contexts. Of course, reading a landscape intelligently requires an awareness of authorship, and one of the earliest and most critical contributions associated with the cultural turn was the questions raised about the emphasis placed on cultures as groups and the apparent rejection of human agency. Since about 1980, approaches to the study of cultural landscape have diversified even further, reflecting in part the proliferation of theoretical bases, and as a result there are today concerns with symbolic landscapes, with landscapes linked to prevailing social power relations, with landscapes of production and consumption, and with landscapes as texts.

Perhaps the most meaningful difference between the Sauerian approach to landscape and the more recently developed approaches to landscape lies in the transition from a concern with *culture* to a concern with *cultures*. Not all cultural geographers have welcomed this conceptual broadening of interest in cultural landscape. Hart (1995:23), for instance, writes: 'I am quite content with the simple vernacular definition of landscape as "the things we see", and I am saddened by the way the meaning of the word has been transmogrified. Some people have endowed the concept of landscape with magical, mystical, or symbolic significance, or have loaded it with metaphysical significance.' In turn, Henderson (1997:95–6) criticizes Hart's point of view as one part of a more general criticism of those landscape studies that stress the richness and diversity of landscape at the expense of such considerations as human inequalities and related social struggles.

Although particular understandings of culture are central to most work in both traditions, two qualifications of this statement are worth noting. First, with reference to landscape school geographers through to the 1970s, Mikesell (1977:460) observed that 'most cultural geographers have adopted a laissez-faire attitude towards the meaning of culture.' Second, some cultural geographers have questioned the intent of the new cultural geography and have sought a major reorientation. These cultural geographers have argued that there is no such thing as culture, but rather only the idea of culture (Mitchell 1995). Further, they have claimed that there is no 'need of a concept of culture as such', but only a need for 'a much deeper, clearer, more operational conception of human behavior and development' (Wagner 1994:5).

Regardless of perspective, one key point emerges from this ongoing and diversifying interest in cultural landscape: landscapes, as the places we create, are important in any attempt to understand who we are—our human identity—and also how we live—our social, political, and economic worlds. Inevitably, then, such studies also help us appreciate our place in the larger world. The interest being shown by other disciplines in this traditional geographic concern speaks clearly to the growing recognition of this point.

Global Cultural Geographies

As noted in Chapter 1, Wagner (1990:41) asserts that 'a theoretically well-grounded, intellectually vigorous, and practically effective social and cultural geography might well assume, in time, a major role in guiding and guarding the evolution of humanity's environments.' Are cultural geographers responding to this ambitious claim?

Given the concern with understanding people and place, cultural geography is certainly well positioned to contribute in various applied areas, and the ecological analyses and studies of cultural and regional inequalities discussed in this book attest to the contributions being made by cultural geographers. Cultural geography continues to embrace a diverse set of ideas and to address issues with multiple layers of relevance. But is this enough? Possibly not, as the following three observations suggest.

First, cultural geographers are saying little about questions of global environmental and human concern. Some studies within cultural ecology offer analyses of particular regional circumstances, but studying these issues at the global scale seems quite another matter. Not much is being written about the all too evident gross inequalities of the human condition.

Consider that in the early twenty–first century, about 1 billion people are malnourished, and that of these, about 650 million people live on the very edge of existence, lack a reliable supply of food, access to clean water, and sustained medical care. Facts such as these identify urgent global issues and are but suggestive of a long and depressing list of problems. While all scholarly work can yield valuable contributions, studying the lives and landscapes of the undernourished and the malnourished in the less developed world with the aim of identifying problems and proposing solutions is one aspect of the geography of consumption that merits much greater effort from cultural geographers. Certainly, major global differences of wealth and power are being largely neglected. Recall the following quote from a political geographer: 'We need to construct a new global geography that focuses on the world map of global inequalities. We should very consciously proclaim this to be the most important map in the world and focus our Geography accordingly' (Taylor 1992:20). Cultural geographers might profitably pursue this line of argument.

Second, cultural geographers have had relatively little to say about the formation of cultural regions at the global scale, despite the suggestion that this is, as Meinig (1976:35) put it, 'the greatest topic in historical geography'. Are there reasons to suggest that cultural geographers might pursue this topic further? One answer comes from an insightful and challenging commentary on Meinig's ambitious series of North American studies (discussed in Box 6.2), in which Harris (1999:10) notes that 'there is awe from a dwindling group of historical geographers, but little response from the rest' because the work 'is not tuned to theory currently in vogue'. This comment must not be taken lightly. Just as much contemporary cultural geography highlights weaknesses, inadequacies, and omissions in earlier work, so that same cultural geography might neglect avenues not favored by the current body of conceptual work. The key point made by Harris is that cultural geographers need to pursue a variety of conceptual inspirations, not simply those prompted by the cultural turn.

Third, there is much discussion today about the possible emergence of a global culture and about the reassertion of some regional and local cultural identities in explicit opposition to globalization. In this context, cultural geographers might usefully evaluate, for example, the six possible global cultural scenarios outlined by Masini (1994:20–5). A first possibility is that cultures, particularly living community cultures, might be reduced to just a relict, museum, or tourist role and be unable to challenge the dominant global identity that reflects Western technology and values. Second, there might be a combination of cultural continuity and change that allows the core elements of cultures to remain while involving some acceptance of global trends. A third option is that many cultures might resist the dominant global trend, with any cultural change being internally generated. A fourth option is that there might be a widespread acceptance by most cultures that no one culture is complete in itself. Fifth, it might be that all cultures are influenced by information from other cultures, with Western culture as the most influential because of advances in information technologies. Finally, perhaps a non–Western culture, most likely an Asian culture, becomes globally dominant. Certainly, cultural geographers today would do well to consider Masini's views on the future of both a global culture and of the myriad of local cultures. On the one hand, the seeming impossibility of isolation means that differences disappear, while on the other hand, local cultural differences seem able to resist many homogenizing forces. Perhaps the future is to be one of a global culture comprising key shared values that coexists with detailed local differences.

Difference and Others

The material included in the discussions of ethnicity, gender, sexuality, and the disadvantaged reflects concerns about identity, place, and relations between identity and place. Although it is common to note that these interests reflect a concern with difference, they are primarily a concern with others—understood as those who are excluded in some way from larger society.

There are three closely related matters of concern regarding the current interest in difference and others. First, there is more than one way to interpret claims about the emergence of alternative cultures, such as those based on a particular sexual preference. In a discussion of the globalization of culture, Albrow (1997:150) states: 'The multiplication of worlds means that individuals can inhabit several simultaneously, but secondly that each severally can only make a small selection from the many which coexist. The result of a plurality of individuals making their own selections is that each builds a different repertoire, and its total scope is obscure to everyone else.' This is a challenging claim implying that the membership of individuals in particular groups is both far from fixed and, more critically, not necessarily a meaningful marker of their identity (see Box 8.2).

Second, an emphasis on human difference rather than human unity often has the unfortunate consequence of implying not just difference but also inequality. Europeans during the period of colonial expansion saw numerous others as different, and this recognition was accompanied or soon followed by inequality. Similarly, any contemporary expression of pride in a particular identity, a particular difference, can easily—although not necessarily—lead to ideas of superiority. This simple point is consistently evident: the need to listen to others is clear, but the need to stress difference is considerably less so.

Third, when discussing questions of power and authority as these relate to human differences, cultural geographers need to be more explicit about the values that underlie critical analyses of, for example, racist and sexist attitudes and practices. Indeed, it is possible that the current preoccupation with those who are different might be a case of overcompensating for the earlier emphasis placed on the dominant culture.

Subdiscipline or Heterotopia?

The cultural geography discussed in this book balances the rich heritage of the subdiscipline with more recent innovations. However, most recent cultural geography books favor a more limited focus by excluding the Sauerian landscape school and focusing almost exclusively on concepts developed and work accomplished since about 1970. This book contends that such a perspective does cultural geography a real disservice by denying a scholarly history and numerous past *and* ongoing achievements.

While it is clear that there are many differences between older and newer approaches, as well there ought to be, it is equally clear that the new did not appear from nowhere—indeed, as suggested throughout this textbook, the new is both a continuation of and a reaction to the older approaches. But not all those working in cultural

geography interpret matters in this way, and Box 10.1 reviews one debate that highlights some of the tensions in cultural geography today.

Two Views of Cultural Geography

So what is cultural geography? This may seem like a strange question to be asking in this final chapter, but it is a necessary one. To repeat, this textbook sees contemporary cultural geography as comprising two principal and related bodies of work. On the one hand, the approach first introduced in the 1920s continues to be applied; on the other hand, the cultural turn and all that this implies is an integral part of the subdiscipline. This view of cultural geography as including both traditional and new content is evident also in the results of a survey of members of the Cultural Geography Specialty Group of the Association of American Geographers (Smith 2003).

But it is important to note that there is another and very different perspective on cultural geography that is evident in several books, including the *Handbook of Cultural Geography* (Anderson, Domosh, Pile, and Thrift 2003a) and *A Companion to Cultural Geography* (Duncan, Johnson, and Schein 2004), both of which aim to provide overviews of cultural geography. Consider the following.

One result of the survey reported in Smith (2003:24–5) refers to the request that survey respondents identify the top five most outstanding living practitioners of cultural geography. The top five, in order of votes received, were Wilbur Zelinsky, Pierce Lewis, Donald Meinig, Yi-Fu Tuan, and Denis Cosgrove. The work of these five cultural geographers is reflected often in the contents of this textbook. But what role do these five outstanding cultural geographers play in the 31-chapter *Handbook* and the 32-chapter *Companion*? Reference to the *Handbook* index shows Zelinsky and Lewis with zero entries, Meinig with two, Tuan with two, and Cosgrove with eleven. Reference to the *Companion* index shows Zelinsky, Lewis, and Meinig with zero entries, Tuan with three, and Cosgrove with fifteen. A second result reported in Smith (2003:25–6) identifies Zelinsky's *Cultural Geography of the United States* as the most outstanding work in cultural geography, a book that is referenced once in the *Companion* and not at all in the *Handbook*.

How are we to explain the discrepancy between the perceptions of the members of the Cultural Geography Specialty Group of the Association of American Geographers (and also the contents of this textbook) on the one hand and the editors of and contributors to the *Handbook* and *Companion* on the other hand?

Expressed simply, there are currently two different ideas about what cultural geography is and what cultural geographers do. Both the survey respondents and this textbook understand work in cultural geography as being informed both by the Sauerian tradition and by more recent conceptual contributions. Both the *Handbook* and the *Companion* elect to exclude the Sauerian tradition and thus almost all of the cultural geography practiced prior to about 1970 as well as much practiced since that time. Indeed, although one stated purpose of the *Handbook* is to 'reflect the varieties of cultural geography being undertaken' (Anderson, Domosh, Pile, and Thrift 2003b:xviii),

the great volume and variety of much past and present American cultural geography is simply excluded. To cite just one example, there is no discussion of the homeland concept. What appears to be happening is that the label 'cultural geography' is being used to encompass only some of the work that, in the view of this textbook and many other cultural geographers, is appropriately included under that label. In light of these different understandings, it will be interesting to see how cultural geography unfolds in the early twenty-first century.

The Past is Prologue

From the perspective of this textbook, ignoring the Sauerian tradition is a mistake not only because it overlooks an extensive body of work in cultural geography but also because it fails to acknowledge substantive precedents to some current work. Admittedly, some cultural geographers may not wish either to lionize Sauer or to institutionalize a landscape school, but to largely omit these matters from meaningful discussions seems

Box 10.1 Debating Approaches

An exchange of views concerning the categories of 'nature' and 'culture' highlights some issues involved in the ongoing debate concerning an appropriate conceptual basis for cultural geography.

Focusing on the West Coast of Canada, Willems-Braun (1997a) explored ways that colonial power relations were established through the construction of nature as separate from culture and how such relations are reproduced in the present. Thus, discussions of nature tend to stress the differences of opinion between those who see nature as a resource to be exploited (such as logging companies) and those who see nature as a wilderness to be preserved (such as conservationist groups). But more important, it can be argued that neither of these options allows for a meaningful appreciation of Native understandings of nature, and Willems-Braun approached the matter by discussing some current conflicts concerning the temperate rainforest ecosystem as these reflect colonial relations established at an earlier time, and by attempting to understand these through a cautious use of selected postcolonial theory.

Commenting on this work, Sluyter (1997:700) noted the absence of any reference to the Sauerian tradition that typically criticized the idea that North America was a primordial wilderness prior to European involvement:

> in his avid digging-up of the colonizers' taken-for-granted epistemologies, Willems-Braun indiscriminately buries other epistemologies. In concert with many smitten by the uptown coolality of postmodernism, he negates the work of other geographers who have striven to understand the ways in which colonizers rhetorically and materially invented colonized peoples and natures.

Responding to these comments, Willems-Braun (1997b:706) expressed concern at the suggestion that the Sauerian tradition was 'a benchmark against which all others must situate their work', and justified excluding the Sauerian tradition on the grounds that it was not an appropriate conceptual basis for work that emphasized questions relating to power and politics, and also on the grounds that the tradition incorporated an imperialist nostalgia.

not just an inappropriate rewriting of the disciplinary past but a rejection of much current work.

An appraisal of current research, based on a survey of current books and journal articles, confirms that both the traditional and the newer cultural geography are with us today, and that both are changing. Newson (1996:279) observes that 'in the USA traditional cultural geography remains an active, and largely distinct, research field that is unlikely to be totally overtaken by new approaches.' On the other hand, the new—much of which is no longer quite so new, of course—has certainly proven to be an enrichment, and most of the recent statements applauding the vitality of cultural geography reflect what can be seen as a 'transformation' (McDowell 1994:146) or a 'reinvention' (Price and Lewis 1993a) of the earlier tradition. As evident from the contents of this textbook, the use of both terms—*transformation* and *reinvention*—is understandable, as the new certainly reflects some significantly different conceptual content and includes a different set of empirical issues. In this sense, the new is not something that has evolved easily from the traditional; rather, as a significantly different set of ideas and practices, it has either been added on to or perhaps even replaced previous ideas and practices. Nevertheless, many links can be identified between the traditional and the new, including

- continuing interest in cultural landscape;
- a mutual mistrust of scientific method;
- an interest in human-and-land relationships;
- a predilection for fieldwork;
- a concern for regional identity;
- a concern for group identity; and
- a willingness to conduct research at various spatial, temporal, and social scales.

To deny these continuing concerns seems unfortunate.

An Integrated Human Geography?

Clearly, a key question to ask about cultural geography—really a question about the larger discipline of geography, or indeed about the way in which knowledge is structured—concerns the viability of identifying a subdiscipline labeled 'cultural geography'. Described in Chapter 1 as a compositional subdiscipline—one concerned with some particular subject matter—cultural geography is always being questioned, as are all other compositional subdisciplines. Certainly, owing largely to the cultural turn, the links with social, political, and economic geographic subdisciplines are increasingly close. For example, in a discussion of economic geography, Crang (1997:4) considers five options for relating the economic and the cultural: a continued opposition of the economic and the cultural, effectively resisting the cultural turn; an export of the economic to the cultural; a view of the economic as embedded in a culturally constructed context; a view of the economic as represented through the cultural, acknowledging that places are cultural constructions; and an understanding of the cultural as materialized in the economic.

Similarly, one study of the discipline of geography was structured around themes rather than subject matter, using four principal headings as follows (Mabogunje 1997): conceptual fundamentals relating to space, time, theory, and methodology; environment and society, including human transformation of earth, climate change, and human response to hazards; spatial structures and social processes, including market and cultural forces, particular modes of production, transportation, and trade; and spatial organization and globalization processes, including the role played by the state, transnational corporations, and megacities. Such a framework for writing about geography effectively incorporates the cultural into a larger geographic context.

Looking Forward

In light of the evident disagreements and uncertainties, what might be the future of cultural geography? As this textbook has often had occasion to stress, cultural geography is contested intellectual terrain, meaning that describing where we are heading is well nigh impossible. In most cases, textbook authors outline the way ahead as a continuation of current trends; at best, such predictions are but a partial truth. Certainly, it is reasonable to assume that the principal current trends will continue, but it is also reasonable to assume that new trends will appear.

It is worth repeating that this book has an ambitious and, for some, a questionable goal, namely that of integrating, or at least incorporating under some common headings, work that might be variously labeled traditional or new cultural geography. There is one principal reason—in addition, of course, to authorial limitations—why this goal is at best only partly attained: cultural geography is all too clearly a moving target, with new, often imported, concepts informing increasingly diverse analyses, and with ever-closer connections to such other traditional compositional subdivisions of human geography as social, political, and economic geography. These are certainly healthy signs, but just as certainly, it makes accommodating the contents of cultural geography within a few key themes—chapters in the context of this book—an unenviable task. Indeed, there is a tendency for reviewers of cultural geographic literature to focus on selections of material rather than on addressing some perceived totality.

For some cultural geographers the goal of integration, as it might appear to imply unity, is questionable because as Duncan (1994a:402) argues, it 'reflects the modernist will to order and discipline, to be governed under a master narrative'. Indeed, although the *Handbook* is one of a series intended to 'represent the "state of the art" in their specific fields', the editors assert that 'if there is one thing about cultural geography that we know for sure, it is that it is not a field' (Anderson, Domosh, Pile, and Thrift 2003b:xviii). To deny the existence of a field and yet at the same time to edit a text that identifies its contents as cultural geography may seem problematic to some practitioners. For the editors of the *Handbook*, cultural geography is characterized by 'its disruption of the usual academic boundaries and by its insatiable enthusiasm for engaging new issues and ideas' (Anderson, Domosh, Pile, and Thrift 2003b:xviii), a description that might easily apply to many other areas of scholarly interest.

With these comments in mind, cultural geography might be described as a *heterotopia*—a site of incompatible discourses—and with this view in mind, Duncan (1994a:402) writes approvingly of the fact that cultural geography 'is no longer as much an intellectual site in the sense of sharing a common intellectual project as it is an institutional site containing significant epistemological differences'.

Whatever the merits of the claim that integration is a questionable goal, this text accepts that there is value in discussing cultural geography—all cultural geography as broadly conceived—within a single organizational context, and this value is especially clear when the needs of student geographers are being considered. Organizational frameworks such as the one adopted in this book aid in understanding diverse ideas and analyses without in any sense imposing unwavering identities on ideas and analyses or suggesting that the framework used is in some sense correct and uncontestable, a point made in the opening chapter of this textbook.

To conclude, this text is sympathetic to the interpretation of cultural geography as 'one of the most dynamic fields within geography today' and to the claim that the 'great heterogeneity' of work being produced is 'surely a sign of strength in a subdiscipline that was regarded by many as marginal as recently as the 1980s' (Murphy and Johnson 2000b:1). Indeed, it seems likely that cultural geographers will continue to move the subdiscipline forward in several directions, employing both traditional and new concepts to address issues of environments, landscapes, identities, and inequalities at both global and local scales.

Further Reading

The following are useful sources for further reading on specific issues.

Dowling (1997) identifies some issues relating to cultural planning in Australia in the context of contemporary cultural geography

Sayer and Storper (1997) discuss the need to understand the values that inform critical cultural geographic analyses.

Schein (2004) comments on the tradition of cultural geography, and Scott (2004) reviews recent developments.

More generally, students interested in pursuing cultural geography further will benefit from appraising the contents of Murphy and Johnson (2000a), Anderson, Domosh, Pile, and Thrift (2003), and Duncan, Johnson, and Schein (2004).

Glossary

ableism   Discrimination on the basis of physical and/or mental abilities. Ableism typically involves the assumption that persons with disabilities are not 'normal.'

acculturation   The process by which a minority or politically weaker cultural group undergoes gradual cultural change to become closer in character to a majority or dominant cultural group. Acculturation does not involve a complete cultural change.

acid rain   An environmental problem caused by the burning of coal, oil, and natural gas, which releases the oxides of sulfur and nitrogen into the atmosphere, resulting in acidified rain that damages plant and animal life.

adaptation   Human adjustment to environmental conditions, both physical and cultural, in order to ensure that the needs of the group are met and that conflict is reduced. It is common to refer to the particular way in which a group meets those needs as an *adaptive strategy*. The term *cultural adaptation* refers to the idea that culture is an adaptive system.

ageism   Discrimination or prejudice on the basis of age. Ageism is often based on an inappropriate assumption that people of similar age have other characteristics in common. Unlike racism and sexism, it is likely that all people suffer from ageism, as identities based on age necessarily change through time.

agoraphobia   A fear of public places and of people in those places.

agricultural revolution   A series of changes that began about 12,000 years ago (ya), near the end of the Pleistocene, involving a movement away from hunting, fishing, and gathering to food production based on the domestication of animals and plants. The agricultural revolution marks the end of the cultural period known as the Mesolithic and the beginning of the Neolithic.

alienation   The circumstance in which a person feels indifferent to or estranged from nature or the means of production. Alienation is based on the sense that our human abilities are taken over by other entities.

anthropocentric   Regarding humankind as the central or most important element of existence, especially as opposed to God or animals (compare *ecocentric*).

apartheid   A system of policies governing the spatial separation of groups of people distinguished on the basis of a racial classification; also, the social distinctions made between those groups. The policies and practices of apartheid—an Afrikaans term meaning 'apartness'—were applied in South Africa between 1948 and 1994.

artificial selection   The process by which humans chose specific members of an animal or plant species to reproduce and to live with (compare *natural selection*).

assimilation   The loss by members of a cultural group of all their previous cultural traits through the adoption of the traits of some other dominant group with which they are in contact. Assimilation involves a loss of identity.

authority   The right, usually by mutual recognition, to require and receive submission by others.

behavior analysis   A comprehensive approach to the study of behavior pioneered by Skinner and informed by the philosophy of behaviorism. The traditional concepts as developed by Skinner include operant conditioning and contingencies.

body   Especially in feminist geography, *body* is understood both as a personal space that contributes to identity formation and as a basis for distinguishing the self from others. In addition, the body is, literally, the basis for our interactions with physical and built environments. Increasingly, the body is being interpreted as one of the concerns of cultural geographers, being variously interpreted as a site of resistance (as in some feminist and postcolonial theory), as a producer of meanings, and as a declaration of subjectivity. Several social theorists, including Bourdieu and Foucault, have provided influential conceptual discussions.

capitalism   A social and economic system for the production of goods and services that is based on private enterprise, that involves a separation of the producer from the means of production, and that allows relatively few individuals to have access to resources.

carrying capacity   The number of people that a given region is able to support, given its specific technological and other cultural characteristics.

Cartesianism   A term that refers to the need for mechanical and mathematical explanations in physical science. Descartes was a key figure in the emergence of modern philosophical and scientific thinking. The Cartesian theory of the mind, *dualism*, maintained that the mind is entirely separate from the body, a view that was famously referred to by Gilbert Ryle as a 'ghost in the machine'.

catastrophism   The idea that the world has experienced and results from a series of sudden violent and unusual events (compare *uniformitarianism*).

civilization   A term with various contested meanings that is often used to refer to a culture with an agricultural surplus, a stratified social system, some specialization of labor, a form of central authority, and a system of keeping records. More generally, it is used to refer to any culture of global significance.

class   A term in frequent use but with disputed meanings. Generally, it refers to a large group of individuals with similar social status, income, and culture. The presence of classes in a society explicitly reflects divisions based on the unequal distribution of economic goods, political power, and cultural status.

cognition   Human thought processes, including perception, reasoning, and remembering.

cognitive map   The model or mental representation of the world in which a person lives. Every individual's cognitive map is different.

colonialism   The process or policy, by a state or a people, of establishing and maintaining authority over another state or people.

conservation (or **constructivism**)   A general term referring to any form of environmental protection, including preservation.

constructionism   A position that claims that all of our conceptual underpinnings, such as ideas about identity, are necessarily contingent and dynamic, not given or absolute (compare *essentialism*). See *social construction*.

consumer culture   A term that, as discussed by Baudrillard, is understood with reference to institutions of consumption, such as shopping malls and advertisements. Consumption is a symbolic practice that needs to be interpreted.

contextualism   Broadly, the claim that it is necessary to take into account the specific discourse, or *context*, within which any account of ideas and facts takes place.

cornucopian thesis   The argument that advances in science and technology will continue to create resources sufficient to support the growing world population.

counterfactual   Used to designate a statement that asks what might have been the consequences if some detail of history had been different. Counterfactual statements may be used to analyze presumed cause–and–effect relationships through a comparison of the observed world and some hypothetical world derived from contemplating a different history.

creation myths   Stories, often invoking the supernatural, that explain both the origin of the world and the origin and early history of people in the world.

cultural ecology   The study of the interactions between humans and nature; more specifically, the analysis of culture as an adaptive system. The term is sometimes used to refer to the discipline of cultural anthropology.

cultural hearth (or **cultural core**)   The place of origin, the heartland, of a culture, from which it diffuses outwards.

cultural realm   An extensive, often continentally based, area within which some uniformity of cultural practice is evident. A cultural realm may be a cluster of related cultural regions. Necessarily, it is less internally homogeneous than is a region.

cultural region   An area that is occupied by a cultural group and that reflects that occupance both in the visible material and symbolic landscapes.

cultural studies   An area of academic interest concerned with various types of texts as these help us understand how values, meanings, and identities are produced. Cultural studies focuses primarily on the political meaning of culture—on gender, sexuality, ethnicity, nationality, and class.

cultural trait   One element in the normal practice of a cultural group. A *cultural complex* is some related set of traits.

cultural turn   A set of changes in social science, prompted by developments in philosophy and social theory, that recognize that culture is too important to be reduced to economics and politics.

culture   Generally, the way of life of the members of a society as evident in their values, norms, and material goods. Note that there are many specific interpretations of this word.

Darwinism   The view that the development of a species results from competition among and within species, gradually eliminating the least fit and permitting the fittest to survive. The fittest are those having genetically based features that give them some competitive advantage over fellow members of their species. Darwinism is non–teleological and materialist.

deconstruction   The principal method used in the postmodernist tradition to study texts. Deconstruction is based on the assumption that a text is not a self-contained entity whose meaning is determined by the intent of the author, but rather that a text exists in a context with both the production (writing) and receiving (reading) being affected by other texts. Deconstruction exposes the unquestioned, taken–for–granted, philosophical assumptions that underpin a text— assumptions that are inherited and typically contain uncritically accepted dualisms.

dependency theory   The idea that European overseas expansion produced a situation in which countries at the core of the world system could maintain the underdevelopment of countries at the periphery in order to maintain the dependence of these countries on the core. Dependency theory involves the belief that in order to become more developed, the more developed world had to create a less developed world.

determinism   The general idea that everything that happens is an effect, that events are necessitated by earlier events. Philosophers such as Hobbes and Hume were determinists. There are many versions of determinism, mostly in physical science, with the principal example being the mechanistic clockwork universe of Newtonian physics. A dominant view in the nineteenth century, determinism was liberally applied in the social sciences in a series of varyingly effective versions. Philosophically, a principal concern with determinism is its application to the human world, with the implication that humans are unable to exercise free will and with related concerns about moral responsibility.

development   A term typically interpreted to mean a process of becoming larger, more mature, and better organized, often as measured by economic criteria. This is a term to be used cautiously, since it often involves an implicit general acceptance of a particular perspective (ethnocentrism).

developmentalism   An analysis of cultural and economic change that treats each country or region of the world separately in an evolutionary manner; it assumes that all areas follow the same stages and that they are autonomous.

dialectic   The resolution of contradictions in the pursuit of truth. A dialectic is a method of reasoning that proceeds from thesis to antithesis to synthesis.

difference   A term that recognizes the need to understand identities in relative terms. Studies of identity frequently acknowledge that identities are socially constructed in opposition to other identities, with power relations playing an important role.

diffusion   The process of spread over space and growth through time of cultural traits, ideas, disease, or people, from a center or centers of origin.

discourse   Language in which the meanings of words are specific to a community of users; for example, the technical vocabulary that distinguishes academic disciplines from each other. A widely used and often confusing term, discourse refers more generally to the various social practices that enable the world to be made intelligible. Although sometimes used interchangeably with the term *text*, discourse is more appropriately seen as a way of connecting texts that share a common point of view, as in sexist discourse. Discourse is often interpreted in terms of the relationships between knowledge, language, and power.

disorganized capitalism   A new form of capitalism characterized by a process of disorganization and industrial restructuring (compare **organized capitalism**).

dualism   In the classic Cartesian formulation, based on earlier Greek ideas, the idea that mind and matter are separate. More generally, dualism refers to the separation of two things that might alternatively be seen as one—humans and nature, for example. It is often contrasted with **holism**.

ecocentric   Emphasizing the value of all parts of an ecosystem rather than, for example, placing humans at the center (compare **anthropocentric**).

ecosystem   An ecological system comprising interacting and interdependent organisms and their physical, chemical, and biological environment.

egalitarian   Denoting a society in which all people are essentially similar in terms of wealth and power, with the only major distinctions being those based on age and sex.

empiricism   An empiricist epistemology asserts that all factual knowledge is based on experience, with the human mind being a blank tablet (*tabula rasa*) before encountering the world. Empiricism is a fundamental component of the philosophy of positivism. It is also related to pragmatism.

energy   The capacity of a physical system for doing work.

Enlightenment   A European intellectual movement, the Age of Reason, of the seventeenth and eighteenth centuries, which is contrasted with the earlier period of irrationality and superstition. Generally, two assumptions reflect the philosophical ideas of the period: an empiricist epistemology and the idea that the aim of social science was to enhance social progress through revealing truths about ourselves.

environmental determinism   The argument that the physical environment is the principal cause of human behavior and human landscape creation.

episteme   The world views, or structures of thought, that a society holds at a particular time and that impose the same standards on all branches of knowledge. The concept is associated with Foucault.

epistemology   The branch of philosophy concerned with the nature, sources, and justification of knowledge. Epistemological questions are about how we know knowledge is knowledge (compare **ontology**).

essentialism   A belief in the existence of fixed, unchanging properties; the attribution of essential characteristics to a group instead of seeing such characteristics as being constructed socially (compare **constructionism** and **social construction**).

ethnic   Denoting a group whose members perceive themselves as different from others because of a common ancestry and shared culture.

ethnic cleansing   The forced removal of an ethnic group by another, more dominant group. Ethnic cleansing often involves extermination (see **genocide**).

ethnic religion   A religion that is associated with a particular group of people and that accordingly does not seek to convert others.

ethnocentric   Making judgments about a culture based on the values of one's own culture and, as a consequence, misrepresenting that culture. For example, Eurocentrism is an ethnocentric view that fails to acknowledge that other ethnic groups do not necessarily share the priorities of European discourse.

evolution   The idea that organisms have developed from primitive forms, through natural processes, to more complex forms. In Europe, belief in evolution was a product of Enlightenment thinking; prior to this, the prevailing view was that of creation. Lamarck was the first important evolutionist, but the critical contribution was from Darwin. There are many applications of an evolutionary epistemology to cultures, with Spencer as the major figure.

existentialism   A humanistic philosophy concerned with human existence, which often emphasizes human estrangement from the larger world with the aim of reunification.

feminism   Generally, the advocacy of the rights of women to equality with men, recognizing that sexism prevails, is wrong, and needs to be eliminated; patriarchy is usually seen as the fundamental issue. There are numerous versions of feminism. Increasingly, feminism is concerned not only with the oppression or subordination of women but also with other oppressions, such as those based on social class, skin color, income, religion, age, culture, and geographic location.

fengshui   A Chinese folk–belief system that ascribes auspicious qualities to both the physical and human environments.

feudalism   A social and economic system, prevalent in Europe prior to the Industrial Revolution, that involved two principal groups: the land was controlled by lords, and peasants were bound to the land and subject to the lords' authority. The clergy were also a distinct group within the feudal system.

first effective settlement   A phrase popularized by Zelinsky to refer to the group that first establishes a viable, self–perpetuating community in an area undergoing settlement from outside. It is similar to the concept of initial occupance noted by Kniffen.

flexible accumulation   Industrial technologies, labor practices, relations between firms, and consumption patterns that are increasingly flexible.

folk–culture landscape   The landscape associated with a group that is relatively unchanging and usually small in number, often characterized by traditional cultural traits associated with clothing, food, architecture, and religion (compare ***popular-culture landscape***).

forces of production   A Marxist term that refers to the raw materials, tools, and workers that actually produce goods.

Fordism   A group of industrial and broader social practices initiated by Henry Ford and dominant until recently in most industrial countries. Fordism is characterized by standardized products of mass production in large factories.

formal region   A cultural region occupied by a relatively distinct cultural group and displaying a relatively uniform landscape.

frontier   A zone of advance penetration by an incoming group. A frontier is often an area of conflict between an existing culture and an incoming culture.

frontier thesis   An environmental determinist view, introduced by the American historian Turner, that sees the American frontier as the place where civilization encountered savagery and where civilization was continually, over a 300–year period, conditioned. The result of this extended experience was a new American, not European, culture.

functionalism   A concern with the analysis of functions; a form of teleological philosophy that explains social situations through an account of roles.

functional region   A cultural region that functions politically or economically as an integrated unit and may be related to some homogeneity in people and/or landscape.

gaze   A particular way of looking at things that may influence the conduct and outcome of research. There is a particular need for researchers to recognize and then avoid a gaze that reflects power differences such as those associated with racism or sexism.

Gemeinschaft   A form of human association based on loyalty, informality, and personal contact, assumed to be characteristic in traditional village communities. The term was introduced by Tönnies.

gender   The social aspect of the relations between the sexes. The term does not refer to physical attributes but rather to the socially formed traits associated with masculine and feminine categories. Masculinity and femininity are not, therefore, naturally occurring; rather, they are the consequences of human history.

gender relations   The idea that gender roles are explained in terms of power relations between women and men, especially that of patriarchy.

gender roles   A set of behaviors traditionally assigned to women, characterized by passivity and relationship behavior (actions that facilitate human interaction)—femininity; and a set of behaviors traditionally associated with men, characterized by activity and instrumental behavior (goal-oriented actions)—masculinity.

genocide   A form of one-sided mass killing in which a state or other authority intends to destroy a particular group, as that group and membership are defined by the perpetrator group.

Gesselschaft   A form of human association based on rationality and depersonalization, assumed to be characteristic in urban areas. The term was introduced by Tönnies.

ghetto   A residential district in an urban area with a concentration of a particular ethnic group.

globalization   A process whereby the population of the world is increasingly bonding into a single culture and economy. It is often related to the emergence of a world system.

greenhouse effect   A term commonly used to refer to the human contribution to the naturally occurring greenhouse effect; the consequences of human activities, especially fossil fuel burning, that add carbon dioxide and other gases to the atmosphere, resulting in the earth's retaining more of the warmth that comes from the sun than it would naturally be able to.

group   Perhaps the most general term used to refer to some collection of human beings. It usually implies that there is a set of relations existing between individual members of the group, and also that each member is conscious of the existence of the group.

hegemony   The ability of a group to exercise control over others without needing to rely on laws or the use of force. Hegemony involves an acceptance by others of fundamentally unequal circumstances.

hermeneutics   The study and interpretation of meaning; the interpretation of texts. Hermeneutics is the process of uncovering cultural meaning in everyday life by understanding the signs and symbols of one's own group in context with those of other groups. There are many versions; all assign a key role to the mental quality of humans (compare ***naturalism***).

heterosexuality   The orientation of emotions and/or sexual activity toward those of the opposite sex.

historical materialism   A method, associated with Marxism and centering on the material basis of society, of attempting to understand social change by reference to historical changes in social relations.

holism   The idea that the properties of individual elements in some complex arrangement are affected by relations and interactions with other elements. Holism emphasizes the value of studying groups of things together rather than apart; ecology is holistic. (Compare ***dualism***).

Holocene   The modern geological epoch, which began some 10,000 ya.

homeland   A type of cultural region that involves interaction with the physical environment to evoke emotional attachment and bonding. It is usually associated with a particular group defined in ethnic terms.

homosexuality   The orientation of emotions and/or sexual activity toward those of the same sex.

humanism   An approach to the study of humans and human behavior that gives priority to the fact of being human. Although the term has many connotations, modern humanism flowered in the nineteenth century as part of the conflict between science, as exemplified by Darwinism, and fundamentalist interpretations of religion. Humanists emphasize our ability to make choices and our responsibility for the actions we take. The humanistic philosophical tradition includes such figures as Vico and Berkeley, both of whom questioned the idea that mathematics and mechanics are keys to understanding the world, and Hegel, whose work was an inspiration for various idealist perspectives.

iconography The description and interpretation of visual images, including landscape, in order to uncover their symbolic meanings; the identity of a region as expressed through symbols.

idealism **1.** A group of philosophies that suggest that what is real is related to and created by the contents of the human mind. Idealism is therefore opposed to both materialism and realism. According to Berkeley, nothing exists outside the mind. According to the transcendental idealism of Kant, the categories of terms that are used to describe the world are not objective characteristics of the world but are, rather, structures imposed by the mind. According to Hegel, history comprises a progressive realization of a single spirit (*Geist*). Generally, idealism is a metaphysical view that only minds and ideas exist. **2.** As introduced into cultural geography, a specific version of the larger idealist perspective, derived from the historical idealism proposed by the historian Collingwood, explicitly concerned with rethinking the thoughts behind human actions.

identity A term that refers to sameness and continuity, and that became popular with the rise of mass society and the related quest for understanding who we are. Contemporary accounts of identity derive from both sociological and psychoanalytic inspirations and focus on who people think they are, or their sense of self. It is recognized that identity is socially constructed and intimately associated with prevailing power relations. Most cultural geographic discussions of identity emphasize one or more of the myth of race, ethnicity, language, religion, class, gender, and sexuality, and typically discuss identity formation in terms of difference.

ideology A socially ordered system of cultural symbols; a body of ideas or a way of thinking. Ideology is closely related to the concept of power, as ideological systems serve to legitimize the differential power that groups possess.

idiographic Denoting a method that stresses the individuality and uniqueness of phenomena rather than the similarities between phenomena (compare **nomothetic**).

image The perception of reality held by an individual or group.

imperialism A relationship between states in which one is dominant over the other; the process of empire establishment that took place during the period of political colonialism.

Industrial Revolution A movement that converted a fundamentally rural society into an industrial society beginning in mid-eighteenth century England. The Industrial Revolution was primarily a technological revolution associated with new energy sources.

infrastructure (or **base**) A Marxist term that refers to the economic structure of a society, especially as it gives rise to political, legal, and social systems.

innovation An idea that leads to change, often increasing individual and/or group productivity, or a cultural trait that is new to a group. An innovator is a person who leads change.

intertextuality The idea that meaning is produced from one text to another rather than being produced between the world, including the author(s) and the historical context, and any given text.

Lamarckianism An evolutionary process proposed by Lamarck, which was essentially abandoned after the acceptance of natural selection. Lamarckianism states that characteristics acquired by individual members of a species through their experiences can be inherited by their offspring. For some, it is appropriate to view cultural evolution, as opposed to biological evolution, in Lamarckian terms.

landscape For many cultural geographers, the basic object of study. Landscape has both material characteristics and symbolic identity.

landscape school   The principal approach to studies in cultural geography, emphasizing landscape evolution, cultural regions, and ecology. Introduced by Sauer in the 1920s, the landscape school has been influential through to the present, although there have been challenges to it, beginning in the 1980s.

language family   A group of related languages derived from a single common ancestral language.

late (or taken-for-granted) capitalism   An economic system characterized by multinational capitalism and by the commodification of culture. It is associated with postmodernity.

lebensraum   Literally, living space. As employed by Ratzel, it is the argument that a political state is similar to a living organism in that it might require space to grow.

limits-to-growth thesis   The argument that, in the future, both world population and world economy may collapse because available world resources will be inadequate.

locale   A setting or context for social interaction. This term, which may be applied at a range of scales, was introduced in structuration theory and has become popular in human geography as an alternative to *place*. The term may be employed at a range of scales.

logical positivism   A philosophical approach, initiated by the Vienna Circle group of scholars in the 1920s, that rejected all metaphysics in favor of a scientific approach.

maladaptation   An adaptive strategy that either fails to achieve the desired goals of the group or involves damage to the environment.

material culture   The visible physical objects that are made or used by a group, including clothing, buildings, and tools.

materialism   The idea that humans are dependent on the natural world. Forms of materialism include the mechanistic materialism of early science and the dialectical materialism of Marxism. The term is sometimes used as a synonym for naturalism (compare *idealism*).

mechanistic   A term closely related to Cartesianism and determinism and more generally to the scientific approach emphasizing cause and effect. A mechanistic world is one that functions like clockwork: it exhibits regularity and predictability and is subject to the operation of laws.

mental map   The already constructed images that humans have in their minds and that affect behavior.

mercantilism   A school of economic thought, dominant in Europe in the seventeenth and early eighteenth centuries, that argued for the involvement of the state in economic life in order to increase national wealth and power. The acquisition of precious metals and a favorable trade balance were important aspects of mercantilism.

Mesolithic   Denoting cultures in early Holocene Europe between the end of the Upper Paleolithic and the start of the agricultural revolution of the Neolithic. The Mesolithic period is characterized by a number of cultural adaptations related especially to changes in plant and animal communities as ice sheets retreated. The term literally means 'middle stone', referring to a transitional stone tool technology between the Paleolithic ('old stone'), and Neolithic ('new stone') periods.

metaphysics   The branch of philosophy that attempts to explore the world of the suprasensible, that is, the world beyond experience. Its proponents claim that metaphysics is able to deal with issues beyond the reach of science.

minority language   A language spoken by a minority group in a state in which the majority of the population speaks some other language. A minority language may or may not be an official language.

mode of production   A Marxist term that refers to the organized social relations through which a human society organizes productive activity. In Marxist theory, human societies are seen as passing through a series of these modes.

modernism   A view that assumes the existence of a reality characterized by structure, order, pattern, and causality.

multiculturalism   A policy that endorses the right of ethnic groups to remain distinct rather than being assimilated into a dominant society. The term is used to refer also to the idea that all cultures are equal.

multilinear evolution   A view of evolution claiming that there are various paths of development that cultures may follow.

nation   In general terms, a group of people sharing a common culture and an attachment to some territory. *Nation* is a difficult term to define objectively. It is now commonly understood to refer to imagined rather than real communities, and has unfortunate associations with the myth of race.

nationalism   The political expression of nationhood or aspiring nationhood. Nationalism reflects a consciousness of belonging to a nation.

nation state   A political unit that contains one principal national group and that identifies itself and its territory with that group.

naturalism   The view that all things are natural; thus, the idea that human behavior can be explained by reference to mechanistic laws similar to those that apply in physical science. A fundamental decision that any social scientist must make is whether or not to adopt a naturalist philosophy. *Naturalism* also refers to the belief that what is studied by the physical and human sciences is all there is. It includes acceptance of natural selection. (Compare **hermeneutics**.)

natural selection   The process whereby organisms better adapted to their environment tend to survive and produce more offspring. This refers to the 'survival of the fittest' (a term coined by Spencer in 1852) and is the principal contribution of Darwin to the question of evolution. It is a materialist explanation for evolutionary change.

nature   It is not helpful to suggest that there is a neat and clear definition of this term that is generally agreed to by cultural geographers and others. There are numerous specific meanings depending on the context in which the term is used. To suggest a contrast of nature with culture is a reasonable general understanding, given the cultural geographic tradition.

neocolonialism   Colonial economic and cultural circumstances in a different guise from the original political colonialism.

Neolithic   A cultural phase characterized by the disappearance of foraging and the start of animal and plant domestication as the principal means of subsistence. Animal and plant domestication may have begun as early as about 12,000 ya. *Neolithic* means literally 'new stone', referring to the production of new types of stone tools, which, along with animal and plant domestication, is a distinguishing characteristic of this period.

new cultural geography   An agenda for cultural geography, first proposed in the 1980s, that stresses the need to understand rather than merely describe cultures and places. Proponents of new cultural geography reject the Sauerian concept of culture as cause, emphasizing instead the diversity of cultures, the role of power relations, and markers of human identity such as 'race', gender, and sexuality. In addition, this approach to cultural geography stresses the constructive power of language, reflecting a shift in focus from modern to postmodern.

niche   An ecological address; the space occupied by an organism and the activities that allow it to survive.

nomothetic   Denoting a method that stresses the similarities between phenomena. It is concerned with seeking laws and is associated with positivism (compare **idiographic**).

non-material culture   The oral components of culture, such as beliefs, customs, songs, paintings, and poetry.

norms   Rules of conduct that identify appropriate and inappropriate behavior for members of a group. Norms are reinforced by sanctions of some form.

occupance   The inhabiting and modification of an area by humans.

ontology   The assumptions about existence that underlie any particular system of ideas. It is common to distinguish between the ontology and the epistemology of knowledge.

ordinary landscape   The landscapes of everyday urban space—rooms, buildings, backyards, streets, and neighborhoods—and of everyday rural space—fields, fences, barns, and farmhouses.

organized capitalism   A form of capitalism that developed after the Second World War, characterized by increased growth of major (often multinational) corporations and increased involvement by the state (often in the form of public ownership) in the economy.

Orientalism   The particular perspective that Western scholars of the Orient have of that area. The term implies a view of the periphery from the center. Although Orientalism is now associated with postcolonial theory, primarily with the work of Said, it is clear that the idea of the Orient as a foil for the West dates back to classical Greece.

other   Philosophically, an ambiguous term that derives from the writings of Hegel and that has been used especially by Lacan; it is a prominent concept in feminist and postcolonial theory. *Other* is usually defined in opposition to the same or to self and is based on the assumption that identities are defined not autonomously but by reference to something that can be either excluded or contradicted. The term may be capitalized when used in this specific context. In cultural geography, it is often used to refer to subordinate groups as these are viewed by and contrasted to dominant groups. Thus, it implies both difference and inferiority; for example, masculine identity is defined in terms of the exclusion of the feminine other.

pair bonding   The formation of a close relationship through courtship and sexual activity with another of the same species.

palimpsest   A writing material on which writing has been removed or partly removed to allow for subsequent writings. The term has been adopted by historical geographers to refer to cultural landscapes that contain features from a series of occupations. There is a suggestive association with the idea of a landscape as a text than can be read.

paradigm   A term used to describe the stable pattern of scientific activity prevailing within a discipline.

patriarchy   Literally, the rule of the fathers. The term refers to a social system in which men dominate, oppress, and exploit women. The system is often seen as comprising six factors: the family household, employment, sexuality, violence, culture, and the state. Most feminist movements seek to combat patriarchy.

perception   The process by which humans acquire information about physical and social environments.

performativity   A feminist-inspired term referring to the idea that bodily performance creates both identities and places.

phenomenology   The science of phenomena as distinct from that of the nature of being. *Phenomenology* refers to various philosophies that provide non-empirical descriptions of phenomena. The modern version, based on the work of Husserl, tries to reveal phenomena as intuited essences through direct awareness.

place   Location. In humanistic geography, this term has acquired a particular meaning as a context for human action that is rich in human significance and meaning. Use of this term usually implies a rejection of various scientific approaches, including positivism.

placelessness   Homogeneous and standardized landscapes that lack local variety and character. Placelessness is sometimes the result of the spread of popular culture at the expense of local cultures. Globalization is an extreme version of this.

Pleistocene   A geological epoch characterized by a series of about eighteen glacial periods. The Pleistocene began about 1.6 million ya and ended about 10,000 ya; it was followed by the Holocene.

plural society   A culture comprising several ethnic groups, each living in a community largely separate from the others.

political ecology   An approach to the study of human-and-nature relationships that employs Marxist concepts to stress political dimensions of the relationship and that typically focuses on rural and agrarian issues in the less developed world.

popular-culture landscape   The landscape associated with a dynamic culture. Popular-culture landscapes are often urban areas possessing traits that reflect recent developments in ideas, values, and preferences (compare *folk-culture landscape*).

positivism   A movement introduced by Comte, related to both empiricism and naturalism, that organized knowledge and technology into a consistent whole. It posits that the history of human thought evolved through three stages—religious, metaphysical, and scientific—and that the sciences form a natural hierarchy ranging from mathematics through to the human science of sociology. The philosophy of logical positivism is often equated with a scientific approach.

postcolonialism   An aspect of the cultural studies perspective, more specifically of postmodernism, that explicitly opposes the ethnocentrism seen as a fundamental component of the European cultural tradition. Postcolonialism sees national cultures of previously colonial areas as defined by the tensions related to the history of colonial domination.

post-Fordism   A group of industrial and broader social practices evident in industrial countries since about 1970. Post-Fordism involves more flexible production methods than those associated with Fordism. It involves the decentralized use of information technologies.

postmodernism   A movement in philosophy, social science, and the arts, arguing that reality cannot be studied objectively and stressing that multiple interpretations are both possible and legitimate.

poststructuralism   A complex body of philosophical thought that is sometimes loosely equated with postmodernism or even with deconstruction. It is, however, better regarded as a set of ideas that expand upon the logic of structuralism to insist that meaning is produced within language and that human actions are constrained by structures.

power   In general, the capacity to affect outcomes; more specifically, the capacity to dominate others by means of violence, force, manipulation, or authority. It is characteristic of an oligarchic society in which power is in the hands of an elite group. Interpreting culture in terms of power enables us to consider what are often called cultural politics and the politics of identity.

pragmatism   The one original American philosophy, founded by Charles Saunders Peirce, William James, and John Dewey, which assumes that truth is to be determined by reference to practical outcomes.

preadaptation   The condition of being adapted to life in conditions not previously encountered. A *preadapted* culture already possesses the necessary cultural traits to allow successful occupation of a new environment prior to movement to that environment; groups with these characteristics have a competitive advantage.

prejudice   Preconceived ideas about an individual or group, which foster resistance to change even in the light of contrary evidence. The term means literally 'prejudging'.

queer theory   A group of ideas developed in gay and lesbian studies and concerned with oppressed sexualities in terms of both social rights and cultural politics.

race   Subspecies; a physically distinguishable population within a species. The term is not applicable to the human species, which comprises one and only one species.

racism   A particular form of prejudice attributing characteristics of superiority or inferiority to a group of people who share some physically inherited characteristics.

realism   A philosophical view holding that material objects exist independently of sense experiences. Realism aims to reveal the causal mechanisms through which events are situated within underlying structures (compare *idealism*).

reductionism   Any doctrine that claims to be able to make seemingly complex matters comprehensible in more simple and limited terms.

reflexivity   The need for researcher self-reflection during the research process in order to ensure that the positionality of the researcher is not unduly affecting any aspect of the research activity.

relations of production   A Marxist term that refers to the ways in which the production process is organized, specifically to the relationships of ownership and control.

renewable resources   Resources that naturally regenerate to provide a new supply within a human lifespan (compare *stock resources*).

representation   Traditionally, the claim that there is a real world that can be mirrored by cultural geographers in their writings. This assumes a neutrality on the part of the author. It is now widely acknowledged that such a mode of representation is not possible; rather, cultural geographers interpret the world through particular lenses, especially in the context of prevailing power relations and discourses.

resource   Something material or abstract that can be used by humans to satisfy a need or perceived deficiency.

restructuring   In a capitalist economy, changes in or between the various component parts of an economic system, resulting from economic change.

sacred space   A landscape that is particularly esteemed by an individual or a group, usually but not necessarily for religious reasons.

satisficing behavior   A model of human behavior that rejects the rationality assumptions of the economic operator, assuming instead that the objective is to reach a level of satisfaction that is acceptable.

semiotics   The study of signs. Semiotics assumes that the meaning of cultural materials is to be understood through an analysis of signs in the context of a particular discourse. It is possible to study many different visual materials in this way, including the built environment.

sense of place   The deep attachments that humans have to specific locations, such as home, and also to particularly distinctive locations.

sequent occupance   An approach to evolutionary landscape analysis that recognizes a series of stages during which the cultural landscape is essentially unchanging, with periods of rapid and profound change occurring between stages.

sexism   Attitudes or beliefs that serve to justify sexual inequalities by incorrectly attributing or denying certain capacities either to women or to men.

sexuality   In some feminist and psychoanalytic theory, a cultural construct rather than a biological given that is aligned with power and control.

simplification   A process of cultural change involving the creation of a less complex culture. Simplification is often experienced by groups moving to a different environment.

simulation   A method of representing a real process in an abstract form for the purposes of experimentation.

situated knowledge   The idea that knowledge is not neutral and cannot be acquired in some detached and disembodied manner, but that all knowledge is partial and located somewhere. Work that emphasizes the situated character of knowledge often highlights the positionality of the researcher and the subjects of research. This is an idea that rejects notions of researcher authority and impartiality.

slavery   Labor that is controlled through compulsion and does not involve remuneration. In Marxist terminology, slavery is one example of a mode of production.

social construction   The recognition that all knowledge reflects the fact that we are born into an existing society that precedes any individual development, with that social knowledge becoming a part of our world view and ideology (compare *essentialism*). The idea of social construction is often applied in the context of identity and has long been a fundamental part of much sociological thinking.

social Darwinism   An interpretation of Darwinian concepts that is applied to human societies. Initially proposed by Spencer, social Darwinism represents an evolutionary and naturalist conception of society.

socialism   A social and economic system that involves common ownership of the means of production and distribution.

society   The system of interrelationships that connect individuals as members of a culture.

sociobiology   A modern growth area within what some call 'neo–Darwinism'. Sociobiology is a concern with the evolutionary interpretation of species behavior, especially concerning social interaction. Behavior is interpreted in terms of strategies that have selective advantage in that they increase the chances of survival. Applied to humans, sociobiology has proven controversial because of the implication that human behavior is genetically determined.

spectacle   Any place or event that is an example of carefully created mass leisure and consumption.

stereotype   A collection of expectations and beliefs about a particular group of people that effectively serves as a blueprint affecting how they are perceived and understood by others.

stock resources   Resources that have evolved over a geological time span and that cannot therefore be used by humans without depleting the total available (compare ***renewable resources***).

Stoicism   A Greek philosophical tradition that was materialist, viewing the earth as designed and fit for human life.

strategic essentialism   The decision to emphasize gender at the expense of other markers of identity on the grounds that such a position is justified for practical and/or political reasons.

stratification   The division of society into social classes that constitute a hierarchy. In a capitalist society, one class, a powerful elite, possesses capital and controls the means of production, while the majority of the population is engaged in the productive process.

structuralism   A range of philosophies sharing the view that the empirical world of observable phenomena results from underlying structures.

structuration theory   A social theory, developed by Anthony Giddens, that aims to integrate knowledgeable human agents and the social structures of which they are a part.

subaltern   A term used in the cultural studies perspective to refer to groups that are considered, on the basis of class, caste, gender, race, or culture, to be socially inferior to other groups. The term is associated with the concept of hegemony, with its focus on relationships between dominant and subordinate groups.

superorganic   Denoting an interpretation of culture that sees it as being above both nature and individuals and therefore as the principal cause of the human world. The superorganic view of culture represents a form of cultural determinism.

superstructure   A Marxist term that refers to the political, legal, and social systems of a society.

surrogate   Substitute data used to represent a variable when precise data pertaining to the desired variable are not available.

sustainability   An adaptive strategy involving the conservation of natural resources to ensure that the environment continues to provide for future generations of the population.

taken-for-granted world   The world of everyday living and thinking, sometimes called the lifeworld and most closely associated with phenomenology; the intersubjective world of lived experience and shared meanings.

technology   The means to convert energy into forms useful to humans. *Technology* refers also to the tools and procedures used by humans to meet their needs.

teleology   The doctrine that everything in the world has been designed by God. The term also refers to the study of purposiveness in the world, the idea that some phenomena are best explained in terms of ends (what they have become or what they achieve) rather than in terms of causes. *Teleology* is sometimes used to refer to a recurring theme in history, such as progress or class conflict.

text   A term that originally referred to the written page but that has broadened to include all human activities, products, and representations that can be read, for example maps and landscape. Texts can be regarded as indicators of deeper cultural realities. Postmodernists recognize that there are any number of realities depending on how a text is read. The term is sometimes seen as synonymous with discourse.

topophilia   The affective ties that humans have with particular places; literally, love of place.

tourist gaze   A concept introduced by the sociologist John Urry to refer to how peoples and places are viewed and consumed by tourists, and to how viewings and consumptive preferences and practices are directed by those responsible for constructing and marketing the tourist site.

transculturation   Cultural borrowing, related to the meeting of two cultures that have similar levels of technology and complexity.

transnational corporations (or **transnational companies**)   Large business organizations that operate in two or more countries.

uniformitarianism   The idea that those physical processes that affected the earth in the past continue to operate today, and vice versa (compare ***catastrophism***).

unilinear evolution   A view of cultural evolution that claims that all cultures pass through the same sequence of stages.

universalizing religion   A religion that does not have a restricted domain because of its claim that its beliefs are appropriate for all people.

Upper Paleolithic   A cultural phase, lasting from about 40,000 ya to about 10,000 ya, characterized by the human use of primitive stone implements. *Paleolithic* means literally 'old stone', referring to a stone tool technology that is contrasted with the subsequent Neolithic ('new stone') phase.

values   Ideas held by an individual or a group concerning what is good, bad, appropriate, and inappropriate. Different cultures possess different values.

vernacular region   A vernacular cultural region is identified as such on the basis of the perceptions held by those both inside and outside the region. A vernacular region usually has a generally accepted name.

verstehen   A research method, associated primarily with phenomenology, that involves the researcher adopting the perspective of the individual or group under investigation. *Verstehen* is a German term that is best translated as 'sympathetic or empathetic understanding'.

world system   A cultural system of global dimensions linking different cultures in some key respects.

References

Abler, R., J.S. Adams, and P. Gould. 1971. *Spatial Organization: A Geographer's View of the World*. Englewood Cliffs, NJ: Prentice Hall.

Abram, S. 1997. 'Performing for Tourists in Rural France'. In *Tourists and Tourism: Identifying with People and Places*, edited by S. Abram, J. Waldren, and D.V.L. Macleod, 29–49. New York: Berg.

Agnew, J.A. 1989. 'The Devaluation of Place in Social Science'. In *The Power of Place: Bringing Together Sociological and Geographical Imaginations*, edited by J.A. Agnew and J.S. Duncan, 9–29. Boston: Unwin Hyman.

———. 1997a. 'Geographies of Nationalism and Ethnic Conflict'. In *Political Geography: A Reader*, edited by J.A. Agnew, 317–24. New York: Arnold.

———. 1997b. 'Places and the Politics of Identities'. In *Political Geography: A Reader*, edited by J.A. Agnew, 249–55. New York: Arnold.

———, and J.S. Duncan. 1989. 'Introduction'. In *The Power of Place: Bringing Together Sociological and Geographical Imaginations*, edited by J.A. Agnew and J.S. Duncan, 1–8. Boston: Unwin Hyman.

Ahrentzen, S. 1997. 'The Meaning of Home Workplaces for Women'. In *Thresholds in Feminist Geography: Difference, Methodology, Representation*, edited by J.P. Jones III, H.J. Nast, and S.M. Roberts, 77–92. Lanham, MD: Rowman and Littlefield.

Aitchison, J., and H. Carter. 1990. 'Battle for a Language'. *Geographical Magazine* 42, no. 3:44–6.

Aitken, S.C. 2001. *Geographies of Young People: The Morally Contested Spaces of Modernity*. New York: Routledge.

Albrow, M. 1997. *The Global Age: State and Society Beyond Modernity*. Stanford: Stanford University Press.

Althusser, L. 1969. *For Marx*, translated by B. Brewster. New York: Penguin.

Alvarez, A. 2001. *Governments, Citizens and Genocide: A Comparative and Interdisciplinary Approach*. Indianapolis: Indiana University Press.

Amedeo, D., and R.G. Golledge. 1975. *An Introduction to Scientific Reasoning in Geography*. New York: Wiley.

Anderson, B. 1983. *Imagined Communities: Reflections on the Origins and Spread of Nationalism*. London: Verso.

Anderson, K. 1995. 'Culture and Nature at the Adelaide Zoo: At the Frontiers of "Human" Geography'. *Transactions of the Institute of British Geographers NS* 20:275–94.

———. 1997. 'A Walk on the Wild Side: A Critical Geography of Domestication'. *Progress in Human Geography* 21:463–85.

———. 2000. '"The Beast Within": Race, Humanity, and Animality'. *Environment and Planning D: Society and Space* 18:301–20

———, M. Domosh, S. Pile, and N. Thrift, eds. 2003a. *Handbook of Cultural Geography*. Thousand Oaks, CA: Sage Publications.

———, M. Domosh, S. Pile, and N. Thrift. 2003b. 'Preface'. In *Handbook of Cultural Geography*, edited by K. Anderson, M. Domosh, S. Pile, and N. Thrift, xviii–xix. Thousand Oaks, CA: Sage Publications.

———, and F. Gale, eds. 1999. *Cultural Geographies*, 2nd edn. Harlow: Longman.

Andrews, H.F. 1984. 'The Durkheimians and Human Geography: Some Contextual Problems in the Sociology of Knowledge'. *Transactions of the Institute of British Geographers NS* 9:315–36.

Appleton, J. 1975a. *The Experience of Landscape*. New York: Wiley.

———. 1975b. 'Prospect and Refuge in the Landscapes of Britain and Australia'. In *Geographical Essays in Honour of Gilbert J. Butland*, edited by I. Douglas, J.E. Hobbs, and J.J. Pigram, 1–20. Armidale: University of New England, Department of Geography.

———. 1990. *The Symbolism of Habitat: An Interpretation of Landscape in the Arts*. Seattle: University of Washington Press.

———. 1996. *The Experience of Landscape*, rev. edn. New York: Wiley.

Arendt, H. 1951. *The Origins of Totalitarianism*. New York: Harcourt Brace.

Arkell, T. 1991. 'Geography on Record'. *Geographical Magazine* 63, no. 7:30–4.

Arnold, D. 1996. *The Problem of Nature: Environment, Culture and European Expansion*. Cambridge, MA: Blackwell.

Aschmann, H. 1965. 'Athapaskan Expansion in the Southwest'. *Association of Pacific Coast Geographers Yearbook* 32:79–97.

————. 1987. 'Carl Sauer, A Self Directed Career'. In *Carl O. Sauer: A Tribute*, edited by M.S. Kenzer, 137–43. Corvallis: Oregon State University Press.

Atkins, P., M. Simmons, and B. Roberts. 1998. *People, Land and Time: An Historical Introduction to the Relations Between Landscape, Culture and Environment*. New York: Arnold.

Augelli, J.P. 1958. 'Cultural and Economic Changes of Bastos, a Japanese Colony on Brazil's Paulista Frontier'. *Annals of the Association of American Geographers* 48:3–19.

Auster, M. 1997. 'Monument in a Landscape: The Question of "Meaning"'. *Australian Geographer* 28:219–27.

Avery, H. 1988. 'Theories of Prairie Literature and the Woman's Voice'. *Canadian Geographer* 32:270–2.

Axford, B. 1995. *The Global System: Economics, Politics and Culture*. New York: St Martin's Press.

Azaryahu, M., and A. Kellerman. 1999. 'Symbolic Places of National History and Revival: A Study in Zionist Mythical Geography'. *Transactions of the Institute of British Geographers, NS* 24:109–23.

Back, L. 2003 'Deep Listening: Researching Music and the Cartographies of Sound'. In *Cultural Geography in Practice*, edited by A. Blunt, et al., 272–85. New York: Arnold.

Badcock, B. 1996. '"Looking-glass" Views of the City'. *Progress in Human Geography* 20:91–9.

Baker, A.R.H. 1992. 'Introduction: On Ideology and Landscape'. In *Ideology and Landscape in Historical Perspective*, edited by A.R.H. Baker and G. Biger, 1–14. New York: Cambridge University Press.

————. 2003. *Geography and History: Bridging the Divide*. New York: Cambridge University Press.

————, and R.A. Butlin, eds. 1973. *Studies of Field Systems in the British Isles*. New York: Cambridge University Press.

Baker, R. 1997. 'Landcare: Policy, Practice and Partnerships'. *Australian Geographical Studies* 35:61–73.

Bale, J. 2003. *Sports Geography*, 2nd edn. New York: Routledge.

Barber, B.R. 1995. *Jihad vs McWorld: How Globalism and Tribalism Are Reshaping the Modern World*. New York: Ballantine Books.

Barker, R.G. 1968. *Ecological Psychology: Concepts and Methods for Studying the Environment and Behavior*. Stanford: Stanford University Press.

Barnes, T.J. 1996. 'Political Economy II: Compliments of the Year'. *Progress in Human Geography* 20:521–8.

————. 2003. 'Introduction: "Never Mind the Economy. Here's Culture"'. In *Handbook of Cultural Geography*, edited by K. Anderson, M. Domosh, S. Pile, and N. Thrift, 89–97. Thousand Oaks, CA: Sage Publications.

————, and M. Curry. 1983. 'Towards a Contextualist Approach to Geographical Knowledge'. *Transactions of the Institute of British Geographers NS* 8:467–82.

————, and J.S. Duncan. 1992. 'Introduction: Writing Worlds'. In *Writing Worlds: Discourse, Text and Metaphor in the Representation of Landscape*, edited by T.J. Barnes and J.S. Duncan, 1–17. New York: Routledge.

Barnett, C. 1998a. 'The Cultural Turn: Fashion or Progress in Human Geography?' *Antipode* 30:379–94.

————. 1998b. 'Guest Editorial: Cultural Twists and Turns'. *Environment and Planning D: Society and Space* 16:631–4.

————. 2005. 'Ways of Relating: Hospitality and the Acknowledgement of Otherness'. *Progress in Human Geography* 29:5–21.

Barrell, J. 1972. *The Idea of Landscape and the Sense of Place, 1730–1840: An Approach to the Poetry of John Clare*. New York: Cambridge University Press.

————. 1982. 'Geographies of Hardy's Wessex'. *Journal of Historical Geography* 8:347–61.

Barrows, H.H. 1923. 'Geography as Human Ecology'. *Annals of the Association of American Geographers* 13:1–14.

Barth, F. 1956. 'Ecologic Relationships of Ethnic Groups in Swat, North Pakistan'. *American Anthropologist* 58:1079–89.

————, ed. 1969. *Ethnic Groups and Boundaries*. Boston: Little Brown.

Bassett, T.J. 1988. 'The Political Ecology of Peasant–Herder Conflicts in the Northern Ivory Coast'. *Annals of the Association of American Geographers* 78:453–72.

Bastian, R.W. 1975. 'Architecture and Class Segregation in Late Nineteenth-Century Terre Haute, Indiana'. *Geographical Review* 65:166–79.

Bate, J. 1991. *Romantic Ecology: Wordsworth and the Ecological Tradition*. New York: Routledge.

Batterbury, S. 2001. 'Landscapes of Diversity: A Local Political Ecology of Livelihood Diversification in South-Western Nigeria'. *Ecumene* 8:437–64.

Bauer, R. 2003. *The Cultural Geography of Colonial American Literatures*. New York: Cambridge University Press.

Baumann, G. 1996. *Contesting Culture: Discourses of Identity in Multi-ethnic London*. New York: Cambridge University Press.

Bayly, C.A. 2003. *The Birth of the Modern World,*

1780–1914: Global Connections and Comparisons. New York: Blackwell.

Bebbington, A.J., and S. Batterbury. 2001. 'Transnational Livelihoods and Landscapes: Political Ecologies of Globalization'. *Ecumene* 8:369–80.

Bell, D., and G. Valentine. 1995. 'Introduction: Orientations'. In *Mapping Desire: Geographies of Sexualities*, edited by D. Bell and G. Valentine, 1–27. New York: Routledge.

———, and ———. 1997. *Consuming Geographies: We Are Where We Eat*. New York: Routledge.

———, and ———. 1999. 'Geographies of Sexuality—a Review of Progress'. *Progress in Human Geography* 23:175–87.

Bell, J. 1999. 'Redefining National Identity in Uzbekistan: Symbolic Tensions in Tashkent's Official Public Landscape'. *Ecumene* 6:183–211.

Bell-Fialkof, A. 1993. 'A Brief History of Ethnic Cleansing'. *Foreign Affairs* 72, no 3:110–21.

Bender, B. 1993. *Landscape: Politics and Perspectives*. Oxford: Berg.

———. 2001. 'Introduction'. In *Contested Landscapes: Movement, Exile and Place*, edited by Barbara Bender and Margot Winer, 1–17. New York: Berg.

Benko, G., and U. Strohmayer. 1997. 'Preface'. In *Space and Social Theory: Interpreting Modernity and Postmodernity*, edited by G. Benko and U. Strohmayer, xiii–xvi. Malden, MA: Blackwell.

Bennett, J.W. 1976. *The Ecological Transition: Cultural Anthropology and Human Adaptation*. New York: Pergamon.

———. 1993. *Human Ecology as Human Behavior: Essays in Environmental and Developmental Anthropology*. New Brunswick, NJ: Transaction Publishers.

Bennion, L.C. 2001. 'Mormondom's Desert Homeland'. In *Homelands: A Geography of Culture and Place across America*, edited by R.L. Nostrand and L.E. Estaville, 184–209. Baltimore: Johns Hopkins University Press.

Berdoulay, V. 1978. 'The Vidal–Durkheim Debate'. In *Humanistic Geography*, edited by D. Ley and M. Samuels, 77–90. Chicago: Maaroufa Press.

———. 1989. 'Place, Meaning, and Discourse in French Language Geography'. In *The Power of Place: Bringing Together Sociological and Geographical Imaginations*, edited by J.A. Agnew and J.S. Duncan, 124–39. Boston: Unwin and Hyman.

Beresford, M.W. 1957. *History on the Ground: Six Studies in Maps and Landscapes*. London: Lutterworth.

Berg, L.D. 1993. 'Between Modernism and Postmodernism'. *Progress in Human Geography* 17:490–507.

———, and R.A. Kearns. 1996. 'Naming as Norming: "Race", Gender, and the Identity Politics of Naming Places in Aotearoa/New Zealand'. *Environment and Planning D: Society and Space* 14:99–122.

———, and R.A. Kearns. 1997. 'Constructing Cultural Geographies of Aotearoa'. *New Zealand Geographer* 53, no. 2:1–2.

Berger, P.L., and T. Luckman. 1966. *The Social Construction of Reality: A Treatise in the Sociology of Knowledge*. Garden City, NY: Doubleday.

———, and S.P. Huntington, eds. 2002. *Many Globalizations: Cultural Diversity in the Contemporary World*. New York: Oxford University Press.

Berger, T.R. 1977. *Northern Frontier, Northern Homeland, Volume 1*. Ottawa: Supply and Services Canada.

Bernard, F., and D. Thom. 1981. 'Population Pressure and Human Carrying Capacity in Selected Locations in Machakos and Kitui Districts'. *Journal of Developing Areas* 5:381–406.

Berry, B.J.L. 1995. 'Editorial: The Postmodernist Pursuit of Pragna'. *Urban Geography* 16:95–7.

———, E.C. Conkling, and D.M. Ray. 1997. *The Global Economy in Transition*, 2nd edn. Upper Saddle River, NJ: Prentice Hall.

Berry, J.W. 1984. 'Cultural Ecology and Individual Behavior'. In *Human Behavior and Environment, Advances in Theory and Research, Volume 4, Environment and Culture*, edited by I. Altman, A. Rapaport, and J.F. Wohlwill, 83–106. New York: Plenum Press.

———. 1997. 'Immigration, Acculturation, and Adaptation'. *Applied Psychology: An International Review* 46:5–68.

Bhaskar, R. 1989. *The Possibility of Naturalism: A Philosophical Critique of the Contemporary Human Sciences*, 2nd edn. Brighton, England: Harvester Press.

Binnie, J. 1997. 'Coming Out of Geography: Towards a Queer Epistemology'. *Environment and Planning D: Society and Space* 15:223–37.

———, and G. Valentine. 1999. 'Geographies of Sexuality—A Review of Progress'. *Progress in Human Geography* 23:175–87.

Birch, B.P. 1981. 'Wessex, Hardy and the Nature Novelists'. *Transactions of the Institute of British Geographers*, NS 6:348–58.

Birks, H.H., et al. 1988. *The Cultural Landscape: Past, Present and Future*. New York: Cambridge University Press.

Biswas, L. 1984. 'Evolution of Hindu Temples in Calcutta'. *Journal of Cultural Geography* 4, no. 2:73–84.

Bjorklund, E.M. 1964. 'Ideology and Culture Exemplified in Southwestern Michigan'. *Annals of the Association of American Geographers* 54:227–41.

Blaikie, P. 1978. 'The Theory of Spatial Diffusion of Innovations: A Spacious Cul-de-Sac'. *Progress in Human Geography* 2:270–95.

———. 1985. *The Political Economy of Soil Erosion in Developing Countries*. London: Longman.

———, and H. Brookfield. 1987. *Land Degradation and Society*. London: Methuen.

Blainey, G. 1983. *A Land Half Won*. Melbourne: Sun Books.

Blau, P.M., and J.W. Moore. 1970. 'Sociology'. In *A Reader's Guide to the Social Sciences*, 2nd edn, edited by B.F. Hoselitz, 1–40. New York: Free Press.

Blaut, J.M. 1977. 'Two Views of Diffusion'. *Annals of the Association of American Geographers* 67:343–9.

———. 1980. 'A Radical Critique of Cultural Geography'. *Antipode* 12:25–9.

———. 1984. 'Commentary on Nostrand's "Hispanos" and Their "Homeland"'. *Annals of the Association of American Geographers* 74:157–64.

———. 1993a. 'Mind and Matter in Cultural Geography'. In *Culture, Form, and Place: Essays in Cultural and Historical Geography*, edited by K. Mathewson. *Geoscience and Man* 32:345–56. Baton Rouge: Louisiana State University, Department of Geography and Anthropology, Geoscience Publications.

———. 1993b. *The Colonizer's Model of the World: Geographical Diffusionism and Eurocentric History*. New York: Guilford.

Blumler, M.A. 1996. 'Ecology, Evolutionary Theory and Agricultural Origins'. In *The Origins and Spread of Agriculture and Pastoralism in Eurasia*, edited by D.R. Harris, 25–50. Washington, DC: Smithsonian Institution Press.

Blunt, A., and C. McEwan, eds. 2002. *Postcolonial Geographies*. New York: Continuum.

———, and J. Wills. 2000. *Dissident Geographies: An Introduction to Radical Ideas and Practice*. New York: Prentice Hall.

———, et al. 2003. *Cultural Geography in Practice*. New York: Arnold.

Boas, F. 1928. *Anthropology and Modern Life*. New York: Norton.

Bondi, L. 1997. 'In Whose Words? On Gender Identities, Knowledge and Writing Practices'. *Transactions of the Institute of British Geographers NS* 22:245–58.

———, and J. Davidson. 2003. 'Troubling the Place of Gender'. In *Handbook of Cultural Geography*, edited by K. Anderson, M. Domosh, S. Pile, and N. Thrift, 325–43. Thousand Oaks, CA: Sage Publications.

———, and M. Domosh. 1992. 'Other Figures in Other Landscapes: On Feminism,

Postmodernism and Geography'. *Environment and Planning D: Society and Space* 10:199–213.

Bonnett, A. 2000. *White Identities*. New York: Prentice Hall.

———, and A. Nayak. 2003. 'Cultural Geographies of Racialization—The Territory of Race'. In *Handbook of Cultural Geography*, edited by K. Anderson, M. Domosh, S. Pile, and N. Thrift, 300–12. Thousand Oaks, CA: Sage Publications.

Boone, J.A. 1996. 'Queer Sites in Modernism: Harlem/The Left Bank/Greenwich Village'. In *The Geography of Identity*, edited by P. Yaeger, 243–72. Ann Arbor: University of Michigan Press.

Boserup, E. 1965. *The Conditions of Agricultural Growth: The Economics of Agrarian Change Under Population Pressure*. London: Allen and Unwin.

Botkin, D. 1990. *Discordant Harmonies: A New Ecology for the Twenty-First Century*. New York: Oxford University Press.

Boulding, K.E. 1950. *A Reconstruction of Economics*. New York: Wiley.

———. 1956. *The Image*. Ann Arbor: University of Michigan Press.

Bourdieu, P. 1977. *Outline of a Theory of Practice*, translated by R. Nice. New York: Cambridge University Press.

———. 1994. *Distinction: A Social Critique of the Judgement of Taste*. New York: Routledge.

Bowden, M.J. 1976. 'The Great American Desert in the American Mind: The Historiography of a Geographical Notion'. In *Geographies of the Mind: Essays in Historical Geosophy in Honor of John Kirtland Wright*, edited by D. Lowenthal and M.J. Bowden, 119–47. New York: Oxford University Press.

Bowen, D.S. 1996. 'Carl Sauer, Field Exploration, and the Development of American Geographical Thought'. *Southeastern Geographer* 36:176–91.

Bowen, E. 1981. *Empiricism and Geographical Thought*. New York: Cambridge University Press.

Bowen, E.H., and H. Carter. 1975. 'The Distribution of the Welsh Language in 1971'. *Geography* 60:1–15.

Bowler, P.J. 1992. *The Fontana History of the Environmental Sciences*. London: Fontana Press.

Bowman, I. 1934. *Geography in Relation to the Social Sciences*. New York: Charles Scribner.

Boyd, W., M. Cotter, and J. Gardiner. 2001. 'Dreaming the Homeland: The Big Scot, Power Poles and the Imagining of Scotland in Australia'. In *2001, Geography—A Spatial Odyssey. Proceedings of the Third Joint Conference of the New Zealand Geographical Society and the Institute of*

Australian Geographers, edited by P. Holland, F. Stephenson, and A. Wearing, 313–20. Hamilton: New Zealand Geographical Society Conference Series, no. 21.

Brace, C. 1999. 'Finding England Everywhere: Regional Identity and the Construction of National Identity, 1840–1940'. *Ecumene* 6:90–109.

Brack, D., and I. Dale, eds. 2003. *Prime Minister Portillo—And Other Things That Never Happened*. London: Politico's.

Breitbart, M.M. 1981. 'Peter Kropotkin: The Anarchist Geographer'. In *Geography, Ideology and Social Concern*, edited by D.R. Stoddart, 134–53. Totowa, NJ: Barnes and Noble.

Broek, J.O.M. 1932. *The Santa Clara Valley, California: A Study in Landscape Change*. Utrecht: Oosthoek.

———. 1965. *Compass of Geography*. Columbus: Merrill.

———, and J.W. Webb. 1978. *A Geography of Mankind*, 2nd edn. New York: McGraw-Hill.

Brookfield, H.C. 1964. 'Questions on the Human Frontiers of Geography'. *Economic Geography* 40:283–303.

———. 1969. 'On the Environment as Perceived'. In *Progress in Geography, Volume 1*, edited by C. Board et al., 51–80. New York: Arnold.

Browett, J. 1980. 'Development, the Diffusionist Paradigm, and Geography'. *Progress in Human Geography* 4:57–79.

Brown, M. 1995. 'Sex, Scale, and the 'New Urban Politics': HIV Prevention Strategies From Yaletown, Vancouver'. In *Mapping Desire: Geographies of Sexuality*, edited by D. Bell and G. Valentine, 245–63. New York: Routledge.

———, and L. Knopp. 2003. 'Queer Cultural Geographies—We're Here! We're Queer! We're Over There Too'. In *Handbook of Cultural Geography*, edited by K. Anderson, M. Domosh, S. Pile, and N. Thrift, 313–24. Thousand Oaks, CA: Sage Publications.

Brunn, S.D. 1963. 'A Cultural Plant Geography of the Quince'. *Professional Geographer* 15:16–18.

Bryant, R.L. 1997. 'Beyond the Impasse: The Power of Political Ecology in Third World Environmental Research'. *Area* 29:5–19.

———. 2001. 'Political Ecology: A Critical Agenda for Change'. In *Social Nature: Theory, Practice, and Politics*, edited by N. Castree and B. Brown, 151–69. Malden, MA: Blackwell.

Buchanan, R.H., E. Jones, and D. McCourt, eds. 1971. *Man and His Habitat: Essays Presented to Emyr Estyn Evans*. London: Routledge and Kegan Paul.

Buckingham-Hatfield, S. 2000. *Gender and Environment*. New York: Routledge.

Bunkse, E.V. 1996. 'Humanism: Wisdom of the Heart and Mind'. In *Concepts in Human Geography*, edited by C. Earle, K. Mathewson, and M.S. Kenzer, 355–81. Lanham, MD: Rowman and Littlefield.

Bunting, T.E., and L. Guelke. 1979. 'Behavioral and Perception Geography: A Critical Appraisal'. *Annals of the Association of American Geographers* 69:448–62.

Burton, I. 1963. 'The Quantitative Revolution and Theoretical Geography'. *Canadian Geographer* 7:151–62.

Buruma, I., and A. Margalit. 2004. *Occidentalism: The West in the Eyes of Its Enemies*. New York: Penguin.

Butler, J. 1990. *Gender Trouble: Feminism and the Subversion of Identity*. New York: Routledge.

———. 1993. *Bodies that Matter: On the Discursive Limits of 'Sex'*. New York: Routledge.

Butler, R., and H. Parr, eds. 1999. *Mind and Body Spaces: Geographies of Disability, Illness and Impairment*. New York: Routledge.

Butler, W.F. 1872. *The Great Lone Land*. London: Sampson Low.

Butlin, R.A. 1993. *Historical Geography: Through the Gates of Space and Time*. New York: Arnold.

———, and N. Roberts. 1995. 'Ecological Relations in Historical Times: An Introduction'. In *Ecological Relations in Historical Times: Human Impact and Adaptation*, edited by R.A. Butlin and N. Roberts, 1–14. Cambridge, MA: Blackwell.

Buttimer, A. 2003. 'Cultural Geographers Forum: What are the Five Most Important Principles that Should be Covered in an Introductory Course in Cultural Geography and Why?' *Place and Culture, The Newsletter of the Cultural Geography Specialty Group of the Association of American Geographers* Spring, 3–4.

Butzer, K.W. 1980. 'Civilizations: Organisms or Systems?' *American Scientist* 68:517–23.

———. 1989a. 'Hartshorne, Hettner, and *The Nature of Geography*'. In *Reflections on Richard Hartshorne's The Nature of Geography*, edited by J.N. Entrikin and S.D. Brunn, 35–52. Washington, DC: Association of American Geographers, Occasional Publication.

——— 1989b. 'Cultural Ecology'. In *Geography in America*, edited by G.L. Gaile and C.J. Willmott, 192–208. Columbus, OH: Merrill.

Buxton, G.L. 1967. *The Riverina: 1861–1891: An Australian Regional Study*. Melbourne: Melbourne University Press.

Calder, A. 1998. '"The Nearest Approach to a Desert": Implications of Environmental Determinism in the Criticism of Canadian Prairie Writing'. *Prairie Forum* 23:171–82.

Callard, F. 1998. 'The Body in Theory'. *Environment and Planning D: Society and Space* 16:387–400.

Cameron, J.M.R. 1974. 'Information Distortion in Colonial Promotion: The Case of Swan River Colony'. *Australian Geographical Studies* 12:57–76.

———. 1977. *Coming to Terms: The Development of Agriculture in Pre-Convict Western Australia*. Perth: University of Western Australia, Department of Geography, Geowest 11.

Carlson, A.W. 1990. *The Spanish-American Homeland: Four Centuries of Change in New Mexico's Río Arriba*. Baltimore: Johns Hopkins University Press.

Carlstein, T. 1982. *Time, Resources, Society and Ecology: On the Capacity for Human Interaction in Space and Time. Volume 1: Preindustrial Societies*. London: Allen and Unwin.

Carmichael, D.L., J. Hubert, B. Reeves, and A. Schanche, eds. 1994. *Sacred Sites, Sacred Places*. New York: Routledge.

Carneiro, R. 1960. 'Slash and Burn Agriculture: A Closer Look at its Implications for Settlement Patterns'. In *Men and Cultures*, edited by A.F.C. Wallace, 229–34. Philadelphia: University of Pennsylvania Press.

Carney, G.O. 1990. 'Geography of Music: Inventory and Prospect'. *Journal of Cultural Geography* 10, no. 2:35–48.

———, ed. 1994a. *The Sounds of People and Places: A Geography of American Folk and Popular Music*, 3rd edn. Lanham, MD: Rowman and Littlefield.

———. 1994b. 'Branson: The New Mecca of Country Music'. *Journal of Cultural Geography* 14, no. 2:17–32.

———. 1995a. 'Introduction: Culture: A Workable Definition'. In *Fast Food, Stock Cars, and Rock 'n' Roll*, edited by G.O. Carney, 1–14. Lanham, MD: Rowman and Littlefield.

———. 1995b. 'Part III: Food'. In *Fast Food, Stock Cars, and Rock 'n' Roll*, edited by G.O. Carney, 95. Lanham, MD: Rowman and Littlefield.

———. 1996. 'Western North Carolina: Culture Hearth of Bluegrass Music'. *Journal of Cultural Geography* 16, no. 1:65–87.

———, ed. 1998. *Baseball, Barns, and Bluegrass: A Geography of American Folklife*. Lanham, MD: Rowman and Littlefield.

Carroll, W.K. 1992. 'Introduction: Social Movements and Counter-Hegemony in a Canadian Context'. In *Organizing Dissent: Contemporary Social Movements in Theory and Practice*, edited by W.K. Carroll, 1–19. Toronto: Garamond Press.

Carson, R. 1962. *Silent Spring*. Boston: Houghton Mifflin.

Carter, E., J. Donald, and J. Squires, eds. 1995. *Cultural Remix: Theories of Politics and the Popular*. London: Lawrence and Wishart.

Carter, G.F. 1948. 'Clark Wissler: 1870–1947'. *Annals of the Association of American Geographers* 38:145–6.

———. 1968. *Man and the Land: A Cultural Geography*, 2nd edn. New York: Holt, Rinehart and Winston.

———. 1977. 'A Hypothesis Suggesting a Single Origin of Agriculture'. In *Origins of Agriculture*, edited by C.A. Reed, 99–109. The Hague: Mouton.

———. 1978. 'Context as Methodology'. In *Diffusion and Migration: Their Roles in Cultural Development*, edited by P.G. Duke et al., 55–64. Calgary: University of Calgary, Archaeological Association.

———. 1980. *Earlier Than You Think: A Personal View of Man in America*. College Station: Texas A&M University Press.

———. 1988. 'Cultural Historical Diffusion'. In *The Transfer and Transformation of Ideas and Material Culture*, edited by P.J. Hugill and D.B. Dickson, 3–22. College Station: Texas A&M University Press.

Carter, H., and J.G. Thomas. 1969. 'The Referendum on the Sunday Opening of Licensed Premises in Wales as a Criterion of a Cultural Region'. *Regional Studies* 3:61–71.

Carter, P. 1987. *The Road to Botany Bay: An Essay in Spatial History*. London: Faber and Faber.

Cartier, C. 1997. 'Symbolic Landscape in High Rise Hong Kong'. *Focus* 44, no. 3:13–21.

Carvel, J. 1997. 'Global Study Finds the World Speaking in 10,000 Languages'. *The Guardian* (22 July).

Castells, M. 1983. *The City and the Grassroots: A Cross-cultural Theory of Urban Social Movements*. Berkeley: University of California Press.

———. 2004. *The Power of Identity*, 2nd edn. Malden, MA: Blackwell.

Castree, N. 2003. 'Strange Natures: Geography and the Study of Human–Environment Relationships'. In *The Student's Companion to Geography*, 2nd edn, edited by A Rogers, H. Viles, and A Goudie, 82–7. Cambridge, MA: Blackwell.

Cater, J., and T. Jones. 1989. *Social Geography: An Introduction to Contemporary Issues*. New York: Arnold.

Cavalli-Sforza, L.L., and M.W. Feldman. 1981. *Cultural Transmission and Evolution: A Quantitative Approach*. Princeton: Princeton University Press.

Chalk, F., and K. Jonassohn. 1990. *The History and Sociology of Genocide*. New Haven: Yale University Press.

Chambers, I. 1994. *Migrancy, Culture, Identity*. New York: Routledge.

Chapman, G.P. 1977. *Human and Environmental Systems: A Geographer's Appraisal*. New York: Academic Press.

Chapple, C.K., and M.E. Tucker. 2000. *Hinduism and Ecology: The Intersection of Earth, Sky, and Water*. Cambridge, MA: Harvard University Press.

Charlesworth, A. 1992. 'Towards a Geography of the Shoah'. *Journal of Historical Geography* 18:464–9.

———. 1994. 'Contesting Places of Memory: The Case of Auschwitz'. *Environment and Planning D: Society and Space* 12:579–93.

———. 2003. 'Landscapes of the Holocaust: Schindler, Authentic History and the Lie of Landscape'. In *Studying Cultural Landscapes*, edited by I. Robertson and P. Richards, 93–107. New York: Arnold.

Childe, V.G. 1936. *Man Makes Himself*. London: Watts and Co.

Chisholm, M. 1975. *Human Geography: Evolution or Revolution?* New York: Penguin.

Chorley, R.J. 1973. 'Geography as Human Ecology'. In *Directions in Geography*, edited by R.J. Chorley, 155–69. London: Methuen.

Chouinard, V. 1997. 'Guest Editorial. Making Space for Disabling Differences: Challenging Ableist Geographies'. *Environment and Planning D: Society and Space* 15:379–87.

Christensen, K. 1982. 'Geography as a Human Science: A Philosophic Critique of the Positivist–Humanist Split'. In *A Search for Common Ground*, edited by P. Gould and G. Olsson, 37–57. London: Pion.

Christopher, A.J. 1973. 'Environmental Perception in Southern Africa'. *South African Geographical Journal* 55:14–22.

———. 1982. 'Towards a Definition of the Nineteenth Century South African Frontier'. *South African Geographical Journal* 64:97–113.

———. 1984. *Colonial Africa*. Totowa: Barnes and Noble.

———. 1994. *The Atlas of Apartheid*. New York: Routledge.

Church News. 1979. *Special Edition: The Era of Mormon Colonization*. Salt Lake City: Deseret News (26 May).

Clark, A.H. 1954. 'Historical Geography'. In *American Geography: Inventory and Prospect*, edited by P.E. James and C.F. Jones, 70–105. Syracuse: Syracuse University Press.

———. 1959. *Three Centuries and the Island*. Toronto: University of Toronto Press.

———. 1975. 'The Conceptions of "Empires" of the St Lawrence and the Mississippi: An Historico-Geographical View With Some Quizzical Comments on Environmental Determinism'. *American Review of Canadian Studies* 5:4–27.

Clarkson, J.D. 1970. 'Ecology and Spatial Analysis'. *Annals of the Association of American Geographers* 60:700–16.

Clay, G. 1980. *Close Up: How to Read the American City*. Chicago: University of Chicago Press.

———. 1987. *Right Before Your Eyes*. Washington, DC: APA Planners Press.

———. 1994. *Real Places: An Unconventional Guide to America's Generic Landscape*. Chicago: University of Chicago Press.

Clayton, D. 2003. 'Critical Imperial and Colonial Geographies'. In *Handbook of Cultural Geography*, edited by K. Anderson, M. Domosh, S. Pile, and N. Thrift, 354–68. Thousand Oaks, CA: Sage Publications.

Clements, F.E. 1905. *Research Methods in Ecology*. Lincoln: University of Nebraska Publishing Company.

Clifford, N., and G. Valentine, eds. 2003. *Key Methods in Geography*. Thousand Oaks, CA: Sage.

Cloke, P., C. Philo, and D. Sadler. 1991. *Approaching Human Geography: An Introduction to Contemporary Theoretical Debates*. New York: Guilford Press.

Coates, P. 1998. *Nature: Western Attitudes Since Ancient Times*. Berkeley: University of California Press.

Cobb, J.C. 1992. *The Most Southern Place on Earth: The Mississippi Delta and the Roots of Regional Identity*. New York: Oxford University Press.

Cohen, S. 1997. 'More Than The Beatles: Popular Music, Tourism, and Urban Regeneration'. In *Tourists and Tourism: Identifying With People and Places*, edited by S. Abram, J. Waldren, and D.V.L. Macleod, 71–90. New York: Berg.

Cohen, S.B. 2003. *Geopolitics of the World System*. New York: Rowman and Littlefield.

Cole, T., and G. Smith. 1995. 'Ghettoization and the Holocaust: Budapest 1944'. *Journal of Historical Geography* 21:300–16.

Connell, J., and C. Gibson. 2003. *Sound Tracks: Music, Identity and Place*. New York: Routledge.

———, and ———. 2004. 'World Music: Deterritorializing Place and Identity'. *Progress in Human Geography* 28:342–61.

Conzen, M.P., ed. 1990a. *The Making of the American Landscape*. Boston: Unwin Hyman.

———. 1990b. 'Ethnicity on the Land'. In *The Making of the American Landscape*, edited by M.P. Conzen, 221–48. Boston: Unwin Hyman.

————. 1993. 'The Historical Impulse in Geographical Writing About the United States, 1850–1900'. In *A Scholar's Guide to Geographical Writing on the American and Canadian Past*, by M.P. Conzen, T.A. Rumney, and G. Wynn, 3–90. Chicago: University of Chicago Press.

————. 2001. 'American Homelands: A Dissenting View'. In *Homelands: A Geography of Culture and Place Across America*, edited by R.L. Nostrand and L.E. Estaville, 238–71. Baltimore: Johns Hopkins University Press.

————, T.A. Rumney, and G. Wynn. 1993. *A Scholar's Guide to Geographical Writing on the American and Canadian Past*. Chicago: University of Chicago Press.

Cook, D., and M. Okenimpke. 1997. *Ngugi wa Thiong'o: An Exploration of His Writings*, 2nd edn. Portsmouth, NH: Heinemann.

Cook, I., D. Crouch, S. Naylor, and J. Ryan. 2000. *Cultural Turns/Geographical Turns: Perspectives on Cultural Geography*. New York: Prentice Hall.

————, and M. Harrison. 2003. 'Cross Over Food: Re-materializing Postcolonial Geographies'. *Transactions of the Institute of British Geographer NS* 28:296–317.

Cook, M. 2004. *A Brief History of the Human Race*. London: Granta.

Cook, N.D., and W.G. Lovell, eds. 1992. *'Secret Judgments of God': Old World Disease in Colonial Spanish America*. Norman: University of Oklahoma Press.

Coon, C. 1962. *The Origin of Races*, 2nd edn. New York: Knopf.

Coones, P., and J. Patten. 1986. *The Penguin Guide to the Landscape of England and Wales*. New York: Penguin.

Cooper, R. 2003. *The Breaking of Nations: Order and Chaos in the Twenty-first Century*. London: Atlantic Books.

Cosgrove, D. 1978. 'Place, Landscape and the Dialectics of Cultural Geography'. *Canadian Geographer* 22:66–72.

————. 1983. 'Towards a Radical Cultural Geography: Problems of Theory'. *Antipode* 15:1–11.

————. 1984. *Social Formation and Symbolic Landscape*. London: Croom Helm.

————. 1985. 'Prospect, Perspective and the Evolution of the Landscape Idea'. *Transactions of the Institute of British Geographers NS* 10:45–62.

————. 1997a. 'Spectacle and Society: Landscape as Theater in Premodern and Postmodern Cities'. In *Understanding Ordinary Landscapes*, edited by P. Groth and T.W. Bressi, 99–110. New Haven: Yale University Press.

————. 1997b. *Social Formation and Symbolic Landscape*, 2nd edn. Madison: University of Wisconsin Press.

————. 2003. 'Landscape and the European Sense of Sight—Eyeing Nature'. In *Handbook of Cultural Geography*, edited by K. Anderson, M. Domosh, S. Pile, and N. Thrift, 249–68. Thousand Oaks, CA: Sage Publications.

————, and S. Daniels, eds. 1988. *The Iconography of Landscape*. New York: Cambridge University Press.

————, and P. Jackson. 1987. 'New Directions in Cultural Geography'. *Area* 19:95–101.

Costello, L., and S. Hodge. 1999. 'Queer/Clear/Here: Destabilizing Sexualities and Spaces'. In *Australian Cultural Geographies*, edited by E. Stratford, 131–52. New York: Oxford University Press.

Cowen, T. 2000. *Creative Destruction: How Globalization is Changing the World's Cultures*. Princeton: Princeton University Press.

Crampton, A. 2001. 'The Voortrekker Monument, the Birth of Apartheid, and Beyond'. *Political Geography* 20:221–46.

————. 2003. 'The Art of Nation-building: (Re)presenting Political Transition at the South African National Gallery'. *Cultural Geographies* 10:218–42.

Crang, M. 1998. *Cultural Geography*. New York: Routledge.

————. 2003. 'On Display: The Poetics, Politics and Interpretations of Exhibitions'. In *Cultural Geography in Practice*, edited by A. Blunt, et al., 255–71. New York: Arnold.

Crang, P. 1994. 'It's Showtime: On the Workplace Geographies of Display in a Restaurant in Southeast England'. *Environment and Planning D: Society and Space* 12:675–704.

————. 1997. 'Cultural Turns and the (Re)constitution of Economic Geography: Introduction to Section One'. In *Geographies of Economies*, edited by R. Lee and J. Wills, 3–15. New York: Arnold.

Cream, J. 1995. 'Re-solving Riddles: The Sexed Body'. In *Mapping Desire: Geographies of Sexuality*, edited by D. Bell and G. Valentine, 31–40. New York: Routledge.

Cresswell, T. 1993. 'Mobility as Resistance: A Geographical Reading of Kerouac's "On the Road"'. *Transactions of the Institute of British Geographers NS* 18:249–62.

————. 1996a. *In Place/Out of Place: Geography, Ideology, and Transgression*. Minneapolis: University of Minnesota Press.

————. 1996b. 'Writing, Reading, and the Problem of Resistance: A Reply to McDowell'. *Transactions of the Institute of British Geographers NS* 21:420–4.

Crick, B. 1996. 'Foreword'. In *Race: The History of an Idea in the West*, by I. Hannaford, xi–xvi. Baltimore: Johns Hopkins University Press.

Cronk, L., N. Chagnon, and W. Irons. 2000. *Adaptation and Human Behavior: An Anthropological Perspective*. New York: Aldine de Gruyter.

Cronon, W. 1983. *Changes in the Land: Indians, Colonists, and the Ecology of New England*. New York: Hill and Wang.

————. 1992. 'A Place for Stories: Nature, History, and Narrative'. *Journal of American History* 78:1347–76.

————. 1995. 'Introduction: In Search of Nature'. In *Uncommon Ground: Toward Reinventing Nature*, edited by W. Cronon, 23–56. New York: W.W. Norton.

Crosby, A.W. 1978. 'Ecological Imperialism: The Overseas Migration of Western Europeans as Biological Phenomenon'. *Texas Quarterly* 21:10–22.

————. 1986. *Ecological Imperialism: The Biological Expansion of Europe, 900–1900*. New York: Cambridge University Press.

————. 1995. 'The Past and Present of Environmental History'. *American Historical Review* 100:1177–89.

Croucher, S.L. 2004. *Globalization and Belonging: The Politics of Identity in a Changing World*. Lanham, MD: Rowman and Littlefield.

Crowley, W.K. 1978. 'Old Order Amish Settlement: Diffusion and Growth'. *Annals of the Association of American Geographers* 68:249–64.

Crumley, C.L. 1994. 'Historical Ecology: A Multidimensional Ecological Orientation'. In *Historical Ecology: Cultural Knowledge and Changing Landscapes*, edited by C.L. Crumley, 1–16. Santa Fe: School of American Research Press.

Crystal, D. 1997. *English as a Global Language*. New York: Cambridge University Press.

Cuba, L., and D. Hummon. 1993. 'A Place to Call Home: Identification with Dwelling, Community, and Region'. *Sociological Quarterly* 34:111–31.

Cullen, I.G. 1976. 'Human Geography, Regional Science and the Study of Individual Behavior'. *Environment and Planning A* 8:397–409.

Cumberland, K.B. 1949. 'Aotearoa Maori: New Zealand About 1780'. *Geographical Review* 39:401–24.

da Costa, M.H.B.V. 2003. 'Cinematic Cities: Researching Films as Geographic Texts'. In *Cultural Geography in Practice*, edited by A. Blunt, et al., 191–201. New York: Arnold.

Dalby, S. 1991. 'Critical Geopolitics: Discourse, Difference, and Dissent'. *Environment and Planning D: Society and Space* 9:261–83.

————. 1996. 'Reading Robert Kaplan's "Coming Anarchy"'. *Ecumene* 3:472–96.

Daniels, S. 1985. 'Arguments for a Humanistic Geography'. In *The Future of Geography*, edited by R.J. Johnston, 143–58. New York: Methuen.

————. 1989. 'Marxism, Culture and the Duplicity of Landscape'. In *New Models in Geography, 2*, edited by R. Peet and N. Thrift, 196–220. Boston: Unwin Hyman.

————. 1994. *Fields of Vision: Landscape Imagery and National Identity in England and the United States*. Cambridge: Polity Press.

————. 2004. 'Landscape and Art'. In *A Companion to Cultural Geography*, edited by J.S. Duncan, N.C. Johnson, and R.H. Schein, 430–46. Malden, MA: Blackwell.

Darby, H.C. 1940. *The Draining of the Fens*. New York: Cambridge University Press.

————. 1956. 'The Clearing of the Woodland in Europe'. In *Man's Role in Changing the Face of the Earth*, edited by W.L. Thomas Jr, C.O. Sauer, M. Bates, and L. Mumford, 183–216. Chicago: University of Chicago Press.

————. 1973. *The New Historical Geography of England*. New York: Cambridge University Press.

————. 1977. *Domesday England*. New York: Cambridge University Press.

Darby, W.J. 2000. *Landscape and Identity: Geographies of Nation and Class in England*. New York: Berg.

Davidson, W.V. 1974. *Historical Geography of the Bay Islands, Honduras*. Birmingham, AL: Southern University Press.

Davis, T. 1995. 'The Diversity of Queer Politics and the Redefinition of Sexual Identity and Community in Urban Spaces'. In *Mapping Desire: Geographies of Sexualities*, edited by D. Bell and G. Valentine, 284–303. New York: Routledge.

Dawson, G. 1994. *Soldier Heroes: British Adventure, Empire and the Imagining of Masculinities*. New York: Routledge.

Day, K. 1999. 'Embassies and Sanctuaries: Women's Experiences of Race and Fear in Public Space'. *Environment and Planning D: Society and Space* 17:307–28.

Dear, M. 1988. 'The Postmodern Challenge: Reconstructing Human Geography'. *Transactions of the Institute of British Geographers NS* 13:262–74.

————, and S. Flusty, eds. 2000. *The Spaces of Postmodernity: Readings in Human Geography*. Malden, MA: Blackwell.

————, and J.R. Wolch. 1987. *Landscapes of Despair*. Princeton: Princeton University Press.

Dei, G.S. 1996. *Antiracism Education: Theory and Practice*. Halifax: Fernwood Publications.

Demeritt, D. 1994. 'The Nature of Metaphors in Cultural Geography and Environmental History'. *Progress in Human Geography* 18:163–85.

Denevan, W.M. 1983. 'Adaptation, Variation and Cultural Geography'. *Professional Geographer* 35:399–406.

————, ed. 1992. *The Native Population of the Americas in 1492*. Madison: University of Wisconsin Press.

————. 2001. *Cultivated Landscapes of Native Amazonia and the Andes*. New York: Oxford University Press.

Dennis, R.J., and H. Clout. 1980. *A Social Geography of England and Wales*. New York: Pergamon.

De Oliver, M. 1996. 'Historical Preservation and Identity: The Alamo and the Production of a Consumer Landscape'. *Antipode* 28:1–23.

de Steiguer, J.E. 1997. *The Age of Environmentalism*, 3rd edn. New York: McGraw-Hill.

Des Jardins, J.R. 2001. *Environmental Ethics: An Introduction to Environmental Philosophy*. Belmont, CA: Wadsworth.

Diamond, J. 1997. *Guns, Germs, and Steel: A Short History of Everybody for the Last 13,000 Years*. New York: W.W. Norton.

Dicken, P. 1992. *Global Shift: The Internationalization of Economic Activity*, 2nd edn. New York: Guilford Press.

Dicken, S., and E. Dicken. 1979. *The Making of Oregon: A Study in Historical Geography*. Portland: Oregon Historical Society.

Dickens, P. 1996. *Reconstructing Nature: Alienation, Emancipation and the Division of Labour*. New York: Routledge.

Dickinson, R.E. 1969. *The Makers of Modern Geography*. London: Routledge and Kegan Paul.

Diesendorf, M., and C. Hamilton, eds. 1997. *Human Ecology, Human Economy: Ideas for an Ecologically Sustainable Future*. St Leonard's, Australia: Allen and Unwin.

Dieterich, M., and J. van der Straaten, eds. 2004. *Cultural Landscapes and Land Use: The Nature Conservation—Society Interface*. Boston: Kluwer.

Dijkink, G. 1996. *National Identity and Geopolitical Visions: Maps of Pride and Pain*. New York: Routledge.

Dodson, B. 2000. 'Dismantling Dystopia: New Cultural Geography for a New South Africa'. In *The Geography of South Africa in a Changing World*, edited by R. Fox and K. Rowntree, 138–57. New York: Oxford University Press.

Doel, M. 1999. *Poststructuralist Geographies: The Diabolical Art of Spatial Science*. Edinburgh: Edinburgh University Press.

————. 2004. 'Poststructuralist Geographies: The Essential Selection'. In *Envisioning Human Geographies*, edited by P. Cloke, P. Crang, and M. Goodwin, 146–71. New York: Arnold.

————, and D.B. Clarke. 1998. 'Figuring the Holocaust: Singularity and the Purification of Space'. In *Rethinking Geopolitics*, edited by G. Ó Tuathail and S. Dalby, 39–61. New York: Routledge.

Dohrs, F.E., and L.M. Sommers, eds. 1967. *Cultural Geography: Selected Readings*. New York: Crowell.

Dollar, D., and A. Kraay. 2002. 'Spreading the Wealth'. *Foreign Affairs* 81, no. 1:120–33.

Dominy, M.D. 2001. *Calling the Station Home: Place and Identity in New Zealand's High Country*. Lanham, MD: Rowman and Littlefield.

Domosh, M. 1989. 'A Method for Interpreting Landscape: A Case Study of the New York World Building'. *Area* 21:347–55.

————. 1992. 'Urban Imagery'. *Urban Geography* 13:475–80.

————. 1996. 'Feminism and Human Geography'. In *Concepts in Human Geography*, edited by C. Earle, K. Mathewson, and M.S. Kenzer, 411–27. Lanham, MD: Rowman and Littlefield.

————. 1998. 'Geography and Gender: Home Again'. *Progress in Human Geography* 22:276–82.

————. 1999. 'Corporate Cultures and the Modern Landscape of New York City'. In *Cultural Geographies*, 2nd edn, edited by K. Anderson and F. Gale, 95–111. Harlow: Longman.

————, and J. Seager. 2001. *Putting Women in Place: Feminist Geographers Make Sense of the World*. New York: Guilford.

Donaldson, D.P. 2001. 'Teaching Geography's Four Traditions With Poetry'. *Journal of Geography* 100:24–31.

Donkin, R.A. 1997. 'A "Servant of Two Masters"'. *Journal of Historical Geography* 23:247–66.

Doob, L.W. 1996. 'Foreword'. In *Nations Without States: A Historical Dictionary of Contemporary National Movements*, edited by J. Minahan. Westport, CT: Greenwood Press.

Doolittle, W.E. 2001. *Cultivated Landscapes of Native North America*. New York: Oxford University Press.

Douglas, I., R. Huggett, and M. Robinson. 1996. 'Preface'. In *Companion Encyclopedia of Geography: The Environment and Humankind*, edited by I. Douglas, R. Huggett, and M. Robinson, ix–xi. New York: Routledge.

Dowling, R. 1997. 'Planning for Culture in Urban Australia'. *Australian Geographical Studies* 35:23–31.

———, and P.M. McGuirk. 1998. 'Gendered Geographies in Australia, Aotearoa/New Zealand and the Asia–Pacific'. *Australian Geographer* 29:279–91.

———, and G. Pratt. 1993. 'Home Truths: Recent Feminist Constructions'. *Urban Geography* 14:464–75.

Downs, R.M. 1979. 'Critical Appraisal or Determined Philosophical Skepticism?' *Annals of the Association of American Geographers* 69:468–71.

———, and D. Stea. 1973. 'Cognitive Maps and Spatial Behavior: Process and Products'. In *Image and Environment: Essays on Cognitive Mapping*, edited by R.M. Downs and D. Stea, 8–26. New York: Arnold.

Dunbar, G.S. 1974. 'Geographic Personality'. In *Man and Cultural Heritage: Papers in Honor of Fred B. Kniffen*, edited by H.J. Walker and W.G. Haag. *Geoscience and Man* 5:25–33. Baton Rouge: Louisiana State University, Department of Geography and Anthropology, Geoscience Publications.

———. 1977. 'Some Early Occurrences of the Term "Social Geography"'. *Scottish Geographical Magazine* 93:15–20.

Duncan, J.S. 1978. 'The Social Construction of Unreality: An Interactionist Approach to the Tourist's Cognition of Environment'. In *Humanistic Geography*, edited by D. Ley and M.S. Samuels, 269–82. Chicago: Maaroufa Press.

———. 1980. 'The Superorganic in American Cultural Geography'. *Annals of the Association of American Geographers* 7:181–98.

———. 1990. *The City as Text: The Politics of Interpretation in the Kandyan Kingdom*. New York: Cambridge University Press.

———. 1993. 'Commentary on "The Reinvention of Cultural Geography"'. *Annals of the Association of American Geographers* 83:517–19.

———. 1994a. 'After the Civil War: Reconstructing Cultural Geography as Heterotopia'. In *Re-reading Cultural Geography*, edited by K.E. Foote, P.J. Hugill, K. Mathewson, and J.M. Smith, 401–8. Austin: University of Texas Press.

———. 1994b. 'The Politics of Landscape and Nature, 1992–93'. *Progress in Human Geography* 18:361–70.

———. 1998. 'Author's Response'. *Progress in Human Geography* 22:571–3.

———, and N. Duncan. 1988. '(Re)reading the Landscape'. *Environment and Planning D: Society and Space* 6:117–26.

———, N.C. Johnson, and R.H. Schein, eds. 2004. *A Companion to Cultural Geography*. Malden, MA: Blackwell.

———, and D. Ley. 1993. 'Introduction: Representing the Place of Culture'. In *Place/Culture/Representation*, edited by J.S. Duncan and D. Ley, 1–21. New York: Routledge.

———, and J.P. Sharp. 1993. 'Confronting Representation(s)'. *Environment and Planning D: Society and Space* 11:473–86.

Duncan, N. 1996. 'Postmodernism in Human Geography'. In *Concepts in Human Geography*, edited by C. Earle, K. Mathewson, and M.S. Kenzer, 429–58. Lanham, MD: Rowman and Littlefield.

———, and S. Legg. 2004. 'Social Class'. In *A Companion to Cultural Geography*, edited by J.S. Duncan, N.C. Johnson, and R.H. Schein, 250–64. Malden, MA: Blackwell.

Dunn, K.M. 1997. 'Cultural Geography and Cultural Policy'. *Australian Geographical Studies* 35:1–11.

———, and A. McDonald. 2001. 'The Geographies of Racisms in NSW: A Theoretical Exploration and Some Preliminary Findings from the mid 1990s'. *Australian Geographer* 32:29–44.

Durham, W.H. 1976. 'The Adaptive Significance of Cultural Behavior'. *Human Ecology* 4:89–121.

Dwivedi, O.P. 1990. 'Satyagraha for Conservation: Awakening the Spirit of Hinduism'. In *Ethics of Environment and Development: Global Challenge, International Response*, edited by J.R. Engel and J.G. Engel, 201–12. Tucson: University of Arizona Press.

Earle, C. 1995. 'Review Article: Historical Geography in Extremis? Splitting Personalities on the Postmodern Turn'. *Journal of Historical Geography* 21:455–9.

Earley, J. 1997. *Transforming Human Culture: Social Evolution and the Planetary Crisis*. Albany: State University of New York Press.

Eden, S. 2001. 'Environmental Issues: Nature Versus the Environment?' *Progress in Human Geography* 25:79–85.

Eder, K. 1996. *The Social Construction of Nature*. London: Sage.

Edwards, C.H., E.W. Brabble, Q.J. Cole, and O.F. Westney. 1991. *Human Ecology: Interactions of Man with His Environments*. Dubuque, IA: Kendall/Hunt.

Edwards, G. 1996. 'Alternative Speculations on Geographical Futures: Towards a Postmodern Perspective'. *Geography* 81:217–24.

Ehrenreich, B., and A. R. Hochschild, eds. 2003. *Global Woman: Nannies, Maids and Sex Workers in the New Economy*. New York: Metropolitan Books.

Eichengreen, B. 1998. 'Geography as Destiny: A Brief History of Economic Growth'. *Foreign Affairs* 77, no. 2:128–33.

Eigenheer, R.A. 1973–4. 'The Frontier Hypothesis and Related Spatial Concepts'. *California Geographer* 14:55–69.

Elazar, D.J. 1984. *American Federalism: A View From the States*. New York: Harper and Row.

———. 1994. *The American Mosaic: The Impact of Space, Time and Culture on American Politics*. Boulder, CO: Westview Press.

Eliade, M. 1959. *The Sacred and the Profane: The Nature of Religion*. New York: Harcourt, Brace and World.

Eliades, D.K. 1987. 'Two Worlds Collide: The European Advance into North America'. In *A Cultural Geography of North American Indians*, edited by T.E. Ross and T.G. Moore, 33–44. Boulder, CO: Westview.

Ellen, R. 1982. *Environment, Subsistence, and System: The Ecology of Small-Scale Social Formations*. New York: Cambridge University Press.

———. 1988. 'Persistence and Change in the Relationship Between Anthropology and Human Geography'. *Progress in Human Geography* 12:229–61.

Eller, J.D. 1999. *From Culture to Ethnicity to Conflict: An Anthropological Perspective on International Ethnic Conflict*. Ann Arbor: University of Michigan Press.

Emanuelsson, U. 1988. 'A Model for Describing the Development of the Cultural Landscape'. In *The Cultural Landscape: Past, Present and Future*, edited by H.H. Birks, H.J.B. Birks, P.E. Kaland, and D. Moe, 111–21. New York: Cambridge University Press.

England, K.V.L. 1994. 'Getting Personal: Reflexivity, Positionality, and Feminist Research'. *Professional Geographer* 46:80–9.

English, P.W., and R.C. Mayfield. 1972. 'Ecological Perspectives'. In *Man, Space and Environment*, edited by P.W. English and R.C. Mayfield, 115–20. New York: Oxford University Press.

Ennals, P., and D. Holdsworth. 1981. 'Vernacular Architecture and the Cultural Landscape of the Maritime Provinces: A Reconnaissance'. *Acadiensis* 10:86–105.

Entrikin, J.N. 1976. 'Contemporary Humanism in Geography'. *Annals of the Association of American Geographers* 66:615–32.

———. 1980. 'Robert Park's Human Ecology and Human Geography'. *Annals of the Association of American Geographers* 70:43–58.

———. 1984. 'Carl Sauer: Philosopher in Spite of Himself'. *Geographical Review* 74:387–407.

———. 1988. 'Diffusion Research in the Context of the Naturalism Debate in Twentieth-Century Social Thought'. In *The Transfer and Transformation of Ideas and Material Culture*, edited by P.J. Hugill and D.B. Dickson, 165–78. College Station: Texas A&M University Press.

———. 1991. *The Betweenness of Place*. Baltimore: Johns Hopkins University Press.

Ericksen, E.G. 1980. *The Territorial Experience: Human Ecology as Symbolic Interaction*. Austin: University of Texas Press.

Estaville, L.E. 1993. 'The Louisiana-French Homeland'. *Journal of Cultural Geography* 13, no. 2:31–45.

———. 2001. 'Nouvelle Acadie: The Cajun Homeland'. In *Homelands: A Geography of Culture and Place across America*, edited by R.L. Nostrand and L.E. Estaville, 83–100. Baltimore: Johns Hopkins University Press.

Evans, D.J. 2001. 'Spatial Analyses of Crime'. *Geography* 86:211–23.

———, and M. Fletcher. 2000. 'Fear of Crime: Testing Alternative Hypotheses'. *Applied Geography* 20:395–411.

Evans, E.E. 1973. *The Personality of Ireland: Habitat, Heritage, and History*. New York: Cambridge University Press.

———. 1996. *Ireland and the Atlantic Heritage: Selected Writings*. Dublin: Lilliput Press.

Evernden, N. 1992. *The Social Creation of Nature*. Baltimore: Johns Hopkins University Press.

Eyles, J., and W. Peace. 1990. 'Signs and Symbols in Hamilton: An Iconology of Steeltown'. *Geografiska Annaler* 72B:73–88.

———, and D.M. Smith. 1978. 'Social Geography'. *American Behavioral Scientist* 22:41–58.

Fairhead, J., and M. Peach. 1996. *Misreading the African Landscape: Society and Ecology in a Forest-Savanna Mosaic*. New York: Cambridge University Press.

Featherstone, M. 1995. *Undoing Culture: Globalization, Postmodernism and Identity*. London: Sage.

Febvre, L. 1925. *A Geographical Introduction to History*. New York: Knopf.

Feder, K.L., and M.A. Park. 1997. *Human Antiquity: An Introduction to Physical Anthropology and Archaeology*, 3rd edn. Mountain View, CA: Mayfield.

Ferguson, N. 1997. *Virtual History: Alternatives and Counterfactuals*. London: Picador.

Fernandez-Armesto, F. 2001. *Civilizations: Culture, Ambition, and the Transformation of Nature*. New York: The Free Press.

————. 2003. *The Americas: The History of a Hemisphere*. London: Weidenfeld and Nicolson.

Firey, W. 1945. 'Sentiment and Symbolism as Ecological Variables'. *American Sociological Review* 10:140–8.

————. 1947. *Land Use in Central Boston*. Cambridge, MA: MIT Press.

Fitzsimmons, M. 1989. 'The Matter of Nature'. *Antipode* 21:106–20.

————. 2004. 'Emerging Ecologies'. In *Envisioning Human Geographies*, edited by P. Cloke, P. Crang, and M. Goodwin, 30–47. New York: Arnold.

Flannery, T. 2001. *The Eternal Frontier: An Ecological History of North America and its Peoples*. New York: Atlantic Monthly Press.

Fleure, H.J. 1919. 'Human Regions'. *Scottish Geographical Magazine* 35:94–105.

————. 1951. *Natural History of Man in Britain*. London: Collins.

Flew, A. 1978. *A Rational Animal and Other Philosophical Essays on the Nature of Man*. Oxford: Clarendon Press.

Fogel, R.W. 1964. *Railroads and American Economic Growth: Essays in Econometric History*. Baltimore: Johns Hopkins University Press.

Foley, M.A. 1976. 'Culture Area'. In *Encyclopedia of Anthropology*, edited by D.E. Hunter and P. Whitten, 104. New York: Harper and Row.

Foote, D.C., and B. Greer-Wootten. 1968. 'An Approach to Systems Analysis in Cultural Geography'. *Professional Geographer* 20:86–91.

Foote, K.E., A. Toth, and A. Arvay. 2000. 'Hungary After 1989: Inscribing a New Past on Place'. *Geographical Review* 90:301–34.

Ford, R. 1998. 'Blacks and Asians "Imprisoned" by Fear of Violence'. *London Times* (13 April).

Forsyth, T. 2003. *Critical Political Ecology: The Politics of Environmental Science*. New York: Routledge.

Foucault, M. 1970. *The Order of Things: An Archaeology of the Human Sciences*, translated by A. Sheridan. New York: Random House.

————. 1977. *Discipline and Punish: The Birth of the Prison*, translated by A. Sheridan. New York: Pantheon Books.

————. 1978. *The History of Sexuality, Volume 1: An Introduction*, translated by R. Hurley. New York: Pantheon Books.

————. 1980. *Power/Knowledge*. New York: Pantheon.

Fox, E.W. 1991. *The Emergence of the Modern European World: From the Seventeenth to the Twentieth Century*. Cambridge, MA: Blackwell.

Francaviglia, R.V. 1978. *The Mormon Landscape*. New York: AMS Press.

————. 1991. *Hard Places: Reading the Landscape of America's Historic Mining Districts*. Iowa City: University of Iowa Press.

————. 1994. 'Elusive Land: Changing Geographic Images of the Southwest'. In *Essays on the Changing Images of the Southwest*, edited by R.V. Francaviglia and D. Narrett, 8–39. College Station: Texas A&M University Press.

————. 1996. *Main Street Revisited: Time, Space, and Image Building in Small-Town America*. Iowa City: University of Iowa Press.

————. 2003. *Believing in Place: A Spiritual Geography of the Great Basin*. Reno: University of Nevada Press.

Francis, R.D. 1988. 'The Ideal and the Real: The Image of the Canadian West in the Settlement Period'. In: *Rupert's Land: A Cultural Tapestry*, edited by R.C. Davis, 253–73. Waterloo, ON: Wilfrid Laurier Press.

Franklin, A. 2002. *Nature and Social Theory*. Thousand Oaks, CA: Sage.

Fredrickson, G.M. 1981. *White Supremacy: A Comparative Study in American and South African History*. New York: Oxford University Press.

Freeman, D.B. 1985. 'The Importance of Being First: Preemption by Early Adopters of Farming Innovations in Kenya'. *Annals of the Association of American Geographers* 75:17–28.

————. 1999. 'Hill Stations or Horticulture? Conflicting Imperial Visions of the Cameron Highlands, Malaysia'. *Journal of Historical Geography* 25:17–35.

Frenkel, S., and J. Walton. 2001. 'Bavarian Leavenworth and the Symbolic Economy of a Theme Town'. *Geographical Review* 90:559–84.

Friesen, G. 1987. *The Canadian Prairies: A History*. Toronto: University of Toronto Press.

Gade, D.W. 1982. 'The French Riviera as Elitist Space'. *Journal of Cultural Geography* 3, no. 1:19–28.

————. 1997. "Germanic Towns in Southern Brazil'. *Focus* 44, no. 1:1–6.

————. 1999. *Nature and Culture in the Andes*. Madison: University of Wisconsin Press.

————. 2004. 'Tradition, Territory, and Terroir in French Viniculture: Cassis, France, and Appellation Côntrolée'. *Annals of the Association of American Geographers* 94:848–67.

Gagen, E.A. 2004. 'Landscapes of Childhood and Youth'. In *A Companion to Cultural Geography*, edited by J.S. Duncan, N.C. Johnson, and R.H. Schein, 404–19. Malden, MA: Blackwell.

Gallagher, S.V. 1991. *A Story of South Africa: J.M.Coetzee's Fiction in Context*. Cambridge, MA: Harvard University Press.

Garrison, J.R. 1990. 'Review of *The American Backwoods Frontier: An Ethnic and Ecological Interpretation* by T.G. Jordan and M. Kaups'. *Annals of the Association of American Geographers* 80:639–41.

Garst, R.D. 1974. 'Innovation Diffusion Among the Gusii of Kenya'. *Economic Geography* 50:300–12.

Gastil, R.D. 1975. *Cultural Regions of the United States*. Seattle: University of Washington Press.

Gates, H.L., Jr, ed. 1986. *'Race', Writing, and Difference*. Chicago: University of Chicago Press.

Gay, J. 1971. *Geography of Religion in England*. London: Duckworth.

Geertz, C. 1963. *Agricultural Involution: The Processes of Ecological Change in Indonesia*. Berkeley: University of California Press.

———. 1973. *The Interpretation of Cultures*. New York: Basic Books.

———. 1996. 'Afterword'. In *Senses of Place*, edited by S. Feld and K.H. Basso, 259–62. Santa Fe: School of American Research Press.

———. 2003. 'Off the Menu: A Review of T. Cowen, *Creative Destruction: How Globalization is Changing the World's Cultures*'. *The New Republic* 17 February, 228:27–30.

Gellner, E. 1997. 'Knowledge of Nature and Society'. In *Nature and Society in Historical Context*, edited by M. Teich, R. Porter, and B. Gustafsson, 9–17. New York: Cambridge University Press.

Gerber, J. 1997. 'Beyond Dualism—The Social Construction of Nature and the Natural and Social Construction of Human Beings'. *Progress in Human Geography* 21:1–17.

Gerlach, J., and J. Janke. 2001. 'The Mall of America as a Tourist Attraction'. *Focus* 46, no. 3:32–6.

Gerlach, R.L. 1976. *Immigrants in the Ozarks: A Study in Ethnic Geography*. Columbia: University of Missouri Press.

Gertler, M.S. 2003. 'A Cultural Economic Geography of Production'. In *Handbook of Cultural Geography*, edited by K. Anderson, M. Domosh, S. Pile, and N. Thrift, 131–46. Thousand Oaks, CA: Sage Publications.

Getz, D. 1995. 'Event Tourism and the Authenticity Dilemma'. In *Global Tourism: The Next Decade*, edited by W.F. Theobald, 313–29. Boston: Butterworth Heinemann.

Geyer, F. 1996. 'Introduction: Alienation, Ethnicity, and Postmodernism'. In *Alienation, Ethnicity, and Postmodernism*, edited by F. Geyer, ix–xxviii. Westport, CT.: Greenwood.

Giddens, A. 1984. *The Constitution of Society: Outline of the Theory of Structuration*. Cambridge: Polity Press.

———. 1987. *Sociology: A Brief But Critical Introduction*, 2nd edn. New York: Harcourt Brace Jovanovich.

———. 1991. *Introduction to Sociology*. New York: W.W. Norton.

———, and J.H. Turner. 1987. 'Introduction'. In *Social Theory Today*, edited by A. Giddens and J.H. Turner, 1–10. Stanford: Stanford University Press.

Gilbert, A. 1988. 'The New Regional Geography in English and French-Speaking Countries'. *Progress in Human Geography* 12:208–28.

Giliomee, H. 2003. *The Afrikaners: Biography of a People*. Charlottesville: University of Virginia Press.

Gill, W.G. 1993. 'Region, Agency, and Popular Music: The Northwest Sound'. *Canadian Geographer* 37:120–31.

Gills, B.K. 1995. 'Capital and Power in the Processes of World History'. In *Civilizations and World Systems: Studying World-Historical Change*, edited by S.K. Sanderson, 136–62. Walnut Creek, CA: Altamira Press.

Gil-White, F.J. 2001. 'Are Ethnic Groups Biological "Species" in the Human Brain'? *Current Anthropology* 42:515–54.

Ginsburg, N. 1970. Geography. In *A Reader's Guide to the Social Sciences*, 2nd edn, edited by B.F. Hoselitz, 293–318. New York: Free Press.

Glacken, C. 1967. *Traces on the Rhodian Shore: Nature and Culture in Western Thought from Ancient Times to the End of the Eighteenth Century*. Berkeley: University of California Press.

———. 1985. 'Culture and Environment in Western Civilization During the Nineteenth Century'. In *Environmental History*, edited by K.E. Bailes, 46–57. New York: University Press of America.

Glasscock, R.E. 1991. 'Obituary: E. Estyn Evans, 1905–1989'. *Journal of Historical Geography* 17:87–91.

Gleeson, B. 1999. *Geographies of Disability*. New York: Routledge.

Gold, J.R. 1980. *An Introduction to Behavioral Geography*. New York: Oxford University Press.

Goldhagen, D.J. 1996. *Hitler's Willing Executioners: Ordinary Germans and the Holocaust*. New York: Knopf.

Goldschmidt, W. 1965. 'Theory and Strategy in the Theory of Cultural Adaptability'. *American Anthropologist* 67:402–8.

Golledge, R.G. 1969. 'The Geographical Relevance of Some Learning Theories'. In *Behavioral Problems in Geography: A Symposium*, edited by K.R. Cox and R.G. Golledge, 101–45. Evanston, Ill.: Northwestern University Press, Studies in Geography, no. 17.

———. 1987. 'Environmental Cognition'. In *Handbook of Environmental Psychology, Volume 1*, edited by D. Stokols and I. Altman, 131–74. New York: Wiley.

————. 1993. 'Geography and the Disabled: A Survey with Special Reference to Vision Impaired and Blind Populations'. *Transactions of the Institute of British Geographers* NS 18:63–85.

————. 2003. 'Reflections on Recent Cognitive Behavioral Research with an Emphasis on Research in the United States of America'. *Australian Geographical Studies* 41:117–30.

————, and R.J. Stimson. 1997. *Spatial Behavior: A Geographic Perspective*. New York: Guilford.

Goode, P. 1926. *The Geographic Background of Chicago*. Chicago: University of Chicago Press.

Gore, A. 1992. *Earth in the Balance: Ecology and the Human Spirit*. Boston: Houghton Mifflin.

Goss, J. 1993. 'The "Magic of the Mall": An Analysis of Form, Function, and Meaning in the Contemporary Built Environment'. *Annals of the Association of American Geographers* 83:18–47.

————. 1999a. 'Modernity and Postmodernity in the Retail Landscape'. In *Cultural Geographies*, 2nd edn, edited by K. Anderson and F. Gale, 199–219. Harlow: Longman.

————. 1999b. 'Once-upon-a-Time in the Commodity World: An Unofficial Guide to the Mall of America'. *Annals of the Association of American Geographers* 89:45–75.

————. 'Geographies of Consumption, 1'. *Progress in Human Geography* 28:369–40.

Gottlieb, R.S. 1996. *This Sacred Earth: Religion, Nature, Environment*. New York: Routledge.

Gould, P. 1963. 'Man Against His Environment: A Game Theoretic Framework'. *Annals of the Association of American Geographers* 53:290–7.

————. 1969. *Spatial Diffusion*. Washington, DC: Association of American Geographers, Resource Paper, no. 4.

————, and R. White. 1986. *Mental Maps*, 2nd edn. Boston: Allen and Unwin.

Gould, S.J. 1990. 'The Golden Rule—a Proper Scale for Our Environmental Crisis'. *Natural History* 99, no. 9:24–30.

Graham, B.J., ed. 1997. *In Search of Ireland: A Cultural Geography*. New York: Routledge.

————, G.J. Ashworth, and J.E. Tunbridge. 2000. *A Geography of Heritage: Power, Culture and Economy*. New York: Arnold.

————, and L.J. Proudfoot. 1993. 'A Perspective on the Nature of Irish Historical Geography'. In *An Historical Geography of Ireland*, edited by B.J. Graham and L.J. Proudfoot, 1–18. New York: Academic Press.

Graham, E. 1997. 'Philosophies Underlying Human Geographic Research'. In *Methods in Human Geography: A Guide for Students Doing a Research Project*, edited by R. Flowerdew and D. Martin, 6–30. Harlow: Longman.

Granö, J.G. [1929] 1997. *Pure Geography*, edited by O. Granö and A. Paasi, translated by M. Hicks. Baltimore: Johns Hopkins University Press.

Gregory, D. 1981. 'Human Agency and Human Geography'. *Transactions of the Institute of British Geographers* NS 6:1–18.

————. 1994. *Geographical Imaginations*. Cambridge, MA: Blackwell.

Gregson, N. 1987. 'Structuration Theory: Some Thoughts on the Possibilities for Empirical Research'. *Environment and Planning D: Society and Space* 5:73–91.

————. 1993. '"The Initiative": Delimiting or Deconstructing Social Geography'. *Progress in Human Geography* 17:525–30.

————. 1995. 'And Now It's All Consumption?' *Progress in Human Geography* 19:135–41.

————, and M. Lowe. 1995. '"Home"-making: On the Spatiality of Daily Social Reproduction in Contemporary Middle-class Britain'. *Transactions of the Institute of British Geographers* NS 20:224–35.

————, and G. Rose. 2000. 'Taking Butler Elsewhere: Performativities, Spatialities and Subjectivities'. *Environment and Planning D: Society and Space* 18:433–52.

————, G. Rose, J. Cream, and N. Laurie. 1997. 'Conclusions'. In *Feminist Geographies: Explorations in Diversity and Difference*, edited by Women and Geography Study Group of the Royal Geographical Society with the Institute of British Geographers, 191–200. Harlow: Longman.

————, et al. 1997. 'Gender in Feminist Geography'. In *Feminist Geographies: Explorations in Diversity and Difference*, edited by Women and Geography Study Group of the Royal Geographical Society with the Institute of British Geographers, 49–85. Harlow: Longman.

Grey, A. 1994. *Aotearoa and New Zealand: A Historical Geography*. Christchurch: Canterbury University Press.

Griffiths, T. 1997. 'Introduction. Ecology and Empire: Towards an Australian History of the World'. In *Ecology and Empire: Environmental History of Settler Societies*, edited by T. Griffiths and L. Robin, 1–16. Seattle: University of Washington Press.

Griggs, R. 2000. 'Philosophy and Methodology in Geography'. In *The Geography of South Africa in a Changing World*, edited by R. Fox and K. Rowntree, 9–30. New York: Oxford University Press.

Gritzner, C.F. 1966. 'The Scope of Cultural Geography'. *Journal of Geography* 65:4–11.

———. 2002. 'What is Where, Why There, and Why Care?' *Journal of Geography* 101:38–40.

Grossman, L. 1977. 'Man–Environment Relationships in Anthropology and Geography'. *Annals of the Association of American Geographers* 67:126–44.

———. 1984. *Peasants, Subsistence, Ecology, and Development in the Highlands of Papua New Guinea.* Princeton: Princeton University Press.

———. 1993. 'The Political Ecology of Banana Exports and Local Food Production in St Vincent, Eastern Caribbean'. *Annals of the Association of American Geographers* 83:347–67.

Groth, P. 1997. 'Frameworks for Cultural Landscape Study'. In *Understanding Ordinary Landscapes*, edited by P. Groth and T.W. Bressi, 1–21. New Haven: Yale University Press.

———, and T.W. Bressi, eds. 1997. *Understanding Ordinary Landscapes.* New Haven: Yale University Press.

Grove, R.H. 1992. 'Origins of Western Environmentalism'. *Scientific American* 267:42–7.

Guelke, L. 1971. 'Problems of Scientific Explanation in Geography'. *Canadian Geographer* 15:38–53.

———. 1974. 'An Idealist Alternative in Human Geography'. *Annals of the Association of American Geographers* 64:193–202.

———. 1975. 'On Rethinking Historical Geography'. *Area* 7:135–8.

———. 1982. *Historical Understanding in Geography.* New York: Cambridge University Press.

———. 1987. 'Frontier Settlement and Human Values: A Comparative Look at North America and South Africa'. In *Abstract Thoughts: Concrete Solutions. Essays in Honour of Peter Nash*, edited by L. Guelke and R. Preston, 181–99. Waterloo: University of Waterloo, Department of Geography, Publication Series, no. 29.

Habermas, J. 1972. *Knowledge and Human Interests.* London: Heinemann.

Hage, G. 1996. 'The Spatial Imagery of National Practices: Dwelling-Domesticating/Being-Exterminating'. *Environment and Planning D: Society and Space* 14:463–85.

Hagen, J. 2004. 'The Most German of Towns: Creating an Ideal Nazi Community in Rothenburg ob der Tauber'. *Annals of the Association of American Geographers* 94:207–27.

Hägerstrand, T. 1951. *Migration and the Growth of Culture Regions.* Lund: Gleerup, Lund Studies in Geography, Series B, no. 3.

———. 1952. *The Propagation of Innovation Waves.* Lund: Gleerup, Lund Studies in Geography, Series B, no. 4.

———. 1967. *Innovation Diffusion as a Spatial Process.* Translated by A. Pred. Chicago: University of Chicago Press.

Haggett, P. 2000. *The Geographical Structure of Epidemics.* Oxford: Clarendon Press.

———. 2001. *Geography: A Global Synthesis.* New York: Prentice Hall.

Hague, E., and J. Mercer. 1998. 'Geographical Memory and Urban Identity in Scotland: Raith Rovers FC and Kirkcaldy'. *Geography* 83:105–11.

Hale, R.F. 1971. 'A Map of Vernacular Regions in America'. Ph.D. dissertation. Minneapolis: University of Minnesota.

———. 1984. 'Vernacular Regions of America'. *Journal of Cultural Geography* 5, no. 1:131–40.

Halfacree, K.H. 2003. 'Landscapes of Rurality: Rural Others/Other Rurals'. In *Studying Cultural Landscapes*, edited by I. Robertson and P. Richards, 141–64. New York: Arnold.

———, and R.M. Kitchin. 1996. '"Madchester Rave On": Placing the Fragments of Popular Music'. *Area* 28:47–55.

Hall, C.S., and G. Lindzey. 1978. *Theories of Personality*, 3rd edn. New York: Wiley.

Hall, S. 1996. 'Introduction: Who Needs Identity?' In *Questions of Cultural Identity*, edited by S. Hall and P. duGay, 1–17. London: Sage.

Hall, T. 2003. 'Art and Urban Change: Public Art in Urban regeneration'. In *Cultural Geography in Practice*, edited by A. Blunt, et al., 221–37. New York: Arnold.

Hallman, B., and A. Joseph. 1999. 'Getting There: Mapping the Gendered Geography of Caregiving to Elderly Relatives'. *Canadian Journal on Aging* 18:397–414

Hamnett, C., ed. 1996. *Social Geography: A Reader.* New York: Arnold.

———. 2003. 'Editorial. Contemporary Human Geography: Fiddling While Rome Burns'. *Geoforum* 34:1–3.

Hannaford, I. 1996. *Race: The History of an Idea in the West.* Baltimore: Johns Hopkins University Press.

Hanson, S. 1992. 'Geography and Feminism: Worlds in Collision'. *Annals of the Association of American Geographers* 82:569–86.

———, and G. Pratt. 1995. *Gender, Work, and Space.* New York: Routledge.

Haraway, D. 1991. *Simians, Cyborgs, and Women: The Reinvention of Nature.* New York: Routledge.

Hardesty, D.L. 1977. *Ecological Anthropology.* New York: Wiley.

———. 1986. 'Rethinking Cultural Adaptation'. *Professional Geographer* 38:11–8.

Hardill, I., D.T. Graham, and E. Kofman. 2001. *Human Geography of the UK*. New York: Routledge.

Hardin, G. 1968. 'The Tragedy of the Commons'. *Science* 162:1243–8.

Harris, D.R. 1996. 'Introduction: Themes and Concepts in the Study of Early Agriculture'. In *The Origins and Spread of Agriculture and Pastoralism in Eurasia*, edited by D.R. Harris, 1–9. Washington, DC: Smithsonian Institution Press.

Harris, L. 2004. *Civilization and Its Enemies*. New York: Free Press.

Harris, M. 1968. *The Rise of Anthropological Theory*. London: Routledge and Kegan Paul.

———. 1979. *Cultural Materialism: The Struggle for a Science of Culture*. New York: Random House.

Harris, R.C. 1977. 'The Simplification of Europe Overseas'. *Annals of the Association of American Geographers* 67:469–83.

———. 1978. 'The Historical Geography of North American Regions'. *American Behavioral Scientist* 22:115–30.

———. 1999. 'Comments on "The Shaping of America, 1850–1915"'. *Journal of Historical Geography* 25:9–11.

Harrison, L.E. 2000. 'Why Culture Matters'. In *Culture Matters: How Values Shape Human Progress*, edited by L.E. Harrison and S.P. Huntington, xvii–xxxiv. New York: Basic Books.

Harrison, R., and D.N. Livingstone. 1979. 'There and Back Again: Towards a Critique of Idealist Human Geography'. *Area* 11:75–9.

Hart, J.F. 1975. *The Look of the Land*. Englewood Cliffs, NJ: Prentice Hall.

———. 1982. 'The Highest Form of the Geographer's Art'. *Annals of the Association of American Geographers* 72:1–29.

———. 1995. 'Reading the Landscape'. In *Landscape in America*, edited by G.F. Thompson, 23–42. Austin: University of Texas Press.

———. 1998. *The Rural Landscape*. Baltimore: Johns Hopkins University Press.

Hartig, K.V., and K.M. Dunn. 1998. 'Roadside Memorials: Interpreting New Deathscapes in Newcastle, New South Wales'. *Australian Geographical Studies* 36:5–20.

Hartshorne, R. 1939. *The Nature of Geography*. Lancaster, PA: Association of American Geographers.

———. 1959. *Perspective on the Nature of Geography*. Chicago: Rand McNally.

Harvey, D.W. 1969. *Explanation in Geography*. New York: Arnold.

———. 1973. *Social Justice and the City*. Baltimore: Johns Hopkins University Press.

———. 1979. 'Monument and Myth'. *Annals of the Association of American Geographers* 69:362–81.

———. 1989. *The Condition of Postmodernity: An Enquiry into the Origins of Cultural Change*. Cambridge, MA: Blackwell.

———. 1993. 'Class Relations, Social Justice and Politics of Difference'. In *Place and the Politics of Identity*, edited by M. Keith and S. Pile, 41–66. New York: Routledge.

Hauptman, L.M., and R.G. Knapp. 1977. 'Dutch–Aboriginal Interaction in New Netherland and Formosa: An Historical Geography of Empire'. *Proceedings of the American Philosophical Society* 121:166–82.

Hausladen, G.J., ed. 2003. *Western Places, American Myths: How We Think About the West*. Reno: University of Nevada Press.

Hawley, A.H. 1950. *Human Ecology: A Theory of Community Structure*. New York: Ronald Press Co.

———. 1968. 'Human Ecology'. In *International Encyclopedia of the Social Sciences, Volume 4*, edited by D.L. Sills, 328–37. New York: Free Press.

———. 1998. 'Human Ecology, Population, and Development'. In *Continuities in Sociological Human Ecology*, edited by M. Micklin and D.L. Poston Jr, 11–25. New York: Plenum Press.

Hayden, D. 1995. *The Power of Place: Urban Landscapes as Public History*. Cambridge, MA: MIT Press.

Haynes, R.M. 1980. *Geographical Images and Mental Maps*. New York: MacMillan.

Head, L. 1993. 'Unearthing Prehistoric Cultural Landscapes: A View From Australia'. *Transactions of the Institute of British Geographers* NS 18:481–99.

———. 2000. *Second Nature: The History and Implications of Australia as Aboriginal Landscape*. Syracuse: Syracuse University Press.

Headland, T. 1997. 'Revisionism in Ecological Anthropology'. *Current Anthropology* 38:605–9.

Heathcote, R.L. 1972. 'The Visions of Australia, 1770–1970'. In *Australia as Human Setting*, edited by A. Rapoport, 77–98. Sydney: Angus and Robertson.

Heffernan, M. 1998. *The Meaning of Europe: Geography and Geopolitics*. New York: Arnold.

Hemmings, C. 2002. *Bisexual Spaces: A Geography of Sexuality and Gender*. New York: Routledge.

Henderson, G. 1997. '"Landscape is Dead, Long Live Landscape": A Handbook for Sceptics'. *Journal of Historical Geography* 24:94–100.

Henderson, J.R. 1978. 'Spatial Reorganization: A Geographical Dimension in Acculturation'. *Canadian Geographer* 12:1–21.

Henderson, M.L. 1996. 'Geography, First Peoples, and Social Justice'. *Geographical Review* 86:278–83.

Herb, G.H. 2004. 'Double Vision: Territorial Strategies in the Construction of National Identities in Germany, 1949–1979'. *Annals of the Association of American Geographers* 94:140–64.

Herbert, D. 1991. 'Place and Society in Jane Austen's England'. *Geography* 76:193–208.

Heyer, P. 1982. *Nature, Human Nature, and Society: Marx, Darwin, Biology and the Human Sciences.* Westport, CT: Greenwood.

Hilliard, S.B. 1972. *Hog Meat and Hoecake: Food Supply in the Old South, 1840–1860.* Carbondale: Southern Illinois University Press.

Hirsch, E., and M. O'Hanlon, eds. 1995. *The Anthropology of Landscape: Perspectives on Space and Place.* Oxford: Clarendon.

Hobsbawm, E. 1993. 'The New Threat to History'. *The New York Review of Books* (16 December):62–4.

Hodge, S. 1996. 'Disadvantage and "Otherness" in Western Sydney'. *Australian Geographical Studies* 34:32–44.

Hoggart, R. 1957. *The Uses of Literacy: Changing Patterns in English Mass Culture.* Fair Lawn, NJ: Essential Books.

Hoke, G.W. 1907. 'The Study of Social Geography'. *Geographical Journal* 29:64–7.

Holloway, L., and P. Hubbard. 2001. *People and Place: The Extraordinary Geographies of Everyday Life.* New York: Prentice Hall.

Holloway, S.L. 2003. 'Outsiders in Rural Society? Constructions of Rurality and Nature–Society Relations in the Racialisation of English Gypsy-Travellers, 1869–1934'. *Environment and Planning D: Society and Space* 21:695–715.

———, and G. Valentine, eds. 2002. *Children's Geographies: Living, Playing, Learning.* New York Routledge.

Homans, G. 1987. 'Behaviorism and After'. In *Social Theory Today*, edited by A. Giddens and J.H. Turner, 58–81. Stanford: Stanford University Press.

Honderich, T. 1995. *The Oxford Companion to Philosophy.* New York: Oxford University Press.

Hooson, D.J.M. 1981. 'Carl O. Sauer'. In *The Origins of Academic Geography in the United States*, edited by B.W. Blouet, 165–74. Hamden, CT: Archon Books.

Hopkins, J.S.P. 1990. 'West Edmonton Mall: Landscape of Myths and Elsewhereness'. *Canadian Geographer* 34:2–17.

———. 1991. 'West Edmonton Mall as a Centre for Social Interaction'. *Canadian Geographer* 35:261–7.

Hopkins, T.K., and I. Wallerstein. 1996. 'The World-System: Is There a Crisis?' In *The Age of Transition: Trajectory of the World System, 1945–2025*, coordinated by T.K. Hopkins and I. Wallerstein, 1–10. Atlantic Highlands, NJ: Zed Books.

Hornbeck, D., C. Earle, and C.M. Rodrigue. 1996. 'The Way We Were: Deployments (and Redeployments) of Time in Human Geography'. In *Concepts in Human Geography*, edited by C. Earle, K. Mathewson, and M.S. Kenzer, 33–61. Lanham, MD: Rowman and Littlefield.

Horton, A. 2003. 'Reel Landscapes: Cinematic Environments Documented and Created'. In *Studying Cultural Landscapes*, edited by I. Robertson and P. Richards, 71–92. New York: Arnold.

Hoskins, W.G. 1955. *The Making of the English Landscape.* London: Hodder and Stoughton.

Hough, M. 1990. *Out of Place: Restoring Identity to the Regional Landscape.* New Haven: Yale University Press.

Howe, S. 1998. *Afrocentrism: Mythical Pasts and Imagined Homes.* New York: Verso.

Howitt, R., and S. Suchet-Pearson. 2003. 'Ontological Pluralism in Contested Cultural Landscapes'. In *Handbook of Cultural Geography*, edited by K. Anderson, M. Domosh, S. Pile, and N. Thrift, 557–69. Thousand Oaks, CA: Sage Publications.

Hsiung, D.C. 1996. *Two Worlds in the Tennessee Mountains: Exploring the Origins of Appalachian Stereotypes.* Lexington: University of Kentucky Press.

Hubbard, P. 2000. 'Desire/Disgust: Mapping the Moral Contours of Heterosexuality'. *Progress in Human Geography* 24:191–217.

———, Kitchin, R., Bartley, B., and D. Fuller. 2002. *Thinking Geographically: Space, Theory and Contemporary Human Geography.* New York: Continuum.

———, and K. Lilley. 2000. 'Selling the Past: Heritage-tourism and Place Identity in Stratford-upon-Avon'. *Geography* 85:221–32.

Huckle, J. 2002. 'Reconstructing Nature: Towards a Geographical Education for Sustainable Development'. *Geography* 87:64–72.

Hudson, B.J. 1992. 'Arnold Bennett's Five Towns: "Internal and External Vision"'. *Journal of Cultural Geography* 11, no. 2:21–30.

Hudson, J.C. 1977. 'Theory and Methodology in Comparative Frontier Studies'. In *The Frontier*, edited by D.H. Miller and J.O. Steffen, 11–32. Norman: University of Oklahoma Press.

———. 1994. *Making the Corn Belt: A Geographical History of Middle-Western Agriculture.* Bloomington: Indiana University Press.

Hudson, W., and G. Bolton. 1997. 'Creating Australia'. In *Creating Australia: Changing Australian*

History, edited by W. Hudson and G. Bolton, 1–11. St Leonards, NSW: Allen and Unwin

Hufferd, J. 1980. 'Toward a Transcendental Human Geography of Places'. *Antipode* 12, no. 3:18–23.

Huggett, R. 1980. *Systems Analysis in Geography*. Oxford: Clarendon Press.

Hughes, J.D. 2001. *An Environmental History of the World: Humankind's Changing Role in the Community of Life*. New York: Routledge.

Hugill, P.J. 1997. 'Review Article: World-System Theory: Where's the Theory?' *Journal of Historical Geography* 23:344–9.

———. 1999. 'Imperialism and Manliness in Edwardian Boys' Novels'. *Ecumene* 6:318–40.

———, and D.B. Dickson, eds. 1988. *The Transfer and Transformation of Ideas and Material Culture*. College Station: Texas A&M University Press.

Hunter, L.M., and M.B. Toney. 2005. 'Religion and Attitudes Toward the Environment: A Comparison of Mormons and the General US Population'. *The Social Science Journal* 42(1): 25–38.

Huntington, E. 1915. Civilization and Climate. New Haven: Yale University Press.

———. 1920. *Principles of Human Geography*, 5th edn. New York: Wiley.

———. 1945. *Mainsprings of Civilization*. New York: Wiley.

Huntington, S.P. 1993. 'The Clash of Civilizations'. *Foreign Affairs* 73, no. 2:22–49.

———. 1996. *The Clash of Civilizations and the Remaking of World Order*. New York: Simon and Schuster.

———. 2000. 'Culture Counts'. In *Culture Matters: How Values Shape Human Progress*, edited by L.E. Harrison and S.P. Huntington, xiii–xvi. New York: Basic Books.

Hurt, D.A. 2003. 'Defining American Homelands: A Creek Nation Example, 1828–1907'. *Journal of Cultural Geography* 21, no. 1:19–43.

Hutcheon, P.D. 1994. 'Is There a Dark Side to Multiculturalism?' *Humanist in Canada* 27, no. 2:26–9.

———. 1995. 'Defining Modern Humanism'. *Humanist in Canada* 28, no. 1:30–3.

———. 1996. *Leaving the Cave: Evolutionary Naturalism in Social-Scientific Thought*. Waterloo, ON: Wilfrid Laurier University Press.

Hutchinson, J., and A.D. Smith, eds. 1996. *Ethnicity*. New York: Oxford University Press.

Huxley, J.S., and A.C. Haddon. 1936. *We Europeans*. London: Jonathan Cape.

Ignatieff, M. 2000. *The Rights Revolution*. Toronto: Anansi.

Imrie, R. 1996. *Disability and the City: International Perspectives*. London: Paul Chapman.

Inglehart, R. 1990. *Culture Shift in Advanced Industrial Society*. Princeton: Princeton University Press.

Innis, H.A. 1930. *The Fur Trade in Canada*. New Haven: Yale University Press.

Isajiw, W. 1974. 'Definitions of Ethnicity'. *Ethnicity* 1:111–24.

Ittelson, W.H., H.M. Proshansky, L.G. Rivlin, and G.H. Winkel. 1974. *An Introduction to Environmental Psychology*. New York: Holt, Rinehart and Winston.

Jackson, J.B. 1974. *A Sense of Place: A Sense of Time*. New Haven: Yale University Press.

———. 1984. *Discovering the Vernacular Landscape*. New Haven: Yale University Press.

———. 1997a. *Landscape in Sight: Looking at America*, edited by H.L. Horowitz. New Haven: Yale University Press.

———. 1997b. 'Foreword'. In *The Evolving Landscape: Homer Aschmann's Geography*, edited by M.J. Pasqualetti, vii–viii. Baltimore: Johns Hopkins University Press.

Jackson, P. 1980. 'A Plea for Cultural Geography'. *Area* 12:110–13.

———. 1987. 'The Idea of "Race" and the Geography of Racism'. In *Race and Racism: Essays in Social Geography*, edited by P. Jackson, 3–21. Boston: Allen and Unwin.

———. 1989. *Maps of Meaning: An Introduction to Cultural Geography*. Boston: Unwin Hyman.

———. 1990. 'The Cultural Politics of Masculinity: Towards a Social Geography'. *Transactions of the Institute of British Geographers* NS 16:199–213.

———. 1992. 'The Politics of the Streets: A Geography of Caribana'. *Political Geography* 11:130–51.

———. 2000a. 'Rematerializing Social and Cultural Geography'. *Social and Cultural Geography* 1:9–14.

———. 2000b. 'Cultures of Difference'. In *The Changing Geography of the United Kingdom*, 3rd edn, edited by V. Gardiner and H. Matthews, 276–95. New York: Routledge.

———. 2002. 'Geographies of Diversity and Difference'. *Geography* 87:316–23.

———, and J. Penrose, eds. 1994. *Constructions of Race, Place, and Nation*. Minneapolis: University of Minnesota Press.

———, and S.J. Smith. 1984. *Exploring Social Geography*. London: Allen and Unwin.

———, and J. Taylor. 1996. 'Geography and the Cultural Politics of Advertising'. *Progress in Human Geography* 20:356–71.

Jackson, R.H. 1978. 'Mormon Perception and Settlement'. *Annals of the Association of American Geographers* 68:317–34.

———, and R. Henrie. 1983. 'Perception of Sacred Space'. *Journal of Cultural Geography* 3, no. 2:94–107.

———, and R.L. Layton. 1976. 'The Mormon Village: Analysis of a Settlement Type'. *Professional Geographer* 28:136–41.

Jacobs, W.R. 1997. 'Perspectives on Scarcity, Population, Ecology, and the American West'. *Journal of the West* 36, no. 3:3–5.

Jakle, J.A. 1985. *The Tourist: Travel in Twentieth Century North America*. Boston: Unwin Hyman.

———. 1987. *The Visual Elements of Landscape*. Amherst: University of Massachusetts Press.

———. 1994. *The Gas Station in America*. Baltimore: Johns Hopkins University Press.

———, S.D. Brunn, and C.C. Roseman. 1976. *Human Spatial Behavior*. North Scituate, MA: Duxbury.

———, and K.A. Sculle. 1999. *Fast Food: Roadside Restaurants in the Automobile Age*. Baltimore: Johns Hopkins University Press.

James, P.E. 1964. *One World Divided: A Geographer Looks at the Modern World*. New York: Blaisdell Publishing Co.

Jameson, F. 1991. *Postmodernism, or The Cultural Logic of Late Capitalism*. Durham, NC: Duke University Press.

Jamieson, D. 1985. 'Against Zoos'. In *In Defense of Animals*, edited by P. Singer, 108–17. Cambridge, MA: Blackwell.

———. 1997. 'Zoos Revisited'. In *The Philosophy of the Environment*, edited by T.D.J. Chappell, 180–92. Edinburgh: Edinburgh University Press.

———, ed. 2001. *A Companion to Environmental Philosophy*. Malden, MA: Blackwell.

Jarosz, L. 1993. 'Defining and Explaining Tropical Deforestation: Shifting Cultivation and Population Growth in Colonial Madagascar (1846–1940)'. *Economic Geography* 69:366–79.

Jeans, D.N. 1974. 'Changing Formulations of the Man–Environment Relationship in Anglo-American Geography'. *Journal of Geography* 73:36–40.

———. 1981. 'Mapping the Regional Patterns of Australian Society: Some Preliminary Thoughts'. *Australian Historical Geography Bulletin* 2:1–6.

———. 1988. 'The First World War Memorials in New South Wales: Centres of Meaning in the Landscape'. *Australian Geographer* 19:259–67.

Jensen, R., and H. Burgess. 1997. 'Mythmaking: How Introductory Psychology Texts Present B.F. Skinner's Analysis of Cognition'. *The Psychological Record* 47:221–32.

Johnson, L.C. 1999. 'Powerlines: A Cultural Geography of Domestic Open Space'. In *Australian Cultural Geographies*, edited by E. Stratford, 87–108. New York: Oxford University Press.

Johnson, N. 1995. 'Cast in Stone: Monuments, Geography, and Nationalism'. *Environment and Planning D: Society and Space* 13:51–65.

Johnson, R. 1996. 'Revising Culture Through 'Women and Nature': An Introduction to the Special Issue'. *Women's Studies* 25:v–xi.

Johnston, J.A. 1979. 'Image and Reality: Initial Assessments of Soil Fertility in New Zealand, 1839–55'. *Australian Geographer* 14:160–5.

———. 1981. 'The New Zealand Bush: Early Assessments of Vegetation'. *New Zealand Geographer* 37, no. 1:19–24.

Johnston, L. 1997. 'Queen(s') Street or Ponsonby Poofters? Embodied Hero Parade Sites'. *New Zealand Geographer* 53, no. 2:29–33.

Johnston, R.J. 1985. 'Introduction: Exploring the Future of Geography'. In *The Future of Geography*, edited by R.J. Johnston, 3–26. New York: Methuen.

———. 1990. 'The Challenge for Regional Geography: Some Proposals for Research Frontiers'. In *Regional Geography: Current Developments and Future Prospects*, edited by R.J. Johnston, J. Hauer, and G.A. Hoekveld, 122–39. New York: Routledge.

———. 1991. *A Question of Place: Exploring the Practice of Human Geography*. Cambridge, MA: Blackwell.

———. 1997. *Geography and Geographers: Anglo-American Geography Since 1945*, 5th edn. New York: Wiley.

———, D. Gregory, and D.M. Smith, eds. 2000. *The Dictionary of Human Geography*, 4th edn. Cambridge, MA: Blackwell.

Jones, E.L. 1987. *The European Miracle: Environments, Economics and Geopolitics in the History of Europe and Asia*, 2nd edn. New York: Cambridge University Press.

Jones, R. 2004. 'What Time Human Geography?'. *Progress in Human Geography* 28:287–304.

Jordan, T.G. 1966. *German Seed in Texas Soil*. Austin: University of Texas Press.

———. 1978. 'Perceptual Regions in Texas'. *Geographical Review* 68:293–307.

———. 1983. 'A Reappraisal of Fenno-Scandian Antecedents for Midland American Log Construction'. *Geographical Review* 73:58–94.

———. 1985. *American Log Buildings: An Old World Heritage*. Chapel Hill: University of North Carolina Press.

———, M. Domosh, and L. Rowntree. 1997. *The Human Mosaic: A Thematic Introduction to Cultural Geography*, 7th edn. New York: Harper and Row.

———, and M. Kaups. 1989. *The American Backwoods Frontier: An Ethnic and Ecological Interpretation.* Baltimore: Johns Hopkins University Press.

———, J.T. Kilpinen, and C.F. Gritzner. 1997. *The Mountain West: Interpreting the Folk Landscape.* Baltimore: Johns Hopkins University Press

Jordan-Bychkov, T.G. 2003. *The Upland South: The Making of an American Folk Region and Landscape.* Santa Fe: Center for American Places.

———, and B. Bychkova Jordan. 2002. *The European Culture Area: A Systematic Geography,* 4th edn. Lanham: Rowman and Littlefield.

Judd, C.M., and A.J. Ray, eds. 1980. *Old Trails and New Directions: Papers of the Third North American Fur Trade Conference.* Toronto: University of Toronto Press.

Juergensmeyer, M. 1995. 'Religious Nationalism: A Global Threat?' *Current History* 95, no. 604:372–6.

Kalman, H. 1994. *A History of Canadian Architecture,* 2 vols. Toronto: Oxford University Press.

Kamin, B. 2005. 'Lincoln Land'. *Chicago Tribune* (10 April). Retrieved online from <*http://www.chicagotribune.com/news/specials/ chi-0504100474apr10,1,2472954.story?coll= chi-homepagenews2-utl&ctrack=1&cset=true*>.

Kaplan, R.D. 1994. 'The Coming Anarchy'. *Atlantic Monthly* 273, no. 2:44–76.

———. 1996. *The Ends of the Earth: A Journey at the Dawn of the 21st Century.* New York: Random House.

———. 2000. *The Coming Anarchy: Shattering the Dreams of the Post-Cold War.* New York: Random House.

Kasperson, J.X., R.E. Kasperson, and B.L. Turner II. 1995. *Regions at Risk: Comparisons of Threatened Environments.* New York: United Nations University Press.

Kates, R.W. 1971. 'Natural Hazard in Human Ecological Perspective: Hypotheses and Models'. *Economic Geography* 47:438–51.

Katz, C. 2003. 'Social Formations: Thinking About Society, Identity, Power and Resistance'. In *Key Concepts in Geography,* edited by S.L. Holloway, S.P Rice, and G. Valentine. Thousand Oaks, CA: Sage.

Kaups, M. 1966. 'Finnish Places Names in Minnesota: A Study in Cultural Transfer'. *Geographical Review* 56:377–97.

Kay, G. 2000. 'On Wordsworth, the Lake District, Protection and Exclusion'. *Area* 32:345–6.

Kearns, K.C. 1974. 'Resuscitation of the Irish Gaeltacht'. *Geographical Review* 64:82–110.

Keil, R., D. Bell, P. Penz, and L. Fawcett, eds. 1998. *Political Ecology: Global and Local.* New York: Routledge.

Keith, M., and S. Pile, eds. 1993. *Place and the Politics of Identity.* New York: Routledge.

Keith, W.J. 1980. *The Poetry of Nature: Rural Perspectives in Poetry from Wordsworth to the Present.* Toronto: University of Toronto Press.

Kelly, K. 1974. 'The Changing Attitudes of Farmers to Forest in Nineteenth Century Ontario'. *Ontario Geography* 8:67–77.

Kingsbury, P. 2004. 'Psychoanalytic Approaches'. In *A Companion to Cultural Geography,* edited by J.S. Duncan, N.C. Johnson, and R.H. Schein, 108–20. Malden, MA: Blackwell.

Kinsley, D. 1994. *Ecology and Religion.* Englewood Cliffs, NJ: Prentice Hall.

Kirk, W. 1951. 'Historical Geography and the Concept of the Behavioral Environment'. *Indian Geographical Journal* 25:152–60.

———. 1963. 'Problems in Geography'. *Geography* 48:357–71.

Kitchin, R.M. 1996. 'Increasing the Integrity of Cognitive Mapping Research: Appraising Conceptual Schemata of Environment– Behavior Interaction'. *Progress in Human Geography* 20:56–84,

———. 1998. '"Out of Place". "Knowing One's Place": Space, Power and the Exclusion of Disabled People'. *Disability and Society* 13:343–56.

———, M. Blades, and R.G. Golledge. 1997. 'Relations Between Geography and Psychology'. *Environment and Behavior* 29:554–73.

Kjekshus, H. 1977. *Ecology Control and Economic Development in East Africa: The Case of Tanganyika.* Berkeley: University of California Press.

Klare, M.T. 1996. 'Redefining Security: The New Global Schisms'. *Current History* 95:353–8.

Knapp, G.W. 1991. *Andean Ecology: Adaptive Dynamics in Ecuador.* Boulder, CO: Westview.

Kneafsey, M. 1998. 'Tourism and Place Identity: A Case-study in Rural Ireland'. *Irish Geography* 31:111–23.

———, and B. Ilbery. 2001. 'Regional Images and the Promotion of Specialty Food and Drink in the West Country'. *Geography* 86:131–40.

Kneale, J. 2003. 'Secondary Worlds: Reading Novels as Geographic Research'. In *Cultural Geography in Practice,* edited by A. Blunt, et al., 39–54. New York: Arnold.

Kniffen, F. 1932. 'Lower California Studies III: The Primitive Cultural Landscape of the Colorado Delta'. *University of California Publication in Geography* 5:43–66.

———. 1936. 'Louisiana House Types'. *Annals of the Association of American Geographers* 26:179–93.

———. 1949. 'The American Agricultural Fair: The Pattern'. *Annals of the Association of American Geographers* 39:264–82.

———. 1951a. 'The American Agricultural Fair: Time and Place'. *Annals of the Association of American Geographers* 41:42–57.

———. 1951b. 'The American Covered Bridge'. *Geographical Review* 41:114–23.

———. 1965. 'Folk Housing: Key to Diffusion'. *Annals of the Association of American Geographers* 55:549–57.

———. 1974. 'Material Culture in the Geographic Interpretation of the Landscape'. In *The Human Mirror*, edited by M. Richardson, 252–67. Baton Rouge: Louisiana State University Press.

———, and H. Glassie. 1966. 'Building in Wood in the Eastern United States: A Time–Place Perspective'. *Geographical Review* 56:40–65.

Knight, D. 1982. 'Identity and Territory: Geographical Perspectives on Nationalism and Regionalism'. *Annals of the Association of American Geographers* 72:514–31.

Knopp, L. 1995. 'Sexuality and Urban Space: A Framework for Analysis'. In *Mapping Desire: Geographies of Sexualities*, edited by D. Bell and G. Valentine, 149–61. New York: Routledge.

Knox, P., and J. Agnew. 1994. *The Geography of the World Economy*, 2nd edn. New York: Arnold.

Kobayashi, A. 1989. 'A Critique of Dialectical Landscape'. In *Remaking Human Geography*, edited by A. Kobayashi and S. Mackenzie, 164–83. Boston: Unwin Hyman.

———. 1993. 'Multiculturalism: Representing a Canadian Institution'. In *Place/Culture/Representation*, edited by J.S. Duncan and D. Ley, 205–31. New York: Routledge.

Koelsch, W.A. 1969. 'The Historical Geography of Harlan H. Barrows'. *Annals of the Association of American Geographers* 59:632–51.

Kofman, E., and G. Young, eds. 1996. *Globalization: Theory and Practice*. New York: Pinter.

Kollmorgen, W.M. 1941. 'A Reconnaissance of Some Cultural–Agricultural Islands in the South'. *Economic Geography* 17:409–30.

———. 1943. 'Agricultural–Cultural Islands in the South, Part II'. *Economic Geography* 19:109–17.

Kong, L. 1993a. 'Negotiating Conceptions of Sacred Space: A Case Study of Religious Buildings in Singapore'. *Transactions of the Institute of British Geographers* NS 18:342–58.

———. 1993b. 'Ideological Hegemony and the Political Symbolism of Religious Buildings in Singapore'. *Environment and Planning D: Society and Space* 11:23–45.

———. 1995. 'Music and Cultural Politics: Ideology and Resistance in Singapore'. *Transactions of the Institute of British Geographers* NS 20:447–59.

———. 1996. 'Popular Music in Singapore: Exploring Local Cultures, Global Resources, and Regional Identities'. *Environment and Planning D: Society and Space* 14:273–92.

———. 1997. 'A "New" Cultural Geography: Debates About Invention and Reinvention'. *Scottish Geographical Magazine* 113:177–85.

———. 1999. 'Cemeteries and Columbaria, Memorials and Mausoleums: Narrative and Interpretation in the Study of Deathscapes in Geography'. *Australian Geographical Studies* 37:1–10.

———, and L.Tay. 1998. 'Exalting the Past: Nostalgia and the Construction of Heritage in Children's Literature'. *Area* 30:133–43.

Kroeber, A.L. 1917. 'The Superorganic'. *American Anthropologist* 19:163–213.

———. 1928. 'Native Cultures of the Southwest'. *University of California Publications in American Archaeology and Ethnology* 23, no. 9:375–98.

———. 1939. *Cultural and Natural Areas of Native North America*. Berkeley: University of California Press, Publications in American Archaeology and Ethnology, 38.

———, and C. Kluckhohn. 1952. *Culture: A Critical Review of Concepts and Definitions*. Cambridge, MA: Papers of the Peabody Museum of American Archaeology and Ethnology, Harvard University, 47, 1.

———, and T. Parsons. 1958. 'The Concepts of Culture and of Social System'. *American Sociological Review* 23:582–3.

Kuper, A. 1999. *Culture: The Anthropologists' Account*. Cambridge, MA: Harvard University Press.

Kuznar, L.A. 1997. *Reclaiming a Scientific Anthropology*. Walnut Creek, CA: Altamira Press.

Kymlicka, W. 2001. *Politics in the Vernacular: Nationalism, Multiculturalism and Citizenship*. New York: Oxford University Press.

Lamal, P.A. 1991. 'Preface'. In *Behavioral Analysis of Societies and Cultural Practices*, edited by P.A. Lamal, xiii–xiv. New York: Hemisphere Publishing Corporation.

Lambert, A.M. 1985. *The Making of the Dutch Landscape: An Historical Geography of the Netherlands*, 2nd edn. New York: Academic Press.

Lambert, D. 2002. 'Geography, 'Race' and Education'. *Geography* 87:297–304.

Lamme, A.J., III, and R.K. Oldakowski. 1982. 'Vernacular Areas in Florida'. *Southeastern Geographer* 22:100–9.

Landes, D.S. 1998. *The Wealth and Poverty of Nations: Why Some Are So Rich and Some So Poor.* New York: W.W. Norton.

Landzelius, M. 2004. 'The Body'. In *A Companion to Cultural Geography*, edited by J.S. Duncan, N.C. Johnson, and R.H. Schein, 279–97. Malden, MA: Blackwell.

Langton, J. 1979. 'Darwinism and the Behavioral Theory of Sociocultural Evolution: An Analysis'. *American Journal of Sociology* 85:288–309.

Latham, A. ed. 2002. *The Oxford Companion to Music.* New York: Oxford University Press.

Leaf, M.J. 1979. *Man, Mind and Science: A History of Anthropology.* New York: Columbia University Press.

Leary, R. 1999. 'Youth Cultures of the Hobart Rivulet'. In *Australian Cultural Geographies*, edited by E. Stratford, 153–69. New York: Oxford University Press.

Lee, D-O. 1997. 'Multicultural Education in Geography in the USA: An Introduction'. *Journal of Geography in Higher Education* 21:261–8.

Lee, R. 1990. 'Regional Geography: Between Scientific Geography, Ideology, and Practice, (or What Use is Regional Geography?)'. In *Regional Geography: Current Developments and Future Prospects*, edited by R.J. Johnston, J. Hauer, and G.A. Hoekveld, 103–21. New York: Routledge.

———, and J. Wills, eds. 1997. *Geographies of Economies.* New York: Arnold.

Lefebvre, H. 1991. *The Production of Space.* Cambridge, MA: Blackwell.

Legrain, P. 2002. *Open World: The Truth About Globalisation.* London: Abacus.

Lehr, J.C. 1973. 'Ukrainian Houses in Alberta'. *Alberta Historical Review* 21:9–15.

Leib, J.I. 2002. 'Separate Times, Shared Spaces: Arthur Ashe, Monument Avenue and the Politics of Richmond, Virginia's Symbolic Landscape'. *Cultural Geographies* 9:286–312.

Leighly, J. 1954. 'Innovation and Area'. *Geographical Review* 44:439–41.

———, ed. 1963. *Land and Life: A Selection from the Writings of Carl Ortwin Sauer.* Berkeley: University of California Press.

———. 1976. 'Carl Ortwin Sauer, 1889–1975'. *Annals of the Association of American Geographers* 66:337–48.

———. 1978. 'Town Names of Colonial New England in the West'. *Annals of the Association of American Geographers* 68:233–48.

———. 1987. 'Ecology as Metaphor: Carl Sauer and Human Ecology'. *Professional Geographer* 39:405–12.

Lemon, J.T. 1972. *The Best Poor Man's Country: A Geographical Study of Early Southeastern Pennsylvania.* Baltimore: Johns Hopkins University Press.

Leopold, A. 1949. *A Sand County Almanac: And Sketches Here and There.* New York: Oxford University Press.

Lester, A. 1996. *From Colonization to Democracy: A New Historical Geography of South Africa.* New York: I.B. Tauris Publishers.

———. 2000. 'Historical Geography'. In *The Geography of South Africa in a Changing World*, edited by R. Fox and K. Rowntree, 60–85. New York: Oxford University Press.

Levison, M., R.G. Ward, and J.W. Webb. 1973. *The Settlement of Polynesia. A Computer Simulation.* New York: Oxford University Press.

Lewin, K. 1944. 'Constructs in Psychology and Ecological Psychology'. In *Authority and Frustration*, edited by K. Lewin et al., 17–23. Iowa City: University of Iowa Press.

———. 1951. *Field Theory in Social Science: Selected Theoretical Papers*, edited by D. Cartwright. New York: Harper.

Lewis, K.E. 2002. *West to Far Michigan: Settling the Lower Peninsula, 1815–1860.* East Lansing: Michigan State University Press.

Lewis, M.W. 1991. 'Elusive Societies: A Regional-Cartographical Approach to the Study of Human Relatedness'. *Annals of the Association of American Geographers* 81:605–26.

———, and K.E. Wigen. 1997. *The Myth of Continents: A Critique of Metageography.* Berkeley: University of California Press.

Lewis, P. 1975. 'Common Houses, Cultural Spoor'. *Landscape* 19, no. 2:1–22.

———. 1979. 'Axioms for Reading the Landscape'. In *The Interpretation of Ordinary Landscapes: Geographical Essays*, edited by D.W. Meinig, 11–32. New York: Oxford University Press.

———. 1983. 'Learning from Looking: Geographic and Other Writing About the American Cultural Landscape'. *American Quarterly* 35:242–61.

Ley, D. 1974. *The Black Inner City as Frontier Outpost: Images and Behavior of a Philadelphia Neighborhood.* Washington, DC: Association of American Geographers, Monograph no. 7.

———. 1977. 'Social Geography and the Taken-for-Granted World'. *Transactions of the Institute of British Geographers* NS 2:498–512.

———. 1981. 'Behavioral Geography and the Philosophies of Meaning'. In *Behavioral Problems in Geography Revisited*, edited by K.R. Cox and R.G. Golledge, 209–30. New York: Methuen.

———. 1998. 'Classics in Human Geography Revisited: Author's Response'. *Progress in Human Geography* 22:78–80.

———, and K. Olds. 1988. 'Landscape as "Spectacle": World's Fairs and the Culture of Heroic Consumption'. *Environment and Planning D: Society and Space* 6:191–212.

———, and M.S. Samuels. 1978. 'Contexts of Modern Humanism in Geography'. In *Humanistic Geography*, edited by D. Ley and M.S. Samuels, 1–18. Chicago: Maaroufa Press.

Leyshon, A., D. Matless, and G. Revill. 1995. 'The Place Of Music'. *Transactions of the Institute of British Geographers NS* 20:423–33.

Liepens, R. 1996. 'Reading Agricultural Power'. *New Zealand Geographer* 52, no. 2:3–10.

Llosa, M.V. 2001. 'The Culture of Liberty'. *Foreign Policy* 122:66–71.

Lomborg, B. 2001. *The Skeptical Environmentalist: Measuring the Real State of the World*. New York: Cambridge University Press.

Longhurst, R. 1994. 'Reflections on and a Vision for Feminist Geography'. *New Zealand Geographer* 50, no.1:14–19.

———. 1997. '(Dis)embodied Geographies'. *Progress in Human Geography* 21:486–501.

Louder, D.R., C. Morissonneau, and E. Waddell. 1983. 'Picking up the Pieces of a Shattered Dream: Quebec and French America'. *Journal of Cultural Geography* 4, no. 1:44–56.

Lovato, A.L. 2004. *Santa Fe Hispanic Culture: Preserving Identity in a Tourist Town*. Albuquerque: University of New Mexico Press.

Lovell, W.G. 1985. *Conquest and Survival in Colonial Guatemala*. Kingston: McGill–Queen's University Press.

———. 1992. '"Heavy Shadows and Black Night": Disease and Depopulation in Colonial Spanish America'. *Annals of the Association of American Geographers* 82:426–43.

Lovelock, J. 1982. *Gaia: A New Look at Life on Earth*. New York: Oxford University Press.

Low, N., and B. Gleeson. 1998. *Justice, Society and Nature: An Exploration of Political Ecology*. New York: Routledge.

Lowenthal, D. 1968. 'The American Scene'. *Geographical Review* 58:61–88.

———. 1997. 'European Landscape Transformations: The Residue'. In *Understanding Ordinary Landscapes*, edited by P. Groth and T.W. Bressi, 180–99. New Haven: Yale University Press.

———. 2000. *George Perkins Marsh: Prophet of Conservation*. Seattle: University of Washington Press.

———, and H.C. Prince. 1964. 'The English Landscape'. *Geographical Review* 54:309–46.

———, and H.C. Prince. 1965. 'English Landscape Tastes'. *Geographical Review* 55:186–222.

Lowie, R.H. 1937. *History of Ethnological Theory*. New York: Farrar and Rinehart.

Luebke, F.C. 1984. 'Regionalism and the Great Plains: Problems of Concept and Method'. *Western Historical Quarterly* 15:19–38.

Lynch, K. 1960. *The Image of the City*. Cambridge, MA: MIT Press.

Lyod, B., and L. Rowntree. 1978. 'Radical Feminists and Gay Men in San Francisco: Social Space in Dispersed Communities'. In *An Invitation to Geography*, 2nd edn, edited by D. Lanegran and R. Palm, 78–88. New York: McGraw-Hill.

Mabogunje, A. 1997. *State of the Earth: Contemporary Geographic Perspectives*. Malden, MA: Blackwell.

McBride, B. 1999. 'The (Post)colonial Landscape of Cathedral Square: Urban Redevelopment and Representation in the "Cathedral City"'. *New Zealand Geographer* 55, no. 1:3–11.

McCann, E.J. 1997. 'Where Do You Draw the Line? Landscape, Texts and the Politics of Planning'. *Environment and Planning D: Society and Space* 15:641–61.

MacDonald, R.H. 1994. *The Language of Empire: Myths and Metaphors of Popular Imperialism, 1880–1918*. Manchester: Manchester University Press.

McDowell, L. 1994. 'The Transformation of Cultural Geography'. In *Human Geography: Society, Space and Social Science*, edited by D. Gregory, R. Martin, and G. Smith, 146–73. New York: MacMillan.

———. 1995. 'Body Work: Heterosexual Gender Performances in City Workplaces'. In *Mapping Desire*, edited by D. Bell and G. Valentine, 75–95. New York: Routledge.

———. 1996. 'Off the Road: Alternative Views of Rebellion, Resistance and "The Beats"'. *Transactions of the Institute of British Geographers NS* 21:412–19.

———. 1997. 'Women/Gender/Feminisms: Doing Feminist Geography'. *Journal of Geography in Higher Education* 21:381–400.

———. 1999. *Gender, Identity and Place: Understanding Feminist Geographies*. Minneapolis: University of Minnesota Press.

McEwan, C. 1996. 'Review Article: Gender, Culture and Imperialism'. *Journal of Historical Geography* 22:489–94.

———. 2003. 'The West and Other Feminisms'. In *Handbook of Cultural Geography*, edited by

K. Anderson, M. Domosh, S. Pile, and N. Thrift, 405–19. Thousand Oaks, CA: Sage Publications.

McIlwraith, T.F. 1997. *Looking for Old Ontario: Two Centuries of Landscape Change*. Toronto: University of Toronto Press.

McKay, I. 1994. *The Quest of the Folk: Antimodernism and Cultural Selection in Twentieth-Century Nova Scotia*. Kingston: McGill–Queen's University Press.

McLeay, C. 1995. 'Musical Words, Musical Worlds: Geographical Imagery in the Music of U2'. *New Zealand Geographer* 51, no. 2:1–6.

———. 1997a. 'Inventing Australia: A Critique of Recent Cultural Policy Rhetoric'. *Australian Geographical Studies* 35:40–6.

———. 1997b. 'Popular Music and Expressions of National Identity'. *New Zealand Journal of Geography* 103:12–17.

McLynn, F. 2004. *1759: The Year Britain Became Master of the World*. New York: Atlantic Monthly Press.

McNeill, W.H. 1995. 'The Changing Shape of World History'. *History and Theory* 34:8–26.

Macphail, C.L. 1997. 'Poetry and Pass Laws: Humanistic Geography in Urban South Africa'. *South African Geographical Journal* 79:35–42.

Macpherson, A. 1987. 'Preparing for the National Stage: Carl Sauer's First Ten Years at Berkeley'. In *Carl O. Sauer: A Tribute*, edited by M.S. Kenzer, 69–89. Corvallis: Oregon State University Press.

McPherson, B.D. 1998. *Aging as a Social Process: An Introduction to Individual and Population Aging*, 3rd edn. New York: Harcourt Brace.

McQuillan, D.A. 1978. 'Territory and Ethnic Identity: Some New Measures of an Old Theme in the Cultural Geography of the United States'. In *European Settlement and North American Development: Essays on Geographical Change in Honour and Memory of Andrew Hill Clark*, edited by J.R. Gibson, 3–24. Toronto: University of Toronto Press.

———. 1990. *Prevailing Over Time: Ethnic Adjustment on the Kansas Prairies, 1875–1925*. Urbana: University of Illinois Press.

———. 1993. 'Historical Geography and Ethnic Communities in North America'. *Progress in Human Geography* 17:355–66.

Malin, J.C. 1947. *The Grasslands of North America: Prolegomena to its History*. Lawrence, KS: J.C. Malin.

Mann, C.C. 2002. '1491'. *Atlantic Monthly* 289, no, 3:41–53.

Mannion, J. 1974. *Irish Settlements in Eastern Canada: A Study of Culture Transfer and Adoption*. Toronto: University of Toronto, Department of Geography, Research Publication no. 12.

Marger, M.N. 2003. *Race and Ethnic relations: American and Global Perspectives*. Belmont, CA: Wadsworth.

Markwick, M. 2001. 'Marketing Myths and the Cultural Commodification of Ireland'. *Geography* 86:37–49.

Marsh, G.P. [1864] 1965. *Man and Nature, or Physical Geography as Modified by Human Action*, edited by D. Lowenthal. Cambridge, MA: Harvard University Press.

Marshall, D. 2004. 'Making Sense of Remembrance'. *Social and Cultural Geography* 5:37–54.

Marshall, H.W. 1995. *Paradise Valley, Nevada: The People and Buildings of an American Place*. Tucson: University of Arizona Press.

Martin, C. 1978. *Keepers of the Game: Indian–Animal Relationships and the Fur Trade*. Berkeley: University of California Press.

Martin, D.G. 1991. *Psychology: Principles and Applications*. Englewood Cliffs, NJ: Prentice Hall.

Martin, G., and J. Pear. 1996. *Behavior Modification: What It Is and How To Do It*, 5th edn. Upper Saddle River, NJ: Prentice Hall.

Martin, G.J. 1987a. 'Foreword'. In *Carl O. Sauer: A Tribute*, edited by M.S. Kenzer, ix–xvi. Corvallis: Oregon State University Press.

———. 1987b. 'The Ecologic Tradition in American Geography'. *Canadian Geographer* 31:74–7.

———, and P.E. James. 1993. *All Possible Worlds: A History of Geographical Ideas*, 3rd edn. New York: Wiley.

Martis, K.C. 2003. 'Review of *Guns, Germs and Steel: The Fates of Human Societies* by J. Diamond. *Political Geography* 22:119–21.

Masini, E. 1994. 'The Futures of Cultures: An Overview'. In *The Futures of Cultures*, compiled by Division of Studies and Programming, UNESCO, 9–28. Paris: UNESCO Publishing.

Mason, O. 1895. 'Influence of Environment upon Human Industries and Arts'. *Annual Report of the Smithsonian Institution* 639–65.

Massey, D.B. 1985. 'New Directions in Space'. In *Social Relations and Spatial Structures*, edited by D. Gregory and J. Urry, 9–19. New York: MacMillan.

Mathewson, K. 1996. 'High/Low, Back/Center: Culture's Stages in Human Geography'. In *Concepts in Human Geography*, edited by C. Earle, K. Mathewson, and M.S. Kenzer, 97–125. Lanham, MD: Rowman and Littlefield.

———. 1999. 'Cultural Landscape and Ecology II: Regions, Retrospects, Revivals'. *Progress in Human Geography* 23:267–81.

———. 2000. 'Cultural Landscapes and Ecology III: Foraging/Farming, Food, Festivities'. *Progress in Human Geography* 24:457–74.

———, and M.S. Kenzer, eds. 2003. *Culture, Land, and Legacy: Perspectives on Carl O. Sauer and the Berkeley School*. Baton Rouge: Geosciences Publication, Louisiana State University.

Matthews, H., and M. Limb. 1999. 'Defining an Agenda for the Geography of Children: Review and Prospect'. *Progress in Human Geography* 23:61–90.

May, J. 1996. 'A Little Taste of Something More Exotic'. *Geography* 81:57–64.

Mayer, J.D. 1996. 'The Political Ecology of Disease'. *Progress in Human Geography* 20:441–56.

Mazrui, A.A. 1997. 'Islamic and Western Values'. *Foreign Affairs* 76, no. 5:118–32.

Mead, W.R. 1981. *An Historical Geography of Scandinavia*. New York: Academic Press.

Meggers, B.J. 1954. 'Environmental Limitation on the Development of Culture'. *American Anthropologist* 56:801–24.

Meigs, P. 1935. 'The Dominican Mission Frontier of Lower California'. *University of California Publication in Geography* no. 7.

Meinig, D.W. 1965. 'The Mormon Culture Region: Strategies and Patterns in the Geography of the American West, 1847–1964'. *Annals of the Association of American Geographers* 55:191–220.

———. 1969. *Imperial Texas: An Interpretive Essay in Cultural Geography*. Austin: University of Texas Press.

———. 1971. *Southwest: Three Peoples in Geographical Change*. New York: Oxford University Press.

———. 1972. 'American Wests: Preface to a Geographical Introduction'. *Annals of the Association of American Geographers* 62:159–84.

———. 1976. 'Spatial Models of a Sequence of Transatlantic Interactions'. In *International Geography, 76, Volume 9, Historical Geography*, 30–5. Toronto: Pergamon.

———. 1978. 'The Continuous Shaping of America: A Prospectus for Geographers and Historians'. *American Historical Review* 83:1186–217.

———, ed. 1979a. *The Interpretation of Ordinary Landscapes: Geographical Essays*. New York: Oxford University Press.

———. 1979b. 'Reading the Landscape: An Appreciation of W.G. Hoskins and J.B. Jackson'. In *The Interpretation of Ordinary Landscapes: Geographical Essays*, edited by D.W. Meinig, 195–244. New York: Oxford University Press.

———. 1979c. 'Symbolic Landscapes: Some Idealizations of American Communities'. In *The Interpretation of Ordinary Landscapes: Geographical Essays*, edited by D.W. Meinig, 164–92. New York: Oxford University Press.

———. 1986. *The Shaping of America: A Geographical Perspective on 500 Years of History. Volume 1: Atlantic America, 1492–1800*. New Haven: Yale University Press.

———. 1993. *The Shaping of America: A Geographical Perspective on 500 Years of History. Volume 2: Continental America, 1800–1867*. New Haven: Yale University Press.

———. 1998. *The Shaping of America: A Geographical Perspective on 500 Years of History. Volume 3: Transcontinental America, 1850–1915*. New Haven: Yale University Press.

Melko, M. 1969. *The Nature of Civilizations*. Boston: Porter Sargent.

Mellor, M. 1997. *Feminism and Ecology*. New York: New York University Press.

Menzies, G. 2002. *1421: The Year China Discovered the World*. London: Bantam Press.

Merchant, C. 1980. *The Death of Nature: Women, Ecology, and the Scientific Revolution*. San Francisco: Harper and Row.

———. 1989. *Ecological Revolutions: Nature, Gender, and Science in New England*. Chapel Hill, NC: University of North Carolina Press.

———. 1990. 'Gender and Environmental History'. *Journal of American History* 76:1117–21.

———. 1995. *Earthcare: Women and the Environment*. New York: Routledge.

Merrens, H.R. 1969. 'The Physical Environment of Early America: Images and Image Makers in Colonial South Carolina'. *Geographical Review* 59:530–56.

Meyer, D.K. 2000. *Making the Heartland Quilt: A Geographical History of Settlement and Migration in Early-Nineteenth-Century Illinois*. Carbondale: Southern Illinois University Press.

Middleton, H., and D. Heater, eds. 1989. *Oxford Atlas of Modern World History*. New York: Oxford University Press.

Mikesell, M.W. 1960. 'Comparative Studies in Frontier History'. *Annals of the Association of American Geographers* 50:62–74.

———. 1961. 'Northern Morocco: A Cultural Geography'. *University of California Publications in Geography* 14:1–136.

———. 1967. 'Geographic Perspectives in Anthropology'. *Annals of the Association of American Geographers* 57:617–34.

———. 1968. 'Landscape'. In *International Encyclopedia of the Social Sciences, Volume 8*, 249–64. New York: MacMillan and Free Press.

————. 1969. 'The Borderlands of Geography as a Social Science'. In *Interdisciplinary Relationships in the Social Sciences*, edited by M. Sherif and C.W. Sherif, 227–48. Chicago: Aldine.

————. 1976. 'The Rise and Decline of Sequent Occupance: A Chapter in the History of American Geography'. In *Geographies of the Mind: Essays in Historical Geosophy in Honor of John Kirtland Wright*, edited by D. Lowenthal and M.J. Bowden, 149–69. New York: Oxford University Press.

————. 1977. 'Cultural Geography'. *Progress in Human Geography* 1:460–4.

————. 1978. 'Tradition and Innovation in Cultural Geography'. *Annals of the Association of American Geographers* 68:1–16.

Miller, D., et al. 1998. *Shopping, Place and Identity*. New York: Routledge.

Miller, E.J.W. 1968. 'The Ozark Culture Region as Revealed by Traditional Materials'. *Annals of the Association of American Geographers* 58:51–77.

Miller, V.P., Jr. 1971. 'Some Observations on the Science of Cultural Geography'. *Journal of Geography* 70:27–35.

Mills, S.F. 1997. *The American Landscape*. Edinburgh: Keele University Press.

Minca, C., ed. 2001. *Postmodern Geography: Theory and Praxis*. Malden, MA: Blackwell.

Mitchell, D. 1995. 'There's No Such Thing as Culture: Towards a Reconceptualization of the Idea of Culture in Cultural Geography'. *Transactions of the Institute of British Geographers NS* 20:102–16.

————. 1996. 'Sticks and Stones: The Work of Landscape (A Reply to Judy Walton's "How Real(ist) Can You Get?")'. *Professional Geographer* 48:94–6.

————. 2000. *Cultural Geography: A Critical Introduction*. Malden, MA: Blackwell.

————. 2003a. 'Cultural Geographers Forum: What are the Five Most Important Principles that Should be Covered in an Introductory Course in Cultural Geography and Why?' *Place and Culture, The Newsletter of the Cultural Geography Specialty Group of the Association of American Geographers* Spring, 2–3.

————. 2003b. 'Cultural Landscapes: Just Landscapes or Landscapes of Justice'. *Progress in Human Geography* 27:787–96.

————. 2004. 'Historical Materialism and Marxism'. In *A Companion to Cultural Geography*, edited by J.S. Duncan, N.C. Johnson, and R.H. Schein, 51–65. Malden, MA: Blackwell.

Mitchell, K. 2004. 'Geographies of Identity: Multiculturalism Unplugged'. *Progress in Human Geography* 28:641–51.

Mitchell, R.D. 1977. *Commercialism and Frontier: Perspectives on the Early Shenandoah Valley*. Charlottesville: University Press of Virginia.

————. 1978. 'The Formation of Early American Cultural Regions'. In *European Settlement and Development in North America*, edited by J.R. Gibson, 66–90. Toronto: University of Toronto Press.

Mithen, S. 2003. *After the Ice: A Global Human History, 20,000–5,000 BC*. London: Weidenfeld and Nicolson.

Mitra, S.K., and R.A. Lewis. 1996. *Subnational Movements in South Asia*. Boulder: Westview Press.

Mogey, J. 1971. 'Society, Man and Environment'. In *Man and His Habitat*, edited by R.H. Buchanan, E. Jones, and D. McCourt, 79–92. New York: Barnes and Noble.

Mohan, G. 1994. 'Deconstruction of the Con: Geography and the Commodification of Knowledge'. *Area* 26:387–90.

Monk, J. 1996. 'Challenging the Boundaries: Survival and Change in a Gendered World'. In *Companion Encyclopedia of Geography: The Environment and Humankind*, edited by I. Douglas, R. Huggett, and M. Robinson, 888–905. New York: Routledge.

————. 1999. 'Gender in the Landscape: Expressions of Power and Meaning'. In *Cultural Geographies*, 2nd edn, edited by K. Anderson and F. Gale, 153–71. Harlow: Longman.

————. 2004. 'Women, Gender, and the Histories of American Geography'. *Annals of the Association of American Geographers* 94:1–22.

Montagu, A. 1942. *Man's Most Dangerous Myth: The Fallacy of Race*. New York: Harper.

————. 1997. *Man's Most Dangerous Myth: The Fallacy of Race*, 6th edn. Walnut Creek, CA: Altamira Press.

Monument Avenue. 2004. Retrieved 9 May 2004, from <*http://xroads.virginia.edu/~UG97/monument/begin.html*>.

Morehouse, B.J. 1996. *A Place Called Grand Canyon: Contested Geographies*. Tucson: University of Arizona Press.

Morgan, L.H. [1877] 1974. *Ancient Society, or Researches in the Lines of Human Progress from Savagery through Barbarism to Civilization*. Gloucester, MA: Peter Smith.

Morgan, W.B., and R.P. Moss. 1965. 'Geography and Ecology: The Concept of the Community and Its Relationship to Environment'. *Annals of the Association of American Geographers* 55:339–50.

Morin, K. 2003. 'Landscape and Environment: Representing and Interpreting the World'. In *Key Concepts in Geography*, edited by S.L. Holloway, S.P. Rice, and G. Valentine, 319–34. London: Sage.

Morley, D., and K. Robins. 2001. *British Cultural Studies: Geography, Nationality, and Identity*. New York: Oxford University Press.

Morrill, R.L. 1965. *Migration and the Spread and Growth of Urban Settlement*. Lund: Gleerup, Lund Studies in Geography, Series B, no. 26.

———. 2002. 'Pausing for Breath'. In *Geographical Voices: Fourteen Autobiographical Essays*, edited by P. Gould and F.R. Pitts, 211–36. Syracuse: Syracuse University Press.

Mosely, C., and R.E. Asher. 1994. *Atlas of the World's Languages*. New York: Routledge.

Moss, P., ed. 2002. *Feminist Geography in Practice: Research and Methods*. Malden, MA: Blackwell.

———, and I. Dyck. 2003. 'Embodying Social Geography'. In *Handbook of Cultural Geography*, edited by K. Anderson, M. Domosh, S. Pile, and N. Thrift, 58–73. Thousand Oaks, CA: Sage Publications.

Muir, R. 1999. *Approaches to Landscape*. Boulder, CO: Barnes and Noble.

Mungall, C., and D.J. McLaren. 1990. *Planet Under Stress: The Challenge of Global Change*. New York: Oxford University Press.

Murdoch, J. 1997. 'Inhuman/Nonhuman/Human: Actor-Network Theory and the Prospects for a Nondualistic and Symmetrical Perspective on Nature and Society'. *Environment and Planning D: Society and Space* 15:731–56.

Murphy, A.B., and D.L. Johnson, eds. 2000a. *Cultural Encounters with the Environment: Enduring and Evolving Geographic Themes*. Lanham, MD: Rowman and Littlefield.

———, and D.L. Johnson. 2000b. 'Introduction: Encounters with Environment and Place'. In *Cultural Encounters with the Environment: Enduring and Evolving Geographic Themes*, edited by A.B. Murphy and D.L. Johnson, 1–13. Lanham, MD: Rowman and Littlefield.

Myrdal, G. 1944. *An American Dilemma: The Negro Problem and Modern Democracy*. New York: Harper.

Myslik, W.D. 1996. 'Renegotiating the Social/Sexual Identities of Places'. In *BodySpace: Destabilizing Geographies of Gender and Sexuality*, edited by N. Duncan, 156–69. New York: Routledge.

Naess, A. 1989. *Ecology, Community and Lifestyle*, translated and edited by D. Rothenberg. New York: Cambridge University Press.

Namaste, K. 1996. 'Genderbashing: Sexuality, Gender, and the Regulation of Public Space'. *Environment and Planning D: Society and Space* 14:221–40.

Nash, C. 1998a. 'Editorial: Mapping Emotion'. *Environment and Planning D: Society and Space* 16:1–9.

———. 1998b. 'Narratives and Names: Irish Landscape Meanings'. In *A European Geography*, edited by T. Unwin, 73–6. New York: Addison Wesley Longman.

———. 1999. 'Irish Placenames: Post-colonial Locations'. *Transactions of the Institute of British Geographers NS* 24:457–80.

———. 2000. 'Performativity in Practice: Some Recent Work in Cultural Geography'. *Progress in Human Geography* 24:653–64.

———. 2002. 'Cultural Geography: Postcolonial Cultural Geographies'. *Progress in Human Geography* 26:219–30.

———. 2003. 'Cultural Geography: Anti-racist Geographies'. *Progress in Human Geography* 27:637–48.

———. 2004. 'Postcolonial Geographies: Spatial Narratives of Inequality and Interconnection'. In *Envisioning Human Geographies*, edited by P. Cloke, P. Crang, and M. Goodwin, 104–27. New York: Arnold.

Nathanson, P., and K.K. Young. 2001. *Spreading Misandry: The Teaching of Contempt for Men in Popular Culture*. Montreal: McGill-Queen's Press.

Naylor, S., and J.R. Ryan. 2003. 'Mosques, Temples and Gurdwaras: New Sites of Religion in Twentieth-Century Britain'. In *Geographies of British Modernity*, edited by D. Gilbert, D. Matless, and B. Short, 168–83. Malden, MA: Blackwell.

Neisser, U. 1967. *Cognitive Psychology*. New York: Appleton Century Crofts.

———. 1976. *Cognition and Reality: Principles and Implications of Cognitive Psychology*. New York: Freeman.

Netting, R.M. 1977. *Cultural Ecology*. Menlo Park, CA: Cummings.

Newcomb, R.M. 1969. 'Twelve Working Approaches to Historical Geography'. *Yearbook of the Association of Pacific Coast Geographers* 31:27–50.

Newson, L.A. 1976. 'Cultural Evolution: A Basic Concept for Human and Historical Geography'. *Journal of Historical Geography* 2:239–55.

———. 1996. 'Review of Culture, Form and Place: Essays in Cultural and Historical Geography edited by K. Mathewson'. *Progress in Human Geography* 20:278–9.

Newton, M. 1974. 'Cultural Preadaptation and the Upland South'. In *Man and Cultural Heritage:*

Papers in Honor of Fred B. Kniffen, edited by H.J. Walker and W.G. Haag. *Geoscience and Man* 5:143–54. Baton Rouge: Louisiana State University, Department of Geography and Anthropology, Geoscience Publications.

Nietschmann, B.Q. 1973. *Between Land and Water: The Subsistence Ecology of the Miskito Indians, Eastern Nicaragua*. New York: Seminar Press.

Niezen, R. 2004. *A World Beyond Difference: Cultural Identity in the Age of Globalization*. New York: Blackwell.

Nisbett, J. 2003. *The Geography of Thought: How Asians and Westerners Think Differently . . . and Why*. New York: Free Press.

Noble, A.G. 1984. *Wood, Brick, and Stone: The North American Settlement Landscape*, 2 vols. Amherst, MA: The University of Massachusetts Press.

———, and R. Dhussa. 1990. 'Image and Substance: A Review of Literary Geography'. *Journal of Cultural Geography* 10, no 2:49–65.

Norberg, J. 2001. *In Defence of Global Capitalism*. Stockholm: Timbro.

North, D.C. 1977. 'The New Economic History After Twenty Years'. *American Behavioral Scientist* 21:187–200.

Norton, W. 1984. *Historical Analysis in Geography*. London: Longman.

———. 1987. 'Humans, Land and Landscape: A Proposal for Cultural Geography'. *Canadian Geographer* 31:21–30.

———. 1988. 'Abstract Cultural Landscapes'. *Journal of Cultural Geography* 8, no. 2:67–80.

———. 1989. *Explorations in the Understanding of Landscape: A Cultural Geography*. Westport, CT: Greenwood.

———. 1995. 'State Boundaries and Agricultural Change in the South Eastern Australian Wheat Belt: Counterfactual Analyses, 1891–1911'. *Australian Geographical Studies* 33:228–41.

———. 1997a. 'Behavior Analysis and Cultural Geography'. *Journal of Cultural Geography* 16, no. 2:1–19.

———. 1997b. 'Human Geography and Behavior Analysis: An Application of Behavior Analysis to the Explanation of the Evolution of Human Landscapes'. *The Psychological Record* 47:439–60.

———. 2001. 'Following Rules in the Intermontane West: Nineteenth Century Mormon Settlement'. *The Behavior Analyst* 24:57–73.

———. 2003a. '"After-the-fact Causality": A Different Direction for Cultural Geography'. *Area* 35:418–26.

———. 2003b. 'Competing Identities and Contested Places: Mormons in Nauvoo and Voree'. *Journal of Cultural Geography* 21, no. 1:95–119.

———. 2004. *Human Geography*, 5th edn. Toronto: Oxford University Press.

Norwine, J., and T.D. Anderson. 1980. *Geography as Human Ecology*. Lanham, MD: University Press of America.

Nostrand, R.L. 1992. *The Hispano Homeland*. Norman, OK: University of Oklahoma Press.

———. 2001. 'The Highland Hispano Homeland'. In *Homelands: A Geography of Culture and Place Across America*, edited by R.L. Nostrand and L.E. Estaville, 155–167. Baltimore: Johns Hopkins University Press.

———, and L.E. Estaville. 1993a. 'Introduction: The Homeland Concept'. *Journal of Cultural Geography* 13, no. 2:1–4.

———, and L.E. Estaville, eds. 1993b. *Special Issue*. *Journal of Cultural Geography* 13, no. 2.

———, and L.E. Estaville. 2001a. 'Introduction: Free Land, Dry Land, Homeland'. In *Homelands: A Geography of Culture and Place Across America*, edited by R.L. Nostrand and L.E. Estaville, xiii–xxiii. Baltimore: Johns Hopkins University Press.

———, and L.E. Estaville, eds. 2001b. *Homelands: A Geography of Culture and Place Across America*. Baltimore: Johns Hopkins University Press.

———, and S.B. Hilliard, eds. 1988. 'The American South'. *Geoscience and Man* 25. Baton Rouge: Louisiana State University, Department of Geography and Anthropology, Geoscience Publications.

Nostrand, R.L. 2003. *El Cerrito: Eight Generations in a Spanish Village*. Norman: University of Oklahoma Press.

Oakes, J., R. Riewe, K. Kinew, and E. Maloney. 1998. *Sacred Lands: Aboriginal World Views, Claims, and Conflicts*. Occasional Publication no. 43. Edmonton, University of Alberta, Canadian Circumpolar Institute.

O'Brien, P.K., ed. 1999. *Oxford Atlas of World History*. New York: Oxford University Press.

O'Connor, J. 1998. *Natural Causes: Essays in Ecological Marxism*. New York: Guilford.

O'Dwyer, B. 1997. 'Pathways to Homelessness: A Comparison of Gender and Schizophrenia in Inner-Sydney'. *Australian Geographical Studies* 35:294–307.

Oelschlaeger, M. 1991. *The Idea of Wilderness*. New Haven: Yale University Press.

Ohmae, K. 1993. 'The Rise of the Region State'. *Foreign Affairs* 76, no 2:78–87.

Olsen, D.H., and D.J. Timothy. 2002. 'Contested Religious Heritage: Differing Views of Mormon Heritage'. *Tourism Recreation Research* 27, no. 2:7–15

Olwig, K. 1980. 'Historical Geography and the Society/Nature 'Problematic': The Perspective of J.F. Schouw, G.P. Marsh and E. Reclus'. *Journal of Historical Geography* 6:29–45.

———. 1996a. 'Nature—Mapping the Ghostly Traces of a Concept'. In *Concepts in Human Geography*, edited by C. Earle, K. Mathewson, and M.S. Kenzer, 63–96. Lanham, MD: Rowman and Littlefield.

———. 1996b. 'Recovering the Substantive Nature of Landscape'. *Annals of the Association of American Geographers* 86:630–53.

———. 2002. *Landscape, Nature, and the Body Politic: From Britain's Renaissance to America's New World.* Madison: University of Wisconsin Press.

———. 2003. 'Landscape: The Lowenthal Legacy'. *Annals of the Association of American Geographers* 93:871–7.

Ortner, S.B. 1972. 'Is Female to Male as Nature Is to Culture?' *Feminist Studies* 1:5–31.

———, and H. Whitehead. 1981. 'Introduction: Accounting for Sexual Meaning'. In *Sexual Meanings: The Cultural Construction of Gender and Sexuality*, edited by S.B. Ortner and H. Whitehead, 1–27. New York: Cambridge University Press.

Osborne, B.S. 1988. 'The Iconography of Nationhood in Canadian Art'. In *The Iconography of Landscape*, edited by D. Cosgrove and S. Daniels, 162–78. New York: Cambridge University Press.

———. 1992. 'Interpreting a Nation's Identity: Artists as Creators of National Consciousness'. In *Ideology and Landscape in Historical Perspective*, edited by A.R.H. Baker and G. Biger, 230–54. New York: Cambridge University Press.

Ostergren, R.C. 1988. *A Community Transplanted: The Trans-Atlantic Experience of a Swedish Immigrant Settlement in the Upper Midwest, 1835–1915.* Madison: University of Wisconsin Press.

———, and J.G. Rice. 2004. *The Europeans: A Geography of People, Culture, and Environment.* New York: Guilford Press.

Ó Tuathail, G. 1995. 'Political Geography I: Theorizing History, Gender and World Order Amidst Crises of Global Governance'. *Progress in Human Geography* 19:260–72.

———. 1996. *Critical Geopolitics: The Politics of Writing Global Space.* Minneapolis: University of Minnesota Press.

Paehlke, R., ed. 1995. *Conservation and Environmentalism: An Encyclopedia.* New York: Garland.

Pahl, R.E. 1965. 'Trends in Social Geography'. In *Frontiers in Geographical Teaching*, edited by R.J. Chorley and P. Haggett, 81–100. London: Methuen.

Pain, R.H. 1997. 'Social Geographies of Women's Fear of Crime'. *Transactions of the Institute of British Geographers* NS 22:231–44.

———. 1999. 'The Geography of Fear'. *Geography Review* 12, no. 5:22–5.

———, and C. Bailey. 2004. 'British Social and Cultural Geography: Beyond Turns and Dualisms?'

Social and Cultural Geography 5:319–29.

———, et al. 2001. *Introducing Social Geographies.* New York: Oxford University Press.

Panelli, R. 2004. *Social Geographies: From Difference to Action.* Thousand Oaks, CA: Sage.

Park, C.C. 1994. *Sacred Worlds: An Introduction to Geography and Religion.* New York: Routledge.

Park, R.E. 1915. 'The City: Suggestions for the Investigation of Human Behavior in the City Environment'. *American Journal of Sociology* 20:577–612.

———. 1952. *Human Communities: The City and Human Ecology.* The Collected Papers of Robert Ezra Park, Volume II. Glencoe, IL: Free Press.

———, and E.W. Burgess. 1921. *Introduction to the Science of Sociology.* Chicago: Chicago University Press.

Parr, H., and C. Philo. 1995. 'Mapping "Mad" Identities'. In *Mapping the Subject: Geographies of Cultural Transformation*, edited by S. Pile and N. Thrift, 199–225. New York: Routledge.

Parsons, J.J. 1979. 'The Later Sauer Years'. *Annals of the Association of American Geographers* 69:9–15.

Paulson, S., and L.L. Gezon, eds. 2005. *Political Ecology Across Spatial Scales and Social Groups.* New Brunswick, NJ: Rutgers University Press.

Pawson, E. 1999. 'Postcolonial New Zealand'. In *Cultural Geographies*, 2nd edn, edited by K. Anderson and F. Gale, 25–50. Harlow: Longman.

———, and G. Banks. 1993. 'Rape and Fear in a New Zealand City'. *Area* 25:55–63.

Peach, C. 1999. 'Social Geography'. *Progress in Human Geography* 23:282–8.

———. 2002. 'Social Geography: New Religions and Ethnoburbs—Contrasts with Cultural Geography'. *Progress in Human Geography* 26:252–60.

Peckham, R.S. 2004. 'Landscape in Film'. In *A Companion to Cultural Geography*, edited by J.S. Duncan, N.C. Johnson, and R.H. Schein, 420–9. Malden, MA: Blackwell.

Peet, R. 1977a. *Radical Geography.* Chicago: Maaroufa Press.

———. 1977b. 'The Development of Radical Geography in the United States'. *Progress in Human Geography* 1:240–63.

———. 1985. 'The Social Origins of Environmental Determinism'. *Annals of the Association of American Geographers* 75:309–33.

———. 1996a. 'Structural Themes in Geographical Discourse'. In *Companion Encyclopedia of Geography: The Environment and Humankind*, edited by I. Douglas, R. Huggett, and M. Robinson, 860–87. New York: Routledge.

———. 1996b. 'Discursive Idealism in the "Landscape-as-Text" School'. *Professional Geographer* 48:96–8.

———, and N. Thrift. 1989. 'Political Economy and Human Geography'. In *New Models in Geography: Volume 1*, edited by R. Peet and N. Thrift, 3–29. Boston: Unwin Hyman.

———. and M. Watts. 1996. 'Liberation Ecology: Development, Sustainability and Environment in an Age of Market Triumphalism'. In *Liberation Ecologies: Environment, Development, Social Movements*, edited by R. Peet and M. Watts, 1–45. New York: Routledge.

Pelto, P.J. 1966. *The Nature of Anthropology*. Columbus, OH: Merrill.

Peluso, N.L., and M. Watts, eds. 2001. *Violent Environments*. Ithaca: Cornell University Press.

Penn, M., and F. Lukermann. 2003. 'Chorology and Landscape: An Internalist Reading of "The Morphology of Landscape"'. In *Culture, Land, and Legacy: Perspectives on Carl O. Sauer and the Berkeley School*, edited by K. Mathewson and M.S. Kenzer, 233–60. Baton Rouge: Geosciences Publication, Louisiana State University

Penrose, J. 1997. 'Construction, De(con)struction and Reconstruction. The Impact of Globalization and Fragmentation on the Canadian Nation-State'. *International Journal of Canadian Studies* 16:15–49.

Peters, B.C. 1972. 'Oak Openings or Barrens: Landscape Evaluation on the Michigan Frontier'. *Proceedings of the Association of American Geographers* 4:84–6.

Peterson, J., and J. Anfison. 1985. 'The Indian and the Fur Trade: A Review of Recent Literature'. *Manitoba History* 10 (Autumn):10–18.

Phillips, R. 2004. 'Sexuality'. In *A Companion to Cultural Geography*, edited by J.S. Duncan, N.C. Johnson, and R.H. Schein, 265–78. Malden, MA: Blackwell.

Philo, C. 1988. 'Conference Report: New Directions in Cultural Geography'. *Journal of Historical Geography* 14:178–81.

———. 1989. Enough to Drive One Mad: The Organization of Space in 19th Century Lunatic Asylums'. In *The Power of Geography*, edited by J. Wolch and M. Dear, 258–90. Boston: Unwin Hyman.

———, compiler. 1991. *New Words, New Worlds: Reconceptualising Social and Cultural Geography*. Aberystwyth, Wales: Cambrian Printers.

———, and H. Parr. 2003. 'Introducing Psychoanalytic Geographies'. *Social and Cultural Geography* 4:283–93.

———, and C. Wilbert, eds. 2000. *Animal Spaces, Beastly Places: New Geographies of Human–Animal Relations*. New York: Routledge.

Pickles, J. 1985. *Phenomenology, Science, and Geography*. New York: Cambridge University Press.

Pile, S. 1993. 'Human Agency and Human Geography Revisited: A Critique of "New Models" of the Self'. *Transactions of the Institute of British Geographers NS* 18:122–39.

———. 1996. *The Body and the City: Psychoanalysis, Space and Subjectivity*. New York: Routledge.

———, and M. Keith, eds. 1997. *Geographies of Resistance*. New York: Routledge.

———, and N. Thrift, eds. 1995. *Mapping the Subject: Geographies of Cultural Transformation*. New York: Routledge.

Pillsbury, R. 1970. 'The Urban Street Pattern as a Cultural Indicator'. *Annals of the Association of American Geographers* 60:428–46.

———. 1974. 'Carolina Thunder: A Geography of Southern Stock Car Racing'. *Journal of Geography* 73:39–47.

———. 1989. 'A Mythology at the Brink: Stock Car Racing in the American South'. *Sport Place: An International Journal of Sports Geography* 3:3–12.

Pinkerton, J.P. 1997. 'Enviromanticism: The Poetry of Nature as Political Force'. *Foreign Affairs* 76, no. 3:2–7.

Pittock, M.G.H. 1999. *Celtic Identity and the British Image*. New York: Manchester University Press.

Platt, R.S. 1962. 'The Rise of Cultural Geography in America'. In *Readings in Cultural Geography*, edited by P.L. Wagner and M.W. Mikesell, 35–43. Chicago: University of Chicago Press.

Plummer, V. 1993. *Feminism and the Mastery of Nature*. New York: Routledge.

Pocock, D.C.D., ed. 1981. *Humanistic Geography and Literature*. London: Croom Helm.

Porter, P.W. 1965. 'Environmental Potentials and Economic Opportunities: a Background for Cultural Adaptation'. *American Anthropologist* 67:409–20.

————. 1978. 'Geography as Human Ecology'. *American Behavioral Scientist* 22:15–39.

————. 1979. *Food and Development in the Semi-Arid Zone of East Africa*. Syracuse: Syracuse University, Maxwell School of Citizenship and Public Affairs.

————. 1987. 'Ecology as Metaphor: Sauer and Human Ecology'. *Professional Geographer* 39:414.

Powell, J.M. 1977. *Mirrors of the New World: Images and Image Makers in the Settlement Process*. Folkestone, England: Dawson.

————. 1980. 'Taylor, Stefansson and the Arid Centre: An Historic Encounter of "Environmentalism" and Possibilism'. *Journal of the Royal Australian Historical Society* 66:163–83.

Pratt, G. 2004. 'Feminist Geographies: Spatialising Feminist Politics'. In *Envisioning Human Geographies*, edited by P. Cloke, P. Crang, and M. Goodwin, 128–45. New York: Arnold.

Pred, A. 1984. 'Place as Historically Contingent Process: Structuration and the Time Geography of Becoming Places'. *Annals of the Association of American Geographers* 74:279–97.

————. 1985. 'The Social Becomes the Spatial, The Spatial Becomes the Social: Enclosures, Social Change and the Becoming of Place in the Swedish Province of Skåne'. In *Social Relations and Spatial Structures*, edited by D. Gregory and J. Urry, 336–75. London: Macmillan.

Price, M., and M. Lewis. 1993a. 'The Reinvention of Cultural Geography'. *Annals of the Association of American Geographers* 83:1–17.

————. 1993b. 'Reply: On Reading Cultural Geography'. *Annals of the Association of American Geographers* 83:520–2.

Prince, H.C. 1971. 'Real, Imagined and Abstract Worlds of the Past'. In *Progress in Geography, Volume 3*, edited by C. Board et al., 1–86. New York: Arnold.

Proctor, J.D. 1998. 'The Social Construction of Nature: Relativist Accusations, Pragmatist and Critical Responses'. *Annals of the Association of American Geographers* 88:352–76.

Pryce, W.T.R. 1975. 'Migration and the Evolution of Culture Areas: Cultural and Linguistic Frontiers in North-East Wales, 1750 and 1851'. *Transactions of the Institute of British Geographers NS* 65:79–107.

Pudup, M.B. 1988. 'Arguments Within Regional Geography'. *Progress in Human Geography* 12:369–90.

Pulvirenti, M. 1997. 'Unwrapping the Parcel: An Examination of Culture Through Italian Australian Home Ownership'. *Australian Geographical Studies* 35:32–9.

Pyle, G.F. 1969. 'The Diffusion of Cholera in the United States in the Nineteenth Century'. *Geographical Analysis* 1:59–75.

Quani, M. 1982. *Geography and Marxism*. Cambridge, MA: Blackwell.

Quinn, J.A. 1950. *Human Ecology*. Englewood Cliffs, NJ: Prentice Hall.

Raby, S. 1973. 'Indian Land Surrenders in Southern Saskatchewan'. *Canadian Geographer* 17:36–52.

Radcliffe, S. 1994. '(Representing) Post-colonial Women: Authority, Difference and Feminisms'. *Area* 26:25–32.

Radding, C. 1997. *Wandering Peoples: Colonialism, Ethnic Spaces, and Ecological Frontiers in Northwestern Mexico, 1700–1850*. Durham: Duke University Press.

Raento, P., and C.J. Watson. 2000. 'Gernika, Guernica, *Guernica*? Contested Meanings of a Basque Place'. *Political Geography* 19:707–36.

Raglon, R. 1996. 'Women and the Great Canadian Wilderness: Reconsidering the Wild'. *Women's Studies* 25:513–31.

Raitz, K.B. 1973a. 'Ethnicity and the Diffusion and Distribution of Cigar Tobacco Production in Wisconsin and Ohio'. *Tijdschrifte voor Economische en Sociale Geografie* 64:293–306.

————. 1973b. 'Theology on the Landscape: A Comparison of Mormon and Amish-Mennonite Land Use'. *Utah Historical Quarterly* 41:23–34.

————. 1979. 'Themes in the Cultural Geography of European Ethnic Groups in the United States'. *Geographical Review* 69:79–94.

————, and R. Ulack. 1981a. 'Appalachian Vernacular Regions'. *Journal of Cultural Geography* 2, no. 1:106–19.

————, and R. Ulack. 1981b. 'Cognitive Maps of Appalachia'. *Geographical Review* 71:201–13.

Raivo, P.J. 1997. 'The Limits of Tolerance: The Orthodox Milieu as an Element in the Finnish Cultural Landscape'. *Journal of Historical Geography* 23:327–39.

Rankin, K.N. 2003. 'Anthropologies and Geographies of Globalization'. *Progress in Human Geography* 27:708–34.

Rapoport, A. 1969. *House Form and Culture*. Englewood Cliffs, NJ: Prentice Hall.

Rappaport, R.A. 1963. 'Aspects of Man's Influence on Island Ecosystems: Alteration and Control'. In *Man's Place in the Island Ecosystem*, edited by F.R. Fosberg, 155–74. Honolulu: Bishop Museum Press.

Ray, A.J. 1974. *Indians in the Fur Trade: Their Role as Hunters, Trappers and Middlemen in the Lands*

Southwest of Hudson Bay, 1660–1870. Toronto: University of Toronto Press.

———. 1996. *I Have Lived Here Since the World Began: An Illustrated History of Canada's Native People*. Toronto: Lester Publishing, Key Porter Books.

Redekop, C. 2000. *Creation & the Environment: An Anabaptist Perspective on a Sustainable World*. Baltimore: Johns Hopkins University Press.

Rees, J. 1989. 'Natural Resources, Economy and Society'. In *Horizons in Human Geography*, edited by D. Gregory and R. Walford, 364–94. New York: MacMillan.

Rees, R. 1984. *Land of Earth and Sky: Landscape Painting of Western Canada*. Saskatoon: Western Producer Prairie Books.

———. 1988. *New and Naked Land: Making the Prairies Home*. Saskatoon: Western Producer Prairie Books.

Rehder, J.B. 2004. *Appalachian Folkways*. Baltimore: Johns Hopkins University Press.

Relph, E. 1970. 'An Inquiry into the Relations Between Phenomenology and Geography'. *Canadian Geographer* 14:193–201.

———. 1976. *Place and Placelessness*. London: Pion.

———. 1981. *Rational Landscapes and Humanistic Geography*. Totowa, NJ: Barnes and Noble.

———. 1984. 'Seeing, Thinking, and Describing Landscapes'. In *Environmental Perception and Behavior: An Inventory and Prospect*, edited by T.F. Saarinen, D. Seamon, and J.L. Sell, 209–23. Chicago: University of Chicago, Department of Geography, Research Paper, no 209.

———. 1985. 'Geographical Experiences and Being-in-the-World: The Phenomenological Origins of Geography'. In *Dwelling, Place and Environment: Towards a Phenomenology of Person and World*, edited by D. Seamon and R. Mugeraur, 15–31. New York: Columbia University Press.

———. 1997. 'Sense of Place'. In *Ten Geographic Ideas That Changed the World*, edited by S. Hanson, 205–26. New Brunswick, NJ: Rutgers University Press.

Renfrew, C. 1988. *Archaeology and Language: The Puzzle of Indo-European*. New York: Cambridge University Press.

Renner, M. 1996. *Fighting for Survival: Environmental Decline, Social Conflict, and the New Age of Insecurity*. New York: Norton.

Reynolds, H. 1982. *The Other Side of the Frontier: Aboriginal Resistance to the European Invasion of Australia*. New York: Penguin.

Rice, J.G. 1977. 'The Role of Culture and Community in Frontier Prairie Farming'. *Journal of Historical Geography* 3:155–75.

Richards, J.F. 2003. *The Unending Frontier: An Environmental History of the Early Modern World*. Berkeley: University of California Press.

Ringer, G., ed. 1998. *Destinations: Cultural Landscapes of Tourism*. New York: Routledge.

Robbins, P. 2004. 'Cultural Ecology'. In *A Companion to Cultural Geography*, edited by J.S. Duncan, N.C. Johnson, and R.H. Schein, 180–93. Malden, MA: Blackwell.

Roberts, N. 1989. *The Holocene: An Environmental History*. Cambridge, MA: Blackwell.

Robinson, G.M. 1998. *Methods and Techniques in Human Geography*. New York: Wiley.

Rodgers, T. 1997. 'Empires of the Imagination: Rider Haggard, Popular Fiction and Africa'. In *Writing and Africa*, edited by M-H. Msiska and P. Hyland, 103–21. New York: Longman.

Rogers, E.M. 1962. *Diffusion of Innovations*. New York: Free Press of Glencoe.

Rogerson, R.J., and A. Gloyer. 1995. 'Gaelic Cultural Revival or Language Decline'. *Scottish Geographical Magazine* 111:46–53.

Rolston, H., III. 1997. 'Nature for Real: Is Nature a Social Construct'. In *The Philosophy of the Environment*, edited by T.D.J. Chappell, 38–64. Edinburgh: Edinburgh University Press.

Rooney, J.R., Jr. 1974. *A Geography of American Sport: From Cabin Creek to Anaheim*. Reading, MA: Addison Wesley.

———. 1993. 'The Golf Construction Boom'. *Sport Place: An International Journal of Sports Geography* 7:15–22.

———, and P.L. Butt. 1978. 'Beer, Bourbon, and Boone's Farm: A Geographical Examination of Alcoholic Drink in the United States'. *Journal of Popular Culture* 11:832–55.

———, and R. Pillsbury. 1992. *Atlas of American Sport*. New York: Macmillan.

———, W. Zelinsky, and D.R. Loudon, eds. 1982. *This Remarkable Continent: An Atlas of United States and Canadian Society and Cultures*. College Station: Texas A&M University Press.

Rose, A.J. 1972. 'Australia as a Cultural Landscape'. In *Australia as Human Setting*, edited by A. Rapoport, 58–74. Sydney: Angus and Robertson.

Rose, C. 1981. 'William Dilthey's Philosophy of Historical Understanding: A Neglected Heritage of Contemporary Humanistic Geography'. In *Geography, Ideology and Social Concern*, edited by D.R. Stoddart, 99–133. Totowa, NJ: Barnes and Noble.

Rose, G. 1993. *Feminism and Geography: The Limits of Geographical Knowledge*. Minneapolis: University of Minnesota Press.

———. 1997. 'Situating Knowledges: Positionality, Reflexivities and Other Tactics'. *Progress in Human Geography* 21:305–20.

———. 2001. *Visual Methodologies*. Thousand Oaks, CA: Sage.

———, V. Kinnaird, M. Morris, and C. Nash. 1997. 'Feminist Geographies of Environment, Nature and Landscape'. In *Feminist Geographies: Explorations in Diversity and Difference*, edited by Women and Geography Study Group of the Royal Geographical Society with the Institute of British Geographers, 146–90. Harlow: Longman.

———, et al. 1997. 'Introduction'. In *Feminist Geographies: Explorations in Diversity and Difference*, edited by Women and Geography Study Group of the Royal Geographical Society with the Institute of British Geographers, 1–12. Harlow, England: Longman.

Rosenau, P.M. 1992. *Post-Modernism and the Social Sciences: Insights, Inroads, and Intrusions*. Princeton: Princeton University Press.

Rosenberg, M. 2004. 'Ethic and Race Relations'. In *Sociology: A Canadian Perspective*, edited by L. Tepperman and J. Curtis, 402–32. Toronto: Oxford University Press.

Rossi, I., and E. O'Higgins. 1980. 'Unit 1: Theories of Culture and Anthropological Methods'. In *People in Culture: A Survey of Cultural Inquiry*, edited by I. Rossi, 31–78. New York: Praeger.

Rostlund, E. 1956. 'Twentieth-Century Magic'. *Landscape* 5, no. 3:23–6.

Routledge, P. 1992. 'Putting Politics in its Place: Baliapal, India, as a Terrain of Resistance'. *Political Geography* 11:588–611.

———. 1997. 'The Imagineering of Resistance: Pollok Free State and the Practice of Postmodern Politics'. *Transactions of the Institute of British Geographers NS* 22:359–76.

Rowles, G. 1978. *Prisoners of Space: Exploring the Geographical Experience of Older People*. Boulder, CO: Westview Press.

Rowntree, L.B. 1996. 'The Cultural Landscape Concept in American Cultural Geography'. In *Concepts in Human Geography*, edited by C. Earle, K. Mathewson, and M.S. Kenzer, 127–59. Lanham, MD: Rowman and Littlefield.

———, K. Foote, and M. Domosh. 1989. 'Cultural Geography'. In *Geography in America*, edited by G. Gaile and C. Wilmott, 209–17. New York: Merrill.

Roxby, P.M. 1930. 'The Scope and Aims of Human Geography'. *Scottish Geographical Magazine* 46:276–99.

Ruggie, J.G. 1998. *Constructing the World Polity: Essays on International Institutionalization*. New York: Routledge.

Rupesinghe, K. 1996. 'Governance and Conflict Resolution in Multi-ethnic Societies'. In *Ethnicity and Power in the Contemporary World*, edited by K. Rupesinghe and V.A. Tishkov, 10–31. New York: United Nations University Press.

Rushton, G. 1979. 'Commentary: On "Behavioral and Perception Geography"'. *Annals of the Association of American Geographers* 69:463–4.

Russell, R.J., and F.B. Kniffen. 1951. *Culture Worlds*. New York: MacMillan.

Ryan, J.R. 2004. 'Postcolonial Geographies'. In *A Companion to Cultural Geography*, edited by J.S. Duncan, N.C. Johnson, and R.H. Schein, 469–84. Malden, MA: Blackwell.

Ryan, S. 1996. *The Cartographic Eye: How Explorers Saw Australia*. New York: Cambridge University Press.

Saarinen, T.F. 1974. 'Environmental Perception'. In *Perspectives on Environment*, edited by I.R. Manners and M.W. Mikesell, 252–89. Washington: Association of American Geographers, Commission on College Geography Publication, no. 13.

———. 1979. 'Commentary-Critique of the Bunting-Guelke Paper'. *Annals of the Association of American Geographers* 69:464–8.

Sack, R.D. 2003. *A Geographical Guide to the Real and the Good*. New York: Routledge.

Sahlins, M. 1974. *Stone Age Economics*. London: Tavistock.

Said, E. 1978. *Orientalism*. New York: Columbia University Press.

———. 1993. *Culture and Imperialism*. London: Chatto and Windus.

Salerno, R.A. 2003. *Landscapes of Abandonment: Capitalism, Modernity, and Estrangement*. Albany: SUNY Press.

Salter, C.L. 1971a. 'The Mobility of Man'. In *The Cultural Landscape*, edited by C.L. Salter, 1–4. Belmont, CA: Duxbury.

———. 1971b. 'Introductory Comments to 'Some Curious Analogies in Explorer's Preconceptions of Virginia' by G. Dunbar'. In *The Cultural Landscape*, edited by C.L. Salter, 18. Belmont: Duxbury.

———. 1972. 'A Speculative Cultural Geography: A Free-for-All Approach to Introductory Cultural Geography'. *Journal of Geography* 71:533–40.

———. 1981. 'John Steinbeck's *The Grapes of Wrath* as a Primer for Cultural Geography'. In *Humanistic Geography and Literature*, edited by D.C.D. Pocock, 142–58. Totowa, NJ: Barnes and Noble.

————, and W.J. Lloyd. 1977. *Landscape in Literature*. Resource Papers for College Geography, 76-3. Washington DC: Association of American Geographers.

Samuels, M.S. 1978. 'Existentialism and Human Geography'. In *Humanistic Geography: Prospects and Problems*, edited by D. Ley and M.S. Samuels, 22–40. London: Croom Helm.

————. 1979. 'The Biography of Landscape: Cause and Culpability'. In *The Interpretation of Ordinary Landscapes*, edited by D.W. Meinig, 51–88. New York: Oxford University Press.

————. 1981. 'An Existential Geography'. In *Themes in Geographic Thought*, edited by M.E. Harvey and B.P. Holly, 115–32. London: Croom Helm.

Sanderson, S.K. 1990. *Social Evolutionism: A Critical History*. Cambridge, MA: Blackwell.

————, ed. 1995. *Civilizations and World Systems: Studying World-Historical Change*. Walnut Creek, CA: Altamira Press.

————, and T.D. Hall. 1995. 'World System Approaches to World-Historical Change'. In *Civilizations and World Systems: Studying World-Historical Change*, edited by S.K. Sanderson, 95–108. Walnut Creek, CA: Altamira Press.

Sauer, C.O. 1924. 'The Survey Method in Geography and its Objectives'. *Annals of the Association of American Geographers* 14:17–33.

————. 1925. 'The Morphology of Landscape'. *University of California Publications in Geography* 2:19–53.

————. 1927. 'Recent Developments in Cultural Geography'. In *Recent Developments in the Social Sciences*, edited by E.C. Hayes, 154–212. Philadelphia: J.B. Lippincott.

————. 1931. 'Cultural Geography'. *In Encyclopedia of the Social Sciences, Volume 6*, 621–4. New York: MacMillan.

————. 1935. Aboriginal Populations of North-western Mexico. *Ibero-Americana* 10. Berkeley: University of California Press.

————. 1940. *Culture Regions of The World: Outline of Lectures*. Berkeley: University of California at Berkeley, Department of Geography.

————. 1941a. 'Foreword to Historical Geography'. *Annals of the Association of American Geographers* 31:1–24.

————. 1941b. 'The Personality of Mexico'. *Geographical Review* 31:353–64.

————. 1952. *Agricultural Origins and Dispersals*. New York: American Geographical Society.

————. 1963. 'Historical Geography and the Western Frontier'. In *Land and Life: A Selection from the Writings of Carl Sauer*, edited by J. Leighly, 45–52. Berkeley: University of California Press.

————. 1968. 'Human Ecology and Population'. In *Population and Economics*, edited by P. DePrez, 207–14. Winnipeg: University of Manitoba Press.

————. 1969. *Seeds, Spades, Hearths, and Herds: The Domestication of Animals and Foodstuffs*. Cambridge, MA: MIT Press.

————. 1970. 'Plants, Animals and Man'. In *Man and His Habitat*, edited by R.H. Buchanan, E. Jones, and D. McCourt, 34–61. London: Routledge and Kegan Paul.

————. 1971. *Sixteenth Century North America*. Berkeley: University of California Press.

————. 1974. 'The Fourth Dimension of Geography'. *Annals of the Association of American Geographers* 64:189–92.

————. 1981. *Selected Essays, 1963–1975*. Berkeley, CA: Turtle Island Foundation.

————. 1985. *North America: Notes on Lectures by Professor Carl O. Sauer at the University of California at Berkeley, 1936*. Northridge, CA.: California State University, Northridge, Department of Geography, Occasional Paper no. 1.

Sayer, A. 1979. 'Epistemology and Conceptions of People and Nature in Geography'. *Geoforum* 10:19–43.

————. 1982. 'Explanation in Economic Geography'. *Progress in Human Geography* 6:68–88.

————. 1992a. *Method in Social Science: A Realist Approach*, 2nd edn. New York: Routledge.

————. 1992b. 'Radical Geography and Marxist Political Economy: Towards a Re-evaluation'. *Progress in Human Geography* 16:343–60.

————, and M. Storper. 1997. 'Guest Editorial Essay—Ethics Unbound: For a Normative Turn in Social Theory'. *Environment and Planning D: Society and Space* 15:1–17.

Scarre, C., and B.M. Fagan. 1997. *Ancient Civilizations*. New York: Longman.

Schaefer, F. 1953. 'Exceptionalism in Geography: A Methodological Examination'. *Annals of the Association of American Geographers* 43:226–49.

Schein, R.H. 1997. 'The Place of Landscape: A Conceptual Framework for Interpreting an American Scene'. *Annals of the Association of American Geographers* 87:660–80.

————. 2004. 'Cultural Traditions'. In *A Companion to Cultural Geography*, edited by J.S. Duncan, N.C. Johnson, and R.H. Schein, 11–23. Malden, MA: Blackwell.

Schmitt, R. 1987. *Introduction to Marx and Engels: A Critical Reconstruction*. Boulder, CO: Westview.

Schnell, S.M. 2003a. 'The Ambiguities of Authenticity in Little Sweden, USA'. *Journal of Cultural Geography* 20(2):43–68.

Schnell, S.M. 2003b. 'Creating Narratives of Place and Identity in "Little Sweden, USA"'. *Geographical Review* 93:1–29.

Schnore, L.F. 1961. 'Geography and Human Ecology'. *Economic Geography* 37:207–17.

Scholte, J.A. 2000. *Globalization: A Critical Introduction*. New York: St Martin's Press.

Schroeder, W.A. 2002. *Opening the Ozarks: A Historical Geography of Missouri's Ste Genevieve District, 1760–1830*. Columbia: University of Missouri Press.

Schutz, A. 1967. *The Phenomenology of the Social World*, translated by G. Walsh and F. Lennert. Evanston, IL: Northwestern University Press.

Schwartz, J.M. 2003. 'Photographs from the Edge of Empire'. In *Cultural Geography in Practice*, edited by A. Blunt, et al., 154–71. New York: Arnold.

Scott, H. 2004. 'Cultural Turns'. In *A Companion to Cultural Geography*, edited by J.S. Duncan, N.C. Johnson, and R.H. Schein, 24–37. Malden, MA: Blackwell.

Seamon, D. 1979. *A Geography of the Lifeworld*. London: Croom Helm.

Seig, L. 1963. 'The Spread of Tobacco: A Study in Cultural Diffusion'. *Professional Geographer* 15:17–21.

Semple, E. 1911. *Influences of Geographic Environment*. New York: Henry Holt.

Senn, P.R. 1971. *Social Science and its Methods*. Boston: Holbrook Press Inc.

Sharp, J. 2004. 'Feminisms'. In *A Companion to Cultural Geography*, edited by J.S. Duncan, N.C. Johnson, and R.H. Schein, 66–78. Malden, MA: Blackwell.

Sharp, J.P. 1994. 'A Topology of "Post" Nationality: (Re)mapping Identity in the Satanic Verses'. *Ecumene* 1:65–76.

———. 1996. 'Locating Imaginary Homelands: Literature, Geography and Salman Rushdie'. *Geojournal* 38:119–27.

———. 2000. 'Towards a Critical Analysis of Fictive Geographies'. *Area* 32:327–34.

Shiva, V. 1988. *Staying Alive: Women, Ecology and Development*. London: Zed Books.

Short, J.R. 2001. *Global Dimensions: Space, Place and the Contemporary World*. London: Reaktion Books.

Shortridge, B.G., and J.R. Shortridge. 1995. 'Cultural Geography of American Foodways: An Annotated Bibliography'. *Journal of Cultural Geography* 15, no. 2:79–108.

———, and ———, eds. 1998. *The Taste of American Place: A Reader on Regional and Ethnic Foods*. Lanham, MD: Rowman and Littlefield.

Shortridge, J.R. 1980. 'Vernacular Regions in Kansas'. *American Studies* 21:73–94.

———. 1995. *Peopling the Plains: Who Settled Where in Frontier Kansas*. Lawrence: University Press of Kansas.

Shurmer-Smith, P., ed. 2002. *Doing Cultural Geography*. Thousand Oaks, CA: Sage.

———, and K. Hannam. 1994. *Worlds of Desire, Realms of Power: A Cultural Geography*. New York: Arnold.

Sibley, D. 1995. *Geographies of Exclusion: Society and Difference in the West*. New York: Routledge.

———. 1999a. 'Outsiders in Society and Space'. In *Cultural Geographies*, 2nd edn, edited by K. Anderson and F. Gale, 135–51. Harlow: Longman.

———. 1999b. 'Creating Geographies of Difference'. In *Human Geography Today*, edited by D. Massey, J. Allen, and P. Sarre, 115–28. Cambridge: Polity Press.

Sidaway, J.D. 2000. 'Postcolonial Geographies: An Exploratory Essays'. *Progress in Human Geography* 24:591–612.

Simmons, C.W. 2004. 'The Political Economy of Land Conflict in the Eastern Brazilian Amazon'. *Annals of the Association of American Geographers* 94:183–206.

Simmons, I.G. 1988. 'The Earliest Cultural Landscapes of England'. *Environmental Review* 12:105–16.

———. 1997. *Humanity and Environment: A Cultural Ecology*. London: Longman.

———. 2003. 'Landscape and Environment: Natural Resources and Social Development'. In *Key Concepts in Geography*, edited by S.L. Holloway, S.P. Rice, and G. Valentine, 305–17. London: Sage.

Simon, J., and H. Kahn, eds. 1984. *The Resourceful Earth*. Cambridge, MA: Blackwell.

Simpson-Housley, P., and A.H. Mallory, eds. 1987. *Geography and Literature: A Meeting of Disciplines*. Syracuse: Syracuse University Press.

Singer, P. 1993. *Practical Ethics*, 2nd edn. New York: Cambridge University Press.

Skelton, T., and G. Valentine, eds. 1998. *Cool Places: Geographies of Youth Cultures*. New York: Routledge.

Skinner, B.F. 1969. *Contingencies of Reinforcement: A Theoretical Analysis*. New York: Appleton Century Crofts.

Slater, D. 2003. 'Cultures of Consumption'. In *Handbook of Cultural Geography*, edited by K. Anderson, M. Domosh, S. Pile, and N. Thrift, 147–63. Thousand Oaks, CA: Sage Publications.

Slater, T., and P. Jarvis, eds. 1982. *Field and Forest: An Historical Geography of Warwickshire and Worcestershire*. Norwich, England: Geobooks.

Sluyter, A. 1997. 'On "Buried Epistemologies: The Politics of Nature in (Post)colonial British Columbia": On Excavating and Burying Epistemologies'. *Annals of the Association of American Geographers* 87:700–2.

Smil, V. 1987. *Energy, Food, Environment: Realities, Myths, Options*. New York: Oxford University Press.

———. 1993. *Global Ecology: Environmental Changes and Social Flexibility*. New York: Routledge.

Smith, A.D. 1986. *The Ethnic Origins of Nations*. Cambridge, MA: Blackwell.

Smith, B.D. 1995. *The Emergence of Agriculture*. New York: Scientific American Library.

Smith, C.T. 1965. 'Historical Geography: Current Trends and Prospects'. In *Frontiers in Geographical Teaching*, edited by R.J. Chorley, 118–43. Methuen: London.

Smith, D.M. 2000. *Moral Geographies: Ethics in a World of Difference*. Edinburgh: Edinburgh University Press.

Smith, J.M., A. Light, and D. Roberts. 1998. 'Introduction: Philosophies and Geographies of Place'. In *Philosophy and Geography III: Philosophies of Place*, edited by A. Light and J.M. Smith, 1–19. Lanham, MD: Rowman and Littlefield.

———. 2004. 'Ethics and the Human Environment'. In *A Companion to Cultural Geography*, edited by J.S. Duncan, N.C. Johnson, and R.H. Schein, 209–20. Malden, MA: Blackwell.

Smith, J.S. 1999. 'Anglo Intrusion on the Old Sangre de Cristo Land Grant'. *Professional Geographer* 51:170–83.

———. 2003. Cultural Geography: A Survey of Perceptions Held by Cultural Geography Specialty Group Members. *Professional Geographer* 55:18–30.

Smith, R. 1997. *The Norton History of the Human Sciences*. New York: W.W. Norton.

———, and B.N. White. 2004. 'Detached from Their Homeland: The Latter-day Saints of Chihuahua, Mexico. *Journal of Cultural Geography* 21 (2):57–76.

Smith, T.R., J.W. Pellegrino, and R.G. Golledge. 1982. 'Computational Process Modeling of Spatial Cognition and Behavior'. *Geographical Analysis* 14:305–25.

Soja, E.W. 1989. *Postmodern Geographies*. London: Verso.

Solot, M. 1986. 'Carl Sauer and Cultural Evolution'. *Annals of the Association of American Geographers* 76:508–20.

Sopher, D.E. 1967. *Geography of Religions*. Englewood Cliffs, NJ: Prentice Hall.

———. 1972. 'Place and Location: Notes on the Spatial Patterning of Culture'. *Social Science Quarterly* 53:321–37.

———. 1981. 'Geography and Religions'. *Progress in Human Geography* 5:510–24.

Southwick, C.H. 1996. *Global Ecology in Human Perspective*. New York: Oxford University Press.

Sowell, T. 1994. *Race and Culture: A World View*. New York: Basic Books.

———. 1996. *Migrations and Cultures: A World View*. New York: Basic Books.

Spate, O.H.K. 1952. 'Toynbee and Huntington: A Study in Determinism'. *Geographical Journal* 118:406–28.

Spencer, C., and M. Blades. 1986. 'Pattern and Process: A Review Essay on the Relationship Between Behavioral Geography and Environmental Psychology'. *Progress in Human Geography* 10:230–48.

Spencer, J.E. 1960. 'The Cultural Factor in Underdevelopment'. In *Geography and Economic Development*, edited by N. Ginsburg, 35–48. Chicago: University of Chicago, Department of Geography, Research Paper, no. 62.

———. 1978. 'The Growth of Cultural Geography'. *American Behavioral Scientist* 22:79–92.

———, and R.J. Horvath. 1963. 'How Does an Agricultural Region Originate?' *Annals of the Association of American Geographers* 53:74–92.

———, and W.L. Thomas Jr. 1973. *Introducing Cultural Geography*. New York: Wiley.

Speth, W.W. 1967. 'Environment, Culture and the Mormon in Early Utah'. *Yearbook of the Association of Pacific Coast Geographers* 29:53–67.

———. 1987. 'Historicism: The Disciplinary World View of Carl O. Sauer'. In *Carl O. Sauer: A Tribute*, edited by M.S. Kenzer, 11–39. Corvallis: Oregon State University Press.

Spooner, D. 2000. 'Reflections on the Place of Larkin'. *Area* 32:209–16.

Spring, D., and R. Spring. 1974. *Ecology and Religion in History*. New York: Harper and Row.

Squire, S.J. 1988. 'Wordsworth and Lake District Tourism: Romantic Reshaping of Landscape'. *Canadian Geographer* 32:237–47.

———. 1994. 'Accounting for Cultural Meanings: The Interface Between Geography and Tourism Studies Re-Examined'. *Progress in Human Geography* 18:1–16.

Stanfield, J.H. 1997. 'Preface to the Sixth Edition'. In *Man's Most Dangerous Myth: The Fallacy of Race*, 6th edn, by A. Montagu, 25–9. Walnut Creek, CA: Altamira Press.

Stanislawski, D. 1946. 'The Origin and Spread of the Grid Pattern Town'. *Geographical Review* 36:105–20.

Steiman, L.B. 1998. *Paths to Genocide: Antisemitism in Western History*. New York: St Martin's Press.

Stein, H.F., and G.L. Thompson. 1993. 'The Sense of Oklahomaness: Contributions of Psychogeography to American Culture'. *Journal of Cultural Geography* 13, no. 2:63–91.

Stein, M. 2000. *City of Sisterly and Brotherly Loves: Lesbian and Gay Philadelphia, 1945–1972*. Chicago: University of Chicago Press.

Steward, J. 1936. 'The Economic and Social Basis of Primitive Bands'. In *Essays in Anthropology Presented to A.L. Kroeber*, edited by R.H. Lowie, 331–45. Berkeley: University of California Press.

———. 1938. *Basin-Plateau Aboriginal Sociopolitical Groups*. Washington, DC: Smithsonian Institution.

———. 1955. *Theory of Culture Change: The Methodology of Multilinear Evolution*. Urbana: University of Illinois Press.

Stilgoe, J.R. 1982. *Common Landscape of America: 1580–1845*. New Haven: Yale University Press.

Stoddart, D.R. 1965. 'Geography and the Ecological Approach'. *Geography* 50:242–51.

———. 1966. Darwin's Impact on Geography'. *Annals of the Association of American Geographers* 56:683–98.

———. 1967. 'Organism and Ecosystem as Geographical Models'. In *Models in Geography*, edited by R.J. Chorley and P. Haggett, 511–48. London: Methuen.

Straussfogel, D. 1997. 'World Systems Theory: Toward a Heuristic and Pedagogic Conceptual Tool'. *Economic Geography* 73:118–30.

Straw, R.A., and H.T. Blethen, eds. 2004. *High Mountains Rising: Appalachia in Time and Place*. Urbana: University of Illinois Press.

Strohmayer, U. 2003. 'Cultural Geographers Forum: What are the Five Most Important Principles that Should be Covered in an Introductory Course in Cultural Geography and Why?' *Place and Culture, The Newsletter of the Cultural Geography Specialty Group of the Association of American Geographers* Spring, 5–6.

Stump, R.W. 1986. 'Introduction'. *Journal of Cultural Geography* 7, no. 1:1–3.

Sumartojo, R. 2004. 'Contesting Place: Anti-gay and lesbian Hate Crime in Columbus, Ohio'. In *Spaces of Hate: Geographies of Discrimination and Intolerance in the USA*, edited by C. Flint, 87–108. New York: Routledge.

Sutter, R.E. 1973. *The Next Place You Come To: A Historical Introduction to Communities in North America*. Englewood Cliffs, NJ: Prentice Hall.

Svobodová, H., ed. 1990. *Cultural Aspects of Landscape*. Wageningen, Netherlands: Pudoc.

Symanski, R. 1974. 'Prostitution in Nevada'. *Annals of the Association of American Geographers* 64:357–77.

———. 1981. *The Immoral Landscape: Female Prostitution in Western Societies*. Toronto: Butterworths.

Tarrow, S. 1992. 'Mentalities, Political Cultures, and Collective Action Frames'. In *Frontiers in Social Movement Theory*, edited by A. Morris and C. Mueller, 174–202. New Haven: Yale University Press.

Tatham, G. 1937. *Environment, Race, and Migration. Fundamentals of Human Distribution: With Special Sections on Racial Classification and Settlement in Canada and Australia*. Toronto: University of Toronto Press.

———. 1951. 'Environmentalism and Possibilism'. In *Geography in the Twentieth Century*, edited by G. Taylor, 128–62. New York: Philosophical Library.

Taylor, A. 2000. "'The Sun Always Shines in Perth': A Postcolonial Geography of Identity, Memory and Place'. *Australian Geographical Studies* 38:27–35.

Taylor, P.J. 1991. 'The English and Their Englishness: "A Curiously Mysterious, Elusive and Little Understood People"'. *Scottish Geographical Magazine* 107:146–61.

———. 1992. 'Understanding Global Inequalities: A World-Systems Approach'. *Geography* 77:10–21.

———. 1994. 'The State as Container: Territoriality in the Modern World-System'. *Progress in Human Geography* 18:151–62.

———. 1996. *The Way the Modern World Works: World Hegemony to World Impasse*. New York: Wiley.

———, and C. Flint. 2000. *Political Geography: World-economy, Nation-state, and Locality*, 4th edn. New York: Prentice Hall.

Taylor, T.G. 1928. *Australia in Its Physiographic and Economic Aspects*, 5th edn. Oxford: Clarendon.

———. 1937. *Environment, Race, and Migration*. Toronto: University of Toronto Press.

———. 1951. *Australia: A Study of Warm Environments and Their Effect on British Settlement*, 6th enlarged edn. New York: Dutton.

Teather, E.K. 1998. 'Themes from Complex Landscapes: Chinese Cemeteries and Columbaria in Urban Hong Kong'. *Australian Geographical Studies* 36:21–36.

———., ed. 1999. *Embodied Geographies: Spaces, Bodies and Rites of Passage*. New York: Routledge.

Teich, M., R. Porter, and B. Gustafsson, eds. 1997. *Nature and Society in Historical Context*. New York: Cambridge University Press.

Tenbrunsel, A.E., K.A. Wade-Benzoni, D.M. Messick, and M.H. Bazerman. 1997. 'Introduction'. In *Environment, Ethics, and Behavior*, edited by M.H. Bazerman, D.M. Messick. A.E, Tenbrunsel, and K.A. Wade Benzoni, 1–9. San Francisco: The New Lexington Press.

Terkenli, T.S. 1995. 'Home as a Region'. *Geographical Review* 85:324–34.

Tetlock, P.E., and A. Belkin, eds. 1996. *Counterfactual Thought Experiments in World Politics: Logical, Methodological, and Psychological Perspectives*. Princeton: Princeton University Press.

Thomas, W.L., Jr. 1956. 'Introductory'. In *Man's Role in Changing the Face of the Earth*, edited by W.L. Thomas Jr., C.O. Sauer, M. Bates, and L. Mumford, xxi–xxxviii. Chicago: University of Chicago Press.

———, C.O. Sauer, M. Bates, and L. Mumford, eds. 1956. *Man's Role in Changing the Face of the Earth*. Chicago: University of Chicago Press.

Thompson, E.P. 1968. *The Making of the English Working Class*. London: Gollancz.

Thompson, G.F. 1995a. 'A Message to the Reader'. In *Landscape in America*, edited by G.F. Thompson, xi–xiv. Austin: University of Texas Press.

———, ed. 1995b. *Landscape in America*. Austin: University of Texas Press.

Thomson, G.M. 1975. *The North West Passage*. London: Martin Secker and Warburg.

Thornthwaite, C.W. 1940. 'The Relation of Geography to Human Ecology'. *Ecological Monograph* 10:343–8.

Thrift, N. 1994. 'Taking Aim at the Heart of the Region'. In *Human Geography: Society, Space and Social Science*, edited by D. Gregory, R. Martin, and G. Smith, 200–31. New York: MacMillan.

Toney, M.B., Keller, C., and L.M. Hunter. 2003. 'Regional Cultures, Persistence and Change: A Case Study of the Mormon Culture Region'. *The Social Science Journal* 40:431–45.

Tough, F. 1996. *As Their Natural Resources Fail: Native Peoples and the Economic History of Northern Manitoba, 1870–1930*. Vancouver: University of British Columbia Press.

Touraine, A. 1981. *The Voice and the Eye: An Analysis of Social Movements*. New York: Cambridge University Press.

Toynbee, A. 1934–61. *A Study of History*, 12 vols. New York: Oxford University Press.

Trépanier, C. 1991. 'The Cajunization of French Louisiana: Forging a Regional Identity'. *Geographical Journal* 157:161–71.

Trigger, B.G. 1982. 'Response of Native Peoples to European Contact'. In *Early European Settlement and Exploitation in Atlantic Canada*, edited by G.M. Story, 139–55. St John's: Memorial University of Newfoundland.

———. 1985. *Natives and Newcomers: Canada's 'Heroic Age' Reconsidered*. Kingston: McGill–Queen's University Press.

———. 1998. *Sociocultural Evolution: Calculation and Contingency*. Malden, MA: Blackwell.

Tuan, Y.-F. 1971. 'Geography, Phenomenology and the Study of Human Nature'. *Canadian Geographer* 15:181–92.

———. 1972. 'Structuralism, Existentialism and Environmental Perception'. *Environment and Behavior* 4:319–42.

———. 1974. *Topophilia: A Study of Environmental Perception, Attitudes, and Values*. Englewood Cliffs, NJ: Prentice Hall.

———. 1975. 'Images and Mental Maps'. *Annals of the Association of American Geographers* 65:205–13.

———. 1977. *Space and Place: The Perspective of Experience*. Minneapolis: University of Minnesota Press.

———. 1982. *Segmented Worlds and Self*. Minneapolis: University of Minnesota Press.

———. 1989. *Morality and Imagination: Paradoxes of Progress*. Madison: University of Wisconsin Press.

———. 1991a. 'A View Of Geography'. *Geographical Review* 81:99–107.

———. 1991b. 'Language and the Making of Place: A Narrative-Descriptive Approach'. *Annals of the Association of American Geographers* 81:684–96.

———. 1997. 'Alexander von Humboldt and His Brother: Portrait of an Ideal Geographer in Our Time'. Los Angeles, University of California at Los Angeles, Department of Geography, Alexander von Humboldt Lecture.

Turner, B.L., II., et al. 1990. *The Earth as Transformed by Human Action: Global and Regional Changes in the Biosphere in the Past 300 Years*. New York: Cambridge University Press.

Turner, B.S. 1994. *Orientalism, Postmodernism and Globalism*. New York: Routledge.

Turner, F.J. 1961. *Frontier and Section*, edited by R.A. Billington. Englewood Cliffs, NJ: Prentice Hall.

Turner, T. 1993. 'Anthropology and Multiculturalism: What Is Anthropology That Multiculturalists Should Be Mindful of It?' *Cultural Anthropology* 8:411–29.

Tylor, E.B. 1916. *Anthropology: An Introduction to the Study of Man and Civilization*. New York: D. Appleton.

————. 1924. *Primitive Culture: Researches into the Development of Mythology, Philosophy, Religion, Language, Art, and Custom*. New York: Brentano's.

United Nations Development Programme. 2002. *Arab Human Development Report, 2002: Creating Opportunities for Future Generations*. New York: United Nations Development Programme.

Unstead, J.F. 1922. 'Geography and Historical Geography'. *Geographical Journal* 41:55–9.

Unwin, T. 1992. *The Place of Geography*. New York: Longman.

Valentine, G. 1989. 'The Geography of Women's Fear'. *Area* 21:385–90.

————. 1993a. '(Hetero)sexing Space: Lesbian Perceptions and Experiences of Everyday Spaces'. *Environment and Planning D: Society and Space* 11:395–413.

————. 1993b. 'Negotiating and Managing Multiple Sexual Identities: Lesbian Time-space Strategies'. *Environment and Planning D: Society and Space* 11:237–48.

————. 1996. '(Re)negotiating the Heterosexual Street'. In *Body Space: Destabilizing Geographies of Gender and Sexuality*, edited by N. Duncan, 146–55. New York: Routledge.

————. 1998. '"Sticks and Stones May Break My Bones": A Personal Geography of Harrassment'. *Antipode* 30:305–32.

————. 1999. 'A Corporeal Geography of Consumption'. *Environment and Planning D: Society and Space* 17:329–51.

————. 2001. *Social Geographies: Space and Society*. New York: Prentice Hall.

Vayda, A.P., and R.A. Rappaport. 1968. 'Ecology, Cultural and Noncultural'. In *Introduction to Cultural Anthropology*, edited by J.A. Clifton, 477–97. Boston: Houghton Mifflin.

Veblen, T.T. 1977. 'Native Population Decline in Totoncapán, Guatemala'. *Annals of the Association of American Geographers* 67:484–99.

Vibert, E. 1997. *Trader's Tales: Narratives of Cultural Encounters in the Columbia Plateau, 1807–1846*. Norman: University of Oklahoma Press.

Vidal de la Blache, P. 1926. *Principles of Human Geography*, edited by E.D. Martonne, translated by M.T. Bingham. London: Constable Publishers.

Visser, G. 2003. 'Gay Men, Leisure Space and South African Cities: The Case of Cape Town'. *Geoforum* 34:123–37.

von Maltzahn, K.E. 1994. *Nature as Landscape: Dwelling and Understanding*. Kingston: McGill–Queen's University Press.

Wacker, P.O. 1968. *The Musconetcong Valley of New Jersey: A Historical Geography*. New Brunswick: Rutgers University Press.

————. 1975. *Land and People: A Cultural Geography of Pre-Industrial New Jersey, Origins and Settlement Patterns*. New Brunswick: Rutgers University Press.

————, and P.G.E. Clemens. 1995. *Land Use in Early New Jersey: A Historical Geography*. Newark: New Jersey Historical Society.

Wagner, P.L. 1958a. 'Nicoya: A Cultural Geography'. *University of California Publications in Geography* 12:195–250.

————. 1958b. 'Remarks on the Geography of Language'. *Geographical Review* 48:86–97.

————. 1960. *The Human Use of the Earth*. Glencoe, IL: Free Press.

————. 1972. *Environments and Peoples*. Englewood Cliffs, NJ: Prentice Hall.

————. 1974. 'Cultural Landscapes and Regions: Aspects of Communication'. In *Man and Cultural Heritage: Papers in Honor of Fred B. Kniffen*, edited by H.J. Walker and W.G. Haag. *Geoscience and Man* 5:133–42. Baton Rouge: Louisiana State University, Department of Geography and Anthropology, Geoscience Publications.

————. 1975. 'The Themes of Cultural Geography Rethought'. *Yearbook of the Association of Pacific Coast Geographers* 37:7–14.

————. 1988. 'Why Diffusion'. In *The Transfer and Transformation of Ideas and Material Culture*, edited by P.J. Hugill and D.B. Dickson, 179–93. College Station: Texas A&M University Press.

————. 1990. 'Review of *Explorations in the Understanding of Landscape: A Cultural Geography* by W. Norton'. *Journal of Geography* 89:40–1.

————. 1994. 'Foreword: Culture and Geography: Thirty Years of Advance'. In *Re-reading Cultural Geography*, edited by K.E. Foote, P.J. Hugill, K. Mathewson, and J.M. Smith, 3–8. Austin: University of Texas Press.

————. 1996. *Showing Off: The Geltung Hypothesis*. Austin: University of Texas Press.

————, and M.W. Mikesell. 1962. 'The Themes of Cultural Geography'. In *Readings in Cultural Geography*, edited by P.L. Wagner and M.W. Mikesell, 1–24. Chicago: University of Chicago Press.

Waitt, G. 1997. 'Selling Paradise and Adventure: Representations of Landscape in the Tourist Advertising of Australia'. *Australian Geographical Studies* 35:47–60.

————. 2003. 'A Place for Buddha in Wollongong, New South Wales? Territorial Rules in the

Place-making of Sacred Spaces'. *Australian Geographer* 34:223–38.

Walker, H.J., and R.A. Detro, eds. 1990. *Cultural Diffusion and Landscapes: Selections by Fred B. Kniffen. Geoscience and Man 27.* Baton Rouge: Louisiana State University, Department of Geography and Anthropology, Geoscience Publications.

Walker, P.A. 2003. 'Reconsidering 'Regional' Political Ecologies: Toward a Political Ecology of the Rural American West'. *Progress in Human Geography* 27:7–24.

———. 2005. 'Political Ecology: Where is the Ecology'. *Progress in Human Geography* 29:73–82.

Walker, R. 1997. 'Unseen and Disbelieved: A Political Economist Among Cultural Geographers'. In *Understanding Ordinary Landscapes*, edited by P. Groth and T.W. Bressi, 162–73. New Haven: Yale University Press.

Wallerstein, I. 1974a. *The Modern World-System: Capitalist Agriculture and the Origins of the European World-Economy in the Sixteenth Century.* New York: Academic Press.

———. 1974b. 'The Rise and Future Demise of the World-Capitalist System: Concepts for Comparative Analysis'. *Comparative Studies in Society and History* 16:387–415.

———. 1980. *The Modern World-System II: Mercantilism and the Consolidation of the European World-Economy, 1600–1750.* New York: Academic Press.

———. 1989. *The Modern World-System III: The Second Era of Great Expansion of the Capitalist World-Economy, 1730–1840s.* New York: Academic Press.

———. 1996. 'The Global Picture'. In *The Age of Transition: Trajectory of the World System, 1945–2025*, coordinated by T.K. Hopkins and I. Wallerstein, 209–25. Atlantic Highlands, NJ: Zed Books.

Walmsley, D.J., and G.J. Lewis. 1984. *Human Geography: Behavioral Approaches.* London: Longman.

Walton, J. 1995. 'How Real(ist) Can You Get?' *Professional Geographer* 47:61–5

Warren, K.J. 1997. 'Taking Empirical Data Seriously: An Ecofeminist Philosophical Perspective'. In *Ecofeminism: Women, Culture, Nature*, edited by K.J. Warren, 3–20. Bloomington: Indiana University Press.

———. 2000. *Ecofeminist Philosophy: A Western Perspective on What It Is and Why It Matters.* Lanham, MD: Rowman and Littlefield.

Waterman, S. 1998. 'Carnivals for Élites? The Cultural Politics of Arts Festivals'. *Transactions of the Institute of British Geographers NS* 22:54–74.

Watson, J.B. 1913. 'Psychology as the Behaviorist Views It'. *Psychological Review* 20:158–77.

Watson, J.W. 1951. 'The Sociological Aspects of Geography'. In *Geography in the Twentieth Century*, edited by G. Taylor, 463–99. New York: Philosophical Library.

———. 1983. 'The Soul of Geography'. *Transactions of the Institute of British Geographers NS* 8:385–99.

Watts, M. 1983. *Silent Violence: Food, Famine, and Peasantry in Northern Nigeria.* Berkeley: University of California Press.

Watts, S.J., and S.J. Watts. 1978. 'On the Idealist Alternative in Geography and History'. *Professional Geographer* 30:123–7.

Webb, N.L. 1990–1. 'Deconstruction and Human Geography: Exploring Four Basic Themes'. *South African Geographer* 18:123–33.

Webb, W.P. 1931. *The Great Plains.* New York: Grosset and Dunlap.

———. 1964. *The Great Frontier.* Austin: University of Texas Press.

Webber, M.J., and D.L. Rigby. 1996. *The Golden Age Illusion: Rethinking Postwar Capitalism.* New York: Guilford.

Weitz, E.D. 2003. *A Century of Genocide: Utopias of Race and Nation.* Princeton: Princeton University Press.

Wekerle, G.R., and C. Whitzman. 1995. *Safe Cities: Guidelines for Planning.* New York: Van Nostrand Reinhold.

Werlen, B. 1993. *Society, Action and Space: An Alternative Human Geography.* New York: Routledge.

Wertz, F.J. 1998. 'The Role of the Humanistic Movement in the History of Psychology'. *Journal of Humanistic Psychology* 38:42–70.

West, R.C., and J.P. Augelli. 1966. *Middle America: Its Lands and Peoples.* Englewood Cliffs, NJ: Prentice Hall.

West Edmonton Mall. n.d. 'Map and Directory: The Wonder of It All'.

Western, J. 1981. *Outcast Cape Town.* Minneapolis: University of Minnesota Press.

Whatmore, S. 1999. 'Hybrid Geographies: Rethinking the "Human" in Human Geography'. In *Human Geography Today*, edited by J. Allen and P. Sarre, 22–39. Cambridge: Polity Press.

———. 2002. *Hybrid Geographies: Natures, Cultures, Spaces.* Thousand Oaks, CA: Sage.

———, and L. Thorne. 1998. 'Wild(er)ness: Reconfiguring the Geographies of Wildlife'. *Transactions of the Institute of British Geographers NS* 23:435–54.

Wheeler, J.O. 2002. Editorial: From Urban Economic to Social/Cultural Urban Geography, 1980–2001. *Urban Geography* 23:97–102.

White, C.L., and G.T. Renner. 1936. *Geography: An Introduction to Human Ecology*. New York: Appleton Century.

———, and ———. 1948. *Human Geography: An Ecological Study of Society*. New York: Appleton Century Crofts.

White, G.W. 2000. *Nationalism and Territory: Constructing Group Identity in Southeastern Europe*. Lanham, MD: Rowman and Littlefield.

White, L., Jr. 1967. 'The Historical Roots of our Ecological Crisis'. *Science* 155:1203–7.

White, L.A. 1949. *The Science of Culture: A Study of Man and Civilization*. New York: Grove Press.

White, R. 1990. 'Environmental History, Ecology, and Meaning'. *Journal of American History* 76:1111–16.

Whittlesey, D. 1929. 'Sequent Occupance'. *Annals of the Association of American Geographers* 19:162–5.

Whyte, I. 2000. 'William Wordsworth's *Guide to the Lakes* and the Geographical Tradition'. *Area* 32:101–6.

Widdis, R. 1993. 'Saskatchewan: The Present Cultural Landscape'. In *Three Hundred Prairie Years*, edited by H. Epp, 142–57. Regina: Canadian Plains Research Centre.

Wilkinson, D. 1987. 'Central Civilization'. *Comparative Civilizations Review* 17:31–59.

Wilkinson, H.R. 1951. *Maps and Politics: A Review of the Ethnographic Cartography of Macedonia*. Liverpool: Liverpool University Press.

Willems-Braun, B. 1997a. 'Buried Epistemologies: The Politics of Nature in (Post)colonial British Columbia'. *Annals of the Association of American Geographers* 87:3–31.

———. 1997b. 'Reply: On Cultural Politics, Sauer, and the Politics of Citation'. *Annals of the Association of American Geographers* 87:703–8.

Williams, M. 1974. *The Making of the South Australian Landscape*. New York: Academic Press.

———. 1983. 'The Apple of My Eye: Carl Sauer and Historical Geography'. *Journal of Historical Geography* 9:1–28.

———. 1987. 'Historical Geography and the Concept of Landscape'. *Journal of Historical Geography* 15:92–104.

Williams, R. 1958. *Culture and Society, 1780–1950*. New York: Columbia University Press.

———. 1976. *Keywords: A Vocabulary of Culture and Society*. New York: Oxford University Press.

———. 1977. *Marxism and Literature*. New York: Oxford University Press.

Wilmer, F. 1997. 'Identity, Culture, and Historicity'. *World Affairs* 160, no. 1:3–16.

Wilson, A. 1992. *The Culture of Nature*. Cambridge, MA: Blackwell.

Wilson, A.G. 1981. *Geography and the Environment: Systems Analytical Methods*. New York: Wiley.

Wilson, C., and P. Groth, eds. 2003. *Everyday America: Cultural Landscape Studies After J.B. Jackson*. Berkeley: University of California Press.

Wilson, C.R. 1998. 'Introduction'. In *The New Regionalism*, edited by C.R. Wilson, ix–xxiii. Jackson: University of Mississippi Press.

Wilson, R.K. 1999. '"Placing Nature": The Politics of Collaboration and Representation in the Struggle for La Sierra in San Luis, Colorado'. *Ecumene* 6:1–28.

Wilson, W.J. 1987. *The Truly Disadvantaged. The Inner City, the Underclass, and Public Policy*. Chicago: University of Chicago Press.

Wilton, R.D. 1998. 'The Constitution of Difference: Space and Psyche in Landscapes of Exclusion'. *Geoforum* 29:173–85.

Winchester, H. 1992. 'The Construction and Deconstruction of Women's Roles in the Urban Landscape'. In *Inventing Places: Studies in Cultural Geography*, edited by K. Anderson and F. Gale, 139–56. Melbourne: Longman Cheshire.

Wishart, D. 2004. 'Period and Region'. *Progress in Human Geography* 28:305–19.

———., A. Warren, and R.H. Stoddard. 1969. 'An Attempted Definition of a Frontier Using a Wave Analogy'. *Rocky Mountain Social Science Journal* 6:73–81.

Wisner, B. 1978. 'Does Radical Geography Lack an Approach to Environmental Relations?' *Antipode* 10:84–95.

Wissler, C. 1917. *The American Indian: An Introduction to the Anthropology of the New World*. New York: McMurtrie.

———. 1923. *Man and Culture*. New York: Crowell.

Withers, C.W.J. 1988. *Gaelic Scotland: The Transformation of a Culture Region*. London: Routledge.

———. 1995. 'How Scotland Came to Know Itself: Geography, National Identity and the Making of a Nation'. *Journal of Historical Geography* 21:371–97.

Wolch, J.R., and J. Emel. 1995. 'Bringing the Animals Back In'. *Environment and Planning D: Society and Space* 13:632–6.

———, J. Emel, and C. Wilbert. 2003. 'Reanimating Cultural Geography'. In *Handbook of Cultural Geography*, edited by K. Anderson, M. Domosh, S. Pile, and N. Thrift, 184–206. Thousand Oaks, CA: Sage Publications.

Wolf, E. 1982. *Europe and the People Without History*. Berkeley: University of California Press.

Wolpert, J. 1964. 'The Decision Process in Spatial Context'. *Annals of the Association of American Geographers* 54:537–58.

Women and Geography Study Group of the Royal Geographical Society with the Institute of British Geographers. 1997. *Feminist Geographies: Explorations in Diversity and Difference*. Harlow, England: Longman.

Wood, W.B. 2001. 'Geographic Aspects of Genocide: A Comparison of Bosnia and Rwanda'. *Transactions of the Institute of British Geographers NS* 26:57–75.

Worster, D. 1984. 'History as Natural History: An Essay on Theory and Method'. *Pacific Historical Review* 53:1–19.

———. 1988. 'Doing Environmental History'. In *The Ends of the Earth: Perspectives on Modern Environmental History*, edited by D. Worster, 289–307. New York: Cambridge University Press.

———. 1990. Transformations of the Earth: Toward an Agroecological Perspective in History'. *Journal of American History* 76:1087–106.

Wright, E.O. 1983. 'Gidden's Critique of Marxism'. *New Left Review* 138:11–35.

Wright, J.K. 1947. 'Terra Incognitae: The Place of Imagination in Geography'. *Annals of the Association of American Geographers* 37:1–15.

Wyckoff, W. 1999. *Creating Colorado: The Making of a Western American Landscape, 1860–1940*. New Haven: Yale University Press.

Yaeger, P. 1996. 'Introduction: Narrating Space'. In *The Geography of Identity*, edited by P. Yaeger, 1–38. Ann Arbor: University of Michigan Press.

Yapa, L.S. 1977. 'The Green Revolution: A Diffusion Model'. *Annals of the Association of American Geographers* 67:350–9.

———. 1996. 'Innovation Diffusion and Paradigms of Development'. In *Concepts in Human Geography*, edited by C. Earle, K. Mathewson, and M.S. Kenzer, 231–70. Lanham, MD: Rowman and Littlefield.

Yeoh, B.S.A., and P. Teo. 1996. 'From Tiger Balm Gardens to Dragon World: Philanthropy and Profit in the Making of Singapore's First Cultural Theme Park'. *Geografiska Annaler* 76B:27–42.

Yeung, H.W-C. 1997. 'Critical Realism and Realist Research in Human Geography: A Method or a Philosophy in Search of a Method'. *Progress in Human Geography* 21:51–74.

Yorgason, E.R. 2003. *Transformation of the Mormon Culture Region*. Urbana: University of Illinois Press.

Young, G.L. 1974. 'Human Ecology as an Interdisciplinary Concept: A Critical Inquiry'. *Advances in Ecological Research* 8:1–105.

———. 1983. 'Introduction'. In *Origins of Human Ecology*, edited by G.L. Young, 1–9. Stroudsburg, PA: Hutchinson Press.

Young, I.M. 1993. 'Together in Difference: Transforming the Logic of Group Potential Conflict'. In *Principled Positions*, edited by J. Squires, 121–50. London: Lawrence and Wishart.

Young, M. 1999. The Social Construction of Tourist Places. *Australian Geographer* 30:373–89.

Zdorkowski, R.T., and G.O. Carney. 1985. 'This Land Is My Land: Oklahoma's Changing Vernacular Regions'. *Journal of Cultural Geography* 5, no. 2:97–106.

Zelinsky, W. 1961. 'An Approach to the Religious Geography of the United States: Patterns of Church Membership in 1952'. *Annals of the Association of American Geographers* 51:139–93.

———. 1967. 'Classical Town Names in the United States: The Historical Geography of an American Idea'. *Geographical Review* 57:463–95.

———. 1973. *The Cultural Geography of the United States*. Englewood Cliffs, NJ: Prentice Hall.

———. 1980. 'North America's Vernacular Regions'. *Annals of the Association of American Geographers* 70:1–16.

———. 1983. 'Nationalism in the American Place-Name Cover'. *Names* 30:1–28.

———. 1988. *Nation into State: The Shifting Symbolic Foundations of American Nationalism*. Chapel Hill: University of North Carolina Press.

———. 1992. *The Cultural Geography of the United States*, revised edn. Englewood Cliffs, NJ: Prentice Hall.

———. 1994. *Exploring the Beloved Country: Geographic Forays Into American Society and Culture*. Iowa City: University of Iowa Press.

———. 1995. 'Review of *Re-Reading Cultural Geography* edited by K.E. Foote, P.J. Hugill, K. Mathewson, and J.M. Smith'. *Annals of the Association of American Geographers* 85:750–3.

———. 1997. 'Seeing Beyond the Dominant Culture'. In *Understanding Ordinary Landscapes*, edited by P. Groth and T.W. Bressi, 157–61. z New Haven: Yale University Press.

Zimmerer, K.S. 1994. 'Human Ecology and the "New Ecology": The Prospect and Promise of Integration'. *Annals of the Association of American Geographers* 84:108–25.

———. 1996a. 'Ecology as Cornerstone and Chimera in Human Geography'. In *Concepts in*

Human Geography, edited by C. Earle, K. Mathewson, and M.S. Kenzer, 161–88. Lanham, MD: Rowman and Littlefield.

———. 1996b. *Changing Fortunes: Biodiversity and Peasant Livelihood in the Peruvian Andes.* Berkeley: University of California Press.

———, and T.J. Bassett. 2003. *Political Ecology: An Integrative Approach to Geography and Environment-Development Studies.* New York: Guilford.

———, and K.R. Young, eds. 1998. *Nature's Geography: New Lessons for Conservation in Developing Countries.* Madison: University of Wisconsin Press.

Index